P9-DEG-442

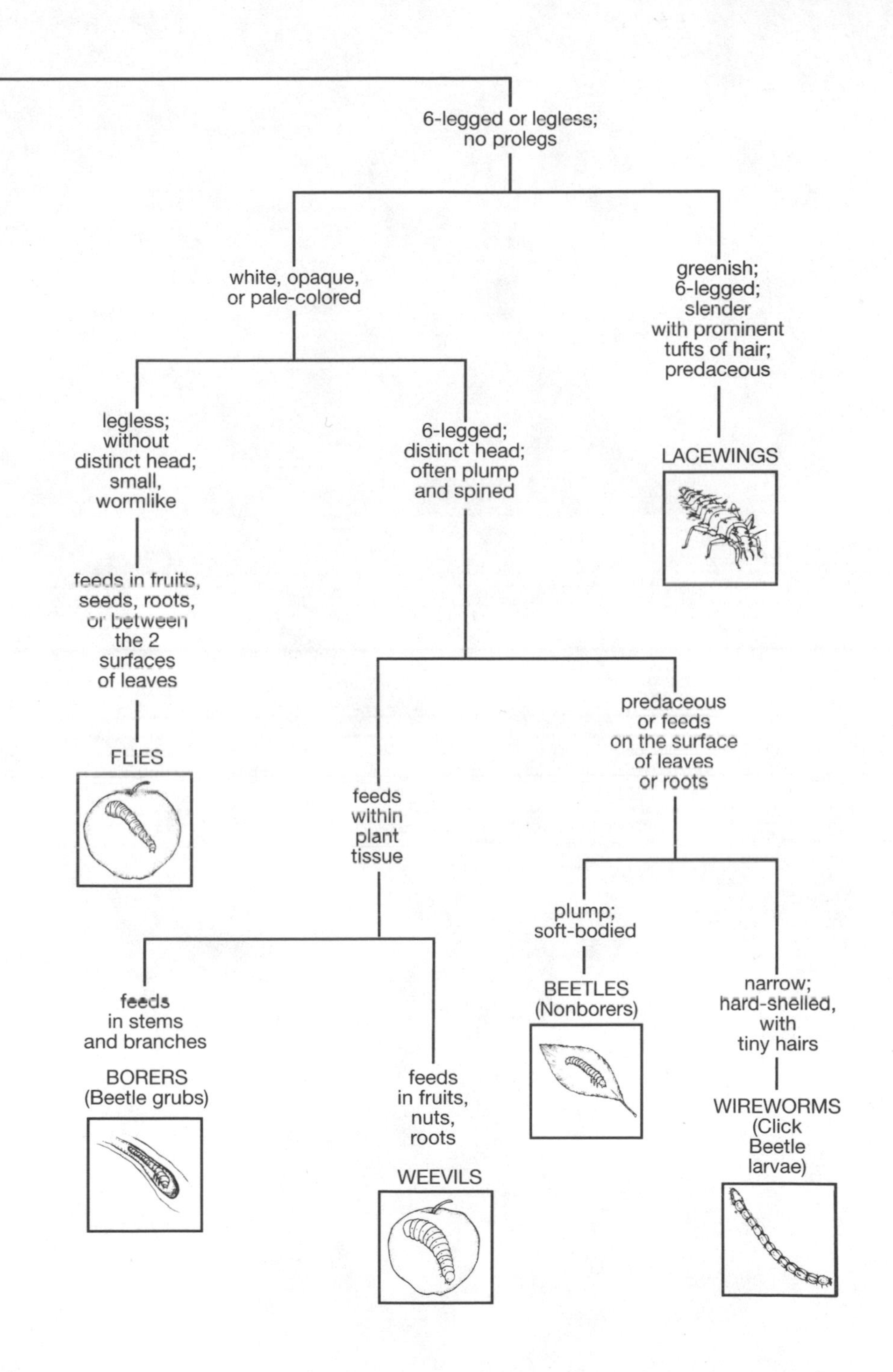
6-legged or legless;
no prolegs
white, opaque,
or pale-colored
greenish;
6-legged;
slender
with prominent
tufts of hair;
predaceous
LACEWINGS
legless;
without
distinct head;
small,
wormlike
6-legged;
distinct head;
often plump
and spined
feeds in fruits,
seeds, roots,
or between
the 2
surfaces
of leaves
FLIES
predaceous
or feeds
on the surface
of leaves
or roots
feeds
within
plant
tissue
plump;
soft-bodied
BEETLES
(Nonborers)
narrow;
hard-shelled,
with
tiny hairs
feeds
in stems
and branches
BORERS
(Beetle grubs)
feeds
in fruits,
nuts,
roots
WEEVILS
WIREWORMS
(Click
Beetle
larvae)

THE ORGANIC GARDENER'S HANDBOOK OF NATURAL INSECT AND DISEASE CONTROL

THE ORGANIC GARDENER'S HANDBOOK OF NATURAL INSECT AND DISEASE CONTROL

A Complete Problem-Solving Guide to Keeping Your Garden & Yard Healthy without Chemicals

EDITED BY

BARBARA W. ELLIS AND FERN MARSHALL BRADLEY

Contributing Writers: Helen Atthowe, Linda A. Gilkeson, Ph.D., L. Patricia Kite, Patricia S. Michalak, Barbara Pleasant, Lee Reich, Ph.D., Alfred F. Scheider

Rodale Press, Emmaus, Pennsylvania

The information in this book has been carefully researched, and all efforts have been made to ensure accuracy. Rodale Press, Inc., assumes no responsibility for any injuries suffered or damages or losses incurred during use of or as a result of following this information. It is important to study all directions carefully before taking any action based on the information and advice presented in this book. When using any commercial product, *always* read and follow label directions. Where trade names are used, no discrimination is intended and no endorsement by Rodale Press, Inc., is implied.

On the Back Cover:

Left: Colorado potato beetle adult, larvae, and eggs. Photo by Ron West.

Right: Mealybug destroyer larva. Photo by Max E. Badgley.

Center: Convergent lady beetle attacking aphids. Photo by Ron West.

Printed in the United States of America
on acid-free (∞), recycled ♻ paper

Senior Managing Editor: Margaret Lydic Balitas
Senior Editor: Barbara W. Ellis
Editor: Fern Marshall Bradley
Production Editor: Nancy J. Ondra
Contributing Editors: Nancy J. Ondra, Deborah L. Martin, Jean M. A. Nick, Paula Dreifus Bakule
Photo Editor: Heidi A. Stonehill
Copy Editor: Lisa D. Andruscavage
Editorial Production Coordinators: Stacy A. Brobst, Susan Nickol
Editorial/Administrative Assistant: Karen Earl-Braymer
Book Designer: Darlene Schneck
Front cover photo: T. L. Gettings
Illustrations: Julia S. Child, Jean Emmons

If you have any questions or comments concerning this book, please write:

Rodale Press
Book Readers' Service
33 East Minor Street
Emmaus, PA 18098

Library of Congress Cataloging-in-Publication Data

The Organic gardener's handbook of natural insect and disease control : a complete problem-solving guide to keeping your garden & yard healthy without chemicals / edited by Barbara W. Ellis and Fern Marshall Bradley ; contributing writers, Helen Atthowe . . . [et al.].
p. cm.
Includes bibliographical references and index.
ISBN 0-87596-124-X hardcover
1. Garden pests—Control. 2. Organic gardening. 3. Plants, Protection of. I. Ellis, Barbara W. II. Bradley, Fern Marshall. III. Atthowe, Helen.
SB974.072 1992
635'.049—dc20 92-3372
CIP

Distributed in the book trade by St. Martin's Press

6 8 10 9 7 5 hardcover

CONTENTS

CONTRIBUTORS

Writers

Helen Atthowe has a master's degree in horticulture (specializing in pomology) from Rutgers University. She has worked as a private pest management consultant and has worked with the New Jersey Cooperative Extension Service as an agent specializing in integrated pest management.

Linda A. Gilkeson, Ph.D., is coauthor of *Rodale's Chemical-Free Yard and Garden* and is an integrated pest management coordinator in the Ministry of Environment for the province of British Columbia. She has a doctorate in entomology from McGill University, Montreal, and has published several articles in scientific journals.

L. Patricia Kite is the author of *Controlling Lawn and Garden Insects, The Home Gardener's Problem Solver,* and *Organic Gardening: Vegetables* and coauthor of *California Cutting Garden.* A resident of Newark, California, she holds a master's degree in journalism and a teaching credential in biology.

Patricia S. Michalak has a master's degree in entomology from Michigan State University. She applies her ten years of research experience in agricultural entomology to her efforts as a freelance writer. She markets herbs and gourmet vegetables organically grown at her farm, Long and Winding Row Farm, in Kempton, Pennsylvania.

Barbara Pleasant is the author of *The Handbook of Southern Vegetable Gardening* and is a contributing editor for *Organic Gardening* magazine.

Lee Reich, Ph.D., is the author of *Uncommon Fruits Worthy of Attention* and *A Northeast Gardener's Year* and a former fruit researcher for Cornell University and the U.S. Department of Agriculture. He currently is a horticultural consultant and writer.

Alfred F. Scheider holds a bachelor's degree from the New York State College of Agriculture at Cornell. He is the author of *Success with Bulbs* and has had articles published in the *New York Times, House Beautiful,* and *House and Garden.*

Editors

Barbara W. Ellis has a bachelor of arts degree from Kenyon College in Ohio and a bachelor's degree in horticulture from The Ohio State University. She is a former publications director/editor for *American Horticulturist,* the publication of the American Horticultural Society, and is the senior editor of garden books at Rodale Press.

Fern Marshall Bradley has a bachelor's degree in plant science from Cornell University and a master's degree in horticulture from Rutgers University. She has managed an organic market garden and is a garden book editor at Rodale Press.

HOW TO USE THIS BOOK

Whether it's spots on the tomatoes, melons that wilt and die mysteriously, or worm-eaten apples, damage from insects and disease organisms is never welcome. How to manage these problems is a top concern of gardeners everywhere. This book is specifically designed to help. Its quick-reference format quickly leads you to complete control information for a wide range of common insect and disease pests.

How to Find It

One look at the contents will illustrate that *The Organic Gardener's Handbook of Natural Insect and Disease Control* is really four books in one. Once you've paged through it to familiarize yourself with the format, flipping to the section you need will become second nature. And for quick and easy access to information, don't forget the index. Here's a rundown of what is included in the pages that follow.

Part 1, Your Healthy Garden: A Plant-by-Plant Guide to Problem Solving and Prevention, contains a plant encyclopedia with symptoms and solutions for major problems of popular plants, including fruits, vegetables, annuals, bulbs, perennials, and trees and shrubs. It also contains preventive information and general entries on major plant groups—plus stunning illustrations of common problems on the most popular garden plants.

Part 2, Insects: Recognizing Your Friends—Eliminating Your Foes, contains information on managing pest insects and attracting beneficials. It also features a photographic "Insect Identification Guide" of more than 100 common pests as well as more than 20 types of beneficial insects. (The names of the beneficial insects are set in green type; the names of the pest insects are in black.)

Part 3, Diseases: Identifying the Causes—Implementing the Cures, is a primer on plant disease that will help you diagnose and control problems caused by fungi, bacteria, and viruses. In this part, you'll also find a photographic "Disease Symptom Guide" to help you identify problems in your own garden.

Part 4, Organic Controls: Using Remedies Safely, is an encyclopedia of organically acceptable control techniques and products. The major types of controls (cultural, physical, biological, and chemical) are presented in order of least to most invasive to make it easy to choose the best control for you.

Remembering the Big Picture

As you use this book, though, it's important that you not lose sight of the big picture of gardening organically. One of the principles that is at the heart of organic gardening is eliminating the need to use chemicals at all—even organically acceptable ones. This may seem impossible at first, even foolhardy, but organic gardeners everywhere will attest to the fact that it makes gardening more enjoyable—and safer—than ever.

So where do you start? Logically enough, at the beginning. The opening pages of part 1 of this book give you the big picture of organic garden management. If you skip right to the controls listed in the plant encyclopedia entries or to the control sections in parts 2 and 3, you'll be missing loads of valuable information that can help prevent pests from reaching a problematic level in your yard and garden.

Surprisingly, most of the best ways to control pests and diseases may not seem like controls at all. Organic gardeners look at their gardens as an overall system that should be kept in balance, so no one pest or disease gets out of hand. They concentrate their efforts on cultural techniques that prevent problems, such as keeping their plants in top-notch health and cultivating rich, organically active soil. They also try to encourage a diverse community of predators to keep pests in check.

While the individual plant, insect, and disease entries do list some preventive measures, they are primarily designed to help you decide what to do once you've encountered specific problems in your garden. But keep in mind that preventive measures are the keys to successful organic gardening. To get the most out of this book—and to develop an effective pest- and disease-control system for your garden—take the time to review the preventive methods discussed in the introduction to part 1, and follow up by studying the descriptions of these methods in part 4. Then, make a plan to begin implementing them in your garden.

Of course, you'll also find that this book is an invaluable reference for finding solutions to garden problems: Throughout these pages, you'll find up-to-date, detailed information on how to handle problems. When you do refer to recommendations for controlling a specific insect or disease, always use the least-invasive method available. Cultural controls are generally the most benign, followed by physical and biological controls. Chemical controls should always be considered a last resort.

Since botanical pesticides are organically acceptable, we include them in our recommendations throughout this book. So there are hundreds of recommendations for applying botanical chemicals as well as sulfur and copper fungicides. This does not mean we are endorsing widespread use of botanical pesticides. It will only be in *exceptional* cases in any individual home garden that the chemical method will be needed. It's critical to remember not to rely on a control approach—you'll be most successful, and safest, if you emphasize practices to promote garden health, and use this book for problem solving in those few cases when a pest population gets out of balance.

If you do decide to use one of the botanical sprays or dusts, *always* remember that they can be dangerous, especially to the person applying them. There is a reason these products kill insects! Never use them casually or carelessly. And whenever you use commercial products, *always* read and follow label directions.

PART 1

YOUR HEALTHY GARDEN

A Plant-by-Plant Guide to Problem Solving and Prevention

Controlling pests and diseases organically means much more than simply changing the types of sprays and dusts you use. Organic gardeners strive to develop a balanced system where problems are regulated naturally and to rely primarily on nonchemical controls to deal with pest problems.

Throughout this book, you'll find hundreds of nonchemical suggestions for stopping insects in their tracks or curbing diseases before they overtake your yard. There are also recommendations for using organically acceptable fungicides and insecticides. These are offered as a last resort—for the exceptional cases when the natural balance has been disturbed and pest problems threaten to ruin the harvest or kill your plants.

To understand why maintaining the natural balance in your garden is important, it helps to think about wild systems like a woodland or meadow. Have you ever wondered why pests and diseases haven't overrun the plants growing there? The answer is that over time, the community has developed a system of checks and balances that prevents devastating outbreaks of pests or diseases. Some insects in a woodland or meadow eat plants, but natural predators and parasites keep their numbers in check. The cycle of plant growth and decay ensures that soil fertility will be maintained to help keep plants healthy and more resistant to insect and disease problems. Plants that can't adapt to the prevailing climate and soil conditions will lose out, and a plant mix that can thrive in those conditions will survive. This natural system is at work in your own garden, and this book is about the many ways you can make these natural checks and balances work for you.

As you peruse the entries in the "Problem-Solving Plant Guide" beginning on page 11—as well as the sections on insects, diseases, and controls that follow—notice that many of the recommendations for controlling pests and diseases don't involve sprays or dusts at all. You'll find suggestions for proper site selection and preferred soil conditions, tips for avoiding or preventing conditions that promote pest and disease outbreaks, and recommendations for steps you can take to prevent problems *before* they occur. That's because keeping plants healthy and preventing problems by proper plant selection and culture are two fundamental principles of gardening organically.

Plan for Healthy Plants

Every minute of every day, the plants in your yard are defending themselves from their natural enemies. When a gust of wind breaks a branch from a tree, chemicals concentrated at the base of that branch mobilize into a protective wall to prevent pests from invading the healthy parts of the tree. When insects nibble on the leaves of a shrub, the plant may respond by changing its leaf chemistry to make itself less appetizing. As earthworms break off pieces of grass roots in their endless tunneling, the grass may turn injury into opportunity by growing two new roots where there used to be one.

Happy Equals Healthy

Maximizing natural resistance is one important way good gardeners keep their gardens flourishing. Pests and disease organisms prefer plants that are weak, injured, or unable to adjust to the site on which they have been planted. For example, when planted in a hot, sunny site, rhododendrons and azaleas are more prone to attack by lace bugs, which aren't a major problem in shady sites. On the other hand, when grown in a suitable climate and provided with healthy soil and an appropriate site, plants generally thrive and are highly resistant to problems. Most plants will have trouble-free lives if provided with growing conditions that closely match their natural preferences. There generally isn't much more the gardener needs to do to keep them happy and healthy. Plants forced to live in conditions that don't suit their needs—with the wrong soil moisture level and sunlight exposure, for example—are doomed from the start. Much of the intrigue of gardening involves discovering—or creating—the perfect growing situation for a special plant.

As you develop your garden, take time to study each site before you decide what to plant there. Then select plants that will thrive in the conditions available. If you match the plant to the site, you're well on the way to a successful garden.

Bear in mind that some pests and diseases can be prevented by site selection alone. For example, a site on which morning sun quickly dries leaves dampened by dew will help limit the conditions many molds and mildews need to flourish. Site selection also can help limit insect-transmitted diseases. When beets are grown close to tomatoes, beet leafhoppers can easily carry a virus called curly top to tomatoes. Flea beetles normally cause only cosmetic damage to potatoes (tiny holes in the leaves), but can move on to corn and infect that crop with bacterial wilt. Whenever you learn that a pest common in your area serves as a vector for disease, separate host plants as much as possible, both physically and by crop timing, to help interrupt the pattern of disease transmission.

Handle with Care

One of the best ways to keep plants healthy and happy is to treat them with care. Bruised or torn leaves, damaged stems, and other plant wounds not only cause unnecessary stress that can weaken plants, they're an open invitation to pests and disease organisms, which find it easy to enter plants through damaged tissue. Keep this in mind whenever you're in your garden, and avoid handling plants whenever possible. Whenever you do need to work in your garden—even walk down the rows between plants—touch the plants gently to keep damage to a minimum. *Never* work around plants when they're wet. Many disease organisms (and some pests, too) travel easily on the film of water that covers foliage on damp days. When you work among wet plants, you're likely not only to help spread disease organisms, but also provide them with easy access to your plants via bruised leaves. Tugging on plants when you harvest fruit or pick flowers can damage roots and stems: Always use a sharp knife or pruning shears to make a clean cut that will heal easily. The same goes for proper pruning.

Handling trees with care is one of the best ways to keep them healthy. Many tree pests can only invade through the exposed tissue at an injury site. For example, flowering dogwood is normally a very insect-resistant tree, its trunk protected by a thick, hard bark. But when bits of bark are broken away, as when you hit a young dogwood with a lawn mower or string trimmer, that injury becomes a point of entry for dogwood borer larvae—a very difficult-to-control pest that eats its way through living wood.

Cultivate Healthy Soil

Cultivating healthy soil is at the foundation of growing healthy plants. Good soil helps plants nurture themselves. Roots flourish in healthy soil. They're able to find and use nutrients as needed, which helps the plant grow strong and resilient.

When grown in poor, compacted soil that is low in nutrients, plants will grow weakly and be stressed by nutrient deficiencies. As a result, they'll be easy prey for insects and

Buying Healthy Plants

Starting out with the healthiest plants available is a great way to ensure a healthy garden. Use these tips to pick the best plants you can find.

■ Look for a sales display where the plants are well cared for. Plants subjected to hot, sunny sidewalks, allowed to wilt frequently, and watered unevenly are never a bargain, no matter how cheap.

■ Look at the entire group of plants being offered for sale. If some of the plants seem to be in poor health, shop somewhere else. Those that look healthy today may be diseased tomorrow.

■ Shop early in the season, when selection is good and plants are young.

■ Examine the roots. Gently shake a plant out of its container. Roots should be plentiful, but not wrapped into a tight spiral. With bare-root plants, look for fungi, lesions, and broken tissue; prune them off before planting.

■ Consider plant color. Pale overall color indicates a need for nutrients, which is easy to correct. However, distinct yellow streaks or brown leaf spots indicate presence of disease.

■ Buy perennials, trees, and deciduous shrubs when they are dormant or just beginning to bud out. Leaf emergence above ground is accompanied by rapid root growth below. Get plants situated before this growth spurt occurs.

■ If bedding plants are already in flower, pinch off the blossoms when you set them out. Pinching will help direct the plant's energy into growing roots, so it will be better able to support more flowers and fruit later in the season.

diseases. In contrast, soil that is fertile, well-drained, and teeming with communities of diverse microbes sharply increases a plant's chance at a healthy, productive life. Also, since many pest and disease organisms spend part or all of their lives below-ground, having a diverse community of organisms to keep them in check is important.

Healthy soil is an intricate mix of tiny rock particles, organic matter, water, air, microorganisms, and other animals. Living things abound in a robust, organically active soil—plant roots, animals, insects, bacteria, fungi, and other organisms. The more organic matter you provide, the livelier the life forms within your soil are likely to be. And, the livelier the soil life becomes, the more heated competition becomes between beneficial and benign soil microorganisms and those capable of doing harm.

Over time, adding organic matter improves soil structure, which in turn improves the soil's ability to absorb and release water. Obviously, without water plants cannot thrive, but too much water can also sabotage plant health. When soil becomes saturated, and water pools up around plant roots, the roots may lose their ability to take up nutrients. Tiny root hairs may begin rotting away, followed by entire root branches. As organisms that cause root rot flourish, the plant may weaken. Above-ground, the plant may be simultaneously attacked by molds and mildews encouraged by damp conditions. In this way, plants can easily die from too much water.

There is tremendous variation in how much water plants can use, and here, too, the question of natural resistance comes into play. Plants that are naturally adapted to wet conditions are resistant to many of the pathogens present in chronically wet soil; those that grow naturally in dry or very well drained settings are easy prey to those same pathogens.

Local soil conditions have an important bearing on what plants are best for your garden. Soil pH affects the availability of certain nutrients to a plant, affecting its overall vigor, thus directly affecting its ability to maintain good health. Where soil tends to be very acidic, plants that like a low pH, such as blueberries and rhododendrons, will be much happier than plants such as asparagus and clematis, which require close monitoring and frequent applications of lime when grown in very acidic soil. Although soil pH is easy to manipulate using organic matter, mineral fertilizers, and mulches, it's wise to consider the natural pH of your soil when choosing long-lived plants.

To help ensure that your soil is healthy and balanced, take the time to learn about its characteristics and fertility levels. It's a good idea to have it tested by the Cooperative Extension Service or by a private laboratory. Use the results as a guide to bring your soil into balance by adding lime, compost, or organic soil amendments and fertilizers as needed. Maintain soil balance by growing green manure crops and adding organic matter each season.

Select Trouble-Free Plants

Some plants are naturally prone to diseases; others are like sparkling beacons to hungry insects. You can eliminate many problems in your garden by avoiding pest- or disease-prone plants. Instead, fill your property with dependable plants that can fend for themselves. In any climate, there are hundreds to choose from. With a solid collection of easy-to-grow plants in place, you can limit your pest-control activities to plants you consider indispensable. Tomatoes are prone to pest problems, but most gardeners agree that vine-ripened tomatoes are well worth the time and trouble they must take to select the best-adapted cultivars, create ideal soil conditions, and patrol the tomato plot regularly to stay on top of insect and disease activity.

Plants differ in their ability to defend

themselves from pests and diseases, and choosing resistant species and cultivars is another basic strategy for limiting pest problems. For example, instead of growing disease-prone hybrid tea roses, you can select from among the new, easier-to-satisfy cultivars of shrub roses, which require little care beyond occasional pruning. In addition to the discussion of plant resistance on page 412, you'll find many suggestions for disease- and insect-resistant cultivars throughout the "Problem-Solving Plant Guide" beginning on page 11. Planting them is often the best way to eliminate a problem altogether. Remember, though, even with resistant plants, proper site selection and preparation are essential.

Feed Regularly

Although rich, well-fed soil is all many plants need to keep them growing vigorously, fertilizing them at the proper time is also beneficial and will help enhance their natural resistance to pests.

As a general rule, make sure annual vegetable and flower crops are well-fed when they are young, and then give them a modest booster feeding when flowering and reproduction commences. Fertilize perennial vegetable and flower crops, along with trees and shrubs, during the first third of their active growth, which is just as they begin growing in spring. You can also fertilize during the second third of the plants' growing season, but not during the last third. Late fertilization can stress the plants by encouraging tender new growth that may be winterkilled. When you apply fertilizer, always keep in mind that more isn't necessarily better. Overfertilizing leads to rank, spindly growth and can even worsen problems with aphids and some other pests.

Encourage Diversity

Any yard or garden is home to countless forms of microscopic life and hundreds of different kinds of insects as well as birds, toads, frogs, mammals, and humans. Relationships within this community are mind-boggling in their complexity, but encouraging a diverse community is a valuable way to ensure a healthy, trouble-free garden.

Rotate Crops

Crop rotation is one way diversity helps prevent problems in your garden. Consider what happens in a monoculture, such as a field planted year after year with corn. In the first year, some corn pests, such as European corn borers and corn earworms, will find the field, eat their fill, and leave eggs or overwintering pupae in the debris that remains after harvest. The following spring, the new generation of pests can get right to work feeding and reproducing themselves. The same is true for disease organisms that develop in the field. Favored host plants are easy to find and are available every year. As a consequence, pest populations can build out of all proportion.

Rotating (essentially taking away) a pest's or disease's preferred host plant hampers its ability to feed and reproduce. If the cornfield were planted with wheat the following year, the borers and earworms would be forced to look someplace else for food. They could not continue to feed and build their populations on that site. Another example of crops that benefit from rotation is members of the cabbage family (Cruciferae). Many diseases such as black leg (a fungus) and black rot (a bacterium) can quickly establish themselves in a cabbage patch. The first year's crop may get by with minimal damage, but if cabbage, broccoli, cauliflower, or other crucifers are planted there again the following year, the disease organisms are ready and waiting to infect defenseless young plants. Rotating cabbage family members with crops that belong to other plant families breaks the disease cycle. For more on crop rotation, see page 415.

Fortunately, rotation generally isn't necessary for long-lived plants such as trees and shrubs, because they have developed their own pest-control strategies that enable them to live long, trouble-free lives. (Unless, of course, you lose a tree to a soilborne diseases such as Verticillium wilt, in which case you shouldn't replace it with another susceptible species.) Below ground, trees and shrubs may release chemicals into the soil that act as natural pesticides to soilborne fungi and bacteria. Above ground, their tough, woody stems help resist bruising and insect attacks. When leaves are lost to disease, a shrub or tree has the ability to produce new leaves several times. When leaf-eating insects dine on tree leaves, the tree may respond by pumping more tannic acid into the leaves that remain, thus deterring further feeding.

Avoid Monocultures

Planting a wide variety of plants is another way to foil pests and diseases by encouraging diversity. Conventional lawns, a type of monoculture, are a good example. Planting a mixture of lawn grasses (especially disease-resistant ones) will help prevent diseases from sweeping through your lawn, because some plants will be susceptible and others won't. Replacing some of your lawn areas with a variety of tough groundcovers will also increase the overall diversity of the home landscape. It's an excellent alternative for sites on which lawn grasses won't naturally thrive.

The importance of using diversity is especially apparent if you look at long-lived plants such as trees and shrubs. Using a large planting of a single species of tree or shrub to screen a busy roadway, or planting the same shade tree that all your neighbors have, are two more examples of monocultures. A devastating pest or disease that appears on the scene can easily move from plant to plant and damage or wipe out the entire planting in a single season. Dutch elm disease, which decimated American elm plantings along thousands of American streets in the mid-1900s, is a perfect example of the dangers of mass planting single species. If you've planted a variety of trees and shrubs for screening or shade, you have built-in protection.

Companion planting is another way to use diversity—and avoid monocultures—to foil pests. Mixing marigolds or strong-smelling herbs in the vegetable garden, as opposed to planting solid blocks or rows of a single crop, is effective, because many pests locate crops by smell. Interplanting can "hide" a crop by masking the odor that attracts pests. Some companion plants work by repelling pests. For example, basil is said to repel tomato hornworms. For more on the many ways companion planting can be used to control pests and diseases, see page 419.

Encourage Natural Predators

Encouraging a diverse community of insect predators is another way organic gardeners manage pests. *Beneficials*, the term usually used for insect-predatory and parasitic insects and arachnids, such as spiders and mites, play a vital role in the complex community that naturally exists in your yard. Many actively help plants reproduce by spreading pollen; others consume or parasitize destructive pests. Beneficials such as lady beetles and honeybees are well-known. Other effective beneficials include lacewings and assassin bugs. For more on beneficials, see pages 258 and 453.

Birds are among the most efficient insect predators. When you consider that chickadees spend much of their time during the winter eating aphid eggs or that Baltimore orioles can eat up to 17 tent caterpillars a minute, you won't wonder why many gardeners want to attract birds to their yards. Planting trees and shrubs that provide food and shelter, along with providing a source of fresh water year-round, will go a long way toward encouraging birds to visit your yard.

And for consuming slugs, cutworms, and

other ground-dwelling pests, nothing beats a resident toad! Be sure to provide shelter and water for these hard-working garden residents.

Develop a System

Keeping plants thriving, improving your soil, planting a variety of plants, and encouraging natural predators are all great ways to prevent pest and disease problems, but planning and record keeping are also essential weapons to add to your arsenal. A garden plan will help you develop a calendar of activities aimed at preventing problems. It will also help you set and achieve goals for your garden, such as eliminating or modifying problem-prone plantings or adding new gardens. Regular record keeping will help you know when to expect problems so you can get a head start on them.

Start your plan by finding out about the plants you're growing as well as the ones you want to grow. Learn what conditions they prefer and which will grow best in your area. Next, draw a sketch of your garden. Mark down existing plantings, but also look for sites that might suit some plants but discourage others. For example, locate frost pockets, sites where air will circulate freely, areas with moist soil, and shady or sunny spots.

Then identify and list the problems you've had in the past or that are common in your area. Use the "Problem-Solving Plant Guide" on the pages that follow to identify problems you've seen in your garden before. Learn about the life cycles of these pests and diseases. How do they overwinter? When and how do they attack your plants?

Once you have all your options—and potential problems—in front of you, make a comprehensive plan. Next, list the ways you could help control each of the problems on your list. Decide what cultural controls will help prevent problems and when they need to be done. Add the biological and physical controls next, and make note of chemical controls for use as a last resort. Be sure to include the life cycle stage at which each control is most effective. It may help to make a chart of the life cycle and control activities.

Use these notes to create a schedule of gardening activities. You'll find that certain chores, such as an annual fall cleanup, are well worth the effort because they'll prevent many types of pests. Planning ahead and following your schedule will enable you to make full use of preventive control measures, and they'll soon become second nature to you. You'll also know what tools and materials to have on hand to fight problems if they arise.

Learn to Look

All the planning in the world won't help if you don't keep an eye out for problems. Begin a garden journal for keeping records throughout the season. Make it a habit to walk through your garden at least once a week—daily is best. Look carefully at your plants, turning over leaves, and note overall appearance.

In your garden records, include pest appearance dates and details of insect behavior along with notes you make on the progress of various plants. This is the only way to discover useful facts, such as when birds are no longer effective cabbageworm predators on cabbage, cauliflower, or broccoli, or which members of the huge daisy family are first to be attacked by four-lined plant bugs.

If you notice plants that don't look healthy or aren't growing the way you'd expected, take time to examine them more closely. Look for signs of the problems you listed, and go over your plan to see what steps you need to take. Use the "Pest Patrol Checklist" on the opposite page to gather the information you need to diagnose the problem. Once you've diagnosed the problem and identified the ways to control it, add the necessary information to your garden plan.

Keep notes on your program during the

Pest Patrol Checklist

In general, plant problems are divided into three main types: pests (including insects, mites, nematodes, even gophers and mice), diseases (such as bacteria, viruses, and fungi), and physiological problems (like nutrient and water imbalances, too much or too little light, or extreme temperatures). These conditions can interact, or one condition might predispose the plants to another. For example, root maggots tunneling in roots directly damage them, but the reason the plants may subsequently die is because rot organisms invade the damaged areas and then spread through the entire root system.

Diagnosing a pest problem on a plant is easiest if you catch the culprit in the act; then it's just a matter of identifying the pest. But reaching a diagnosis often takes patience and careful observation, because the insect that caused the damage may only feed at night or underground. In addition, insect damage often resembles infections from fungi or viruses, nutrient deficiencies might look like a viral disease or vice versa. In these cases use the steps below to collect as much information as you can. Then use the plant, insect, and disease encyclopedia sections of this book to determine the cause of the problem.

1. Examine the entire afflicted plant and note the range of abnormal symptoms. Is the entire plant affected or just a part of it? If a part is affected, what part? Do you see symptoms only on the new growth or old growth, or on a single branch? Is just one plant affected, or do you see symptoms along an entire row? If plants are affected randomly within a row or bed, and the symptoms do not spread to the other plants, the problem is probably not an infectious disease. Clusters of plants showing symptoms may indicate disease, or just a pocket of poor soil.

2. Examine both the affected areas and healthy portions of the plant with a magnifying glass. Inspect the leaves, especially on the undersides, stems, flowers, and roots, if possible. Look for signs of hidden damage, such as borer holes in stems or fine webbing on the undersides of leaves. To avoid damaging roots when you observe them, dig the plants rather than pulling them from the ground. Carefully wash soil off the roots for easier inspection.

3. Check the condition of the soil. Is it poorly aerated, encouraging root rots or at least poor root function? Or is the soil droughty, so plants wilt from lack of water? If soil conditions are poor, root systems may not be able to spread in search of nutrients, and plants in that area may suffer from nutrient deficiencies.

4. Collect any insects associated with the ailing plant, or take samples of damaged leaves. Seal samples in pill bottles or clear plastic bags.

5. Use the information you've gathered to narrow down the possible causes, using the information and photographs in this book. You may also want to refer to other insect, disease, and horticulture handbooks for help in confirming your diagnosis and identifying species and for advice on how to deal with the problem.

6. If you are still stumped, review everything you know about the conditions preferred by the plant, correct any environmental or nutritional deficiencies, and give the plants the best care you can. Many unexplained afflictions are physiological in origin and plants often grow out of them when conditions improve. Bear in mind that nutrient problems will take weeks or even months to correct. If more plants in the row become affected, as might happen if disease is spreading, pull and discard the damaged plants to reduce the source of disease.

Cultural Questions

Use the following questions to help organize your thinking about the causes for garden problems and the possible ways you could prevent them by improving plant culture.

■ Is the plant growing in an appropriate soil type with the kind of exposure and drainage it requires? Is the pH within tolerable limits? Should you move the plant? If so, when is the best time?

■ Are climatic factors limiting the plant's growth? Could windbreaks, winter protection, early planting, or use of a different species solve the problem?

■ Have you given the plant too little or too much fertilizer? Have you fertilized it at the appropriate times?

■ How severe is the damage? Is it cosmetic or is the plant's health really deteriorating? If you have seen this pest before, did the affected plants outgrow the problem?

■ Is the pest soilborne and/or indigenous and likely to appear year after year?

■ Have you followed appropriate crop rotations?

■ Is the plant surrounded by good companions or bad ones?

■ Can you remove affected plant parts to stop the progress of the pest?

■ For insect pests: Does the pest fly, jump, or remain stationary? Are other plants likely to become infested? Are natural predators present?

■ Is the plant too old to adequately defend itself? Could healthy replacement plants solve the problem?

■ Could you replace the plant with one that is resistant to the problem?

■ Have you chosen the control with the most limited environmental impact?

■ Are there routine garden tasks you could use, such as annually raking and removing afflicted leaves, to control the problem?

■ Have you made plans for nurturing follow-up care to help the plant recover from the pest challenge?

season. Write down what works and what doesn't, what controls you used, how much you used, and where. The more information you have, the better you'll be able to refine the plan for the next season.

The more time you spend in your garden, the better you will be able to recognize, understand, and treat pest problems. When you consider pest problems thoughtfully, you may be surprised at how little damage actually has occurred. Many plants can tolerate up to 30 percent loss of foliage without suffering serious cuts in productivity. And, although leaves with little holes in them don't look nice, they may actually help the plant defend itself by triggering the production of pest-repelling chemicals.

You'll also gain a greater appreciation for just how well plants can keep themselves healthy, given the opportunity. Even in the middle of summer, when pest problems are most severe, bear in mind that the crisis will pass and your plants will get another chance with each new season. Plants have been making use of these second chances, devising their own self-defense tricks and strategies, since time began. Learning and using natural pest-control maneuvers gets you involved in these adventures, which are part of the fun and fascination of gardening.

PROBLEM-SOLVING PLANT GUIDE

Use this encyclopedia of over 200 popular garden plants to help diagnose, prevent, and control problems throughout your garden. Individual plant entries include symptoms for the most common garden problems you may encounter, suggestions for cultural measures that help prevent problems, and control options for specific pest and disease problems.

Food plants—vegetables, fruits, and herbs—are listed by common name. However, ornamentals—annuals, bulbs, perennials, and trees, shrubs, and vines—are listed by botanical name. If you don't know the botanical name of a particular plant, there are two quick ways to look it up. For example, suppose you want information on disease problems of lilacs, and can't find a Lilac entry. One option is to look up lilac in the index. It will refer you to the *Syringa* index entry, where you'll find page listings. Or, you can turn to "Common and Botanical Names of Plants" on page 487, look up lilac, find the botanical name *Syringa,* and then go to the Syringa entry in this section of the book.

The entries are designed to help you quickly zero in on specific plant problems. In the "Problems" section of each entry, you'll find short descriptions of symptoms in boldface type, such as **"Leaves with yellow, V-shaped spots on margins."** Scan these descriptions until you find the one that most closely matches the problem on your plant. You can then read that section in the entry to learn the cause of the problem and how to avoid or control it.

In addition to the individual plant entries, you'll also find separate general entries on annuals, bulbs, herbs, lawns, perennials, and vegetables, along with an entry on trees, shrubs, and vines. These general entries highlight problems that attack a wide variety of plants. You'll find them useful for solving some of the most common problems that attack your garden.

If you're using this guide to diagnose a specific plant problem, always be sure your diagnosis is correct before deciding on a control strategy. If you want more help in identifying a specific problem, you'll find photographs and descriptions of specific insects and diseases in the "Insect Identification Guide" beginning on page 268 and in the "Disease Symptom Guide" beginning on page 370.

For more detailed information on how to safely and effectively use specific controls, refer to "Organic Controls" beginning on page 404. Controls are divided into four categories: cultural, physical, biological, and organically acceptable chemical controls. If you're not sure which category a particular control belongs to, your best bet is to look it up in the index.

Descriptions of the controls include information on the protection they offer, toxicity to nonpest organisms, and any possible hazards related to their use. When choosing a control strategy, always start with the lowest-impact control possible, and only use chemical controls as a last resort.

Abies

Fir. Trees.

Firs are cone-bearing evergreens with needlelike foliage and a symmetrical form, especially when young. They make impressive specimens or, in groups, effective dark backgrounds and tall screens.

Firs grow best under cool, moist conditions; they do not thrive in hot, dry climates. Set container-grown or balled-and-burlapped plants into acid, moist but well-drained soil, preferably in spring. Full sun is best, but firs will grow well (with a somewhat more open habit) in partial shade. Fir trees are seldom bothered by insects or diseases in the landscape. Planting healthy trees in the right conditions will help the plants avoid most potential pest problems.

Problems

Needles deformed. Cause: Balsam twig aphids. This 1/8″ long, greenish aphid has a white waxy coating. It feeds primarily on succulent new growth and exudes sticky honeydew, which covers the needles. Most of the damage occurs in late spring to early summer. Control with insecticidal soap, or spray with pyrethrin for severe infestations.

Needles light gray or bronze. Cause: Spruce spider mites. These tiny, spiderlike pests damage needles as they feed; the injury weakens the plant and can kill young trees. Mites start feeding on lower, older needles and progress upward and outward; tiny webs may be visible on needles. Control in spring with dormant oil before growth starts. During active growth, spray plants repeatedly with strong jets of water or make 2 applications of insecticidal soap 7-10 days apart; apply pyrethrin for severe infestations.

Branches defoliated. Cause: Hemlock loopers. The 1″ long, greenish yellow, black-spotted caterpillars usually appear in June. They feed on the needles, starting from the branch tips and working toward the center of the plant. Handpick small populations; control large infestations with BTK.

Branches with brown tips. Cause: Spruce budworms. These brown, 1″ caterpillars have white dots on their backs. They start tunneling into older needles and gradually migrate to opening buds. Damaged needles turn brown and drop; you may see tiny webs on shoot tips. Prune off affected branch tips or handpick if only a few pests are present. Spray severe infestations with BTK.

Plant defoliated; branches bear cocoonlike bags. Cause: Bagworms. See "Plant defoliated; branches bear cocoonlike bags" on page 236.

Acer

Maple. Trees.

Maples are a large group of easy-to-grow, 15′-100′ trees with opposite leaves and distinctive, 2-winged fruit. They are popular as street or specimen trees.

In general, maples prefer rich, moist but well-drained, acidic soil and a sunny position. Some smaller species, such as Japanese maples (*Acer palmatum*), need at least partial shade. Set out maples in spring as container-grown or balled-and-burlapped plants.

Although they can host a range of pests, most maples survive nicely with little or no spraying. Note that oil sprays and insecticidal soaps may damage some maples; before using either product, note restrictions on the label and use with caution. Test on a few leaves before treating the whole plant.

Problems

Leaves with brown, dry edges. Cause: Leaf scorch. Caused by lack of water or reflected heat from surrounding pavement, this damage is often a problem on newly planted trees. Avoid by keeping soil around the tree evenly moist; a 2″-3″ thick layer of mulch keeps roots cool and retains moisture.

Leaves yellow; whole plant weakened. Cause: Maple scales. Both maple phenacoccus and cottony maple scale produce small, fluffy, white egg masses. They feed on leaves and stems, weakening the plant. Spray the trunk and branches with dormant oil before growth starts in spring, or with superior oil or insecticidal soap as buds begin to open. Repeat the soap or superior oil application in midsummer, spraying the leaves and branches.

Leaves and shoots blackened; leaves with moist or brown sunken spots. Cause: Anthracnose. See "Leaves and shoots blackened; leaves with moist or brown sunken spots" on page 238 for more details on this disease.

Leaves distorted or bumpy. Cause: Mites. Several different mites attack maple leaves, causing pointed or wartlike swellings in various colors. Although they are unsightly, galls don't seriously threaten tree health. Pick off damaged leaves, or apply dormant oil in early spring, before growth starts.

Leaves, shoots, or seeds distorted. Cause: Boxelder bugs. These ½″ long bugs have charcoal-colored wings with red veins. They feed on leaves, flowers, and fruit and often congregate on tree trunks. Spray plants with superior oil or insecticidal soap when pests first appear.

Leaves skeletonized or with large holes. Cause: Caterpillars. See "Leaves skeletonized or with large holes; branches may be webbed" on page 236 for details.

Leaves with powdery white coating. Cause: Powdery mildew. See "Leaves with powdery white coating" on page 237 for controls.

Leaves with spots. Cause: Leaf spots. See "Leaves with spots" on page 237 for details.

Leaves wrinkled and discolored. Cause: Aphids. See "Leaves wrinkled and discolored" on page 235 for control information.

Branches wilt. Cause: Verticillium wilt. Parts of trees infected with this fungal disease may wilt suddenly or produce leaves that are yellow or smaller than normal. Cut and destroy or dispose of affected branches. Trees may recover if you provide extra water and fertilizer to promote strong, new, healthy growth. Remove severely infected trees; don't replant another maple in the same spot.

Trunk or branches with oozing lesions; branch tips die back. Cause: Canker. See "Trunks or branches with oozing lesions; branch tips die back" on page 238 for details.

Trunk or branches with small holes; limbs die or break off. Cause: Borers. See "Trunk or branches with small holes; limbs die or break off" on page 238 for controls.

Achillea

Yarrow. Perennials.

These hardy, easy-to-grow, mid-border perennials prosper in well-drained soil in full sun. They thrive on heat, need little water once established, and enjoy poor soil. Overly rich soil or excess fertilization gives poor growth and few flowers. Leaves are feathery and grayish green. Tiny yellow, red, or white flowers form dense, broadly flattened clusters from spring to midsummer. Plants form vigorous clumps that can spread; divide regularly for best performance.

Leaves of crowded plants in moist, humid, shady conditions may develop white or gray powdery patches of powdery mildew; see "Leaves covered with white powder" on page 177 for controls. Thin such plants to improve air circulation; avoid watering late in the day.

Aesculus

Horse chestnut. Trees and shrubs.

These deciduous plants have 3″-10″ long, opposite, compound leaves with 3-9 leaflets. The larger species are used as specimens or street trees; the shrubs are best in mass plantings or shrub borders.

All of these species thrive in moist, well-drained soil rich in organic matter. The tree forms do best in full sun, while the shrub kinds will flourish in conditions ranging from full sun to partial shade. Note that insecticidal soap sprays may damage some horse chestnuts; read the label before applying and use with caution. Test spray on a few leaves before spraying the whole plant.

Problems

Leaves with irregular brown spots. Cause: Leaf blotch. Caused by a fungus, these spots may spread over leaves and lead to defoliation. Clean up and destroy fallen leaves to remove overwintering spores.

Leaves with brown, dry edges. Cause: Leaf scorch. This symptom indicates weak or damaged roots or poor soil conditions; it is caused by lack of water or by heat reflected from surrounding pavement. Avoid by planting in rich, evenly moist soil and irrigating, if necessary, in very dry seasons.

Leaves with powdery white coating. Cause: Powdery mildew. See "Leaves with powdery white coating" on page 237 for controls.

Leaves skeletonized. Causes: Whitemarked tussock moths; Japanese beetles. Whitemarked tussock moths are 1½″ long; yellow caterpillar has 4 clumps of white hair on its back, along with 2 tufts of black hair on the head and 1 at the rear. Birds and natural parasites usually keep this pest under control. If damage occurs, reduce future populations by scraping the frothy white egg masses off trunks and branches. Band trunks with a sticky coating, such as Tanglefoot, to catch caterpillars as they crawl up the tree. Spray leaves with BTK to control serious infestations.

Japanese beetles also skeletonize leaves. See "Leaves skeletonized" on page 236 for control measures.

Leaves and shoots blackened; leaves with moist or brown sunken spots. Cause: Anthracnose. See "Leaves and shoots blackened; leaves with moist or brown sunken spots" on page 238 for details on this fungal disease.

Trunk or branches with oozing lesions; branch tips die back. Cause: Canker. See "Trunk or branches with oozing lesions; branch tips die back" on page 238 for details.

Plant defoliated; branches bear cocoonlike bags. Cause: Bagworms. See "Plant defoliated; branches bear cocoonlike bags" on page 236 for controls.

Ageratum

Ageratum, flossflower. Annuals.

The soft lavender, blue, or pink flower clusters of ageratum are a beautiful addition to the flower garden. Plants grow 6″-36″ high and about 1′ wide. They bloom from early summer to first frost. Ageratums also make good container plants.

Direct-seeding ageratum doesn't always give good results. For better germination, sow seeds indoors 8 weeks before the last frost. Seeds need light, so don't cover them; expect germination in about 10 days. Initial seedling growth is slow. Many gardeners prefer the ease of starting with nursery plants.

Ageratum grows best in fertile, well-drained soil on a sunny site with good air circulation. Poor soil and insufficient water cause plant browning. Too much shade causes decreased flowering and legginess. Remove spent flower heads to promote more blooms. Feed plants with a general-purpose fertilizer once a month. The first frost turns plants black.

Problems

Leaves yellow; plant weakened. Cause: Whiteflies. For control measures, see "Leaves yellow; plant weakened" on page 18.

Leaves stippled with yellow; foliage webbed. Cause: Spider mites. See "Leaves stippled with yellow; foliage webbed" on page 18.

Plant wilts. Cause: Fungal wilt. Various fungi thrive in the moist, fertile soil ageratums require. They attack roots, causing plant stunting and wilting. Wilting may begin at lower leaves and progress toward the upper ones, or it may occur rapidly. If the plant is pulled up, you'll see the roots are dark brown and rotting. Discard infected plants. Improve drainage and lighten the soil with organic matter.

Ajuga

Ajuga, bugleweed. Perennials.

These hardy groundcovers favor partial shade and moist, well-drained soil. Flower spikes appear spring to summer over mats of dark green, bronze, or burgundy leaves. Common bugleweed (*Ajuga reptans*) spreads vigorously via stolons, desirable for a slope, less so in a lawn. Other species are more easily controlled. Set plants 6″ apart for rapid coverage.

Excess moisture, especially in heavy, slow-draining soils, encourages crown and root rots caused by soil-dwelling fungi. Dark spots appear on the crown and leaves of infected plants; roots blacken. Controls are cultural: Avoid planting ajuga in wet areas; cultivate to improve air circulation. Remove infected plants and surrounding soil; do not compost diseased plants. Clean up debris in fall to limit places where fungi overwinter.

Albizia

Albizia, mimosa, silk tree. Trees.

Mimosas are deciduous trees growing to 40′ high, with finely textured foliage and pink blooms. They are easy to transplant. Mimosas flower best in full sun. They can adapt to a range of soil conditions.

Because of their insect and disease problems, mimosas are usually short-lived. Mimosa webworms, 1″ long, brown caterpillars, bind leaves with webs and skeletonize foliage. Remove and destroy nests; spray leaves with BTK, or with pyrethrin as a last resort. Mimosa wilt produces wilted leaves and dead branches. Remove and destroy infected trees.

Alcea

Hollyhock. Biennials.

Hardy hollyhocks create an excellent backdrop for a perennial border and offer blooms in shades of red, pink, purple, yellow, and white. Hollyhocks grow 5′-9′ tall, bearing midsummer spikes of fluffy single or double flowers, up to 4″ wide.

Sun-loving and free-flowering, hollyhocks prosper in rich, heavy, moist, well-drained soils, but will thrive in ordinary soils, too. Plant 2′ apart in a site sheltered from wind. Most cultivars require staking.

Problems

Leaf surfaces pale; powdery orange spots beneath. Cause: Rust. Hollyhock's most common disease also deforms leaves and stems and causes early leaf drop. Its orange spots release fungal spores that spread via wind and rain. Grow resistant cultivars; keep leaves dry and encourage good air circulation. In problem areas, apply wettable sulfur several weeks before rust normally appears. Remove and destroy infected plant parts.

Leaves skeletonized. Cause: Leaf-feeding beetles. Japanese beetles are prime suspects here, although other beetles, such as spotted cucumber beetles and rose chafers, may also attack. Handpick or spray with a solution of 1 tablespoon 70 percent isopropyl alcohol to 1 pint prepared pyrethrin mixture. Repeat every 3-5 days until beetles are gone.

Buds and leaves deformed or dwarfed. Cause: True bugs. Several of the true bugs injure hollyhocks with their piercing-sucking mouthparts and release a toxin that deforms plants. See "Buds and leaves deformed or dwarfed" on page 178 for control information.

Allium

Allium. Bulbs.

Related to onions and garlic, but much prettier, ornamental alliums bloom in shades of blue, purple, pink, white, and yellow. Allium foliage may be rounded and hollow like onion leaves or straplike and solid; when bruised, the leaves release the pungent fragrance associated with culinary alliums. The blossoms, however, may have a pleasant floral smell. Small star- or bell-shaped flowers cluster in 2″-9″ globes atop leafless stems in early summer. Plants range from 6″ to 5′ tall.

Plant allium bulbs in fall at a depth roughly 3 times the bulbs' diameter. Select a site with full sun and well-drained soil; taller species may need some protection from wind. Water regularly throughout the growing season, but avoid excess moisture, which encourages rots. Allium foliage dies back shortly after flowering ends. If division is needed to reduce crowding, separate bulblets after foliage dies back; replant immediately.

Problems

Leaves with large, ragged holes. Cause: Slugs and snails. Slugs and snails may feed on foliage; see "Leaves with large, ragged holes" on page 50.

Leaves with silver-white streaks. Cause: Onion thrips. These 1/50″-1/25″ thrips suck sap from foliage; severe infestations stunt plants and may impair flowering. Thrips are hard to control because they burrow into plant tissue. Remove and destroy infested plant parts; encourage predators such as pirate bugs, lacewings, and lady beetles. Clean up debris where thrips overwinter. Monitor and trap thrips with blue sticky traps. Apply insecticidal soap sprays regularly once pests appear on traps.

Almond

Prunus dulcis var. *dulcis* (Rosaceae)

Almonds are deciduous trees that grow to about 20′ and bloom in early spring. Nuts are borne on short-lived spurs—short branches that elongate only a fraction of an inch per year. Almonds are hardy in Zones 6-9.

Grow almonds where summers are hot and dry, in sunny sites, free from late spring frosts. To set nuts, many cultivars need cross-pollination; others, such as 'All-in-One,' are self-pollinating. The cultivar 'Archedoise' resists many fungal diseases. Almonds and peaches belong to the same genus and are affected by similar insects and diseases. For more information on problems and solutions, see the Peach entry beginning on page 164.

Problems

Blossoms shriveled, covered with fuzzy gray masses. Cause: Brown rot. Nut hulls also may rot. To control this fungal disease, harvest nuts early, just when hulls begin to split. Before spring growth begins, remove infected nuts and twigs with sunken lesions. For greater control, spray with sulfur or copper.

Nuts disappear. Cause: Squirrels and other pests. For more information, see "Stopping Animal Pests" on page 408.

Growing shoots wilted or dying. Causes: Oriental fruit moths; peach twig borers. Both pests tunnel into growing shoots and cause wilting. For information on controls, see "Growing shoots wilted or dying" on page 166.

Leaves with small purple spots, some spots with centers missing. Cause: Shothole disease. Centers of leaf spots often enlarge to about ¼″, then fall out. For more information on this fungal disease, see "Leaves with small purple spots, some spots with centers missing" on page 167.

Amaranthus

Amaranth. Annuals.

Their colorful leaves or drooping flower clusters make these plants ideal focal points in the garden. They grow from 1′ to 6′ tall, depending on the species and cultivar. The leaves or flowers also stand out in cut flower arrangements.

Amaranth seedlings do not transplant well, so direct-seed in early summer when night temperatures are consistently 60°F or above. Water and feed seedlings regularly only until their colors appear. Thin 1′-2′ apart. Amaranth markedly prefers hot, dry, sunny areas and poor to average soil. Excess fertility causes dull foliage and weak stems. Overwatering leads to root rot, causing stems, crowns, and roots to turn brown or black. Once it begins, there is no cure, so the best control is prevention.

Anemone

Anemone, windflower. Tubers.

Tuberous species of anemones include low-growing Grecian windflower (*Anemone blanda*), a 6″-8″ plant with deeply divided leaves and daisylike, blue, pink, or white flowers, and showy, 7″-15″ poppy anemone (*A. coronaria*), a more tender plant (Zone 8) with single or semidouble blooms in deep shades of red, blue, violet, and white. Both species bloom in early spring.

Soak tubers overnight in warm water before planting in fall. Plant tubers 3″ deep in light shade in richly organic, neutral to slightly alkaline, well-drained soil. Windflower tubers often lack an obvious top and bottom; plant them on their sides, so stems grow up and roots grow down. Tubers planted sideways have a better chance of growing than those planted upside down. Poppy anemone tubers resemble claws; plant claw side down. Plantings expand slowly. Divide crowded clumps in late summer after foliage dies down, making sure each piece has a bud. Dust with a copper-based fungicide or sulfur and let dry for 2 days in a shady, airy spot before replanting.

Problems

Foliage disappears. Causes: Animal pests; slugs and snails. Birds and rabbits are quite fond of early anemone leaves. Cover young plants with netting if the problem is severe enough to warrant control. See "Stopping Animal Pests" on page 408 for control information.

Slugs and snails also eat anemone leaves; see "Leaves with large, ragged holes" on page 50 for controls.

Leaves, stems, and buds distorted, sticky; clusters of small insects. Cause: Aphids. These tiny pests occasionally trouble anemones. See "Leaves, stems, and buds distorted, sticky; clusters of small insects" on page 51 for control information.

Annuals

Controlling pests and diseases on annual plants is largely a matter of prevention. Annuals die naturally after 1 season, so insect and disease control is less critical than on more permanent plants. Sometimes, though, pest problems can cut short the bloom season and your enjoyment of the flowering annuals, so you'll need to be aware of the few potential problems. Check the individual entries, or see the text below for common problems.

Leaf Problems

Leaves stippled with yellow; foliage webbed. Cause: Spider mites. These tiny, spiderlike pests generally feed on the undersides of plant leaves. They suck sap from plant leaves, initially causing a yellow flecking on the upper leaf surfaces. Severe infestations can cause leaves to turn yellow or white; damaged leaves will eventually turn brown and drop. Tiny webs may be evident on leaves and stem tips.

Control spider mites by spraying plants thoroughly with water (especially the undersides of the leaves) 2-3 times a day for several days. For severe infestations, spray plants with insecticidal soap twice, 2-3 days apart.

Leaves yellow; plant weakened. Cause: Whiteflies. These tiny, white, mothlike flies and their flattened, scalelike larvae feed on leaves, sucking out the plant juices. The adults often fly up in a cloud when you disturb an infested plant. Whiteflies secrete a sugary substance called honeydew, which makes leaves sticky and may encourage the growth of sooty mold fungus; see "Leaves with black coating" below for more information.

Eliminating garden weeds helps keep whitefly populations down. A few yellow sticky traps in the area may catch whiteflies and other pest bugs, but also some beneficials later in the season. Try introducing green lacewings into the garden. Spray leaves with insecticidal soap every 3 days for 2 weeks, paying special attention to lower leaf surfaces. Use pyrethrin as a last resort.

Leaves with large, ragged holes. Cause: Slugs and snails. Slugs and snails range in size from 1/8" to 8" and may be gray, tan, green, black, yellow, or spotted. A thick layer of mulch and plants with low-growing leaves provide shady hiding places from which these pests emerge to feed at night. All species rasp large holes in leaves and stems; slimy trails of mucus also signal their presence.

To deter these pests, sprinkle bands of coarse, dry, scratchy materials, such as cinders, wood ashes, cedar sawdust, and diatomaceous earth, around plants or beds to irritate the soft bodies of these pests; renew frequently. Push 4″-8″ copper strips into the soil around beds as edging. Lay boards, cabbage leaves, or overturned clay pots—anything that offers a cool, damp daytime haven—around the garden; destroy pests that congregate underneath. Handpick slugs and snails from plants at night; drop them into soapy water or sprinkle them with table salt to dehydrate and kill them. Set out shallow pans of beer; remove drowned pests daily.

Leaves with black coating. Cause: Sooty mold. This fungus grows on the sugary, sticky honeydew produced by aphids, scales, whiteflies, and mealybugs. The black fungal coating doesn't harm leaves directly, but it does shade the leaves and reduce growth.

The best control is to deal with the pests that are producing the honeydew. Determine what pests your plant has and apply the appropriate control. (If the plant itself doesn't show signs of pest damage, the honeydew may be dripping down from an overhanging plant.) On small plants, you can wipe the leaves with a damp cloth to remove the honeydew and the mold.

Leaves with powdery white patches. Cause: Powdery mildew. The powdery leaf patches of this common plant disease consist of fungal strands and spores. The fungus feeds on plant nutrients, causing leaf yellowing. The problem is most common on upper leaf surfaces. Fungus from 1 plant can spread through a crowded area within days.

Encourage good air circulation by thinning plants. Water from below to keep foliage dry. Protect mildew-prone plants, like zinnias, by spraying leaves early in the season with an antitranspirant spray. Pick off infected leaves; remove and destroy seriously infected plants. Spray remaining plants with sulfur or a commercial fungicidal soap spray.

Leaves greenish yellow; growth poor. Cause: Aster yellows. This is a common disease of many annuals. Leafhoppers carry the disease, spreading it through their sap-sucking feeding habits. These tiny, green or brown, wedge-shaped insects move about by running sideways or by hopping. They usually feed on the undersides of plant leaves. Affected plants branch abnormally. Flowers are deformed or absent. Flowers, regardless of original color, turn a yellow-green.

There is no cure. Remove and destroy infected plants. Control leafhoppers by spraying remaining plants with a mixture of insecticidal soap and 70 percent isopropyl alcohol (1 tablespoon alcohol to 1 pint soap solution). Spray 3 times, once every 3 days. Overwatering increases chances of the disease, particularly in heavy soils.

Leaves spotted; buds wilted and dark. Cause: Plant bugs. Tarnished plant bugs are 1/4″ insects with yellow, black, and red mottling on the wings. Four-lined plant bugs are 1/4″, yellow bugs with 4 black stripes on their wings. Both pests attack a wide range of ornamentals. Control by handpicking; dust with rotenone for serious infestations.

Leaves with spots. Cause: Leaf spots. A large number of fungi and bacteria cause spots on plant leaves, in a variety of colors, shapes, and sizes. In some cases, the spots may spread to cover entire leaves, stunting plant growth.

Fortunately, the same controls are effective against many leaf spot diseases. Remove and discard infected leaves. Thin plants and avoid crowding future plantings. Wash your tools and hands after handling infected plants. Avoid overhead watering. Inspect bedding plants on purchase. Clean up plant debris to remove overwintering sites.

Leaves rolled. Cause: Leafrollers. These ½"-1", green-bodied, brown-headed caterpillars web leaf edges together and feed on the enclosed leaves and buds. If the infestation is light, break open the "envelopes" and pick off the caterpillars; for heavier infestations, spray the leaves with BTK.

Whole Plant Problems

Leaves, stems, and buds distorted. Cause: Aphids. Several species of pear-shaped, 1/32"-1/8" aphids attack annuals. Aphids can be green, pink, black, gray, or with a white fluffy coating and have long antennae and 2 short tubes projecting from the rear of their abdomens. These pests cluster under leaves and on growing tips. Aphids suck plant sap, causing leaf and bud distortion and blossom and leaf drop. Their feeding may spread diseases such as aster yellows. As they feed, they excrete sticky honeydew on which sooty mold grows; see "Leaves with black coating" on page 19 for more information.

Wash aphids from plants by spraying plants with water; repeat as needed to control infestations. Encourage natural predators and parasites. For serious infestations, spray with insecticidal soap, or pyrethrin as a last resort.

Seedlings or young plants cut off at soil level. Cause: Cutworms. These brown or gray, 1"-2" caterpillars feed on plants at night, cutting off transplants and sometimes eating entire seedlings. You'll find severed plants lying on the ground the next morning. Most active in May and June, cutworms can destroy several plants each night. These caterpillars remain below the soil surface during the day.

Protect plants by placing a collar, such as a toilet paper roll or an open-ended can, around each young plant. Push the collar into the soil so that about half is below the surface. Apply parasitic nematodes to the soil.

Leaves and stems with white, cottony clusters. Cause: Mealybugs. At first, mealybugs may be hard to see, but as the plant wilts, and mealybugs rapidly multiply, colonies become quite visible. These minute insects are covered with a fluffy white coating. They suck plant sap and cause the plant to look unhealthy. Mealybugs secrete a sugary substance called honeydew, which makes leaves sticky and may encourage the growth of sooty mold fungus; see "Leaves with black coating" on page 19 for more information.

Control small infestations by spraying them off the plant with water. Insecticidal soap will also control mealybugs; be sure to cover the plant thoroughly.

Seedlings die. Cause: Damping-off. Damping-off is caused by various soil fungi. Sometimes seedling stems rot before they even appear above the surface; in other cases, seedlings are affected after they emerge. The weak, blackened stems will collapse, resulting in seedling death.

Damaged seedlings cannot be cured. To prevent damping-off, let the soil surface dry slightly between waterings. Indoors, run a low-speed fan near your seedlings to promote good air circulation. Outdoors, thin seedlings to avoid overcrowding. Do not add nitrogen fertilizers until seedlings have produced their first true leaves.

Antirrhinum

Snapdragon.
Tender perennials grown as annuals.

Most gardeners try snapdragons at least once, and many make them a garden mainstay. Bloom time is winter and spring in mild winter areas, and spring to summer elsewhere. Snapdragons prefer cool weather but tolerate heat. Single or double upright flower spikes offer a variety of color choices, including red, white, yellow, orange, and pink. Height range is 4″-48″. Plants may self-sow.

Start seeds indoors in late summer or early fall in mild climates and in early spring where frosts are common. Do not cover seeds; they require light and warmth for germination. Seedlings appear in 7 days. Move seedlings outside in spring when the planting bed can be worked. Direct-seeding outdoors is possible, but seeds dry out rapidly, so frequent sprinkling is necessary. Purchased bedding plants flower faster than home-grown plants because of the controlled growing conditions.

Place plants 6″-12″ apart in full sun and humus-rich, well-drained, somewhat sandy soil. To keep plants from becoming leggy or spindly, pinch them back when they are 3″-6″ high. Pinching may cause a slight flowering setback at first, but later, flower production will markedly increase. Give ample water from below. Fertilize once a month. Cut flower spikes frequently to maintain bloom. When flowering slows down, cut plants back severely and give liquid fertilizer. In windy areas, staking may be necessary.

Problems

Stems and leaf undersides with dusty, dark brown spots. Cause: Rust. This fungus develops quite rapidly, so leaves must be monitored for initial spotting. Rust-resistant snapdragons are now available and should be used whenever possible. However, they don't eliminate the problem.

Infections are less likely if snapdragons are grown rapidly and vigorously. Keeping the soil evenly moist will encourage strong, healthy growth. If rust spotting appears, or if you have had rust problems in prior years, dust every 2 weeks with sulfur, beginning in very early spring. Fertilize and water regularly, but avoid overhead watering, which helps transfer rust spores. Water in mornings. Remove and destroy badly infected plants. In severe situations, where rust appears in succeeding years despite controls, change planting locations.

Leaves, stems, and buds distorted. Cause: Aphids. See "Leaves, stems, and buds distorted" on the opposite page for controls.

Leaves stippled with yellow; foliage webbed. Cause: Spider mites. For control measures, see "Leaves stippled with yellow; foliage webbed" on page 18.

Apple

Malus spp. (Rosaceae)

Apples are deciduous trees growing from 6′ to more than 30′, depending on soil, rootstock, and cultivar. The showy, pink-tinged white blossoms appear in spring mostly on spurs—short branches that elongate only a fraction of an inch per year. Apples are hardy in Zones 3-9.

Culture

Plant in full sun in well-drained, moderately fertile, slightly acidic soil. Train trees to a framework of well-spaced, wide-angled branches. Prune bearing trees each winter to admit light into the tree and encourage good air circulation. As you prune, remove diseased and spindly wood and crossed branches. Where dense growth blocks out light, remove extra shoots at their bases. To develop growth on spindly shoots, remove the end of the shoot just above an outward-facing bud. For more pruning information, see "Pruning and Training" on page 101. For best results, choose cultivars resistant to a wide range of diseases. These include 'Jonafree', 'Liberty', 'Nova Easy Grow', 'Priscilla', 'Redfree', and 'William's Pride'.

To set fruit, most cultivars need cross-pollination by a second compatible apple or crab apple planted within 40'-50'. Some cultivars, such as 'Jonagold' and 'Mutsu', produce nonviable pollen and can't serve as pollinators. A few, like 'Golden Delicious', are self-pollinating. If you're planting 1 tree, improve fruit set by grafting a branch of a suitable pollinator onto the tree. For more information on setting fruit, see "Setting Fruit" on page 101.

Fruit Problems

Fruit with holes surrounded by brown, crumbly excrement. Cause: Codling moths. Adult codling moths appear in early spring and lay eggs in trees within 2-6 weeks of blossom time. Eggs hatch into larvae within 5-14 days. The fat, white or pinkish, ⅞" caterpillars tunnel through fruit and may have departed by the time you discover the holes, which may be filled with what looks like moist sawdust. For light infestations, kill eggs by spraying superior oil on leaves and twigs within 2-6 weeks of blossoming. For heavy infestations, kill larvae before they tunnel into fruit by spraying the tree canopy with ryania at petal fall and again 10-14 days later. Pyrethrin spray mixed with a synergist (see "The Other Ingredients" on page 469) also kills larvae. For additional details on codling moth's life cycle and controls, see "Coordinated Control" on page 262.

Codling moths often produce several generations per growing season. Trapping the pupating caterpillars aids control. Remove loose bark and wrap the trunk with a band of corrugated cardboard or burlap. Periodically remove the band and destroy pupae. Also inspect harvest containers for pupae.

Since codling moths prefer crowded fruit, you can discourage attack by thinning apples until no fruit touches. Infested apples may drop early; pick up and destroy dropped fruit before larvae emerge to pupate. If you have a single backyard tree and no other trees in your area, try trapping male moths with pheromones. Mating disruption pheromones are also available. Since females rarely fly more than 100 yards looking for host plants, plant your trees well away from other apple, pear, or walnut host trees in your area.

The effectiveness of other controls depends on local conditions. Introduced *Trichogramma* wasps control this pest in some orchards but not in others. Success depends on such factors as weather, optimum timing of parasite release, and using the correct species of wasp. If you'd like to try this technique, be sure to purchase a species of *Trichogramma* that parasitizes codling moths. BTK may also help with control, but you have to apply it during the 3-5 days between the time when the eggs hatch and the larvae enter the fruit. Once the larvae are in the fruit, BTK is ineffective. You may find that BTK is more effective when combined with a feeding attractant (also called an appetite stimulant), such as molasses. See the opposite page for an illustration of this pest.

In the future, gardeners may be able to buy new experimental codling moth control products, such as codling moth granulosis virus. Another experimental product, encapsulated BTK, releases the bacteria through a timed-release capsule for potential long-term control of codling moth larvae. Also, cultivars resistant to codling moths may eventually be available.

Fruit dimpled; brown tunnels through flesh. Cause: Apple maggots. These ¼" larvae of the apple maggot fly ruin fruit by copious tunneling. Adult flies emerge from soilborne pupae in late June and continue to appear until early autumn. Flies puncture fruit skin and deposit eggs, which hatch into fruit-tunneling maggots. Infested apples often drop. To prevent buildup of pupae around trees, collect and destroy dropped fruit at least weekly.

Apple maggot flies are attracted to fruit by sight. You can control them by trapping them on dark red balls coated with a sticky coating, such as Tangle-Trap. Buy commercially made traps or make your own from discarded croquet balls. In mid-June hang 1

APPLE ◆ What Goes Wrong and Why

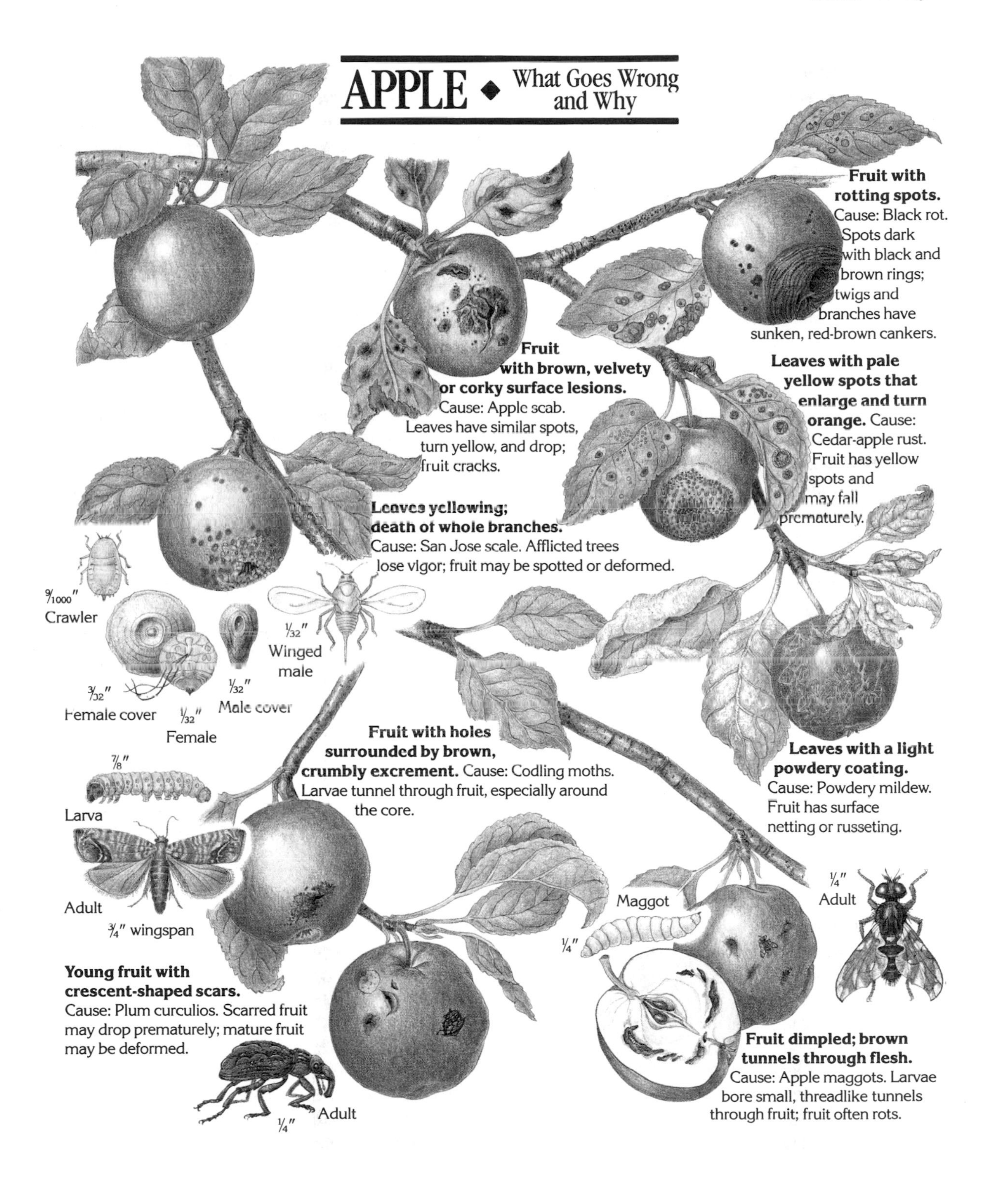

trap per dwarf tree or 4-8 traps per full-size tree. Hang traps at eye-level, 2′-3′ in from branch tips, near fruit but not completely hidden by leaves. Clean traps every few days and reapply the sticky coating. See page 23 for an illustration of this pest.

Young fruit with crescent-shaped scars. Cause: Plum curculios. This beetle, common east of the Rocky Mountains, leaves a characteristic crescent-shaped scar as it lays eggs in fruit. Damaged apples may drop, but frequently will remain on the tree. Since plum curculios cannot mature in hard apple flesh, fruit that doesn't drop will be superficially scarred but otherwise edible.

To control this pest without sprays, spread a dropcloth beneath the tree and jar the trunk and branches with a padded mallet. Collect and destroy curculios that fall onto the sheet. For best results, jar the tree twice a day, beginning as soon as you see the first scarred fruit. In addition, prevent hatching of some curculio eggs by picking up and discarding dropped fruit. Some apples ('Mutsu', for example) are more resistant to plum curculios than others. In the future, breeders may be able to develop other apple cultivars with even greater curculio resistance. For more information, see "Young fruit with crescent-shaped scars" on page 186. See page 23 for an illustration of this pest.

Fruit with brown, velvety or corky surface lesions. Cause: Apple scab. This serious, widespread disease begins when spring warmth and moisture promote the discharge of fungal spores from old, infested apple leaves into the air. These spores can infect the leaves and fruits of susceptible apples growing nearby. To prevent scab from spreading into the tree each spring, destroy dropped leaves in the fall. Either collect and compost leaves in a hot compost pile or hasten their decay on the ground by shredding them with a lawn mower. You may also till leaves into the soil or spray them with a nitrogen source, such as blood meal, to speed decomposition.

Growing-season applications of copper, sulfur, or lime-sulfur sprays will help control scab. If you have susceptible cultivars or if the weather is very warm and wet, spray weekly beginning with the appearance of the first green tips on the tree until disease pressure subsides. Pruning to encourage sunlight penetration and air circulation also helps with control. Scab-resistant cultivars include 'Golden Delicious', 'Grimes Golden', 'Jonagold', 'Jonathon', 'Mutsu', 'Spigold', and 'York'.

In the future gardeners may be able to treat fallen leaves with a fungus such as *Athelia bambacina,* which inhibits development of overwintering apple scab. See page 23 for an illustration of apple scab.

Fruit with rotting spots. Cause: Summer disease. Summer diseases are fungal diseases associated with hot weather. Dark spots with alternating black and brown rings indicate black rot. Other summer diseases include bitter rot (slightly sunken, tan spots) and white rot (watery decay). Black rot prevails where summers are cool; bitter rot and white rot, where summers are warm. Summer disease fungi overwinter in mummified fruit and in cankers on diseased wood. To control, remove and dispose of all cankered wood. Collect and destroy mummified fruit. Black rot and white rot attack weakened or wounded trees, so keep them healthy with good pruning and nutrition. Prune in early spring, when wounds heal most quickly. Sulfur sprays help control black rot (see page 23 for an illustration). Cultivars resistant to bitter rot include 'Akane', 'Blairmont', 'Fuji', 'Hawaii', 'Jonafree', 'Jonalicious', 'Liberty', 'Melrose', 'Priscilla', and 'Spartan'. Cultivars resistant to white rot include 'Akane', 'Arkansas Black', 'Arkansaw', 'Dayton', 'Fuji', 'Hawaii', 'Jonafree', 'Jonalicious', 'Liberty', 'Limbertwig', and 'Melrose'.

Fruit with yellow skin spots that later turn orange. Cause: Cedar-apple rust. Infected fruit is small, deformed, and may fall prematurely. For more information see "Leaves with pale yellow spots that enlarge and turn orange" below.

Fruit with surface netting or russeting. Cause: Powdery mildew. For more information, see "Leaves with a light powdery coating" below.

Fruit with red skin spots bearing white centers. Cause: San Jose scale. For more information, see "Leaves yellow; death of whole branches" on page 26.

Fruit with raised black spots or brown smudges on the surface. Causes: Fly speck; sooty blotch. Black spots are fly speck; brown smudges are sooty blotch. Both of these blemishes are fungal diseases and both can be controlled with sulfur sprays. But since both are superficial, you can just rub them off the fruit.

Leaf and Branch Problems

Leaves with olive-brown, velvety spots that become brown and corky. Cause: Apple scab. Infected leaves may turn yellow and drop prematurely, further weakening the tree. For more information see "Fruit with brown, velvety or corky surface lesions" above.

New leaves twisted or curled and covered with a sticky coating. Cause: Aphids. You may find these tiny green, black, gray, pink, or white fluffy-coated insects on leaf undersides. The leaves may be covered with a black coating, caused by a fungus called sooty mold that feeds on the honeydew exuded by aphids. Aphids weaken trees by sucking sap, but they depart by midsummer. For a light infestation, just wait it out. Also grow nectar-producing flowers, such as dill and buckwheat, near your trees to provide food for aphid predators (parasitic wasps, lady beetles, and hover flies). Or attract aphid predators by spraying commercial or homemade yeast-and-sugar mixtures on your trees. For more information, see "New leaves twisted or curled and covered with a sticky coating" on page 67.

Leaves with pale yellow spots that enlarge and turn orange. Cause: Cedar-apple rust. Leaves infected with cedar-apple rust may drop prematurely. This fungal disease overwinters as a gall on various juniper species—commonly on Eastern red cedar (*Juniperus virginiana*). In the spring the galls swell, push out orange horns, and discharge disease spores that are borne on the wind to infect apple trees. The leaf spots caused by cedar-apple rust won't spread disease into the tree, and you can't make them go away once they appear.

Cedar-apple rust isn't a problem in areas with few Eastern red cedars or other junipers (*Juniperus* spp.). Removing nearby Eastern red cedars won't prevent the arrival of windborne disease coming from trees miles away. Sulfur sprays are only fairly effective. Resistant cultivars include 'Arkansas Black', 'Empire', 'Gala', 'Granny Smith', 'Gravenstein', 'Grimes Golden', 'Jerseymac', 'McIntosh', 'Macoun', 'Tydeman's Red', and 'Winesap'. See page 23 for an illustration of this disease.

Leaves with a light powdery coating. Cause: Powdery mildew. As this fungal disease becomes more severe, leaves may curl lengthwise and drop. Since it overwinters in dormant buds, winter pruning of infected buds, which show white fungal growths, aids control. Spray sulfur for growing-season control. Resistant cultivars include 'Empire', 'Golden Delicious', 'McIntosh', 'Mutsu', 'Red Delicious', 'Rhode Island Greening', and 'Spigold'. See page 23 for an illustration of this disease.

Leaves suddenly blacken, with tips of growing shoots bent over. Cause: Fire blight. Don't confuse this disease with sooty mold, a black fungus that rubs off easily. Fire blight bacteria may travel to roots and kill the entire tree. For growing-season control, remove blighted parts at least 6″ below the infected area. Between cuts, dip tools into isopropyl alcohol or 10 percent bleach solution (1 part bleach to 9 parts water). Resistant cultivars include 'Arkansas Black', 'Baldwin', 'Ben Davis', 'Empire', and 'Winesap'. For more information, see "Leaves suddenly blacken, with tips of growing shoots bent over" on page 170.

Whole Plant Problems

Leaves yellow; death of whole branches. Cause: San Jose scale. Clusters of these sucking insects cling to bark and appear as small gray bumps that can be easily scraped off with a fingernail. Control with late winter application of dormant oil spray. See page 23 for an illustration of this pest.

Fruit disappears; bark gnawed. Causes: Deer; rabbits; mice. For information on controlling these pests, see "Stopping Animal Pests" on page 408.

Tree declines; sawdustlike material on trunk near ground level. Cause: Roundheaded appletree borers. These creamy white, dark-headed larvae bore into trunks near ground level, girdling the tree or tunneling into the heartwood. Kill borers by inserting a thin, flexible wire or by injecting parasitic nematodes into the hole.

Apricot

Prunus armeniaca (Rosaceae)

Apricots are deciduous trees growing 20′-30′. The pink blossoms appear in spring on spurs—short branches that elongate only a fraction of an inch per year—1 year old and older. Apricots are hardy in Zones 4-9.

Plant in a sunny site with well-drained, moderately fertile soil and protected from late frost. Train young trees to a modified central leader; for complete instructions, see the illustration on page 102. Most apricots are self-pollinating, but some cultivars bear more if cross-pollinated. For more information, see "Setting Fruit" on page 101. Apricots and peaches belong to the same genus and have similar problems. For more information on problems and solutions, see the Peach entry beginning on page 164.

Problems

Young fruit with crescent-shaped scars. Cause: Plum curculios. These beetles, common east of the Rockies, leave characteristic scars as they lay eggs in fruit. Damaged fruit usually drops. For information on controls, see "Young fruit with crescent-shaped scars" on page 186.

Fruit with small brown spots that enlarge and grow fuzzy in humid weather. Cause: Brown rot. Blossoms attacked by this fungal disease also may wither, fruit may mummify (dry and shrivel) on the tree, and leaves may turn brown. Resistant cultivars include 'Harcot', 'Hargrand', and 'Harlayne'. For more information, see "Fruit with small brown spots that enlarge and grow fuzzy in humid weather" on page 166.

Fruit with small, dark, sunken spots or cracks on skin. Cause: Bacterial leaf spot. This disease, common in the Southeast, is very difficult to control. Plant resistant cultivars, including 'Alfred', 'Curtis', 'Harcot', 'Hargrand', and 'Harlayne'. For more information, see "Fruit with small, dark, sunken spots or cracks on skin" on page 166.

Fruit with pinkish worms. Cause: Oriental fruit moth larvae. For more information on this pest, which also tunnels into growing shoots, see "Growing shoots wilted or dying" on page 166.

Growing shoots wilted or dying. Causes: Oriental fruit moth larvae; peach twig borers. Both pests tunnel into growing shoots and cause wilting. For more information, see "Growing shoots wilted or dying" on page 166.

Leaves with small purple spots, some spots with centers missing. Cause: Shothole disease. Centers of leaf spots often enlarge to about ¼″, then fall out. This fungal disease, common in the West, spreads rapidly on wet foliage.

For more information, see "Leaves with small purple spots, some spots with centers missing" on page 167.

Tree declines; gummy exudate mixed with sawdustlike material on trunk near ground level. Cause: Peachtree borers. Inspect the trunk near or just below the ground; you may find holes and a gummy exudate made by peachtree borers, which are the larvae of a clear-winged moth that bore the inner bark. For information on controls, see "Tree declines; gummy exudate mixed with sawdustlike material on trunk near ground" on page 168.

Branches wilting and dying, fail to leaf out in spring. Causes: Bacterial canker; Valsa canker. Both diseases may cause amber gum to exude from the bark. The cultivars 'Harcot', 'Hargrand', 'Harlayne', and 'Harogem' are resistant to Valsa. For more information, see "Branches wilting and dying, fail to leaf out in spring" on page 167.

Aquilegia

Columbine. Perennials.

Columbines brighten spring and summer with blossoms of yellow, white, purple, red, or bicolor. Plants reach 1′-3′ and bear 1″-4″, spurred flowers.

Plant these woodland natives in moist, organic, well-drained, slightly acid soil. Overly rich soil causes weak, short-lived plants, lush growth, and few flowers. Plants tolerate full sun if temperatures stay below 80°F but prefer light shade. Set plants 1′ apart in spring after the last frost or in fall when plants are dormant, placing crowns at soil level. Columbines have long roots; only young plants transplant easily. To start from seed, sow outdoors in early summer or indoors in winter. Don't cover the seed; it needs light to germinate. Columbines self-seed, but seedlings may not match the parents. Mulch lightly in cold winter areas.

Problems

Leaves with tan or brown blotches or serpentine tunnels. Cause: Leafminers. These tiny pale green fly larvae feed between the upper and lower leaf surfaces. Prune off and destroy infested leaves. Remove debris in fall to destroy overwintering leafminers. Let parasitic wasps control these pests, or apply weekly insecticidal soap sprays starting when the first tunnel appears.

Stems blacken at base; leaves yellow; plant topples over. Cause: Fungal and/or bacterial rots. Prevent with cultural practices: Plant only in well-drained sites; avoid crown or root injuries from careless digging. Keep winter mulch away from crowns to prevent rotting during dormancy. Remove affected plants and surrounding soil; don't compost them.

Arabis

Rock cress. Perennials.

These low-growing, spreading plants are often used as groundcovers, in rock gardens, and as border edging; few species grow over 1′ tall. Rock cresses prefer soil on the dry side, cool temperatures, and full sun. White, pink, or rose flowers appear in early spring.

Summer heat can cause rock cresses to die out in the center, a condition best prevented with good growing conditions. Divide mature plants in spring in most areas. Space plants 5″ apart. Cut back after flowering to encourage dense growth. Apply a winter mulch in cold areas. If thrips, mites, or aphids cause leaf yellowing, knock them from plants with a strong water spray. Time watering so plants have dried by nightfall; wait until soil is fairly dry before respraying. Overwatering encourages rot, as does poor drainage.

Artemisia

Artemisia, wormwood. Perennials.

This genus includes southernwood (*Artemisia abrotanum*), wormwood (*A. absinthium*), and tarragon (*A. dracunculus*). Durable and sometimes woody, most species are grown for their feathery, gray-green to silver foliage. Plants given full sun and average, well-drained soil suffer few problems. Rich soil promotes rangy growth; excess moisture encourages root rot. Mounding types such as silvermound artemisia (*A. schmidtiana*), often react to high temperatures by dying out in the center; cut stems back to about 1″ tall to encourage overall new growth. Artemisias' aromatic oils repel most insects.

Asparagus

Asparagus officinalis (Liliaceae)

Asparagus is a long-lived perennial. Its tender young shoots are one of the first vegetables ready to harvest in the spring.

Culture

Asparagus is hardy in Zones 2-9. It thrives in any area with winter ground freezes or a dry season to provide a dormant period each year. Asparagus does best in full sun and deep, well-drained soil. Select a permanent location carefully, since plants will produce for 20 years or more. Dig out all weeds and add plenty of compost to the soil before planting. Asparagus requires high levels of phosphorus, potassium, and nitrogen. Do a soil test and add amendments as necessary. If your soil is heavy or poorly drained, plant asparagus in raised beds.

Plant 1-year-old crowns from a reputable nursery that sells fresh, firm, disease-free roots, or start your plants from seed. Soak seeds or crowns in compost tea for 5 minutes before planting to reduce disease problems.

Most seed-grown asparagus plants eventually out-produce those started from crowns. Growing from seed also allows you to eliminate female plants. A bed of all male plants can produce as much as 30 percent more spears than a mixed bed of male and female plants. Plants grown from seed will flower their first summer. When the tiny flowers appear, observe them with a magnifying glass. Female flowers have well-developed, 3-lobed pistils; male blossoms are larger and longer than female flowers. Weed out all female plants. The following spring, transplant the males to the permanent bed.

Harvesting new plantings too soon can stress plants and make them more susceptible to pest problems. Harvest for 2 weeks the second season, 4 weeks the third season, and up to 8 weeks thereafter.

Mulch with a high-nitrogen compost each spring before spears emerge, and again in fall. Leave winter-killed foliage, along with straw or other light mulch, on the bed to provide winter protection. Remove and destroy the foliage before new growth appears in the spring; it can harbor diseases and pest eggs. Over the years, the crowns will push closer to the soil surface, resulting in smaller and less-tender spears. To remedy this, mound 6″ of soil over the rows each spring.

Spear Problems

Spears small. Causes: Young plants; low soil fertility; overharvested plants. Harvest lightly the first few years so plants can grow strong roots. Reduce harvest if established plantings begin to lose vigor. Asparagus is a heavy feeder; add lots of compost to maintain high soil fertility.

Spears small with brown streaks or girdled at soil line. Cause: Asparagus miners. Larvae are 3/16″ long, white maggots. Adults are small black flies. Destroy infested spears to control.

Spears turn brown, may get soft or wither and dry. Cause: Cold injury. Cut and discard damaged spears. Cover spears with mulch or newspaper when freezing nights are predicted.

Spears small, with large lesions at or below soil line. Cause: Fusarium wilt. Leaves and stems yellow, plants wilt, seedlings die. There is no cure; remove and destroy infected plants. To avoid problems, don't put new plantings where asparagus or other Fusarium-susceptible plants have grown in 8 years. Plant disease-free seed and crowns or resistant cultivars such as 'Greenwich', 'Jersey Giant', 'Martha Washington', and 'Viking KB3'. Disinfect seed before planting. Soak it for 2 minutes in a 10 percent bleach solution (1 part bleach to 9 parts water), and rinse for 1 minute under running water.

There are reports that adding rock salt to asparagus soil at 2 pounds per 100 square feet may suppress Fusarium infection and improve asparagus root growth.

Spears turn brown near soil line. Cause: Crown rot. Remove and destroy diseased plants, including roots. Prevent crown rot by planting in raised beds and maintaining good drainage. Keep soil pH above 6.0. Wait 2 years before harvesting new plantings. Some research suggests that not harvesting the first spear of the spring on each plant may help prevent crown rot because the developing frond produces food for the plant.

Spears crooked and deformed, may be brown, or scarred; leaves chewed or missing. Causes: Asparagus beetles; asparagus fern caterpillars; cucumber beetles; mechanical injury. Asparagus beetles are blue-black, 1/4″ long insects with cream-colored spots and red borders on wing covers. Larvae are 1/3″ long, humpbacked, gray grubs with black heads. They are most active in cool weather. Spotted asparagus beetles are reddish orange with 12 black spots on wing covers. Larvae are orange. Fertilize plants to encourage new leaf growth. Spray with a commercial pyrethrin/rotenone mix to control beetles. Destroy tops in late winter to remove overwintering beetles and future problems.

Asparagus fern caterpillar (also known as the beet armyworm) feeds on leaves. It is a dull green, 1½″ long caterpillar with a light-colored stripe along each side of the body. Handpick or spray with BTK to control.

Cucumber beetles also eat asparagus leaves. For description and controls, see "Leaves with chewed holes" on page 215.

Deep cultivation can damage developing spears. Cultivate shallowly or use mulch instead. In windy areas, blowing sand can scar spears; protect plantings with windbreaks.

Spears chewed at soil line. Cause: Cutworms. Check for cutworms at night with a flashlight. Control with BTK or parasitic nematodes. See "Seedlings clipped off at soil line" on page 246 for more controls.

Spear bracts open prematurely (feathering). Cause: Excessive heat. Harvest spears daily when they are about 8″ high, especially when temperatures are above 95°F.

Leaf Problems

Leaves yellow; growth slow. Causes: Nitrogen deficiency; waterlogged soil. Spray foliage with fish-meal tea and side-dress with compost to correct nitrogen deficiency. Waterlogged soil will produce the same symptoms. Make sure soil is well-drained or plant in raised beds.

Leaves yellow; plant dwarfed or rosetted. Cause: Asparagus aphids. These soft-bodied, pale green, powdery-looking insects suck plant juices and cause plants to weaken. Knock aphids off plants with a strong blast of water, or spray with insecticidal soap. If that fails to control the problem, spray with a commercial pyrethrin/rotenone mix.

Leaves turn brown and drop; stems and branches have small reddish blisters. Cause: Rust. Clean up and dispose of tops in late winter to eliminate overwintering spores. Rust weakens plants by reducing the leaf area and the amount of food stored in the roots. Repeated attacks can kill plants. Plant resistant cultivars such as 'California 500', 'Jersey Giant', 'Martha Washington', 'Mary Washington', 'Rutgers Beacon', 'Viking KB3', and 'Waltham Washington'.

Leaves dull gray-green to brown. Cause: Thrips. Adults are tiny, pale, rapidly moving, winged insects. The larvae are smaller, wingless versions barely visible with the naked eye. Trap thrips with sticky traps or spray with insecticidal soap. Research indicates that blue and yellow sticky traps are the most effective thrips catchers in greenhouses. Other studies suggest that thrips are most attracted to white sticky traps outdoors. Hang traps slightly higher than the tops of the plants.

Aster

Aster, Michaelmas daisy. Perennials.

Daisylike aster flowers of purple, pink, or white appear in late summer and fall on plants that range from 9″ to 6′ in height.

Asters require ample water—about 1″ per week—but also need well-drained soil and a sunny location. Plants will self-sow with abandon if fading flowers are not removed; seedlings aren't usually true to parent type. Grow tall cultivars out of wind and stake as needed. Divide in spring every 2-3 years.

Problems

Leaves, stems, and buds distorted, sticky; clusters of small insects. Cause: Aphids. For controls, see "Leaves, stems, and buds distorted, sticky; clusters of small insects" on page 177.

Leaves and/or flowers with holes. Cause: Japanese beetles. See "Leaves and/or flowers with holes" on page 176 for controls.

Leaves covered with white powder. Cause: Powdery mildew. For controls, see "Leaves covered with white powder" on page 177.

Leaves and flowers deformed, yellowish; small, tarlike spots on undersides. Cause: Lace bugs. These small pests with lacy wings cause yellow-brown leaf spots. Dark brown droppings on lower leaf surfaces confirm their activity. Remove debris in spring and fall to deter overwintering. For severe problems, spray superior oil, insecticidal soap, or pyrethrin.

Leaves with pale areas on upper surfaces; "downy" patches underneath. Cause: Downy mildew. This fungus spreads quickly during cool, wet nights and warm, humid days, causing leaves to wilt and die. Remove and destroy infected plant parts, encourage air circulation, and water early in the day to allow plants to dry before nightfall.

Astilbe

Astilbe, false spirea. Perennials.

Astilbes' dark green to bronze, fernlike foliage appears in spring, followed by pink, white, purple, or red plumelike flower spikes. These mid-border plants grow 1′-3′ tall with a similar spread.

Astilbes enjoy the soggy soil conditions shunned by most other perennials, although good drainage is needed if winters are wet. Use a 3″ layer of mulch around plants to conserve water. Never let soil dry completely—this causes brown-edged leaves and poor growth. Set plants 1′ apart in partial shade and rich soil. Astilbes benefit from spring fertilization and light feedings during the growing season. Divide every 3-5 years.

Grown under the conditions described above, astilbes suffer few pest or disease problems beyond those common to herbaceous perennials.

Problems

Leaves and/or flowers with holes. Cause: Japanese beetles. See "Leaves and/or flowers with holes" on page 176 for controls.

Leaves covered with white powder. Cause: Powdery mildew. Crowding under moist conditions invites powdery mildew. For control information, see "Leaves covered with white powder" on page 177.

Plant wilts while soil is moist. Cause: Fusarium wilt. *Fusarium* spp. fungi sometimes infect astilbes, causing wilting unrelated to adequate soil moisture. Young plants wilt quickly, while older ones turn pale green and lower leaves wilt. Stems show brown streaks that darken gradually; grayish pink mold may appear. Remove and destroy infected plants and surrounding soil; do not compost diseased materials. Don't plant susceptible crops in sites where wilt has appeared. Sterilize tools after use around infected plants.

Avocado

Persea americana (Lauraceae)

Avocado trees grow to about 60′ and have long, drooping leaves. Though evergreen, some trees shed their leaves just before a growth flush. Each tree bears many thousands of flowers, but pollination of only a small percentage of flowers ensures a full crop. Avocados are mostly self-pollinating. They are hardy to Zone 10.

Avocados need full sunlight and well-drained soil. Pruning is only necessary to keep a tree within bounds and to remove diseased wood.

Problems

Fruit with brown or purplish, scablike lesions. Cause: Fruit scab. Scab is a superficial skin fungus and not otherwise harmful to the plant. No treatment is necessary unless infection is so severe that fruit is deformed. For severe infection, apply copper sprays in early May and twice more at 4-5 week intervals.

Fruit gnawed or eaten. Causes: Birds; squirrels; rats; other small animals. Use traps or physical barriers to get rid of rodents; netting may be necessary for birds. See "Stopping Animal Pests" on page 408 for more information on controlling pests.

Leaves skeletonized and webbed together. Causes: Avocado caterpillars; omnivorous leafroller caterpillars. The yellowish green avocado caterpillar (also called an amorbia) is the larva of a reddish brown moth. Omnivorous leafrollers, yellowish green caterpillars with a stripe down their backs, are the larvae of dark brown moths. Natural enemies, such as *Trichogramma* wasps, often keep these pests sufficiently in check. For heavy infestations, spray BTK. Also look for and crush caterpillars and masses of eggs.

Leaves wilt and yellow on an entire branch. Cause: Verticillium wilt. A soilborne fungus causes this disease, and there is no way to save a tree once it has this problem. Don't plant avocados in soil that has sustained other Verticillium-susceptible crops such as tomatoes, peppers, and eggplants. If you suspect *Verticillium* fungi may be in your soil, solarize it for 1-2 months over the summer before planting Verticillium-susceptible plants.

Leaves pale green to yellow and dropping. Cause: San Jose scale. Look for the small, ash-colored to sooty black bumps of San Jose scale clinging to the bark. Colonies of these sucking, immobile insects cling to bark and weaken trees by sucking sap. Control scale with dormant oil spray applied in late winter.

Tree declines; stunted pale leaves; no new growth. Cause: Root rot. Root rot can be caused by overwatering or by poorly drained soil, but the disease may take years to develop. There is no cure. Replant in well-drained soil or on a raised mound of soil. 'Grande' and 'Martin' are rootstocks resistant to this problem.

Basil

Ocimum basilicum (Labiatae)

Basil is an annual herb grown for its aromatic leaves. To grow healthy, trouble-free plants, sow seed indoors in 70°F soil mix, or outdoors after the soil has warmed. Basil does best in rich, moist, well-drained soil with a pH between 5.0 and 8.0, and needs at least 4 hours of full sun per day for good growth.

Basil is rarely troubled by pests or diseases. Protect plants with row covers when temperatures drop below 40°F, or dark spots caused by cold injury may appear on leaves. Fungi may also cause dark spots on foliage. Remove spotted leaves and spray foliage with compost tea or sulfur if problem is severe. If leaves are mottled yellow with turned-down edges, plants may have cucumber mosaic virus. Destroy infected plants. See the Herbs entry beginning on page 116 for other problems.

Bean

Phaseolus spp. (Leguminosae)

Beans are annual vegetables grown for their immature pods, immature seeds, and nutritious dry seeds. They are legumes and, with the help of certain soil-dwelling bacteria, can transform nitrogen from the air into nitrogen compounds that plants can absorb. A wide range of bean types and cultivars are available.

Culture

Beans thrive in most soil. Work in plenty of low-nitrogen compost before planting to loosen the soil. For a healthy, trouble-free crop, plant beans after soil has warmed. Optimum soil temperature for germination is 80°F. At soil temperatures below 60°F, most bean cultivars germinate poorly and are more susceptible to pests and root rot. Choose light, well-drained soil for early plantings, if possible, and cover beans with row cover or clear plastic until they emerge. If you use clear plastic, be sure to remove it as soon as the seeds germinate to avoid "cooking" the seedlings.

Beans do best when soil pH is between 5.5 and 6.8. They don't require high soil fertility. In fact, high nitrogen levels will delay maturity. Spray young plants with seaweed extract to prevent micronutrient deficiencies and improve overall plant health.

Soak seed in compost tea for 25 minutes before planting to help prevent disease and speed germination. Treat seed with an inoculant labeled for the type of bean you are planting before sowing to promote nitrogen fixation. Be sure to buy fresh inoculant each year, or check the date on the package for viability.

Don't touch plants when foliage is wet to avoid spreading diseases. Compost plants after harvest. Prevent problems by not planting beans in the same location more often than every 3 years.

Whole Plant Problems

Seedlings die or fail to emerge; plant stunted. Causes: Seedcorn maggots; root rot. Look for 1/4" long, yellow-white maggots feeding on seeds and seedlings. Adults are small flies. Seedlings that do come up are deformed and spindly. Control these soil-dwelling pests by applying parasitic nematodes to the soil before replanting.

Root rot causes reddish black streaks on the roots; plants are stunted and yellow. Cool, wet soil encourages this fungal disease. Destroy wilted plants. Replant with fresh seed in well-drained, warm soil. Soak seed in compost tea before planting.

Plant yellow and stunted, wilts during hot days and recovers at night. Causes: Wireworms; bean leaf beetle larvae; root knot nematodes. If you suspect any of these pests, pull up a plant and examine the roots. Wireworm larvae are up to 1½", yellow to reddish brown, slender, tough-bodied, segmented grubs. Adults are ½" long, dark-colored, elongated click beetles. Apply parasitic nematodes to the soil to control.

Slender white grubs up to 1/3" long feeding on roots are bean leaf beetle larvae. See "Leaves with large holes" on page 36 for more information.

Root knot nematodes cause swollen and darkened enlargements of roots. Destroy infested plants. Control root knot nematodes by applying chitin or parasitic nematodes to the soil. Small, round, pinkish nodules attached to the roots are caused by nitrogen-fixing bacteria and are beneficial.

Leaf Problems

Leaves with yellow, curling margins. Causes: Potato leafhoppers; calcium deficiency; salt injury. Look for leafhoppers on plants. Adults are 1/10", yellow-green, winged insects.

BEAN ◆ What Goes Wrong and Why

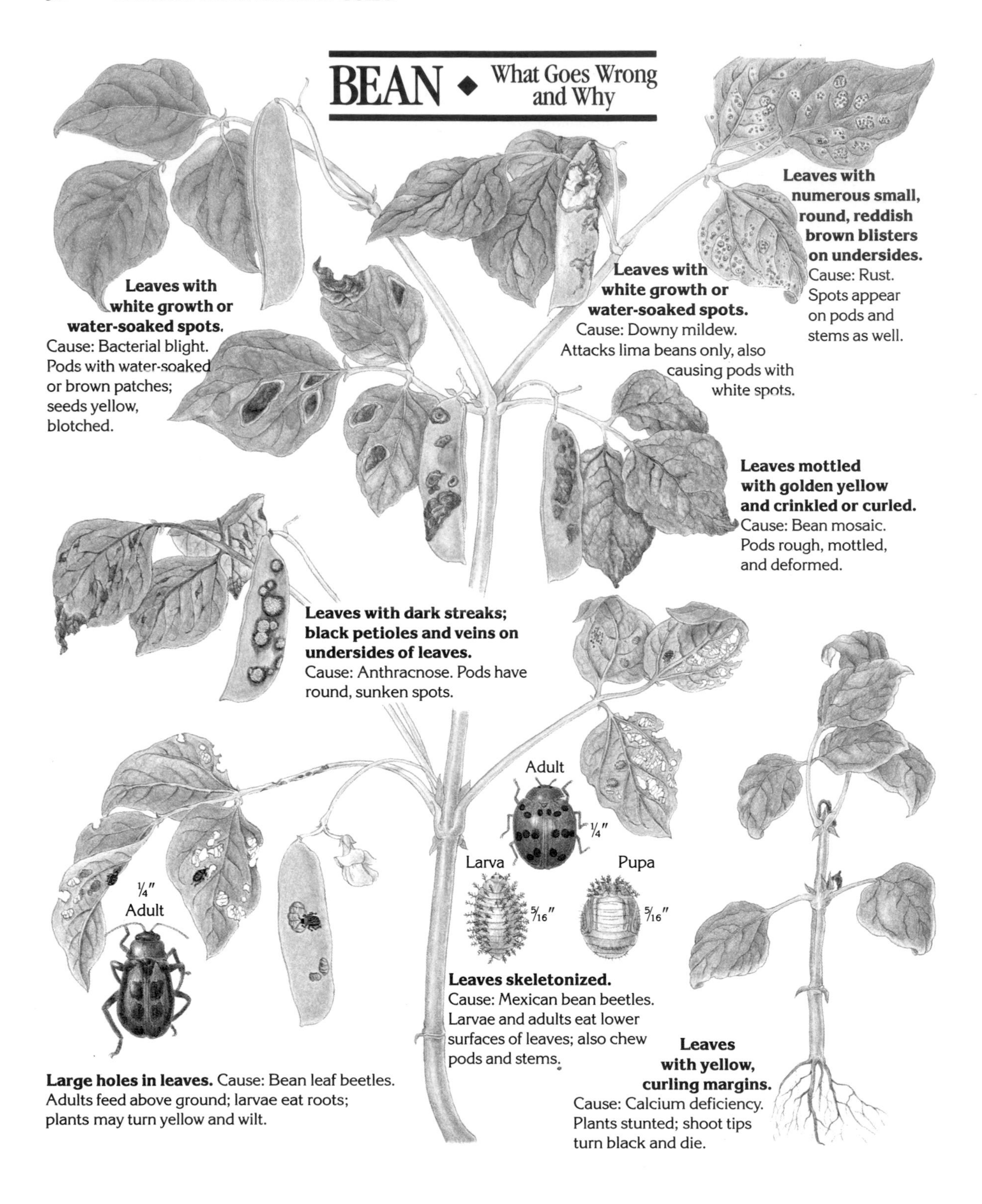

Nymphs are smaller and wingless. Severe infestations cause plants to wilt and flowers and pods to drop. Leafhoppers can spread viruses from plant to plant, so control is important. Cover emerging seedlings with row cover if leafhoppers have been a problem. Spray plants with insecticidal soap in the evening to control. If not effective, spray with a commercial pyrethrin/rotenone mix. Controls are most effective on wingless nymphs since adults fly away when disturbed. Be sure to spray the undersides of leaves where nymphs congregate.

Salt injury is common in areas of the country with saline soils, because beans are very sensitive to sodium. Planting in raised beds (with lots of compost worked in) and watering thoroughly may help.

If plants are also stunted and shoot tips turn black and die, suspect calcium deficiency. See the opposite page for an illustration of this problem. Keep soil evenly moist to help prevent problems. If soil test shows deficiency, add gypsum, or calcitic lime if pH is below 6.2.

New growth distorted and yellow. Cause: Tarnished plant bugs. Adults are oval, light green to brown, 1/4" long bugs. Nymphs are smaller and yellow-green. Trap them with white sticky traps, or spray with sabadilla in the evening to control.

Leaves yellow; growth stunted. Cause: Nitrogen deficiency. This problem is often brought on by waterlogged soil. Spray foliage and drench roots with fish emulsion or fishmeal tea to alleviate symptoms. Prevent problems by adding compost to the soil and providing good drainage.

Leaves yellow and withered. Cause: Bean aphids. Adults are small, black, soft-bodied, sucking insects. For mild infestations, knock pests off plants with a blast of water. Spray plants with insecticidal soap if aphid populations are low, or with a commercial pyrethrin/rotenone mix if more than half the shoot tips have aphids.

Leaves mottled with golden yellow and crinkled or curled. Cause: Bean mosaic. Leaves of infected plants curl downward, and plants are stunted. Seeds are small and shriveled. See the opposite page for an illustration of this disease. Remove and destroy plants infected with this viral disease. Prevent problems by controlling aphids that spread the disease and by planting tolerant cultivars, such as 'E-Z Pick', 'Goldcrop', 'Lake Largo', 'Morgane', 'Provider', 'Roma 2', 'Sungold', 'Tendercrop', 'Topcrop', and 'Venture'.

Leaves puckered and curled downward; plant dwarfed. Cause: Curly top virus. Remove and destroy infected plants. Prevent problems by controlling aphids that spread the disease and by planting resistant or tolerant cultivars such as 'Goldcrop', 'Great Northern', and 'Hystyle'.

Leaves with white growth or water-soaked spots. Causes: Downy mildew; powdery mildew; white mold; bacterial blight. Warm, damp weather encourages these diseases, the first 3 of which are caused by fungi. Downy mildew only attacks lima beans, causing young shoots and flowers of infected plants to develop white growth; leaf veins are twisted and purplish. See the opposite page for an illustration of downy mildew. Powdery mildew and white mold can attack all types of beans. Leaves develop a powdery or fuzzy white coating.

To control fungal diseases, thin plants to increase air movement. If weather is wet, spray sulfur in the evening. Plant cultivars, such as 'Provider' and 'Tendercrop', that tolerate all 3 diseases. Under some conditions, a baking-soda-and-soap spray (1 teaspoon baking soda, 1 teaspoon liquid dish soap, 1 quart water) may prevent further infection.

Water-soaked spots caused by bacterial blight turn brown with a yellow halo and then become dry and brittle. See the opposite page for an illustration of this disease. Spray plants with copper to reduce the spread of the disease if plants are setting fruit. If no new pods are forming, don't bother to spray; remove and destroy infected plants.

Leaves with dark streaks; black petioles and veins on undersides of leaves. Cause: Anthracnose. Spots exude salmon-colored ooze in cool, moist weather. See page 34 for an illustration of this fungal disease. This disease thrives in wet, humid conditions. Spray plants with sulfur to control. Prevent problems by planting tolerant cultivars such as 'Espada', 'Marbel', 'Morgane', and 'Rocdor'.

Leaves with numerous small, round, reddish brown blisters on undersides. Cause: Rust. As this fungal disease progresses, leaves turn yellow and drop. Spots also appear on pods and stems. See page 34 for an illustration of this disease. Rust usually develops in late summer. To control, spray sulfur as soon as you see indications of the disease. Plant cultivars that are rust-tolerant, such as 'Burpee Stringless', 'Kentucky Wonder', 'Roma', 'Spurt', and 'Sungold', to prevent problems.

Leaves pale and stippled or bronzed. Cause: Mites. Hot, dry weather encourages outbreaks. Look for these tiny, spiderlike creatures on the undersides of leaves where they appear as pale or dark specks. There may be a fine webbing on the undersides of leaves. To control, spray plants with water or insecticidal soap in the evening.

Leaves with wandering, white or translucent tunnels. Cause: Leafminers. Adults are tiny black-and-yellow insects. Larvae are pale green and maggotlike and tunnel into leaves. Remove and destroy mined leaves. Apply row cover as soon as plants emerge to prevent problems. Use yellow sticky traps to catch adults. Certain parasitic nematodes can actually attack leafminer larvae inside leaf tunnels.

Leaves riddled with tiny holes. Cause: Flea beetles. These small, shiny, black beetles hop when disturbed. Spray plants with a commercial pyrethrin/rotenone mix if damage is severe. Prevent problems by covering young plants with row cover.

Leaves skeletonized. Cause: Mexican bean beetles. See page 34 for an illustration of this pest. In severe infestations pods also are chewed. Adults are oval, 1/4" long, yellowish brown beetles with 16 black spots on the wing covers. Nymphs are yellowish orange and spined. Eggs are yellow and found in groups on the undersides of leaves. A tiny parasitic wasp, *Pediobius foveolatus,* provides effective control. Release the wasps when there are bean beetle larvae present. In cold climates you must release the wasps each year as they cannot overwinter. Spined soldier bugs are also efficient predators. Use a commercial pyrethrin/rotenone mix as a last resort.

Leaves with large holes. Causes: Bean leaf beetles; cucumber beetles; various caterpillars. Bean leaf beetles are 1/4" long and dark yellow with 6 black dots. The larvae are white, up to 1/3" long, and feed on roots and underground stems. Cucumber beetles are 1/4" long and greenish yellow with black stripes or spots. Prevent problems by covering plants with row cover. If beetle damage is severe, spray leaves with sabadilla or a commercial pyrethrin/rotenone mix. See page 34 for an illustration of bean leaf beetle; see page 214 for an illustration of cucumber beetle.

Many caterpillars also feed on bean leaves. Spray plants with BTK to control.

Flower and Pod Problems

Blossoms appear but no pods form. Causes: Excessive heat; mechanical injury; zinc deficiency. Hot days (85°F or more) or mechanical damage caused by heavy rains or strong water sprays can cause flowers to drop. Wait for new blossoms.

Zinc deficiency can also cause pods to drop. Spray plants with seaweed extract to prevent deficiency problems.

Pods with dark, round, sunken spots with lighter centers. Cause: Anthracnose. Salmon pink ooze appears in wet weather. See "Leaves with dark streaks; black petioles and veins on undersides of leaves" above for controls.

Pods with water-soaked or brown patches; seeds yellow and blotched. Cause: Bacterial blight. See "Leaves with white growth or water-soaked spots" on page 35 for controls.

Pods with white spots. Causes: Powdery mildew; white mold; downy mildew. See "Leaves with white growth or water-soaked spots" on page 35 for more information and controls. Lima bean pods infected with downy mildew shrivel and turn black.

Pods rough, mottled, and deformed. Cause: Bean mosaic. See "Leaves mottled with golden yellow and crinkled or curled" on page 35 for controls.

Pods with wartlike pimples. Cause: Green stink bugs. Adults are large, flat, shield-shaped, green bugs. Nymphs have reddish markings. For control if injury is severe, spray or dust plants with sabadilla or a commercial pyrethrin/rotenone mix.

Pods with chewed holes. Cause: Caterpillars. Cover plants with row cover to exclude egg-laying moths if caterpillars have caused damage in the past. Control feeding worms with BTK spray.

Pods pitted and browned. Cause: Cold injury. Most cultivars may be damaged at temperatures below 45°F. Protect late crops with row cover.

Dry beans tunneled. Cause: Bean weevils. Adults are gray or brown, 1/10″–1/8″ long snout beetles. Larvae are small and light-colored. To control this storage pest, dry seed in a 125°F oven for 25 minutes, or store at 0°F for 3–4 days to kill larvae.

Beet

Beta vulgaris
Crassa group (Chenopodiaceae)

Beets are biennial vegetables grown for their firm, sweet roots and succulent greens. Cultivars with red, yellow, and red-and-white roots are available.

Culture

Beets grow best in deep, well-drained soil with a pH between 6.5 and 7.5. They are a cool-season plant and will tolerate temperatures as low as 40°F. However, plants will bolt if exposed to 2–3 weeks of temperatures below 50°F after the first true leaves have formed. Beets grow poorly above 75°F and are best grown as a spring or fall crop. Keep soil moist, but not soggy since rapid and uninterrupted growth produces the best roots. Prevent problems by not planting beets in the same location more often than every 3 years.

Leaf Problems

Leaves with purplish patches. Cause: Phosphorus deficiency. This is common in cool spring soils. Plants usually outgrow problem when soils warm. Spray leaves with seaweed extract to speed up recovery.

Leaves yellow; plant stunted. Cause: Nitrogen deficiency. This deficiency affects older leaves first. Spray foliage and drench roots with fish emulsion to alleviate symptoms.

Leaves yellow and curled; plant stunted. Cause: Aphids. Adults are soft-bodied, small, green, gray, black, or pinkish sucking insects, sometimes with a white fluffy coating. Knock aphids off the plants with a blast of water, or spray plants with insecticidal soap to control. If soap spray is not effective, spray with a commercial pyrethrin/rotenone mix.

Leaves with brown tips. Cause: Excessive heat. Bright sun and 80°F temperatures may injure beet leaves. Plant in a partially shaded location or provide shade in climates with hot summers.

Leaves with dark-bordered, tan spots. Cause: Cercospora leaf spot. Destroy spotted leaves. Spray plants with copper if this fungal disease is severe to prevent further spread, or before symptoms develop if it has been a problem in the past. To prevent problems, soak seed in 122°F water for 25 minutes before planting. (Be aware that this treatment can damage seed viability; for complete instructions, see page 422.) Eliminate weeds that can harbor the disease. Plant cultivars, such as 'Big Red Hybrid' and 'Red Ace', that are tolerant of this disease.

Leaves with light-colored spots on upper leaf surfaces. Cause: Downy mildew. Leaves infected with this fungal disease have spots covered with white fuzzy growth on the undersides. Spray plants with copper when symptoms first appear.

Leaves stunted and crinkled. Cause: Curly top virus. There is no cure once plants are infected; destroy infected plants. Leafhoppers spread the virus as they feed; control them with insecticidal soap, or for severe infestations, a commercial pyrethrin/rotenone mix. Prevent problems by covering seedbed with row cover after planting.

Leaves with wandering, white or translucent tunnels. Cause: Leafminers. Pale green, maggotlike larvae feed inside leaves, leaving empty tunnels behind them. Adults are tiny black-and-yellow insects. Once larvae have entered leaves, spraying will not control them. Destroy mined leaves. Catch adults with yellow sticky traps. Prevent problems by covering plants with row cover as soon as they emerge. Certain parasitic nematodes can attack leafminer larvae inside leaf tunnels.

Leaves riddled with small holes. Cause: Flea beetles. These small, shiny, black, brown, or bronze beetles hop when disturbed. Prevent problems by covering plants with row cover as soon as they emerge.

Leaves with large, ragged holes. Causes: Beet armyworms; garden webworms. Greenish brown, 1½" long caterpillars with a light-colored stripe on each side are beet armyworms. Pale green to nearly black, 1" long caterpillars with a black or light stripe down their backs and dark spots are garden webworms. Webworms roll leaf edges over. Handpick or spray actively feeding caterpillars with BTK in the evening.

Root Problems

Roots distorted, with rough, cracked skin. Cause: Downy mildew. See "Leaves with light-colored spots on upper leaf surfaces" above for controls.

Roots with raised, rough, brown spots on surface. Cause: Scab. Prevent this fungal disease by adding compost to soil, and plant in raised beds to improve drainage.

Roots with black, dead, hard spots in flesh. Cause: Boron deficiency. Roots may be wrinkled or cracked. Plants can be stunted; leaf edges may be brown and lower leaf surfaces may be reddish purple. Prevent deficiency problems by spraying plants with seaweed extract as soon as the first true leaves appear, and every few weeks thereafter. Check deficiency with a soil test. Correct by adding 1 tablespoon of borax dissolved in 1 gallon water, or 10 pounds of kelp, per 100 square feet of soil.

Roots forked. Cause: Calcium deficiency. Prevent problems by keeping soil evenly moist. If soil test shows deficiency, add gypsum, or if pH is below 6.2, calcitic lime.

Roots small and poorly developed. Causes: Nitrogen deficiency; crowded roots. If leaves are small and yellow, feed plants with fish emulsion or compost tea to boost nitrogen level. Thin plants so roots don't touch to allow further growth.

Roots covered with hairy side roots; flesh woody. Cause: Curly top virus. See "Leaves stunted and crinkled" above for controls.

Roots unusually dark-colored. Cause: Potassium deficiency. Affected roots are prone to rot. If soil test confirms deficiency, amend soil as needed.

Roots light-colored; zoned rings in flesh very evident. Causes: Excessive heat; uneven soil moisture. Prevent problems by keeping soil evenly moist and mulching with straw during hot weather.

Begonia

Begonia. Tender perennials grown as annuals, and tubers.

The genus *Begonia* provides many beautiful species that are well-adapted to garden culture. Outdoors, the most common types are the fibrous-rooted and tuberous-rooted begonias. Each type has its own habits and needs.

The fibrous-rooted wax begonia, with its prolific flowering habit, is among the most popular bedding plants. Clusters of single or double, pink, white, or reddish flowers, up to 2″ across, appear in spring and continue until frost. Wax begonias do not always compete successfully with other annuals. Spotlight them in masses of 1 color. These are good plants to line the front of a sunny or partially shady border, growing only 6″-9″ tall and up to 1′ wide. In mild climates they may overwinter and live for years. They do well in window boxes and are widely grown as flowering and ornamental houseplants. Glossy wax begonia leaves are shades of green or red.

Tuberous-rooted begonias have fewer but larger and more showy blooms. The single or double flowers can grow to 6″ or more across. They bloom on upright plants, growing to 2′ tall, in bright and pastel shades of white, pink, red, yellow, and combinations. Tuberous-rooted begonias brighten up shady beds and borders.

Because powdery wax begonia seeds require much nurturing, many gardeners prefer to buy bedding plants in spring for direct garden placement. If you do decide to grow your own, sow the tiny seeds indoors, 5 months before the final frost. Use a mix of fine peat and sand as the medium. Fluorescent lighting helps keep the desired warm temperature. Water seedlings from the bottom. When they are large enough to handle, transfer to 3″ pots; transplant outdoors when night temperatures remain above 50°F. Set plants 9″ apart.

Start tuberous-rooted begonias indoors, planting them 8-10 weeks before your frost-free date in a loose growing medium. Barely cover tubers with the concave side up (it should have little pink buds coming out of the center) and moisten lightly. Give lots of water and light after the shoots emerge. Move tubers to individual 4″ or 5″ pots when shoots are 1″-3″ tall. After all danger of frost is past, plant in partial shade in fertile, moist but well-drained soil with plenty of organic matter. Water liberally in warm weather. Feed every 3 weeks or so with compost tea or fish emulsion.

Begonias need rich, very well drained, light soil. Work in generous amounts of leaf mold, dehydrated manure, humus, and bonemeal. They tolerate full sun in cool climates only; in hot climates, leaves in full sun may show dry spotting from sunscald. Heavy shade, however, results in legginess and fewer flowers. Partial shade is generally best. Water when soil 1″ below surface is still moist but not wet. Fertilize every 3-4 weeks.

If you want to overwinter tuberous-rooted begonias, lift the plants (with the soil still attached) in fall when the leaves are yellow and withered. After a week or so, cut the stems to within a few inches of the tuber. Once the stem stub dries completely, shake the soil off the tubers and store in dry peat or coarse sand at 45°-55°F. Start them again in spring.

Some gardeners dig up favored garden wax begonias, bringing them indoors before first frost to use as houseplants, then making cuttings in spring for garden use. As houseplants, begonias may bloom most of the year. Pinch back stems by at least half to promote new growth. Keep soil damp but not wet. Mist leaves occasionally. Keep in sun or some shade. When bringing any plant indoors from outdoors, consider that you also may be transferring pests and diseases indoors. These can spread rapidly to other household plants.

Problems

Leaves with powdery white patches. Cause: Powdery mildew. This fungal disease is very common on begonias. For control information, see "Leaves with powdery white patches" on page 19.

Flower buds drop off. Causes: Excess water; high temperatures. Poor drainage and overwatering are the prime causes of early bud drop. Ease up on the watering and plant begonias in humus-rich soil with good drainage next time. Overly high temperatures also cause bud drop. In areas with unexpected hot spells, mulch roots and mist plants in late morning.

Leaves with angular brown blotches. Cause: Leaf nematodes. Microscopic, wormlike creatures, known as leaf nematodes, may be feeding on the affected leaves. They cause brown blotches that enlarge until the leaves curl up and drop off. Symptoms usually appear on lower leaves first and spread to upper leaves. Plant growth is stunted; new leaves may not appear.

Remove and destroy infested leaves and the next 2 leaves directly above them. Do not let affected plants touch unaffected ones. Water from below rather than above; nematodes can move from one plant to another through water. Leaf nematodes can survive 3 or more years in the soil or surrounding debris. If you have a problem, don't replant begonias in that area.

Leaves with fuzzy, brown or gray spots. Cause: Botrytis gray mold. The Botrytis fungus tends to attack weak plants. Once the resultant gray mold takes hold, the disease moves into healthy plant tissue. High humidity and cool temperatures favor the infection process, as do crowded plantings, rain, and overhead watering. Botrytis gray mold fungus can be a year-round problem in mild winter areas. Remove and destroy infected leaves, stems, and flowers. Thin plants to encourage good air circulation.

Leaves with spots. Cause: Leaf spots. Both bacteria and fungi cause leaf spots on begonias. Bacterial leaf spot produces small blisters that are brown with yellow margins; fungi may form brown, black, or transparent spots. For control information, see "Leaves with spots" on page 19.

Leaves, stems, and buds distorted. Cause: Aphids. See "Leaves, stems, and buds distorted" on page 20 for controls.

Leaves yellow; plant weakened. Cause: Whiteflies. For controls, see "Leaves yellow; plant weakened" on page 18.

Leaves and stems with white, cottony clusters. Cause: Mealybugs. See "Leaves and stems with white, cottony clusters" on page 20 for control information.

Stems with soft, water-soaked spots. Cause: Stem rot. Both bacterial and fungal infections can cause begonia stems to rot. If you catch the stem spots when they are small, try cutting them out with a razor blade or sharp knife. Remove and destroy seriously infected leaves, stems, and plants. These diseases can spread quickly, so wash your hands and tools after working with infected plants. Avoid overcrowding and overhead watering; pick a site with good air circulation. Keep mulch a few inches away from the stem. Don't replant begonias in affected areas.

Leaves with reddish brown lines. Cause: Thrips. These tiny, light brown, fast-moving insects generally attack leaves, causing reddish brown lines or spots on upper leaf surfaces and silvery blotches on leaf undersides. Stems, buds, and flowers can also be affected. Remove and destroy severely infested plants.

Leaves with large, ragged holes. Cause: Slugs and snails. See "Leaves with large, ragged holes" on page 18 for controls.

Berberis

Barberry. Shrubs.

Barberries are spiny deciduous or evergreen shrubs commonly used as hedges, barriers, or foundation plants.

Barberries grow in sun or partial shade and prefer well-drained soil. They require little care and tolerate even severe pruning. Set out container-grown or bare-root plants in spring or fall.

Pests and diseases are seldom a problem on cultivated barberries. Many states prohibit growing common barberry (*Berberis vulgaris*), which is an alternate host to a rust fungus that also attacks cereal grains. Check with your extension agent for local restrictions.

Problems

Leaves and shoots blackened; leaves with moist or brown sunken spots. Cause: Anthracnose. See "Leaves and shoots blackened; leaves with moist or brown sunken spots" on page 238 for control suggestions.

Leaves wrinkled and discolored. Cause: Aphids. See "Leaves wrinkled and discolored" on page 235 for control measures.

Leaves chewed; branch tips with webbing. Cause: Barberry webworms. These 1½", black caterpillars tie together leaves and twigs, forming a web nest in midsummer as they feed on the leaves. If not too numerous, remove nests by hand. Control large infestations with 3 applications of BTK made 1 week apart, or spray with pyrethrin.

Leaves shrivel and turn brown. Cause: Barberry wilt. This soilborne fungus attacks water-conducting tissue, causing leaves to wilt and eventually killing the entire plant. Remove the whole affected plant and its surrounding soil; replace with fresh soil.

Leaves with notched leaf margins. Cause: Japanese weevils. The ¼", brown adults attack foliage, and the legless white grubs feed on roots. Apply a sticky substance, such as Tanglefoot, to the lower stems to prevent adults from climbing up the plants. Drench the soil around the base of the plant with a solution of parasitic nematodes to control the larvae. Spray leaves several times with pyrethrin for major infestations of adults.

Twigs covered in small reddish brown bumps. Cause: Barberry scale. Heavy infestations may cause yellowing foliage and stunted growth, leading to the death of the plant. Pests are most active in June and July. Spray twigs with dormant oil in early spring, before growth starts. During the growing season, spray plant with superior oil or repeated applications of insecticidal soap.

Whole plant stunted and lacking vigor. Cause: Nematodes. See "Whole plant stunted and lacking vigor" on page 239 for more information.

Bergenia

Bergenia. Perennials.

Bergenias are sturdy, low-growing border plants with rose-pink flower clusters borne above the foliage in early spring. Broad, glossy leaves are 10″ long and heart-shaped in the case of heartleaf bergenia (*Bergenia cordifolia*), oval in the case of leather bergenia (*B. crassifolia*). The foliage takes on a dark burgundy color in fall; leaves are evergreen but often damaged by cold weather. Space plants 1′-1½′ apart in average, well-drained soil in a partially shaded spot.

Bergenias' large, low leaves form a haven for slugs and snails. See "Leaves with large, ragged holes" on page 177 for control information.

Betula

Birch. Trees.

Birches are deciduous trees with simple leaves. They are popular as single or multiple-stemmed specimen trees; attractive features include their striking trunks, peeling bark, graceful branches, and magnificent fall color.

Most birches grow best in cool climates and almost invariably perform better in the northern portions of their range. Choose a site with light shade and moist but well-drained, acidic soil. Transplant in spring as balled-and-burlapped specimens. Don't prune in winter or early spring, when trees can "bleed" sap; late summer is a better time.

Birches are relatively short-lived, especially when stressed, and they are beset by a number of insect pests that can cause serious damage and even loss. Fortunately, you can prevent most of the common problems by planting trees in the right conditions and keeping them vigorous. Fertilize regularly, and water deeply during drought. Some species, such as river birch (*Betula nigra*), are less prone to common birch pests.

Those species that are noted for their predominantly white bark develop this characteristic slowly; do not be disturbed if the young plant you have acquired is not white when you get it!

Problems

Leaves blister and turn brown. Cause: Birch leafminers. Perhaps the most serious and prevalent of all birch pests, birch leafminers produce ¼″, white larvae that feed between the upper and lower surfaces of the leaves. Several generations occur each year; the first, usually active in May, is most destructive. Remove infested leaves. Repeated applications of insecticidal soap will kill the small, black, sawflylike adults, reducing future populations.

Leaves yellow; branch tips dead or dying; trunk with lumps under bark. Cause: Bronze birch borers. The adults are ½″-1″, blunt-headed, reddish gray beetles that feed on the foliage, and then lay their eggs in slits in the bark. The eggs hatch into light-colored grubs that tunnel through the bark into the sapwood, starting from the top of the tree and working downward. By the time you see the symptoms, it is usually too late to save the tree. If the damage is not yet severe, you could try pruning out and destroying the infested parts. The best control for borers is planting trees in the right conditions and keeping them healthy and vigorous. Avoid planting European white birch (*B. pendula*), which is very susceptible to this pest.

Leaves skeletonized. Causes: Cankerworms; birch skeletonizers; Japanese beetles. Cankerworms are ½″-1″, yellow, green, or brown caterpillars that feed on the foliage of birches, often defoliating the tree. To avoid this problem, apply a band of a sticky substance, such as Tanglefoot, around the base of the tree in spring to keep females from climbing up the tree and laying their eggs. Apply fresh coats in September and February to trap both spring and fall cankerworms. If their feeding becomes a problem, spray with BTK as soon as you notice them, and continue every 2 weeks until the pests are gone.

Birch skeletonizers also feed on foliage, leaving brown leaves with visible leaf veins behind. These ¼″, yellowish green caterpillars do the most damage in July and August. In most cases this pest is a problem only every few years, and the damage happens so late in the season that no control is necessary. If you want to protect young trees from serious damage, apply BTK or superior oil as soon as you spot pests in midsummer.

Japanese beetles are foliage feeders, too. For information on controlling them, see "Leaves skeletonized" on page 236.

Plant defoliated. Cause: Gypsy moths. See "Plant defoliated" on page 236 for control suggestions.

Trunk or branch crotches with swollen, cracked lesions. Cause: Canker. This fungal disease attacks at the forks of branches, causing swellings that crack open to expose the wood. The callus tissue that forms may also be affected. The canker can eventually girdle and kill the branch or tree. If the canker is small, cut out the tissue involved; cut down and destroy seriously infected trees. Healthy trees are less susceptible.

Leaves tunneled. Cause: Leafminers. See "Leaves tunneled" on page 237 for controls.

Leaves wilt or curl and pucker. Cause: Birch aphids. These small, pear-shaped insects attack the foliage, weakening and distorting the growth. They also produce copious quantities of honeydew, which attracts ants and encourages the growth of black-colored, sooty mold. Spray plants with a strong jet of water to knock off the pests. For serious infestations, apply insecticidal soap, or pyrethrin spray as a last resort.

Branches die. Cause: Dieback. This fungal disease produces symptoms very similar to those caused by bronze birch borers, except for the lumpy bark. Trees weakened by drought, low fertility, or borers are most susceptible to infection. Prevent this problem by planting trees in the right spot and watering during drought to keep them growing vigorously. If symptoms do occur, prune out and destroy or dispose of infected branches.

Leaves with powdery white coating. Cause: Powdery mildew. See "Leaves with powdery white coating" on page 237 for controls.

Leaves with spots. Cause: Leaf spots. See "Leaves with spots" on page 237 for control suggestions.

Blackberry

Rubus spp. (Rosaceae)

Blackberries are perennials that bear fruit on second-year canes. Some cultivars grow erect and some bear long, trailing canes. Blackberries are hardy in Zones 5-8.

Like other bramble fruits, blackberries need full sun and well-drained, moisture-retentive soil. Train plants to posts or fences. Prune annually. Pull unwanted suckers. On erect blackberries pinch the tips of 3′ canes to force growth of lateral shoots. During the dormant season, shorten all lateral shoots to about

$1\frac{1}{2}'$. For both types of blackberry, cut away all fruiting canes right after harvest. In winter thin out canes leaving 3 or 4 canes per clump for erect types and 8–12 canes per clump for trailing types.

Blackberries are self-pollinating and require no cross-pollination to set fruit. Blackberries belong to the same genus as raspberries and are affected by similar insects and diseases. For more information on problems and solutions, see the Raspberry entry beginning on page 196.

Problems

Flowers with reddish, twisted petals; no fruit set. Cause: Double blossom. Double blossom, known as rosette, also causes plants to throw out witch's brooms, dense masses of deformed twigs with pale leaves. To control this fungal disease, remove infected canes as soon as you notice them. To help prevent the spread of the fungus, spray the whole planting with copper 2 or 3 times at 10-day intervals. If the disease is rampant, mow down the whole planting and destroy the newly cut canes. Avoid planting sites near wild brambles, which may carry the disease. Resistant cultivars include 'Flordagrand' and 'Humble'.

Fruit covered with light gray fuzz. Cause: Fruit rot. Especially common during wet weather, this fungal disease appears less frequently where plants have good air circulation and proper pruning. To control fruit rot, harvest berries often. Also pick and discard infected fruit far away from plants.

Fruit fails to ripen or remains red and sour. Cause: Redberry mites. Control these microscopic mites with lime-sulfur applied in early spring when new shoots are 1″ long and again when canes have about 1′ of new growth.

Canes and leaves with purple spots. Causes: Anthracnose; leaf and cane spot disease. Controls are the same for both these fungal diseases. Keep plantings well-pruned to promote good air circulation. Remove and destroy infected canes right after harvest. Early in the season, spray with lime-sulfur. Cultivars resistant to both diseases include 'Black Satin', 'Dirksen Thornless', and 'Gem'.

Leaf undersides with bright orange pustules appearing in spring. Cause: Orange rust. This incurable fungal disease infects the entire plant and can't be controlled by spraying. Plants become weak and unfruitful. Diseased plants never recover; infection spreads quickly to neighboring plants. Dig and destroy plants as soon as you notice disease symptoms. Avoid sites near wild blackberries, which may carry the disease. Late in the season, a benign leaf rust appears on some blackberries; disregard it. The cultivars 'Cheyenne', 'Comanche', 'Eldorado', 'Evergreen', and 'Shawnee' are resistant to orange rust.

Canes dying. Cause: Borers. For information identifying and controlling borers, see "Canes dying, break off easily" and "Cane tips wilted and dying" on page 199.

Canes turn yellow and die in midsummer. Cause: Verticillium wilt. This is a common soilborne fungal disease. Typically, leaves yellow, wilt, and fall before the entire cane dies. Avoid planting blackberries following other Verticillium-susceptible crops, such as strawberries, potatoes, tomatoes, and eggplants. Soil solarization may prevent the disease. Verticillium-resistant cultivars include 'Evergreen', 'Logan', 'Marion', and 'Olallie'. See page 197 for an illustration of this disease on raspberries.

Canes with wartlike or corky swellings; canes dying. Cause: Crown gall. Crown gall bacteria live in the soil and enter plants through wounds. For information and controls, see "Canes with wartlike or corky swellings; canes dying" on page 199.

Blueberry

Vaccinium spp. (Ericaceae)

Blueberries are deciduous shrubs ranging in height from 1′ for lowbush blueberries (*Vaccinium angustifolium*) to 6′ for highbush blueberries (*V. corymbosum*) to over 30′ for rabbiteye blueberries (*V. ashei*). The flowers look like dainty white bells and appear in spring on shoots that grew the previous season. Blueberries are hardy in Zones 3-9, depending on species and cultivar.

Culture

Blueberries require full sun and well-drained, moisture-retentive, acidic soil with a pH of 4.0-5.0. Of the 3 species, highbush blueberries are the most finicky about soil. Blueberries generally grow well in soil enriched with acidic organic material, such as peat moss, composted pine needles or oak leaves, or compost made from pine, oak, or hemlock bark. Fertilize with acidic fertilizers, such as cottonseed meal or soybean meal. Blueberries enjoy a thick, organic mulch.

Most blueberries are not wholly self-pollinating. Plant at least 2 different cultivars near each other for adequate cross-pollination.

Prune plants each winter, beginning when bushes are about 4 years old. On highbush and rabbiteye plants, remove drooping or very old branches and thin out branches where growth is too dense. Cut lowbush plants to the ground every third year for a harvest every 2 out of 3 years.

Flower and Fruit Problems

Berries soft and mushy. Cause: Blueberry maggots. If you find 3/8″, white maggots inside berries, you've discovered the larvae of the blueberry maggot fly. This insect, closely related to the apple maggot fly, deposits eggs just under the skin of the fruit from late June to August. Eggs hatch into fruit-devouring maggots that later drop to the soil and pupate over the winter. To reduce the number of maggots that pupate in the soil through the next winter, harvest frequently and destroy any infested berries you find. Another control is to trap adult flies on sticky red spheres like the ones used to trap apple maggot flies. Hang 1 trap per highbush plant or 1 per several lowbush plants before the first berries turn blue. For information on making these maggot fly traps, see "Fruit dimpled; brown tunnels through flesh" on page 22.

Berries shriveled and webbed together with silk. Causes: Cherry fruitworms; cranberry fruitworms. Berries will contain sawdustlike material and either a white caterpillar (cherry fruitworm) or a yellowish green caterpillar (cranberry fruitworm). These are the larvae of 2 different moth species that lay eggs on berries in the spring. Larvae hatching from these eggs bore into berries near the stem end and web berries together as they feed. You cannot spray anything to kill worms inside the fruit. To reduce damage from these pests without sprays, harvest berries frequently and destroy infested fruit to prevent larvae from maturing. If infestation has been heavy in previous years, spray plants with rotenone immediately after bloom.

Berries drop early; mature berries turn gray, shriveled, and hard. Cause: Mummy berry. This fungal disease overwinters on dried, diseased berries called mummies and spreads most rapidly in cold, wet springs. Control mummy berry by picking all mummies off the plant. In the spring cultivate around bushes to bury dropped mummies, or add mulch to cover the remaining fungal spores. Resistant cultivars include 'Bluetta', 'Burlington', 'Collins',

'Darrow', 'Dixie', 'Jersey', and 'Rubel'.

Blossoms covered with brown splotches or brownish gray mold. Cause: Botrytis blight. This fungal disease spreads most rapidly when cool, wet weather extends throughout the bloom period. An effective control for Botrytis is to carry a paper bag into the garden and carefully pick off and discard blighted blossoms and foliage as soon as you find them. After working around infected plants, wash your hands thoroughly before working with healthy plants. Also space and prune plants to encourage good air circulation and rapid drying of plants after rain. Avoid wetting the foliage when watering and avoid excess fertilization, which brings on lush growth and increased susceptibility to this blight.

Fruit disappears. Cause: Birds. These creatures love blueberries so much that unprotected bushes are often stripped clean of berries, even before they are fully ripe! The only sure protection is a net draped over the planting and held to the ground to prevent birds from getting underneath.

Leaf and Branch Problems

New leaves with black centers; growing tips wilted. Cause: Mummy berry. This fungal disease spreads most easily in cold, wet spring weather. For more information, see "Berries drop early; mature berries turn gray, shriveled, and hard" on page 45.

Leaves or twigs covered with brownish gray mold. Cause: Botrytis blight. An effective control for this fungal disease is to remove and discard blighted foliage as soon as you see it. For more information, see "Blossoms covered with brown splotches or brownish gray mold" above.

New leaves yellow with green veins. Cause: Iron deficiency. The soil usually has sufficient iron but is not acidic enough to make iron available to the plant. Spray leaves with iron salts or chelates for quick relief of symptoms. A long-term solution is to reduce the soil pH by mulching with an acidic material, such as pine needles.

Cane dieback. Causes: Blueberry cane canker; Fusicoccum canker. Blueberry cane canker, most prevalent in the South, shows up as reddish, conical stem swellings. The next year these swellings become blisterlike, light gray, and then black and fissured. If this disease is a problem in your area, plant rabbiteye blueberries, which are not susceptible to cane canker. Or plant highbush cultivars resistant to cane canker, including 'Atlantic' and 'Jersey'.

Fusicoccum canker, a fungal disease more common in colder regions, begins as red stem spots that enlarge and develop a bull's-eye pattern. Resistant cultivars include 'Berkeley', 'Blueray', 'Burlington', 'Coville', and 'Rubel'. Other blights also cause cane dieback, and the best course for all such problems is to prune away diseased branches as soon as you spot them. To avoid spreading disease further as you prune, sterilize pruning tools between cuts in isopropyl alcohol or a 10 percent bleach solution (1 part bleach to 9 parts water).

Leaves skeletonized. Cause: Japanese beetles. Beginning in early summer, these metallic blue-green beetles with bronze wing covers like to feed on plants in the sun. The simplest control is to check plants early in the morning while beetles are sluggish and knock them off leaves into jars of soapy water. Traps baited with floral or fruit scents plus pheromones may be effective if placed at some distance from plants, but some gardeners find that a trap placed in the middle of a planting attracts more beetles than normal to the garden. If you use traps, be sure to place them well away from the plantings you're trying to protect.

Populations of this pest naturally decline by midsummer, but for very heavy infestations, spray with rotenone. Milky disease spores, available commercially and applied to lawns to kill Japanese beetle grubs, may not work very well to control this pest. Not all the grubs in your yard will become infected with the disease, and beetles that emerge from neighboring properties may fly to your plants. You can also buy parasitic nematodes to apply to the soil for control of Japanese beetle grubs.

Broccoli

Brassica oleracea
Botrytis group (Cruciferae)

Broccoli is a cool-season vegetable grown for its crisp green heads of flower buds. Broccoli and cabbage require similar culture and are attacked by the same diseases and insects: See the Cabbage entry beginning on page 52 for culture and information on problems.

Prevent problems by planting the following improved cultivars: 'Green Dwarf #36', 'Emperor', and 'Mariner' are black rot-tolerant; 'Premium Crop' is resistant to Fusarium yellows; 'Citation', 'Emperor', 'Esquire', 'Green Dwarf #36', 'Hi-Caliber', and 'Mariner' are tolerant of downy mildew; and 'De Cicco' is tolerant of flea beetles.

In addition to the problems listed below, transplants exposed to cool temperatures (35°-45°F) for 10 days or more may form tiny, useless flower heads prematurely. High temperatures can cause similar tiny head formation. Broccoli grows best at temperatures between 45°-75°F. Harvest heads when buds are still tight and dark green or dusky violet, except for 'Romanesco', which should be yellow-green.

Problems

Heads small and uneven; stems hollow. Cause: Potassium deficiency. Prevent problems by spraying plants with seaweed extract every 2 weeks. Check soil potassium with a soil test and amend as necessary.

Heads with black or discolored centers. Causes: Fungal rot; cold injury. Broccoli heads rot when water collects between the individual flower buds. Avoid wetting heads when watering. Cold weather can also cause black areas in the center of heads. Protect plants with row cover when temperatures below 40°F are predicted.

Browallia

Browallia, sapphire flower. Annuals.

Browallias bear profuse quantities of purple or blue, 1″-2″ flowers. They grow in bush form, 10″-18″ high, or will cascade over container edges. They bloom in summer and into fall. Browallias do well as bedding plants, hanging baskets, or houseplants.

Browallias grow easily from seed. Sow indoors in March. Do not cover seed; they need light to germinate. Seedlings appear in 2 weeks. Pinch when 6″ tall for bushier plants. Transplant when night temperatures remain consistently above 50°F; space 10″ apart.

Browallias need partial shade and moderately rich soil. Keep the soil evenly moist. Fertilize lightly but regularly. Pests and diseases seldom bother browallias. Whiteflies sometimes feed on plants; see "Leaves yellow; plant weakened" on page 18.

Brussels Sprouts

Brassica oleracea
Gemmifera group (Cruciferae)

Brussels sprouts are a cool-season vegetable grown for their small, cabbage-shaped buds. They are one of the hardiest members of the cabbage family and can tolerate lower pH (5.5-6.8) than any of the other brassicas. But, they are less tolerant of heat. In warmer climates grow sprouts in soil with a high clay content if you have a choice, and shade the soil around roots. Harvest sprouts when they are 1″ in diameter or smaller and still tight. Twist them off gently, starting at the base.

Brussels sprouts and cabbage require similar culture and are attacked by the same diseases and insects: See the Cabbage entry beginning on page 52 for culture and information on problems.

In addition to problems outlined in the cabbage entry, a combination of cold injury and viral disease can cause leaves with black specks. Destroy infected plants. Caterpillars will bore small holes into sprouts; see "Leaves with large, ragged holes" on page 56 for more details and controls. Plants deficient in potassium have poorly developed sprouts: Spray with seaweed extract at transplanting and repeat several times to prevent problems. Check potassium levels with a soil test, and amend as necessary.

Buddleia

Butterfly bush. Shrubs.

Butterfly bush (*Buddleia davidii*) is a deciduous shrub producing long spikes of bloom in midsummer. In severe climates it tends to die back to the roots over winter; where this does not occur, cut plants back to near ground level to promote healthy growth and flowering.

Buddleias thrive in rich, loamy soil in full sun. They are remarkably free of insect and disease problems. Japanese beetles have been known to attack them, but rarely severely. Handpicking or inoculating your lawn with milky disease spores will usually suffice; stronger control methods can endanger the butterflies that these plants attract.

Bulbs

The term *bulb* is used to describe a variety of underground storage organs, including corms, rhizomes, and tuberous roots. Some are modified stems, while others are simply enlarged roots. Although structurally different, bulbs all store food for plant growth in subsequent seasons and spread (in the case of rhizomes) or otherwise multiply asexually to increase a plant's population.

Whether they are hardy corms, such as crocuses, that bloom year after year in the same spot, or tender rhizomes, such as cannas, that must be dug up and stored over winter, bulbs generally enjoy similar growing environments and suffer similar problems. Pest and disease prevention is preferable to applying controls after trouble arises. The following bulb-buyer's guidelines will help you avoid most bulb problems.

■ Buy only dormant bulbs that show little, if any, root development and no topgrowth other than a pale, fat bud. (Lilies, however, are never completely dormant; their bulbs often have fleshy roots attached.)

■ Purchase and plant bulbs at the right time. Reputable dealers sell bulbs in defined

BUYING HEALTHY BULBS

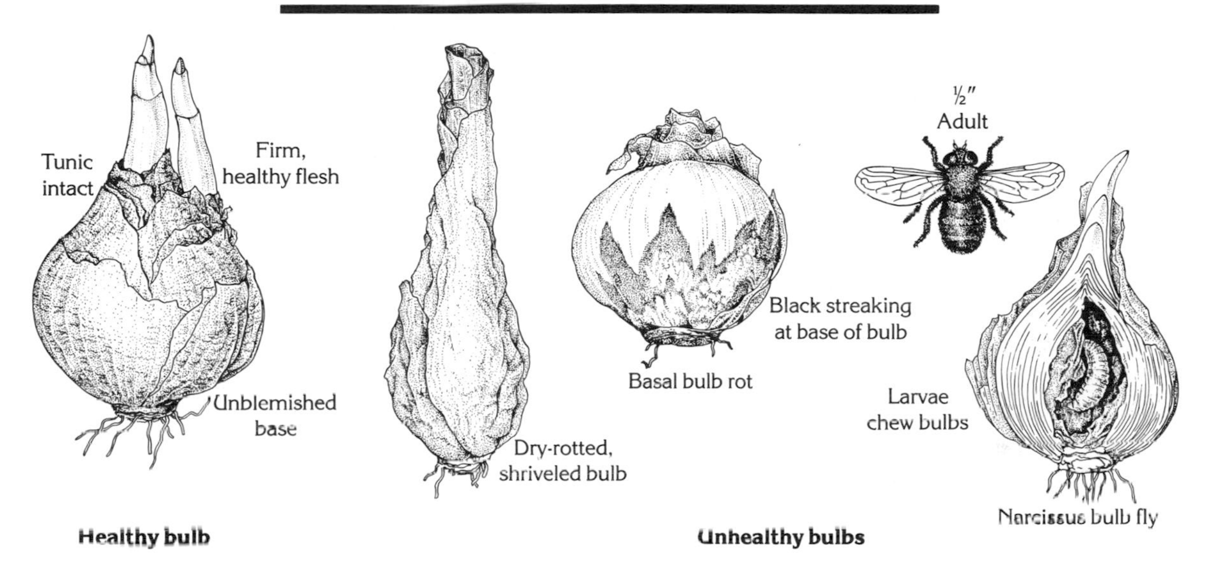

Choosing bulbs carefully is an important step toward a beautiful and care-free flower garden. Select bulbs, such as the healthy daffodil pictured, that seem heavy for their size and that are free of cuts, bruises, and discoloration. Be wary of lightweight or shriveled bulbs.

seasons: fall for spring-flowering bulbs (daffodils, tulips), spring for summer-flowering bulbs (lilies, glads), and summer for fall-flowering species (some crocuses).

■ Look for bulbs that have their papery skins—called tunics—intact. These contain natural compounds that inhibit disease and premature sprouting.

■ Choose bulbs packaged in materials that permit air to enter. Damp bulbs in plastic bags often rot.

■ Select bulbs that are firm with few wrinkles and no soft spots. Healthy bulbs seem somewhat heavier than their size suggests. Avoid unusually lightweight bulbs and those with cuts, dark or water-soaked spots, or discolored or scabby areas.

Growing healthy, trouble-free bulbs also depends on good cultural practices. Fleshy bulbs are prone to bacterial and fungal rot organisms, most of which thrive in wet conditions. Prepare bulb planting beds well and make sure the soil is well-drained. Handle bulbs carefully when planting to avoid injuries that provide access to diseases. Remove problem plants quickly to keep pests and diseases from spreading. Always let bulb foliage die back naturally to allow food production for growth and flowering in subsequent years. Mark sites where bulbs are planted so you can find them after foliage has faded. Clean up flower beds in fall to remove plant debris that offers shelter to many pests and diseases.

Healthy bulbs and good culture go a long way toward successful plantings. A few pests and diseases, however, are prevalent among bulbous plants; controls for these problems do not vary significantly among species.

Problems

Leaves yellow or distorted; bulbs decayed. Cause: Bulb mites. These 1/50″-1/25″, whitish mites feed in groups on bulbs. Infested bulbs have corky, brown spots that become powdery. Bulb mites are attracted to damaged or rotting bulbs, but can move to healthy ones, carrying rot-producing fungi and bacteria as they travel. When buying bulbs, inspect them carefully for damage or signs of infestation; for more information, see the illustration on page 49. Dip bulbs in hot (120°F) water for a few minutes to kill mites. Dig and destroy severely infested bulbs; solarize the soil before planting in previously infested ground.

Leaves and/or flowers with holes. Causes: Japanese beetles; other leaf-eating beetles. A number of hungry beetles, including 3/4″ long black blister beetles, 1/4″ long spotted cucumber beetles, 1/3″ long, reddish brown rose chafers, and nocturnal, 1/3″ long, gray Fuller rose beetles feed on the leaves and sometimes the blossoms of flowering bulbs. Iridescent green-and-bronze, 1/2″ Japanese beetles top this list of troublesome pests; the adults devour the leaves, stalks, and flowers of nearly 300 plant species, while the 3/4″, C-shaped grub larvae feed on roots and are a major pest of turf grasses.

Injury caused by adult beetles ranges from small holes in leaves to skeletonized foliage to complete defoliation; feeding by weevils, such as long-nosed Fuller rose beetles, resembles "ticket punches" around leaf margins. Japanese and spotted cucumber beetles are fond of flowers and often do most of their damage there. Most beetles are active during the day and are large enough to be readily visible while they feed. Weevils more often prefer nighttime feeding.

Handpick adult beetles into a bucket of soapy water or a 5 percent solution of isopropyl alcohol. Wear gloves if your pests are blister beetles; contact with crushed beetles causes burns and blisters on skin. Shake infested plants in early morning to knock beetles onto a dropcloth, then scoop them up and destroy. Pheromone traps are available for Japanese beetles, but they are most helpful if you can convince your neighbors to use them, too; otherwise, your traps may just attract beetles from their yard into yours. Apply milky disease spores to your lawn for long-term larval control. Apply parasitic nematodes to soil to limit all beetle grub populations. As a last resort, spray with neem, pyrethrin, or rotenone.

Leaves with large, ragged holes. Cause: Slugs and snails. Differing only in the presence or absence of a shell, slugs and snails range in size from 1/8″ to 8″ and may be gray, tan, green, black, yellow, or spotted. Broad foliage, arising at or near the soil surface, makes many flowering bulbs appealing to these busy mollusks. Mulch provides shady hiding places from which they emerge to feed nocturnally. All species rasp large holes in leaves, stems, and bulbs; slimy trails of mucus also signal their presence.

Since slugs and snails travel over the ground on a flat, muscular foot, you can impede their progress with barriers of materials that irritate their soft bodies. Sprinkle bands of coarse, dry, scratchy materials such as cinders, wood ashes, cedar sawdust, and diatomaceous earth around plants or beds; renew frequently. Push 4″-8″ copper strips into the soil around beds as edging. Lay boards, cabbage leaves, or overturned clay pots—anything that offers a cool, damp daytime haven—around the garden; destroy pests that congregate underneath. Handpick slugs and snails from plants at night; drop them into soapy water or sprinkle them with table salt to dehydrate and kill them. Set shallow pans into the soil, placing the lip flush with the soil surface and fill with stale beer or any fermenting liquid; remove drowned pests daily. Encourage predatory ground beetles by maintaining clover, sod, or stone walkways.

Leaves, stems, and buds distorted, sticky; clusters of small insects. Cause: Aphids. Several species of pear-shaped, 1/32"-1/8" aphids attack flowering bulbs. Aphids can be green, pink, black, gray, or with a white fluffy coating and have long antennae and 2 short tubes projecting from the rear of their abdomens. These pests gather under leaves and on growing tips. Some, such as tulip bulb aphids, also infest bulbs, clustering beneath the tunic. Aphids suck plant sap, causing leaf and bud distortion and blossom and leaf drop. Their feeding may spread viral diseases. As they feed, they excrete sticky honeydew on which sooty molds grow; see "Leaves with black coating" below for more information.

Wash aphids from plants with a strong spray of water; repeat as needed to control infestations. Encourage natural predators and parasites such as aphid midges, assassin bugs, lacewings, lady beetles, and spiders; apply homemade garlic or tomato-leaf sprays. If water sprays fail, try sprays of alcohol, citrus oil, insecticidal soap, or neem. Use boric acid baits to control ants that herd aphids onto plants. As a last resort, spray aphids with pyrethrin, nicotine, or rotenone.

Leaves with black coating. Cause: Sooty mold. This fungus grows on the sugary, sticky honeydew produced by aphids and other sucking pests such as scales, whiteflies, and mealybugs. The black fungal coating doesn't harm leaves directly, but it does shade the leaves and reduce growth. The best control is to deal with the pests that are producing the honeydew. Identify the pest and apply the appropriate control. (If the plant itself doesn't show signs of pest damage, the honeydew may be dripping down from an overhanging plant.) On small plants, wipe the leaves with a damp cloth to remove the honeydew and the sooty mold.

Buxus

Boxwood. Shrubs.

Boxwoods are evergreen shrubs with glossy, opposite leaves. They are widely used for hedges.

Boxwoods prefer a site partially sheltered from winter winds. They will flourish in a range of conditions, from full sun to deep shade, and can adapt to almost any soil (with the exception of poorly drained sites or heavy clays). Set out as balled-and-burlapped or container-grown plants in spring. Set out plants no deeper than they grew in the nursery. A thick layer of mulch will help keep roots cool and retain moisture. Clean up all fallen leaves and remove debris from branch crotches in autumn.

Problems

Leaves blistered and browned. Cause: Boxwood leafminers. The 1/8", orange larvae tunnel and feed within leaves. After they feed, the maggots change to the adult form and emerge as 1/8", orange, mosquito-like flies. Use yellow sticky traps to monitor the emergence of adults. When they begin to appear (usually late April and early May), spray plants with superior oil.

Leaves curled and cupped around stem tips. Cause: Boxwood psyllids. The eggs and larvae of this tiny green insect overwinter on buds and infest new leaves in spring. Spray plants with insecticidal soap, or pyrethrin for serious infestations, at the first sign of damage.

Leaves stippled with yellow; foliage webbed. Cause: Spider mites. See "Leaves stippled with yellow; foliage webbed" on page 236 for controls.

Leaves brown; twigs die back. Cause: Winter damage. Cold, dry winter winds can seriously injure tender growth. Avoid damage by spraying leaves with an antidesiccant in fall, or provide shelter for plants with burlap windbreaks. Mulch heavily in fall, and water well before the ground freezes. Fertilize plants in early spring only; summer feeding promotes easily damaged, late-season growth. Prune out affected parts.

Leaves yellow; stems and leaves covered with small bumps. Cause: Scales. See "Leaves yellow; stems and leaves covered with small bumps" on page 237 for control suggestions.

Leaves yellow, sparse, distorted, or with brown edges; branches die; growth stunted. Cause: Decline. For more information, see "Leaves yellow, sparse, distorted, or with brown edges; branches die; growth stunted" on page 238.

Whole plant stunted and lacking vigor. Cause: Nematodes. See "Whole plant stunted and lacking vigor" on page 239 for controls.

Cabbage

Brassica oleracea
Capitata group (Cruciferae)

Cabbage is a cool-season vegetable grown for its crisp, dense leaf heads. There are green and purple cultivars and ones with savoyed, or crinkled, leaves.

Culture

Plant cabbages in full sun in a site with fertile, well-drained soil and a pH between 6.0 and 6.8. If you have a choice of sites, spring plantings do best in lighter, sandier soils, while fall plantings do better in soils that contain more clay. Plants grow best at temperatures between 40° and 75°F.

Cabbages are biennial, and transplants exposed to cool temperatures (35°-45°F) for 10 or more days may bolt or go to flower prematurely. High temperatures also cause bolting.

These plants have very shallow roots, so be sure to keep the top few inches of soil from drying out. Fluctuations in soil moisture after the heads have formed may cause them to split. Mulching helps to balance and conserve moisture. Most cabbage diseases need free water to spread, so don't water with overhead sprinklers.

To avoid problems, don't plant cabbages where members of the cabbage family (broccoli, cauliflower, brussels sprouts, or kale) have grown for at least 3 years. Also, avoid areas

CABBAGE ◆ What Goes Wrong and Why

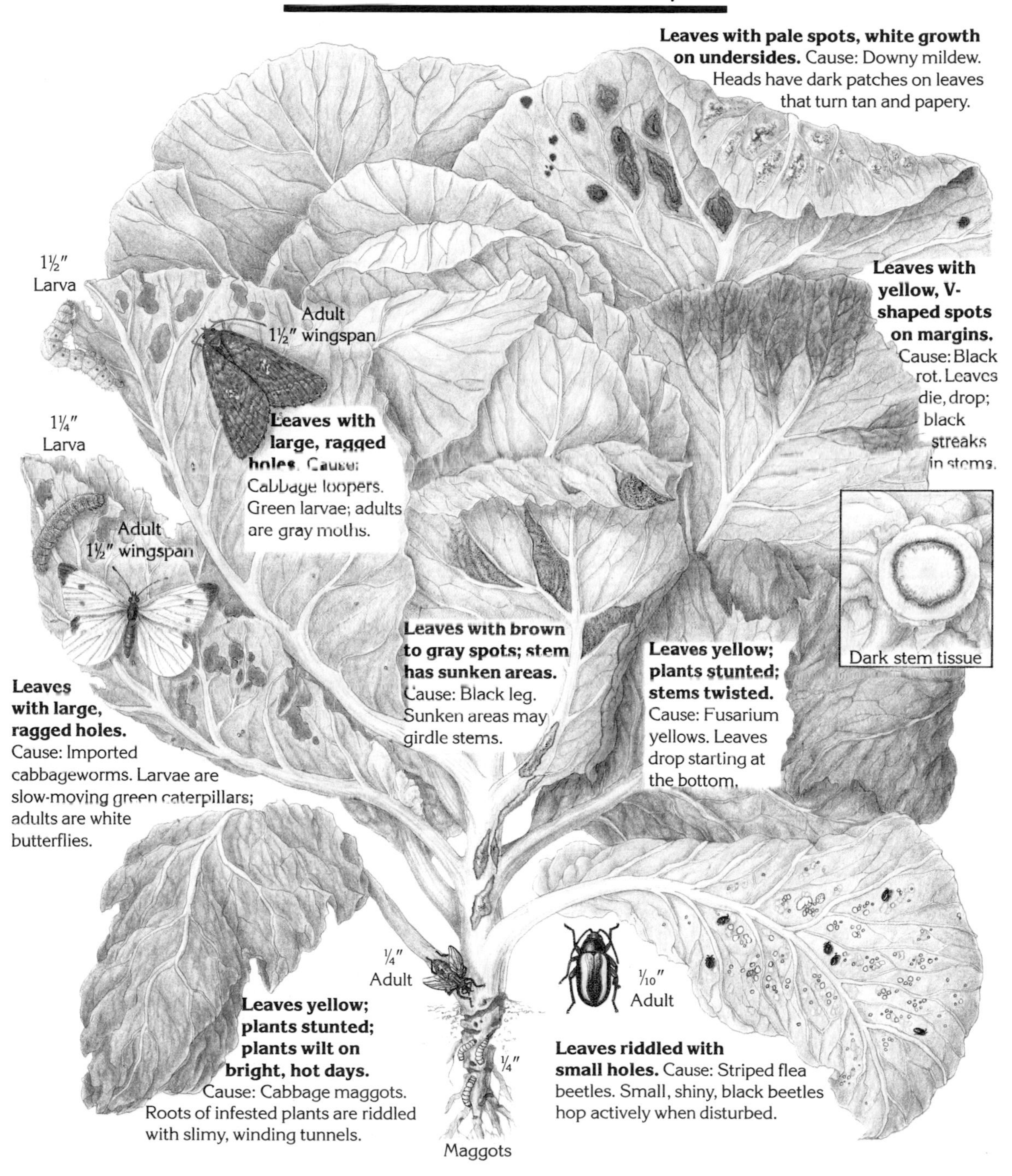

with cabbage family weeds, such as wild mustard. Destroy all crop residues, including roots, after harvest.

If you start your own plants, soak the seed in 122°F water for 25 minutes before planting to eliminate seed-borne diseases. (Be aware that this treatment can damage seed viability; for complete instructions, see page 422.) Once your seeds have germinated, grow seedlings at 60°F to keep them short and stocky.

Cabbages are heavy feeders and are susceptible to several nutrient deficiencies, including boron, calcium, phosphorus, and potassium.

Problems

Stems of transplants shrunken and dark near soil line. Cause: Wirestem. Weak plants may wilt and die as the disease progresses; seedlings that survive yield poorly. Later in the season, the same fungal disease rots the head from the bottom up. Spray plants with copper before symptoms develop if you have had problems in the past.

Leaves of transplants purple. Cause: Phosphorus deficiency. Phosphorus is not readily available in cool spring soils. Spray plants with seaweed extract to correct.

Leaves yellow; plant stunted. Cause: Nitrogen deficiency. When this deficiency develops, older leaves turn yellow first. Spray plants and drench roots with fish emulsion to correct.

Leaves yellow; plant stunted; plant wilts on bright, hot days. Causes: Club root; root knot nematodes; cabbage maggots. If you suspect any of these pests, pull up a plant and examine the roots. Roots that are enlarged and clublike indicate club root, a fungal disease; remove and destroy diseased plants. Prevent problems by rotating crops and improving drainage. Add lime to raise the pH to 7.2 or above. Cabbage can tolerate this pH, but club root fungi are less active.

Cabbage roots covered with irregularly shaped galls are infested with root knot nematodes. Destroy severely infested plants. Reduce the number of nematodes in the soil by solarizing soil, keeping area clear of crops and weeds for a summer, or by planting a winter cover crop of wheat. Applying parasitic nematodes or chitin to the soil can also help control pest nematodes.

Cabbage roots riddled with slimy, winding tunnels are infested with cabbage maggots. Larvae are white, 1/4" maggots that feed on roots and transmit bacterial and fungal diseases. Adults look somewhat like houseflies and lay eggs on the soil near the base of the plant. See page 53 for an illustration of this pest. Remove and destroy infested plants. Frequent, light cultivation when plants are young helps to decrease larval populations. Or plant each transplant through an X-shaped slit in the center of a 6"-8" square of tar paper to prevent the adults from laying eggs near the stems. Solarizing the soil will help reduce maggot populations in problem areas.

Leaves yellow; plant stunted; stems twisted. Cause: Fusarium yellows. Leaves of afflicted plants drop off starting at the bottom and working up, often leaving a bare stem. Cut stems reveal darkened vascular tissue in afflicted areas. This fungal disease usually occurs soon after transplanting. See page 53 for an illustration of this disease. Destroy infected plants. Use at least a 5-year rotation to prevent problems or plant resistant cultivars such as 'Applause', 'Bravo', 'Early Jersey Wakefield', 'Gourmet', 'Grand Slam Hybrid', 'Green Cup', 'Market Prize', 'Market Victor', 'Ranger', 'Red Danish', and 'Stonehead'.

Leaves curled and yellow; plant stunted. Cause: Aphids. Look for small, green, pink, gray, black, or white fluffy-coated, soft-bodied insects feeding on plants. Aphids can also transmit viral diseases. Control by knocking the pests off the plant with a blast of water, or use an insecticidal soap spray. Severe infestations can be controlled by spraying with a commercial pyrethrin/rotenone mix.

Leaves with yellow, V-shaped spots on margins. Cause: Black rot. As this bacterial disease progresses, yellow areas on the leaves enlarge and veins turn black. Infected leaves die and drop off. Stems have internal black streaks. Heads are dwarfed and often 1-sided. See page 53 for an illustration of this disease. Destroy infected plants. Spray copper if weather is wet and if you have had problems in the past. Resistant cultivars include 'A & C #5', 'Blue Boy', 'Bravo', 'Fortuna', 'Grand Slam Hybrid', and 'Guardian'.

Leaves with yellow-brown, concentrically ringed spots. Cause: Alternaria blight. This fungal disease attacks lower leaves first. Spots merge and leaves die as the disease progresses. Plants eventually die. Spray plants with copper at the first sign of symptoms. Plant tolerant cultivars such as 'Hybrid H' to prevent problems.

Leaves with white and yellow blotches. Cause: Harlequin bugs. Leaves turn brown, wilt, and die in severe infestations. Look for flat, ½" long, shield-shaped stink bugs with red and black, spotted markings. Nymphs look like adults but are smaller and more round. Handpick, or spray with a commercial pyrethrin/rotenone mix. Destroy crop residues after harvest to eliminate pests.

Leaves with pale spots, white growth on undersides. Cause: Downy mildew. Infected heads have dark patches on leaves. As the disease progresses, infected areas enlarge and turn tan and papery. See page 53 for an illustration of this fungal disease. Control by spraying plants with a baking-soda-and-soap spray (1 teaspoon baking soda, 1 teaspoon liquid dish soap, 1 quart water) or copper. The cultivar 'Hybrid H' is tolerant of downy mildew.

Leaves with white-bronze spots. Cause: Thrips. Severe infestations cause leaves to wilt. These tiny insects look like grains of pepper. Trap with sticky blue traps or spray insecticidal soap, making sure to cover undersides of leaves where thrips feed.

Leaves with brown to gray spots; stem has sunken areas. Cause: Black leg. Leaf spots are speckled with tiny black dots. As the sunken areas on stem enlarge, they may girdle the plant and cause it to wilt and die. Dead leaves remain attached to the plant. See page 53 for an illustration of this disease. Spray plants, especially the stems, with copper at the first sign of this fungal disease. When spraying stems, try to avoid spraying copper on the soil, because it is toxic to many soil organisms, including earthworms. Prevent black leg by using a 4-year rotation and improving soil drainage.

Leaves with brown tips. Causes: Excessive heat; calcium deficiency. Browning of the tips of outer leaves is caused by extreme heat. Cabbage is sensitive to bright sun and temperatures above 80°F. Plant cultivars that resist tipburn such as 'Hancock Hybrid' and 'Multikeeper'.

Brown leaf tips inside the head are caused by calcium deficiency. Many factors contribute to poor calcium uptake. To prevent problems, keep soil moisture constant and side-dress plants with compost to provide balanced nutrition.

Leaves cracked and corky; stem water-soaked or hollow inside the head. Cause: Boron deficiency. Plants in the cabbage family need lots of boron. Spray plants with seaweed extract when transplanting and repeat several times until head formation to help prevent problems. Raise the boron level of the soil by adding 1 tablespoon of borax dissolved in 1 gallon water, or 10 pounds of kelp, per 100 square feet.

Leaves riddled with small holes. Cause: Striped flea beetles. These small, shiny, black beetles hop when disturbed. They can transmit viral and bacterial diseases. See page 53 for an illustration of this pest. Protect young plants with row cover until they have at least 6 leaves. 'Early Jersey Wakefield' tolerates beetle damage better than most cultivars.

Leaves with large, ragged holes. Cause: Caterpillars. Look for dark green excrement at the base of leaves or plant. Imported cabbageworm is a velvety green, slow-moving caterpillar up to 1¼″ long. Adults are small white butterflies. Cabbage loopers are light green, 1½″ long caterpillars with a white stripe along each side. They travel with a characteristic "looping" movement, like an inchworm. Adults are gray moths. See page 53 for an illustration of these 2 pests. Diamondback moth caterpillars are pale green with light brown heads, 5/16″ long, pointed at both ends, and covered with fine black hairs. Adults are slender, ½″ long moths.

Apply a BTK spray as soon as active caterpillars or feeding are observed; remember to look inside heads for early signs of feeding. Pheromone traps are available for male cabbage looper moths; spray with BTK 1-2 weeks after the first moths are caught in the traps. Help reduce problems by planting cultivars, such as 'Danish Ballhead', 'Early Jersey Wakefield', and 'Red Ace', that are less susceptible to cabbage looper and imported cabbageworm feeding.

These caterpillars are all parasitized by wasps, and other beneficials feed on both eggs and larvae. A native virus may also kill caterpillars later in the season. Infected caterpillars move slowly, become yellow, and die. Let nature work for you and avoid handpicking or spraying if caterpillars are infected or covered with the small, white, spindle-shaped eggs of parasitic wasps.

Heads small and soft; color poor in red cabbage. Cause: Potassium deficiency. Check with a soil test and amend as needed.

Heads split. Cause: Excessive water. Keep soil constantly moist, but not saturated, especially when heads are forming.

Heads rot. Cause: Fungal diseases. A variety of fungi cause cabbage heads to rot. Destroy rotted plants. Spray plants with copper before symptoms develop if you have had problems in the past.

Calendula

Calendula, pot marigold. Annuals.

Calendulas are herbs that have 1″-3″, orange or yellow flowers, giving bouquets of color for borders and indoor use. Plants grow 1′-3′ high with a bushy habit. Calendulas blend well into bedding schemes, herb gardens, and containers.

Sow seeds indoors 4-6 weeks before transplanting outdoors into warm spring soil. Direct-seed in early summer for color in late summer or fall. Plant seeds ¼″ deep; they need darkness to germinate. They should sprout in 1 week. Thin seedlings to 1′.

For earliest color, transplant or seed in full sun. In hot climates plants appreciate some afternoon shade. While calendulas tolerate poor soil, they prefer average soil and must have good drainage. Water regularly, keeping soil slightly moist. When first buds appear, pinch back the main stem to promote maximum flowering. In adverse conditions, such as excess heat, calendulas may still flower, but stems will be shorter.

Problems

Leaves, stems, and buds distorted. Cause: Aphids. For controls, see "Leaves, stems, and buds distorted" on page 20.

Leaves with large, ragged holes. Causes: Imported cabbageworms; slugs and snails. Cabbageworms are green caterpillars with a yellow back stripe and yellow side spots. Handpick cabbageworms or spray leaves with BTK when pests first appear.

Slugs and snails cause similar damage, and leave behind slimy trails of mucus. For controls, see "Leaves with large, ragged holes" on page 18.

Leaves greenish yellow; growth poor.

Cause: Aster yellows. For more information, see "Leaves greenish yellow; growth poor" on page 19.

Callistephus

China aster. Annuals.

Not true asters, China asters provide showy, single or double, daisylike blossoms from midsummer until autumn with successive plantings. There are early, mid-season, and late-flowering cultivars. A range of color choices is available, including blue, purple, pink, red, rose, yellow, or white. Individual flowers can reach 5″ wide; petal types are quilled, spooned, or curled. Plant size varies from 9″ to 24″ high, with a possible 1½′ spread. Low-growing cultivars make good edging plants. Taller cultivars are popular for long-lasting cut flowers. Plants will not rebloom once flowers have been cut.

Direct-sow outdoors a week or two after the last frost, barely covering the seed. Seeds germinate within 2 weeks. Or, for earlier flowers, sow seeds indoors 6-8 weeks before the last frost. Move seedlings to individual 3″ pots when the first true leaves appear. Move plants outdoors when nights are consistently above 50°F. Handle carefully during transplanting to avoid damaging the roots. Space 6″-15″ apart. Each plant blooms for about 4 weeks. Successive plantings 2 weeks apart are the key to a long display season.

China asters tolerate light shade only. They thrive in fertile, moist but well-drained soil in full sun. China asters are shallow-rooted, so cover soil with a 1″ mulch to retain soil moisture and keep roots cool. Pinch growing tips to promote bushiness. Water regularly so soil stays slightly moist. Uneven care that slows growth makes the plant susceptible to disease. Remove old flowers. Tall cultivars may need staking to support heavy flowers on the slim stems.

Problems

Leaves greenish yellow; growth poor. Cause: Aster yellows. For more information, see "Leaves greenish yellow; growth poor" on page 19.

Lower leaves yellow, wilt, and die. Cause: Fusarium wilt. Other symptoms of this common fungal disease include yellowing and wilting of 1 side of plant only; drooping flower heads; stunted mature plants; dark brown areas on stems; and partial or total root decay. Once introduced on diseased plants or transmitted by wind, water, or handling, Fusarium wilt lives on indefinitely in soil. There is no cure. Remove infected plants and dispose of them in sealed containers in household trash. Never plant China asters in the same location 2 years in a row. Wilt-resistant cultivars are available, but they are not completely immune.

Leaves, stems, and buds distorted. Cause: Aphids. For control measures, see "Leaves, stems, and buds distorted" on page 20.

Camellia

Camellia. Shrubs and small trees.

Camellias are evergreen shrubs or small trees that bear spectacular flowers. They are popular as foundation plants or hedges in areas where they are hardy. In cold climates grow these plants in a cool greenhouse.

Camellias require a moist, peaty, acidic soil well-enriched with organic matter. Set out as bare-root plants in partial shade. A thick layer of mulch will protect the shallow roots and help retain soil moisture.

Problems

Leaves yellow; stems and leaves covered with small bumps. Cause: Scales. See "Leaves yellow; stems and leaves covered with small bumps" on page 237 for control suggestions.

Flowers turn brown and fall off. Cause: Flower blight. This soilborne fungus causes small brown spots that expand to cover entire petals. Prevent problems by setting out only bare-root plants and picking off all blooms as they fade. If the disease does strike, remove and destroy all the flowers. Use a preventive spray of bordeaux mix in the spring. Surround plants with a 3″ layer of mulch to keep spores from splashing up from the soil to the flowers.

Leaves with notched edges. Cause: Fuller rose beetles. The adult is a 1/3″, grayish brown beetle with a cream-colored stripe on each side of its body. It feeds at night on foliage. The yellowish, brown-headed larvae feed on plant roots. Handpick adults as they hide in foliage during the day. Place sticky barriers, such as Tanglefoot, around the base of plants to keep adults from climbing up to the leaves. Drench soil around the plant with a solution of parasitic nematodes to control larvae.

Campanula

Bellflower, harebell. Perennials.

The profuse dainty, blue, purple, or white, bell-shaped flowers of *Campanula* spp. brighten gardens from spring through summer. This large genus includes annual, biennial, and perennial species that range in size from low-growing Carpathian harebells (*C. carpatica*) to 2′–3′ tall, peach-leaved bellflowers (*C. persicifolia*).

Mountain natives, bellflowers prefer full sun to light shade; well-drained, organic soil; and regular watering. Serbian bellflower, *C. poscharskyana,* tolerates drought, but most other bellflowers do not. A layer of mulch helps keep the root zone cool and holds in moisture during hot summer weather; a winter mulch protects roots from heaving. Divide plants in spring every 3–4 years.

Problems

Leaves, stems, and buds distorted, sticky; clusters of small insects. Cause: Aphids. See "Leaves, stems, and buds distorted, sticky; clusters of small insects" on page 177 for controls.

Leaves with large, ragged holes. Cause: Slugs and snails. For controls, see "Leaves with large, ragged holes" on page 177.

Leaves and flowers distorted and with brown spots. Cause: Thrips. The same thrips that feed on onions feed on bellflowers. Damage to young plants is most severe. Water stress encourages thrips multiplication and damage. These 1/50″–1/25″, yellow, brown, or black pests are difficult to see with the naked eye, but the streaking, spotting, and distortion caused by a heavy infestation is distinctive. Protect plants from thrips damage with predatory mites and blue or yellow sticky traps; in severe cases apply sprays of alcohol, insecticidal soap, neem, nicotine, or pyrethrin, or dust undersides of leaves and surrounding soil with diatomaceous earth.

Leaves yellow and wilting; stems blackened at base; plant topples over. Cause: Stem, root, and crown rots. These problems are common when bellflowers are kept overly wet or are grown in poorly drained soil. The base of an infected plant may show grayish white mold; dark lesions blacken the stem from the soil up, eventually girdling the plant. Remove and destroy infected plants and surrounding soil. Let soil dry somewhat between waterings, and incorporate organic matter to improve soil drainage. Avoid overwatering and overcrowding; keep winter mulch away from the crowns to prevent excess moisture. Protect crowns from injury, especially when plants are dormant and hard to see. Wash tools after working around infected plants.

Canna

Canna. Rhizomes.

Cannas' large, paddle-shaped leaves and brightly colored red, yellow, orange, pink, white, or variegated flowers add a tropical look to summer and fall gardens. These heat-loving hybrids (*Canna* × *generalis*) grow 1½′–8′ tall with a spread of about 3′.

In Zones 8–10 cannas will survive winter temperatures, but in northern zones, the rhizomes are overwintered indoors. In either case, plant rhizomes 3″–5″ deep and 1′ apart in average, moist but well-drained soil in full sun. Both shade and excess nitrogen fertilizer can reduce bloom. In cold winter areas, start rhizomes indoors in pots as early as February. Move outdoors in spring. If necessary, divide roots at spring planting time. Use a sharp knife, and leave at least 1 good-size bud on an ample piece of rhizome. Let divisions dry overnight before planting. To overwinter, carefully dig up the rhizomes in fall after frost blackens the leaves. Allow them to dry for a few days, then store them in containers of barely moist sand or perlite in a cool place.

Problems

Leaves and/or flowers with holes. Cause: Japanese beetles. See "Leaves and/or flowers with holes" on page 50 for controls.

Leaves with large, ragged holes. Cause: Slugs and snails. For control information, see "Leaves with large, ragged holes" on page 50.

Leaves rolled. Cause: Leafrollers. These pests roll leaves into tubes and fasten them with webbing while feeding. Handpick and destroy rolled leaves; encourage parasitic *Trichogramma* wasps. Sprays of BTK are most effective when pests are young.

Cantaloupe

Cucumis melo
Reticulatus group (Cucurbitaceae)

Cantaloupe is a name commonly used for muskmelons in the United States. See the Melon entry beginning on page 148 for culture and pest information. Harvest when fruit smells ripe and the stem separates easily from the vine.

Carpinus

Hornbeam. Trees.

Hornbeams are bushy, slow-growing, alternate-leaved trees. They are occasionally used as specimen trees, but difficulty in transplanting limits their use.

Set out balled-and-burlapped plants in early spring. Hornbeams are excellent urban trees and, while preferring full sun, will grow satisfactorily in partial shade. They are widely tolerant of various soil types.

Hornbeams are remarkably free of problems. Maple phenacoccus scale may detract from the appearance of the plant, coating the undersides of leaves with fluffy white egg masses. See "Leaves yellow; whole plant weakened" on page 13. Leafminers may also attack; see "Leaves tunneled" on page 237.

Carrot

Daucus carota var. *sativus* (Umbelliferae)

Carrots are a biennial vegetable grown for their crisp, sweet roots. They are a cool-season crop, hardy enough to be undisturbed by light frosts in the spring and fall.

Culture

Carrots grow best in deeply worked, loose soils with a pH between 5.5 and 6.8. No other vegetable is as sensitive to poor soil structure. Misshapen carrots are more often a result of lumpy or compacted soil than any pest problem. Carrots do well in raised beds. Work in a generous amount of compost or well-rotted manure before planting. Cultivars with short roots will tolerate shallow or poor soil better than long, thin cultivars.

Carrots grow best when temperatures are between 60° and 70°F. They grow poorly above 75°F, but will tolerate temperatures as low as 45°F. Most cultivars grow short roots at high temperatures, and longer, more pointed roots at lower temperatures.

Abundant water is necessary for good root development. It is especially important to give emerging seedlings an edge against weeds. Keep soil evenly moist, but not saturated.

To prevent problems with diseases and insects, do not plant carrots where carrots or parsley has grown for 3 years.

Carrots require moderate to high levels of potassium and phosphorus, but only moderate levels of nitrogen, so avoid high-nitrogen fertilizers. Carrots are very sensitive to salt injury and do poorly in soils with high sodium levels.

Leaf Problems

Young leaves yellow and dwarfed; growth bushy. Cause: Aster yellows. Older leaves of plants infected with this disease may be purplish. Once a plant is infected, there is no cure; destroy infected plants. The disease is spread by green or brown, wedge-shaped, 1/10"-1/8" long leafhoppers. Cover emerging seedlings with row cover if leafhoppers have been a problem. Control leafhoppers by spraying plants with insecticidal soap in the evening, or for severe infestations, with a commercial pyrethrin/rotenone mix.

Leaves yellow and curled or distorted. Causes: Aphids; tarnished plant bugs. Aphids are small, soft-bodied, green, black, gray, pink, or white fluffy-coated insects that suck plant sap. For a mild infestation, knock pests off plants with a blast of water. Control them by spraying plants with insecticidal soap, or with a commercial pyrethrin/rotenone mix if infestation is severe.

Tarnished plant bugs are oval, light green to brown, 1/4" long bugs with a lighter triangle on each wing. Nymphs are smaller, yellow-green insects with black spots. Control them with white sticky traps or by spraying plants with a commercial pyrethrin/rotenone mix in the evening. Turn crop residues under after harvest to reduce future problems.

Leaves yellow; plant wilts during bright, hot days but recovers at night. Cause: Root knot nematodes. See "Roots misshapen and covered with numerous hairlike roots" below for controls.

Older leaves yellow; plant stunted. Causes: Nitrogen deficiency; waterlogged soil. Spray foliage and drench roots with fish emulsion or fish-meal tea to alleviate symptoms. Plant carrots in well-drained soil. Keep soil evenly moist, but not saturated.

Leaves with dark, yellow-bordered spots. Causes: Cercospora leaf blight; Alternaria leaf blight; bacterial leaf blight. For the fungal blights, pick off spotted leaves and spray foliage with fish emulsion to encourage new growth. If disease continues to spread, spray foliage with sulfur. To prevent problems, soak seed in 122°F water for 20 minutes before planting. (Be aware that this treatment can damage seed viability; for complete instructions, see page 422.) Plant cultivars, such as 'Huron', 'Orlando Gold Hybrid', and 'Seminole', that are somewhat resistant to Alternaria.

Bacterial blight also causes dark brown streaks on petioles. Destroy the whole plant. Spray plants with copper if bacterial blight has been a problem in the past. To prevent problems, soak seed in 122°F water for 20 minutes before planting.

Root Problems

Roots forked and misshapen. Cause: Lumpy or compacted soil. Prevent problems by working soil deeply and adding lots of compost. Plant shorter-rooted cultivars in rocky or clay soil.

Roots misshapen and covered with numerous hairlike roots. Causes: Root knot nematodes; aster yellows. Root knot nematodes cause tiny round swellings on side roots. Prevent future root knot nematode problems by applying chitin or parasitic nematodes to the soil before planting.

Roots infected with the disease aster yellows are small, woody, hairy, and taste bitter. See "Young leaves yellow and dwarfed; growth bushy" above for controls.

Roots with green shoulders. Cause: Light exposure. Keep carrots covered with 2″ of soil to prevent problems.

Roots with dark tunnels. Causes: Carrot weevils; carrot rust flies; carrot beetles. Carrot weevils usually attack the upper and outer parts of the root. Tunnels are often in a zigzag pattern. Larvae are creamy white, 1/3″ long grubs with brown heads. Adults are coppery brown, 1/6″ long snout beetles. Destroy infested roots; don't put them in the compost. Control larvae by applying parasitic nematodes to the soil. Control adults, which emerge early in the spring to lay eggs, by covering the seedbed with row cover or by spraying plants with a commercial pyrethrin/rotenone mix as soon as you see adult weevils.

Carrot rust fly maggots feed throughout the root, leaving randomly patterned tunnels filled with a rust-colored, sawdustlike material. Larvae are 1/3″ long, white maggots. Adults are small flies with yellow heads. Apply parasitic nematodes to soil to control larvae. Prevent problems by planting after June 1st to avoid the first hatching of the season. Cover seedbed with row cover to prevent adults from laying eggs on soil.

Carrot beetle larvae are 1″, bluish white grubs that feed on the roots. Adults are 1/2″, reddish brown or black beetles. Fall cultivation reduces overwintering populations; crop rotation will also help.

Roots with small, irregular holes; plant may be stunted and yellow. Cause: Wireworms. Damage usually occurs later in the season and is worse in dry years. Larvae are up to 1 1/2″ long, yellow to reddish brown, slender, tough-bodied, segmented grubs with brown heads. Adults are 1/3″–3/4″ long, dark-colored, elongated click beetles. Control wireworms by applying parasitic nematodes to the soil.

Roots or crowns rotted. Cause: Various fungal or bacterial diseases. Root rot is often brought on by soggy, poorly drained soil or previous insect damage to the roots. Plant carrots in loose, well-drained soil and keep soil moist, but not soggy. Use raised beds to improve drainage.

Roots with internal cavities; may split open. Cause: Cavity spot. This condition is caused by a combination of factors: calcium deficiency, high levels of potassium, and possibly various diseases. To prevent problems, keep soil moist, but not soggy. Add gypsum, or high-calcium lime if pH is below 6.2, to raise calcium level, and withhold high-potassium fertilizer.

Roots with jagged cracks. Causes: Freezing injury; uneven soil moisture; cavity spot. Temperatures below 30°F cause cracked roots with a water-soaked appearance. Protect roots with mulch before temperatures fall.

Roots can also crack if soil is wetted after being dry. Prevent problems by keeping soil evenly moist.

Severe cases of cavity spot can cause open cracks. See "Roots with internal cavities; may split open" on page 61 for causes and controls.

Roots poorly formed or pale. Causes: Nutrient deficiency; extreme temperatures. Spindly, short roots can be caused by potassium deficiency or excessive heat. Poor color and taste are caused by magnesium deficiency, phosphorus deficiency, and low or high temperatures. Copper deficiency can also cause poor root development. Spray foliage with seaweed extract to prevent deficiencies. Do a soil test and amend soil as needed.

Cauliflower

Brassica oleracea
Botrytis group (Cruciferae)

Cauliflower is a cool-season crop grown for its dense, white heads of flower buds. Cauliflower and cabbage require similar culture and are attacked by many of the same diseases and insects. See the Cabbage entry beginning on page 52 for culture and more information on problems. Prevent problems by planting the following cultivars: 'Alpha Paloma' and 'White Rock' are tolerant of cabbage root maggots; 'Super Snowball A' tolerates flea beetles better than most cultivars.

Cauliflower is more fussy about temperatures than the other brassicas. Plants will only tolerate a low of 45°F and grow poorly above 75°F. Transplants exposed to cool temperatures (35°-45°F) for 10 days or more may form tiny, unusable flower heads. High temperatures can cause the same problem. Its optimum temperature range is 60°-65°F.

Problems

Heads with black spots on surface. Cause: Downy mildew. Control this fungal disease by spraying plants with a baking-soda-and-soap spray (1 teaspoon baking soda, 1 teaspoon liquid dish soap, 1 quart water) or copper at the first sign of disease. Repeat weekly if weather is wet. Prevent problems by planting 'Cloud Nine', which is somewhat tolerant of downy mildew.

Heads with brown or discolored curds. Causes: Light exposure; boron deficiency; cold injury. Cauliflower heads develop green or purple pigments when exposed to sunlight. To blanch cauliflower, tie the outside leaves loosely up over developing heads so that air can move through, but water cannot drip in. Save time by growing self-blanching cultivars such as 'Montano'.

Boron deficiency turns curds brown and makes them appear water-soaked. Prevent problems by spraying plants with seaweed extract when transplanted and every 2 weeks thereafter. Raise the boron level of the soil by adding 1 tablespoon of borax dissolved in 1 gallon water, or 10 pounds of kelp, per 100 square feet.

Curds will also turn brown in response to cold and freezing injury. Protect cauliflower heads with row cover below 45°F.

Heads with loose or irregular flower curds. Causes: Excessive heat; molybdenum deficiency. If weather has been hot, suspect heat stress. Prevent problems by growing cauliflower as a fall crop when temperatures are cooler. Prevent molybdenum deficiency by spraying plants with seaweed extract at transplanting and every few weeks thereafter.

Cedrus

Cedar. Trees.

Cedars are among the most attractive evergreen trees. They have horizontal branching habits and bear large, upright cones. All are ideal as specimen trees.

Cedars all like a deep, loamy soil, well-drained but enriched with organic matter. Plant them in spring as small, balled-and-burlapped or container-grown plants, setting them in a sunny location. Cedars are remarkably free of serious insect pests. Keep your trees healthy by applying compost mulch and watering thoroughly during periods of drought to minimize damage.

Problems

Leaves reddish brown; branch tips die back. Cause: Deodar weevils. This ½″ long, reddish brown beetle has irregular white spots on its back. It will occasionally gnaw through bark and feed on the growing tissue beneath, girdling small branches. The grubs also feed on branch tips. This pest rarely attacks healthy trees, so fertilize and water trees regularly to keep them vigorous. Drench soil around the plant with a solution of parasitic nematodes. Spray leaves with pyrethrin to control severe infestations.

Trunk or branches with oozing lesions; branch tips die back. Cause: Canker. See "Trunk or branches with oozing lesions; branch tips die back" on page 238 for more information.

Leaves yellow and wilt. Cause: Root rot. See "Leaves yellow and wilt" on page 237 for details.

Leaves yellow; stems and leaves covered with small bumps. Cause: Scales. See "Leaves yellow; stems and leaves covered with small bumps" on page 237 for controls.

Celastrus

Bittersweet. Vines.

Bittersweets are vigorous twining vines with alternate leaves. The orange-red berries are popular for dried arrangements.

Bittersweets flourish in sun or partial shade in virtually any soil. Be sure to obtain both male and female plants to ensure fruiting. Control the aggressive vines by pruning them back severely each winter.

The major insect enemy of bittersweets is scales; see "Leaves yellow; stems and leaves covered with small bumps" on page 237. Powdery mildew may affect leaves; see "Leaves with powdery white coating" on page 237.

Celery

Apium graveolens var. *dulce*
(Umbelliferae)

Celery is a biennial vegetable grown for its crispy leaf stalks. It is in the same family as carrots and is troubled by many of the same pests and diseases. See the Carrot entry beginning on page 60 for more information on culture and problems.

Celery does well if given plenty of moisture, rich soil with a pH between 6.0 and 6.8, and cool temperatures or light shade. Celery will withstand a light frost, but will bolt or flower prematurely and produce only a few small stalks if temperatures are below 55°F for more than 10 days when plants are young.

Celery is a heavy feeder. Boron deficiency causes brown, mottled leaves and horizontal cracks on stalks. 'Florida 683' is tolerant of low boron. Calcium deficiency causes the center of the plant to blacken and die. 'Utah 52-70R Improved' is tolerant of low calcium. Magnesium deficiency causes yellow leaves. Discolored streaks on stalks are a symptom of potassium deficiency. Phosphorus deficiency causes plants to form rosettes. Prevent problems by adding plenty of compost to the soil and by spraying plants with seaweed extract every 2 weeks. Check suspected deficiencies with soil tests and amend soil as needed.

Stalks become tough, bitter, and stringy if plants do not have enough water or nutrients, if temperatures are too high, or if they are not harvested promptly.

Problems

Leaves with dark, yellow-bordered spots. Cause: Blight diseases. Various fungal blight diseases attack celery. Stalks of afflicted plants may also develop spots or dark areas. Destroy spotted leaves. Spray plants with copper if symptoms are serious. Soak seed in 118°F water for 30 minutes before planting to prevent problems. (Be aware that this treatment can damage seed viability; for complete instructions, see page 422.) Plant cultivars, such as 'Emerson Pascal' and 'Golden Self-Blanching', that tolerate certain blights.

Leaves yellow; plant stunted; veins in stalks reddish. Cause: Fusarium yellows. Destroy plants with this fungal disease. Prevent problems by planting tolerant cultivars such as 'Bishop', 'Deacon', 'Emerson Pascal', 'Starlet', 'Summit', 'Ventura', and 'Vicar'.

Leaves yellow and mottled; stalks twisted; plant dwarfed. Cause: Celery mosaic. Destroy infected plants. Control aphids because they can spread viral diseases as they feed. Prevent problems by planting resistant cultivars such as 'Florida 683' and 'Utah 52-70R Improved'.

Leaves and stalks with irregular tan spots. Cause: Brown spot. Control this fungal disease by spraying plants with bordeaux mix when spots appear to prevent further spread.

Leaves with chewed holes or rolled edges. Cause: Caterpillars. Handpick caterpillars or spray plants with BTK to control.

Stalks or crowns rotted. Cause: Bacterial and fungal diseases. Various bacteria and fungi will rot celery stalks and crowns. Destroy infected plants. Spray remaining plants with copper. Prevent problems by controlling insect pests, such as carrot rust flies, that injure roots, creating openings for disease organisms. Also, avoid handling plants while they are wet, and plant in raised beds if possible.

Celosia

Celosia, cockscomb. Annuals.

Celosias form feathery plumes or velvety crests in shades of pink, red, orange, and yellow. Blossoms appear in summer through late fall; plants grow 7"-36" tall. Plumed types are great in beds and borders; crested types are ideal for the cutting garden.

Give celosias a sunny, well-drained site. Plants tolerate drought and poor soils, but they bloom poorly and become leggy if shaded. Sow seeds indoors 4 weeks before last frost, or direct-seed outdoors after the soil warms up. Cover lightly so seeds receive some light but don't dry out. Tall-growing, crested types are often top-heavy and benefit from staking.

Celosias are usually trouble-free. Careless transplanting can injure the terminal bud and reduce flowering. Damping-off may cause seedlings to rot at the soil line; see "Seedlings die" on page 20 for controls.

Centaurea

Cornflower, bachelor's-button. Annuals.

Usually seen in shades of blue, hardy cornflowers, or bachelor's-buttons, are available in red, pink, white, and violet, too. These 12″-26″ high garden favorites provide 1″ flowers all spring. Seeds are favored by birds, particularly finches. Use in beds and borders or for cutting.

Direct-sow outdoors in early spring where growing seasons are short, and in later summer through fall where winters are mild. Indoor seeding in April is an option, but seedlings transplant poorly. Sow evenly to avoid overcrowding, which results in weak flower stalks and slow growth. Seedlings appear in 2 weeks. Thin to 6″-9″ apart.

For best results, give bachelor's-buttons full sun and rich, moist, neutral soil. They will, however, tolerate average soil and drought. Always water early in the day to give the leaves a chance to dry before nightfall. Add organic matter or fertilizer to the planting site before seeding, and do not fertilize again; extra fertilizer promotes the development of leaves rather than flowers. By midsummer, bachelor's-buttons tend to become woody and unsightly. Picking flowers prolongs bloom until July, but plan on pulling out the spent plants and replacing them with a later-blooming annual, such as marigolds or celosias.

Bachelor's-buttons have very few problems or pests. They may reseed too prolifically for some gardeners, but removing spent flowers decreases this possibility. If plants look too leggy, they are not getting enough sun; plant on a sunnier site next year. Rusts may cause orange spots on leaves; see "Stems and leaf undersides with dusty, dark brown spots" on page 21. Aster yellows can cause plants to turn yellow; see "Leaves greenish yellow; growth poor" on page 19.

Chaenomeles

Flowering quince. Shrubs.

Flowering quinces are spring-blooming deciduous shrubs with alternate leaves and thorny branches. The plants are used as deciduous hedges, in foundation plantings, or in shrub borders.

Flowering quinces require full or nearly full sun. While not particular as to soils, they prefer a good, well-drained loam. Renewal pruning (removal of a few of the oldest canes each year) is recommended to keep them flowering freely.

Aphids may feed on young shoots and leaves; see "Leaves wrinkled and discolored" on page 235. Scales may also be a problem; see "Leaves yellow; stem and leaves covered with small bumps" on page 237.

Chard

Beta vulgaris, Cicla group (Chenopodiaceae)

Chard, also known as Swiss chard, is a type of beet grown for its succulent tops. It does not form enlarged roots. Green and red cultivars are available.

Chard grows vigorously from late spring to fall frost. Plants prefer sandy, well-drained soil with a pH of 6.5-7.5. They will tolerate a wide range of temperatures and will withstand light frost. To keep leaves tender, provide chard with plenty of water and nitrogen. Compost worked into the soil at planting followed by alfalfa meal sprinkled on the soil surface after seeding usually provides chard's requirements.

Chard and beets are susceptible to the same diseases and pests. See the Beet entry beginning on page 37 for descriptions and controls.

Cherry

Prunus spp. (Rosaceae)

Cherries are deciduous trees growing to 35′ for standard-size sweet cherries (*Prunus avium*) and to 20′ for standard-size sour cherries (*P. cerasus*). These plants are hardy in Zones 4-9.

Culture

Cherries need full sun and, especially for sweet cherries, well-drained soil. Choose a site free of late spring frosts. Train sweet cherries to a central leader form and sour cherries to an open center form. For more pruning information, see "Pruning and Training" on page 101.

Mature sweet cherries need little or no annual pruning. Sour cherries need enough pruning each winter to thin out branches and stimulate a moderate amount of growth.

To set fruit, most sweet cherries need cross-pollination from a second compatible cultivar planted nearby. Some sweet cherry cultivars need specific cultivars for cross-pollination. A few sweet cherries, such as 'Garden Bing' and 'Stella', are self-pollinating. Sour cherries are self-pollinating. For more information, see "Setting Fruit" on page 101.

Fruit Problems

Fruit malformed, shrunken, and drops early. Causes: Plum curculio larvae; cherry fruit fly maggots. Check affected fruit for crescent-shaped scars and small, brown-headed grubs inside. These are larvae of plum curculio beetles common to the East. The crescent-shaped scar marks the site where these pests lay eggs. To control curculios without sprays, spread a dropcloth beneath the tree and jar the trunk and branches twice a day with a padded mallet. Collect and destroy curculios that fall onto the sheet. For more information see "Young fruit with crescent-shaped scars" on page 186.

Creamy white, 1/4″ maggots inside fruit are the larvae of cherry fruit flies. These flies, which are half the size of houseflies, emerge in late spring from soilborne pupae and lay eggs in fruit. Eggs hatch into fruit-feeding maggots that later drop to the ground to pupate until next spring. To control this pest, trap adult flies in beginning to late May using red sticky spheres just like those for apple maggots. For instructions on making apple maggot traps see "Fruit dimpled; brown tunnels through flesh" on page 22. Rotenone sprayed when fruit begins to color is also effective. See the opposite page for an illustration of this pest.

Fruit with small brown spots that enlarge and become fuzzy in humid weather. Cause: Brown rot. This fungal disease may also cause

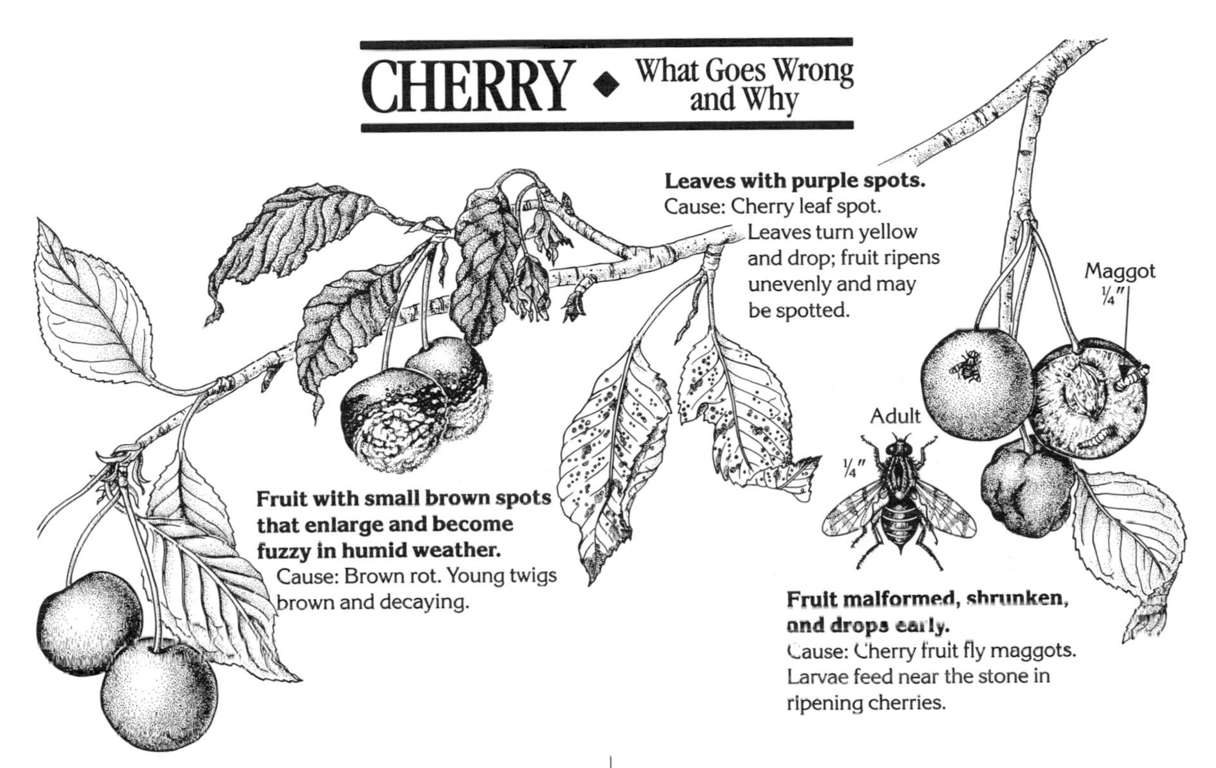

blossoms to turn brown and decay. Infected cherries may drop early, or remain on the tree as dried, shriveled fruit known as mummies. To control brown rot, inspect trees in early spring. Remove and destroy both mummified fruit and twigs with gummy-looking lesions. For further control, spray sulfur early in the season to control the disease on blossoms, then again later in the season to protect fruit. Copper sprays also aid control. Since injured fruit is more susceptible to brown rot, control fruit-damaging insects, such as plum curculios. See above for an illustration of this disease.

Leaf and Branch Problems

New leaves twisted or curled and covered with a sticky coating. Cause: Black cherry aphids. Look on leaf undersides for clusters of these tiny black insects. The sticky coating is honeydew secreted by these pests. Leaves also may be covered with a black fungus, called sooty mold, which feeds on honeydew. Aphids leave by midsummer, so if an infestation isn't severe, just wait it out.

For heavy infestations, knock aphids off with a strong spray of water. Or try insecticidal soap spray, which kills aphids on contact. Rotenone also kills aphids, but reserve this as a last resort since it also kills beneficial insects. Apply dormant oil spray in the winter to suffocate aphid eggs.

Many natural predators keep aphids in check, including parasitic wasps, lady beetles, and hover flies. To attract aphid predators, grow nectar-producing flowers, such as dill and buckwheat, near your trees, or spray commercial or homemade yeast-and-sugar mixtures on your trees. Introduced lacewings often control aphids. Buy these at commercial insectaries or from organic farm supply companies. Ants often introduce an aphid infestation; trap ants with sticky barriers around the trunks. Avoid excessive use of high-nitrogen fertilizer, which favors aphid reproduction.

Leaves covered with lacy, brownish patches. Cause: Pear slugs. These green-black, slimy creatures are not true slugs; they're the larvae of the pear sawfly, a black-and-yellow insect slightly larger than a housefly. In the spring sawflies emerge from soilborne cocoons and lay eggs on leaves. Eggs hatch into sluglike larvae that skeletonize leaves. Remove them by handpicking or with blasts of water. For heavy infestations, use insecticidal soap spray; for a dwarf tree, apply wood ashes or diatomaceous earth. (Caution: Wood ashes may damage leaf tissue.) Spider mites multiply quickly in dry, dusty conditions, so wash ashes off after a few days to avoid encouraging a rise in the spider mite population.

Growing shoots wilt and die. Cause: Oriental fruit moth larvae. Slit the stem below the wilted portion and look for a pinkish white, ½″ caterpillar—the larva of an oriental fruit moth. Larvae tunnel into shoots and remain for 2 weeks before leaving to pupate for 10 days in cocoons suspended in trees. Larvae from later generations bore into and ruin fruit. Use pheromone traps to monitor and control the pests. Mating disruption pheromones, such as Isomate-M, are also effective. Replace them every 90 days throughout the growing season. For heavy infestations, spray superior oil to smother eggs and larvae.

Leaves with purple spots. Cause: Cherry leaf spot. This fungal disease, prevalent during rainy springs in the East and Midwest, causes circular purple leaf spots that later turn brown and drop out leaving small holes. Leaves may yellow and drop early. Infected trees decline in fruit production and winter hardiness. Since the disease overwinters in fallen leaves, control it by thoroughly removing leaves before spring growth begins. For persistent infection, apply lime-sulfur or sulfur spray. Since lime-sulfur discolors fruit, use it only at petal fall, then use sulfur alone for subsequent sprays. Plant leaf spot-resistant cultivars including 'Meteor' and 'Northstar' sour cherries and 'Gold', 'Lambert', 'Hedelfingen', 'Schmidt Bigarreau', and 'Viva' sweet cherries. See page 67 for an illustration of this disease.

Leaves covered with a white powdery coating. Cause: Powdery mildew. Leaves may also be twisted or stunted and a powdery coating may cover fruit. Rainy weather does not cause this fungal disease to spread. It is most common in weather patterns featuring cool nights changing to warm days. For control, apply sulfur or lime-sulfur spray. 'Northstar' sour cherry is resistant to powdery mildew.

Young twigs brown and decaying. Cause: Brown rot. With this fungal disease, limbs may develop gummy-looking lesions. For more information, see "Fruit with small brown spots that enlarge and become fuzzy in humid weather" on page 66.

Leaves wilting and dying on whole branches. Causes: Bacterial canker; Valsa canker; shothole borers. With both cankers, trees may not leaf out in the spring, and branches may develop sunken, elliptical lesions that ooze a reddish gum. With bacterial canker, this gum smells sour, and leaves may have small, angular spots. For both cankers, prune wilted or dying branches off below the infection point, sterilizing tools in isopropyl alcohol or a 10 percent bleach solution (1 part bleach to 9 parts water) between cuts. Valsa canker enters trees through bark injuries; prevent it with cultural practices that avoid bark injury. Cultivars resistant to bacterial canker include 'Bada', 'Corum', 'Early Burlat', 'Sam', and 'Sue' sweet cherries. For more information, see "Branches wilting and dying, fail to leaf out in spring" on page 167.

Dying branches covered with numerous small holes are the work of tiny beetles called shothole borers. Inside the sawdust-filled tunnels, look for white, ⅛″ borer larvae with reddish brown heads. Shothole borers prefer injured or diseased trees. Vigorous, healthy trees are much less susceptible to problems.

Leaves yellowing; limb dieback. Cause: San Jose scale. Colonies of these sucking insects cling to the bark and appear as small gray bumps that can easily be scraped off with a fingernail. Control San Jose scale with dormant oil spray applied in late winter or early spring just before blossoms open.

Twigs and branches bearing tarry, black galls. Cause: Black knot. For more information, see "Twigs and branches bearing tarry, black galls" on page 187.

Whole Plant Problems

Tree lacking vigor. Cause: Peachtree borers. The larvae of a clear-winged moth, these 1", white borers create holes and a gummy exudate on trunks near or just below ground level. Either dig the borers out without damaging the tree excessively or locate borer holes and kill larvae by inserting a wire into holes. Spreading a ring of tobacco dust around the base of the tree discourages these pests. Keeping your tree vigorous and avoiding mechanical damage also discourages peachtree borers.

Tree stripped of fruit. Cause: Birds. Birds especially love red sweet cherries. Plant yellow-fruited cultivars, which are somewhat less attractive to birds. Cover small trees with netting. For more information on controlling birds, see "Stopping Animal Pests" on page 408.

Chestnut

Castanea spp. (Fagaceae)

Chestnuts are deciduous trees with large, alternate leaves. The nuts are enclosed in prickly burrs, a drawback to the use of the trees for ornamental purposes.

At one time, American chestnut (*Castanea dentata*) was the dominant tree in eastern forests. It was a splendid specimen tree, highly valued for its wood and for the nuts it yielded. Unfortunately, the fungus that causes chestnut blight was introduced to the United States in the early 1900s, and virtually all American chestnuts have died back from the disease. Infected trees sometimes produce new sprouts, but they don't grow to maturity. Using these sprouts, researchers are working to develop resistant strains. Until then, blight-resistant species, such as Japanese chestnut (*C. crenata*) and Chinese chestnut (*C. mollissima*), are practical alternatives to American chestnut.

Chestnuts are not especially demanding in their soil requirements, as long as drainage is good. They are remarkably heat-tolerant. Plant in a sunny location; set out in spring or fall as balled-and-burlapped plants.

Problems

Leaves skeletonized. Causes: Cankerworms; Japanese beetles. These ½"-1" long, green, yellow, or brown caterpillars feed on foliage, often defoliating the tree. To avoid this problem, apply a band of a sticky substance, such as Tanglefoot, around the base of the tree in spring to keep females from climbing up the tree and laying their eggs. Apply fresh coats in September and February to trap both spring and fall cankerworms. If their feeding becomes a problem, spray with BTK as soon as you notice them; continue every 2 weeks until pests are gone.

Japanese beetles often feed on chestnut leaves, causing similar symptoms. See "Leaves skeletonized" on page 236 for control suggestions.

Plant defoliated. Cause: Gypsy moths. See "Plant defoliated" on page 236 for controls.

Leaves skeletonized or with large holes; branches may be webbed. Cause: Caterpillars. See "Leaves skeletonized or with large holes; branches may be webbed" on page 236 for more information.

Nuts damaged. Cause: Chestnut weevils. The larvae of this pest feed inside the nuts, and then enter the soil when the nuts fall to the ground. Pick fallen nuts from the ground daily to reduce future populations.

Trunk or branches with small holes; limbs die or break off. Cause: Borers. Several kinds of borers attack chestnuts; see "Trunk or branches with small holes; limbs die or break off" on page 238 for controls.

Chives

Allium spp. (Liliaceae)

Common chives (*Allium schoenoprasum*) are a hardy (Zone 3) perennial herb grown for their mild, onion-flavored leaves and pink flowers. Garlic chives (*A. tuberosum*) are slightly less hardy (Zone 5), and have larger, straplike leaves and white, rose-scented flowers. To grow healthy, trouble-free plants, sow seed indoors in 60°-70°F soil mix, or outdoors after soil has warmed. Chives do best in moderately rich, moist, well-drained soil with a pH between 5.0 and 7.0. Chives need full sun.

Chives are closely related to onions and prone to the same problems, especially rust, downy mildew, and smut. See the Onion entry beginning on page 155 for information.

Chrysanthemum

Chrysanthemum, garden mum. Perennials.

Early wild chrysanthemums, mentioned in 5th century B.C. writings, bore rather plain, yellow blossoms. Later, whites and purples were found and cultivated. Today, hundreds of cultivars are available. Within this genus are Shasta daisies, costmary, painted daisies, feverfew, and marguerites, in addition to the huge array of colors and blossoms found among the popular hardy mums.

Full sun and fertile, well-drained soil promote healthy mums. Fertilize plants in spring with compost or a slow-acting, general-purpose fertilizer, then give supplemental feedings monthly during the growing season; stop fertilizing in August. Start plants from stem cuttings, divisions, or seeds. When buying mums, look for compact, bushy plants, which will flower better than larger specimens. Space 1′ apart, avoiding areas already occupied by tree and shrub roots. When plants reach 6″-8″, pinch growing tips back to stimulate branching and flower bud production. Give these shallow-rooted plants ½″-1″ of water weekly; water stress causes woody stems and lower leaf drop. Overwatering, particularly when combined with poor drainage, causes yellowing leaves that blacken and drop. Cut plants back to within a few inches of the ground after blooming. Protect roots with a winter mulch.

Although the list of pests and diseases that *may* attack chrysanthemums is long, mums are relatively trouble-free. Unless mums are stressed by unfavorable growing conditions, few pests will seek them out; fewer still will cause significant damage.

Problems

Leaves, stems, and buds distorted, sticky; clusters of small insects. Cause: Aphids. See "Leaves, stems, and buds distorted, sticky; clusters of small insects" on page 177 for controls.

Leaves with yellow-brown spots or blotches; leaves die and turn brittle. Cause: Foliar nematodes. These microscopic roundworms overwinter in soil or plant debris, then travel up a plant's stem in the film of water created by spring rains, entering the lower leaves through the stomata, which are minute openings in the leaves. Symptoms move up the plant; foliar nematodes may even infest the petals. Spots eventually cover entire leaves, which die, turn brittle, and drop. Severe infestations can kill entire plants. Remove and destroy infested plants and the surrounding soil; do not compost the debris. There is no cure; clean up debris in fall to destroy overwintering nematodes. Rotate plantings and mulch in spring to keep nematodes from climbing up plants; avoid wetting leaves when watering.

Leaves with tan or brown blotches or serpentine tunnels. Cause: Leafminers. These tiny pale green fly larvae feed between the upper and lower surfaces of leaves. Prune off and destroy infested leaves until only healthy growth remains. Remove debris in fall to destroy overwintering leafminers. Let parasitic wasps control these pests, or repel larvae with weekly insecticidal soap sprays starting when the first tunnel appears.

Leaves and flowers greenish yellow, distorted; new growth spindly. Cause: Viral diseases. Chrysanthemums are prone to several viruses, which are spread by sucking insects such as aphids and leafhoppers. Control such pests to limit virus problems. Remove and destroy infected plants; do not compost them. Wash tools used around infected plants. Viruses overwinter in perennials and weeds such as daisies and plantains.

Cimicifuga

Bugbane, snakeroot. Perennials.

The durable bugbanes are 2′-6′ tall natives that spread to 2′ wide. Small creamy white blooms appear in multiples to cover upright, slim spikes that tower above the dark green, fernlike leaves. Flowering occurs from late summer to early fall, depending on species. Flower aroma has been described as "sickly sweet" and "strange." Whether it deters pests or not remains undetermined, but parts of stinking bugbane (*Cimicifuga foetida*) are being tested as a general insect repellent.

Bugbanes prosper in rich, moist, well-drained, organic soil, similar to their native woodland habitat. These plants prefer light shade; bugbanes grown in deep shade bear fewer flowers. Plants can succeed in sunny areas with ample watering. Clumps expand outward slowly and can be divided in fall or spring, but bugbanes are deep-rooted and may suffer from division. Germination from seed is slow and inconsistent. Provide a winter mulch in colder areas, a protective summer mulch in warmer areas. Fertilize and water regularly.

Problems

Leaves with large, ragged holes. Cause: Slugs and snails. While sturdy bugbanes withstand most pest invasions well, check regularly beneath the dense clumps of foliage for slugs and snails. Differing only in the presence or absence of a shell, slugs and snails range in size from 1/8″ to 8″ and may be gray, tan, green, black, yellow, or spotted. These voracious mollusks rasp large holes in leaves and stems, leaving slimy mucus trails as they travel.

Repel slugs and snails from flower beds with a 4″-8″ strip of copper edging, or surround plant with a band of cinders, wood ashes, cedar sawdust, or diatomaceous earth, and renew frequently. Trap slugs and snails with shallow pans of stale beer set with rims flush with the soil surface; remove drowned pests daily. Or lay boards, cabbage leaves, or overturned clay pots around the garden, and destroy pests that congregate underneath. Handpick slugs and snails from plants at night; drop them into soapy water or sprinkle them with table salt to dehydrate and kill them.

Citrus

Citrus spp. (Rutaceae)

Citrus trees are evergreens that grow in flushes throughout the year. They bear fragrant flowers and fruit that hangs for months without dropping. These plants are hardy in Zones 9-10.

To flourish and bear well, citrus trees require a site with excellent drainage and full sun. Prune to remove dead wood, to keep trees in bounds, and to thin out centers where growth is too dense. Citrus flowers attract bees; don't plant them in high-traffic areas. Citrus trees bear fruit without cross-pollination.

Fruit Problems

Fruit with raised, light brown, corky areas. Cause: Citrus scab. Leaves also may show scabby lesions. This fungal disease, common in Florida, attacks mainly grapefruit, 'Temple' and 'Murcott' honey oranges, lemons, sour oranges, satsumas, and some tangelos. To control citrus scab, prune and site trees so that foliage and fruit remain as dry as possible during spring growth.

Fruit with small, brown, sunken spots that turn dark and are raised. Cause: Melanose. This fungal disease usually attacks trees more than 10 years old and is worst in wet springs. As rain washes spores over fruit, a tear-streaked pattern of infection develops. Copper spray applied just once at fruit set controls this disease, but since fruit damage is only cosmetic, spraying isn't really necessary. Keep melanose in check by pruning out dead wood, which may harbor the disease.

Fruit with firm, light brown spots on the rind. Cause: Brown rot. In humid weather brown rot may appear as a white mold on the rind. Rain splashes the spores of this fungal disease from the ground onto low-hanging fruit. The simplest control is to mulch the soil around the tree and prune away low-hanging branches. Copper spray applied to the ground and low-hanging limbs also works. This disease spreads easily in stored fruit; keep infected fruit away from clean fruit.

Fruit contains webbing. Cause: Navel orangeworms. These reddish orange, brown-headed larvae bore into cracked fruit and spin their cocoons inside. Control pests by picking off damaged fruit. Clean up dropped fruit. Spray BTK to kill larvae before they enter.

Leaf and Branch Problems

Leaves with pale yellow stipples. Cause: Citrus mites. These pests may also attack fruit causing it to turn brown or silver and drop prematurely. Pesticides tend to kill natural mite predators, so unsprayed trees are less prone to mite infestation. You can also introduce predatory mites, such as *Amblyseius* species. The simplest mite remedy is to hose down the tree with plain water or insecticidal soap solution. For more serious infestations, spray superior oil or sulfur.

Leaves with oily brown spots on undersides. Cause: Greasy spot. With this fungal disease, yellowish brown blisters on leaf undersides eventually turn oily brown and coalesce. Superficial pitting may appear on fruit rind. Since the disease overwinters on leaf litter, the simplest control is to collect and destroy fallen leaves. Wet foliage encourages spread of the fungus; keep leaves as dry as possible, especially during the early summer. For recurrent greasy spot, spray with superior oil or copper.

Trunk with brown patches exuding a thick, amber gum. Cause: Brown rot gummosis. Eventually brown patches form dark, sunken cankers. Lower branches above the canker may wither and die. This disease begins when spores of Phytophthora fungus splash up from the ground and enter the trunk through a bark injury. To control this disease, scrape away the canker into healthy bark, allow it to dry, and spray it with copper. To prevent the disease, keep surface water and sprinkler water away from the trunk. Plant only in well-drained soil using trees grafted onto Phytophthora-resistant rootstocks, such as trifoliate orange, sour orange, and 'Troyer' citrange.

Leaves with small, dark, circular depressions bearing yellow margins. Cause: Melanose. For more information, see "Fruit with small, brown, sunken spots that turn dark and are raised" above.

Leaves turn pale green or yellow. Cause: Nutrient deficiency. If old leaves turn yellow and drop off, suspect nitrogen deficiency, a condition that sometimes occurs when cold soil slows root activity. If young leaves turn yellow with green veins, suspect iron deficiency, a condition common to wet or alkaline soils, even if the soil contains enough iron. To correct nitrogen deficiency, apply an organic nitrogen source directly to the soil. To correct iron deficiency, adjust the soil pH to be more acidic. You can also spray either of these nutrients directly on the leaves for immediate help.

Leaves turn yellow and fall; tree declines. Cause: Scales. Various types of scales infest citrus trees, causing decline. Trees may suffer from twig dieback and reduced vigor, and leaves may be covered with honeydew excreted by the scales as well as sooty mold, which feeds on the honeydew. Cottony cushion scale looks like white, cottony masses clustered on leaves, stems, and branches. Red, brown, or black scale may look like crusty, waxy, or smooth bumps on leaves, trunk, stems, or fruit. Many natural enemies keep scale insects in check. If scale infestation is heavy, introduce outside predators for additional control. You must choose the species of predator that feeds on your species of scale. For example, Vidalia beetles (*Rodolia cardinalis*) feed on cottony cushion scale, and the parasitic wasp *Aphytis melinus* preys on California red scale. Ants protect scales by destroying their natural predators, so keep ants out of trees with sticky barriers around trunks. You can also control scale by spraying specially refined superior oil.

Leaves with white, cottony masses. Cause: Mealybugs. Leaves may also be covered with a sticky honeydew excreted by the mealybugs or a black mold, which feeds on the honeydew. Natural enemies, such as lacewings, usually keep mealybugs in check, but you can speed along the process by introducing the mealybug destroyer, an Australian predatory beetle available through commercial insectaries and organic farm suppliers. Ants protect mealybugs by destroying their natural predators. Restrict ants from trees by placing sticky barriers around trunks and removing branches that droop to the ground.

New leaves curled or twisted. Cause: Aphids. Look on leaf undersides for clusters of these tiny sucking insects. Many natural enemies keep aphids in check, including parasitic wasps and hover flies. Provide food for natural predators and parasites by growing nectar-producing flowers, such as dill and buckwheat, near your trees. Or spray commercial or homemade yeast-and-sugar mixtures on your trees to attract predators. Introduced lady beetles may not control aphids effectively, but introduced lacewings often do. Ants protect aphids by destroying their natural predators, so restrict ant movement into your trees with sticky barriers around the trunks. Excessive use of nitrogen fertilizer favors aphid reproduction.

For heavy infestations, apply a strong spray of water or insecticidal soap solution to trees, making sure you spray undersides of leaves as well. Superior oil sprays kill aphid eggs. Rotenone also works, but use it only as a last resort since it also kills beneficial insects.

Clematis

Clematis. Vines.

Clematis are mostly deciduous vines that climb by means of their twining leaf stalks. They bear attractive flowers in a range of sizes and colors.

Clematis prefer neutral to slightly alkaline soils, well-enriched with compost. Set out small, container-grown plants in spring or fall, planting several inches deeper than when grown in the nursery. Provide some sort of support. Clematis perform best in full sun, but they do like to have their roots cool and shaded; a rich, deep, organic mulch is of great benefit. Clematis may be slow to initiate top growth; they must first establish a substantial root system.

Problems

Vines droop and shrivel. Cause: Clematis wilt. This fungal disease is one of the most serious problems for clematis. When it strikes, prune out and destroy all affected parts. If you set the plant a few inches deeper than it was growing in the nursery at transplanting time, new vines may come up from the buried stem portion and produce new, healthy topgrowth. If the whole plant dies, don't try to grow another clematis on that site.

Plant stunted; top dies back. Cause: Clematis borers. This ⅔″ long, white-bodied, brown-headed grub feeds on the roots and crowns of clematis vines. To control, cut out and destroy infested stems, and dig the larvae out of the crowns. Dig out and destroy severely infested plants.

Leaves yellow; stems and leaves covered with small bumps. Cause: Scales. See "Leaves yellow; stem and leaves covered with small bumps" on page 237 for control suggestions.

Colchicum

Autumn crocus. Corms.

These hardy, autumn-flowering corms are not crocuses, despite their common name. Pink, white, or purple, 2″-4″ tall flowers bloom without foliage, which comes up in early spring and dies off by midsummer. Narrow dark green leaves grow 6″-12″ tall.

Plant autumn crocus corms in July and August, setting them 3″ deep in average, well-drained soil in full sun to light shade. Lift and separate corms after foliage fades to reduce crowding or to move plantings; replant immediately.

Almost no pests or diseases bother these corms, which are quite poisonous if eaten.

Coleus

Coleus. Annuals.

Brilliant red, maroon, cream, and green leaves make coleus a colorful addition to the shady garden. Coleus can grow to 3′ but are usually smaller. The flowers are insignificant, so pinch them off to promote leaf growth.

Sow seeds indoors 10 weeks before last spring frost. Seeds need light to germinate, so don't cover them. After the last frost, transplant outdoors into rich, well-drained soil and moderate shade. Too much sun fades leaves and causes drooping. Feed with high-nitrogen fertilizer to encourage leaf development.

Coleus are usually trouble-free. Mealybugs may feed on plants; see "Leaves and stems with white, cottony clusters" on page 20 for controls.

Collard

Brassica oleracea
Acephala group (Cruciferae)

Collards are a cool-season vegetable grown for the leafy, cabbage-type greens. Collards can stand more heat than other members of the cabbage family. They prefer growing temperatures of 65°-70°F and tolerate 40°-80°F. 'Champion' and 'Vates' are cold-hardy cultivars. Soil pH should be between 5.5 and 6.8.

Collards are grown like cabbage and are attacked by the same diseases and insects. See the Cabbage entry beginning on page 52 for culture and information on problems.

Convallaria

Lily-of-the-valley. Perennials.

Fragrant lilies-of-the-valley (*Convallaria majalis*) scent springtime air with small, white, dangling bell-shaped flowers carried on stalks above 6″ long, pointed green leaves. Attractive but poisonous red berries follow the blossoms. Lily-of-the-valley spreads by thick rhizomes. Purple shoots appear in April, followed by buds.

Plants multiply rapidly; in favorable conditions they can be invasive, a trait that makes lily-of-the-valley useful as a groundcover in controlled situations. Late fall division encourages bloom. Plant lilies-of-the-valley about 5″ apart in moist, well-drained, organic soil. The light shade beneath deciduous trees is fine for these woodland natives; heavy shade decreases bloom. Fertilize in fall.

Problems

Leaves stippled, reddish to yellow, with fine webbing. Cause: Spider mites. These 1/50″ arachnids quickly disfigure ornamentals. Their feeding on the undersides of foliage causes it to yellow, dry up, and die; excessive feeding turns foliage almost white and fine webs cover leaves and growing tips. Mites seek out water-stressed plants in hot, dry weather; adequate watering guards against these pests. Hose down plants when symptoms first appear; a strong stream knocks mites off plants and may give sufficient control. Introduce predatory mites (1,000 per 500 square feet); spray severe infestations with avermectins, citrus oils, insecticidal soap, neem, pyrethrin, or rotenone.

Leaves with large, ragged holes. Cause: Slugs and snails. See "Leaves with large, ragged holes" on page 177 for recommended controls.

Coreopsis

Coreopsis, tickseed. Perennials.

Golden daisylike flowers grace plantings of coreopsis from mid-June well into July. Most species of these 1′-3′ plants are hardy perennials, although tickseed (*Coreopsis grandiflora*) performs as a biennial north of Zone 7. Prolific 1″-3″, single or double blooms attract beneficial syrphid flies.

Space coreopsis 15″ apart in well-drained, average to poor soil in full sun. Once established, these natives are very drought-tolerant but bloom better with regular watering. Excess feeding encourages rank, spindly growth; a single spring fertilization is sufficient. Remove spent flowers to prolong bloom and reduce coreopsis' tendency to self-sow. Divide clumps every 3–4 years.

Problems

Leaves, stems, and buds distorted, sticky; clusters of small insects. Cause: Aphids. See "Leaves, stems, and buds distorted, sticky; clusters of small insects" on page 177 for controls.

Buds and leaves deformed or dwarfed. Cause: Four-lined plant bugs. For recommended controls, see "Buds and leaves deformed or dwarfed" on page 178.

Leaves stippled, distorted. Cause: Aster leafhoppers. Greenish yellow pests only 1/8″ long, these insects feed on leaf undersides, giving foliage a finely mottled look; leaves yellow, shrivel, and drop. Besides injury from feeding, leafhoppers transmit the disease aster yellows. Use water or insecticidal soap sprays to dislodge nymphs; encourage enemies such as bigeyed bugs and parasitic wasps. For control of severe infestations, spray with pyrethrin or rotenone.

Corn

Zea mays var. *rugosa* (Gramineae)

Corn is an annual vegetable grown for its large ears of tasty kernels. Sweet corn is harvested when the ears are tender and immature. Sweet corn cultivars are classified in 3 groups: normal sugary (traditional corn flavor, hybrid or open-pollinated), sugar enhanced (increased tenderness and sweetness, hybrid), and supersweet (God's gift to sweet-toothed corn lovers, hybrid). Be sure to plant supersweets at least 25′ away from any other corn cultivars or cross-pollination will cause starchy, tough kernels. Be aware that whether you plant other cultivars or not, supersweet cultivars generally suffer from more than their share of pests.

While most gardeners stick to sweet corn, other types of corn are just as easy to grow. Popcorn, ornamental corn, and field corn are harvested when the seeds are hard and mature and the husks are dry. Popcorn is popped for a tasty, low-calorie snack. Dent (field) corn and flint (Indian) corn cultivars are used for cornmeal and decoration. Flour corn is starchy and used for flour and decoration.

Culture

Corn does best in a rich, sandy, or well-worked soil with a pH between 5.5 and 6.8. Prepare soil by working in a generous amount of compost. Side-dress plants with alfalfa meal when they are 1′ high and again when silk first shows at the end of the ears. Spraying plants with seaweed extract or compost tea periodically also improves your harvest and prevents deficiencies.

Plant corn seed only after the soil is at least 60°F, or 75°F for supersweets. Seed planted in cooler soil is prone to many problems, including poor germination. To help speed soil warming, cover soil with clear plastic at least 2 weeks before you want to plant. After planting, use row cover for about a month to give seedlings a boost.

Corn needs at least 1″ of water a week. Keep soil moist, but not soggy. Mulch plants to conserve moisture and cut down on weed competition.

Plant corn in blocks rather than long single rows to ensure good pollination. Do not plant corn where it has grown in the past 2 years to prevent problems. After harvest cut or mow stalks and let them dry. Then turn them under or collect and compost them. Destroy any diseased or infested material.

Leaf and Whole Plant Problems

Seedlings cut off at ground level; leaf margins ragged. Cause: Cutworms. Young cut-

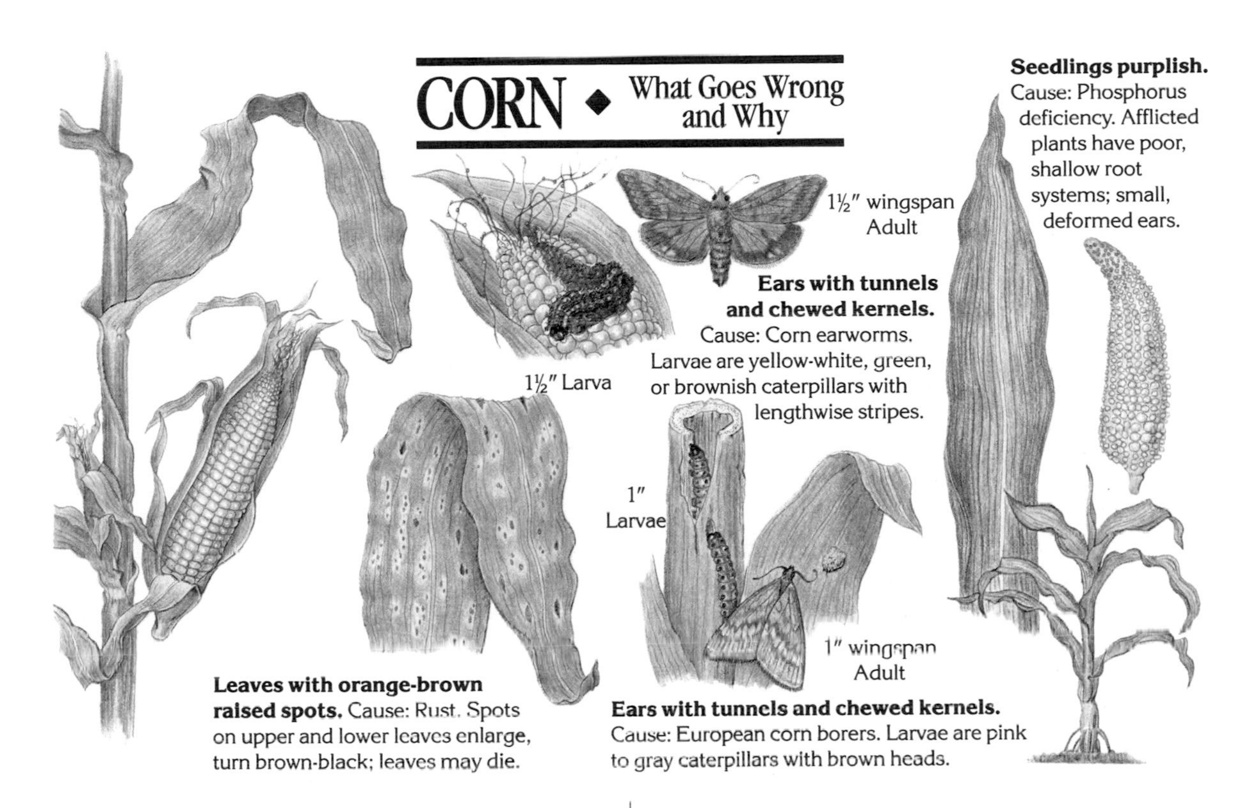

worms feed on leaves. Most cutworms are gray or dull brown and up to 2″ long. Spray seedlings with BTK in the evening if cutworms are still feeding on leaves. Sprinkle moist bran mixed with BTK on the soil surface in the evening, or add parasitic nematodes to the soil to control stem-girdling cutworms. For small plantings, place cutworm collars around plants or groups of plants to prevent damage.

Seedlings fail to emerge or are stunted. Causes: Cool soil; corn rootworms; seedcorn beetles; wireworms; seedcorn maggots; white grubs. Corn seed germinates poorly in cool soil and is more prone to insect damage and rots. Plant after soil is warm to prevent problems. Do not plant corn where grass grew the year before because many pests live in sod.

Corn rootworms are white, slender, ½″ long larvae with brown heads. Adults are beetles. Western rootworm beetles are ¼″ and yellow with 3 black stripes. Northern rootworm adults are yellow to pale green, and southern rootworms (spotted cucumber beetles) are yellow-green with 11 black spots. Rootworm larvae feed on seeds, and adults feed on leaves, silks, and tassels.

Seedcorn beetle larvae are white, slender, ½″ long grubs; adults are ⅓″ long ground beetles with light brown edges. Both larvae and adults feed on seeds.

Wireworms are 1½″ long, yellow to reddish brown, slender, tough-bodied, segmented larvae with brown heads. Adults are dark-colored, elongated click beetles.

Seedcorn maggots are ¼″ long, yellow-white, and spindle-shaped. Adults are small flies. White grubs have thick, soft, ¾″ bodies and curl into a C-shape when disturbed. Adults are Japanese or June beetles. Wireworms, seedcorn maggots, and white grubs eat corn seed.

Apply parasitic nematodes to the soil before planting to control soil-dwelling pests.

Prevent adults from laying eggs on or near seedbed by covering it with a large piece of row cover after planting. Remove row cover when tassels appear.

Seedlings purplish. Cause: Phosphorus deficiency. See page 77 for an illustration of this condition. Phosphorus is not readily available in cool soils. Spray plants with seaweed extract or compost tea to alleviate symptoms.

Plant stunted and yellow. Causes: Nitrogen deficiency; too much or too little water; nematodes. If older leaves turn yellow first, suspect nitrogen deficiency. Drench roots and spray plants with fish emulsion or compost tea to alleviate symptoms.

To avoid root damage, keep the soil evenly moist. Plant in raised beds to ensure good drainage.

If entire plant is yellow and doesn't green up when sprayed with seaweed or compost, and soil is well-drained and not too dry, suspect pest nematodes. Roots may be stubby or have swellings or beadlike galls. Control by applying chitin or parasitic nematodes to soil.

Leaves yellow and curled; plant stunted. Cause: Aphids. These small, soft-bodied, green, black, gray, pink, or white fluffy-coated insects suck plant sap. For mild infestations, knock pests off plants with a blast of water. Spray infested plants with insecticidal soap, or with a commercial pyrethrin/rotenone mix if severe.

Leaves mottled yellow and green. Cause: Maize dwarf mosaic. The stem doesn't grow normally, so new leaves are very close together. Leaves may develop lengthwise stripes. There is no cure; destroy infected plants. Prevent problems by controlling aphids, which spread the virus as they feed, and perennial weeds such as Johnsongrass that can harbor the disease. Plant tolerant cultivars such as 'Bellringer', 'Bi-guard', 'Bunker Hill', 'Earlibelle', 'Enforcer', 'Merit', and 'Silverette'.

Leaves with lengthwise yellow stripes; plant may wilt. Cause: Bacterial wilt. Leaves are dwarfed. Cut off the infected plant near the soil line and look for yellow bacterial slime oozing from the cut stem to confirm diagnosis. Destroy infected plants. Control flea beetles and cucumber beetles that spread the disease. Prevent problems by planting resistant cultivars such as 'Bellringer', 'Bunker Hill', 'Buttersweet', 'Earlibelle', 'Gold Cup', 'Merit', 'Silverado', 'Silver Queen', 'Stylpak', 'Summer Pearl', and 'Sweet Sue'.

Leaves with yellow, tan, or gray spots or blotches. Causes: Northern or southern corn leaf blight; other fungal leaf spots. Spray infected plants with sulfur to prevent blight from spreading. Prevent problems by using a 3-year rotation for corn and by planting cultivars, such as 'Sweet Sal' and 'Ultimate', that are tolerant of northern leaf blight or 'Apache', 'Cherokee', 'Comet', 'Florida Staysweet', and 'Wintergreen', which resist both blights.

Leaves with orange-brown raised spots. Cause: Rust. See page 77 for an illustration of this disease. Spray infected plants with sulfur early in the day and thin plants to maintain good air circulation. If rust is a recurrent problem, try spraying plants with an antitranspirant before symptoms develop to prevent infection. (Do not spray with antitranspirant when corn is silking.) Plant rust-tolerant cultivars such as 'Earlibelle', 'Flavor King', 'Summer Pearl', and 'Sweetie 82'.

Leaves with numerous small holes. Causes: Flea beetles; billbugs. Leaves may develop bleached-out spots or stripes. Flea beetles are tiny black insects that hop when disturbed. They can transmit disease and are likely to be more numerous after mild winters. Cover plants with row cover as soon as they come up to exclude beetles. Remove row cover when tassels form. Spray plants with a commercial pyrethrin/rotenone mix to control severe infestations.

If holes are arranged in rows, look for billbugs. These hard-shelled, nearly black, ¼"–½" snout beetles are usually seen only at night. Spray heavily infested plants with a commercial pyrethrin/rotenone mix.

Leaves with large, ragged holes. Causes: Corn earworms; European corn borers; armyworms; other caterpillars. Handpick or spray plants with BTK if caterpillars are feeding. See "Ears with tunnels and chewed kernels" below for controls.

Leaves skeletonized. Cause: Japanese beetles. Silk and tassels may be chewed. Adults are 1/2" long, metallic blue-green beetles with bronze wing covers. Handpick or trap adults. Reduce beetle population in following years by applying parasitic nematodes or milky disease spores to garden and lawn areas.

Stems girdled at base; stems and ears with tunnels. Cause: Southwestern corn borers. These white, dark-spotted, brown-headed larvae bore into stalks and ears. To reduce future damage, plant corn early and use resistant cultivars. Cut corn stalks down at the soil level and remove them right after harvest.

Ear Problems

Ears with tunnels and chewed kernels. Causes: Corn earworms; European corn borers; fall armyworms. Leaves and silk may be chewed. Stalks and tassels may be tunneled and may snap off. Corn earworm larvae are light yellow, green, pink, or brown caterpillars up to 2" long, with lengthwise stripes. Adults are tan moths. See page 77 for an illustration of this pest.

European corn borer larvae are beige, brown-spotted caterpillars up to 1" long, with brown heads. Adults are pale yellow to tan moths with wavy lines on their wings. See page 77 for an illustration of this pest.

Armyworm larvae are greenish brown caterpillars up to 1½" long, with a characteristic inverted Y on the head. Adults are pale gray moths and have a 1½" wingspan.

Cover plants with row cover until tassels emerge to exclude the first generation of pests, especially if you have had problems in previous years.

Use insect-specific pheromone traps to catch male moths of various pest species, or ask your local extension agent when moths appear in your area. If you are growing large quantities of corn, you may want to try using a blacklight trap to catch all types of moths; be aware that many kinds of beneficial insects are also attracted to and killed by these traps. When moths are present, spray plants with ryania or a commercial pyrethrin/rotenone mix in the evening to prevent egg-laying.

Even before the tassels emerge, check the upright, topmost leaves of your plants every 2–3 days for signs of caterpillar feeding. Spray BTK as soon as any feeding holes are found. Make sure to spray the undersides of leaves and insides of unfolding leaves where pests feed.

Once silks appear, spray them with BTK or sprinkle a few grains of granular BTK directly on each silk. Apply a few drops of mineral oil to each silk 4–5 days after they wilt to discourage any resident pests, or inject a few drops of parasitic nematode suspension around the tip of the ear to kill larvae.

Kernels at tips or upper half of ears hollowed out. Cause: Sap beetles. These small black beetles with yellow spots invade ears after the silks turn brown. Handpick beetles, or spray ear tips with a commercial pyrethrin/rotenone mix for severe infestations. Sap beetles can be trapped in containers baited with fermenting fruit.

Ears with dried tips exposed; ears stripped or missing. Causes: Birds; raccoons. Birds peck at tips of ears, exposed tips turn dry and greenish brown. Raccoons harvest ripe, juicy corn, usually the night before you would pick it yourself. Repel birds with loud noises or visual scare devices. Frustrate raccoons by surrounding your corn with a 3-strand electric fence with wires 3"–4" apart and off the ground. See "Stopping Animal Pests" on page 408 for further information.

Ears misshapen or with areas of undeveloped kernels. Causes: Poor pollination; nutrient deficiency; viral diseases. Insufficient or ineffective pollination can cause undeveloped bare tips, scattered kernels, or entirely bare cobs. Plant corn in a block, rather than a long row to ensure effective wind pollination. Insects feeding on silks before pollination occurs can prevent pollination, as can very dry conditions. Control insects that feed on silk, and keep soil moist, but not wet, to ensure even pollination.

Ears with bare, undeveloped tips can also be caused by potassium deficiency. Phosphorus deficiency also causes small, irregular ears. See page 77 for an illustration of this condition. If ears are misshapen and kernels have corky, brown bands at their bases, suspect boron deficiency. Spray young plants with seaweed extract or compost tea to help prevent deficiencies. Confirm deficiencies with a soil test and amend soil as needed.

Viral diseases such as maize dwarf mosaic can cause poor kernel formation at base of ears or bare ears. See "Leaves mottled yellow and green" on page 78 for more symptoms.

Ears or tassels with enlarged galls. Cause: Corn smut. Young galls are firm and whitish, older ones are spongy and filled with black powder. Remove galls before they split open, and destroy infected plants. Do not compost them. Prevent problems by planting resistant cultivars such as 'Bellringer', 'Gold Cup', 'Merit', 'Sweet Sue', and 'Viking'.

Cornus

Dogwood. Shrubs and small trees.

Dogwoods are deciduous trees or shrubs, usually with opposite leaves. These plants provide year-round landscape interest with their showy flowers, attractive fruit, striking fall color, and interesting bark. Use dogwoods as specimen plants, in woodland plantings, or as informal barriers.

Culture

Spring is the best time to plant dogwoods, as balled-and-burlapped or container-grown plants. Most thrive in well-drained soil enriched with organic matter. The tree forms, in particular, are intolerant of drought, so water them thoroughly in dry periods. A deep organic mulch helps to conserve moisture and keep down weeds. The shrub types will grow in sun or partial shade; the trees are best in dappled shade.

Dogwoods vary widely in their susceptibility to pests and diseases. Some, such as flowering dogwood (*Cornus florida*), tend to have more problems than resilient species like kousa dogwood (*C. kousa*) and cornelian cherry dogwood (*C. mas*). To avoid problems with any dogwood, keep plants growing vigorously with regular watering and fertilization. Avoid damaging the trunk with lawn mowers, string trimmers, or pruning equipment; many dogwood problems, like canker and borers, enter plants through wounds in the bark.

Problems

Trunk with lumpy swellings; limbs die or break off; bark falls off. Cause: Borers. Several species of borers attack dogwoods. The most common type, the dogwood borer, has white-bodied, brown-headed larvae. They hatch from eggs laid on the bark and enter the tree through wounds or scars in the bark. Dogwood borers often attack young, newly transplanted trees.

Although tree wraps have long been recommended to protect dogwoods from borers, research has shown that these guards are not a good idea. They tend to keep bark soft and moist, actually protecting borer larvae. If you want to use a tree wrap to protect plants from winter deer damage, remove the guard during the growing season. Also, avoid wounding the trunk with lawn mowers, string trimmers, or pruning equipment; these injuries provide an easy entrance for borers. Preventive measures are the best control for these pests. If borers do strike, try cutting them out of the trunk with a sharp knife. Remove and destroy seriously infected trees.

Leaves with small, purple-rimmed spots or large brown blotches; twigs die back. Cause: Dogwood anthracnose. This fungal disease can be a serious problem on flowering dogwood. Lower branches die back as the disease progresses; eventually the whole tree may die. The problem is most serious in woodland settings, where the dogwoods may already be weakened by lack of light and competition for food and water. Vigorously growing plants are more resistant. If disease does strike, prune out affected parts; clean up and destroy fallen leaves in autumn. A dormant spray of lime-sulfur or bordeaux mix may give some control.

Leaves yellow, sparse, distorted, or with brown edges; branches die; growth stunted. Cause: Decline. In recent years, flowering dogwoods have been afflicted by decline. The exact cause is unknown. Trees that are stressed by insects, disease, or environmental changes are very susceptible to further pest and disease problems. Dogwood anthracnose (see above) commonly attacks already weakened trees, for example. For more information, see "Leaves yellow, sparse, distorted, or with brown edges; branches die; growth stunted" on page 238.

Twigs die back. Cause: Twig blight. Prune off affected parts back to live wood. Keep the tree growing vigorously with regular watering and fertilization.

Trunk or branches with oozing lesions; branch tips die back. Cause: Canker. See "Trunk or branches with oozing lesions; branch tips die back" on page 238 for more information.

Twigs with clublike galls. Cause: Gall midges. The orange larvae of these small reddish brown midges tunnel into young shoots. One-inch-long swellings form around the developing larvae on the twig; leaves on the branch may wilt or turn brown. Prune off and destroy galls as soon as you see them.

Leaves yellow; stems and leaves covered with small bumps. Cause: Scales. See "Leaves yellow; stem and leaves covered with small bumps" on page 237 for controls.

Leaves tunneled. Cause: Leafminers. See "Leaves tunneled" on page 237 for suggested controls.

Leaves with powdery white coating. Cause: Powdery mildew. See "Leaves with powdery white coating" on page 237 for controls.

Cosmos

Cosmos. Annuals.

The daisylike flowers of these fast-growing plants come in many shades of red, yellow, white, pink, lavender, and magenta. Cosmos bloom on 1½′-4′, slender stems covered with finely cut foliage.

Direct-sow cosmos outdoors from March through July. Seedlings appear in 10 days. Cosmos are also easy to transplant. Space plants 8″-24″ apart, depending on ultimate size. They prefer full sun and light, average soil kept slightly on the dry side. Avoid overwatering, and do not fertilize.

Pinch back seedlings before flower buds form to encourage compact bushy plants, or buy lower-growing cultivars. Staking is necessary for larger cosmos. Plant in groups or among sturdier bedding plants for extra wind protection. Plants may self-sow.

Problems

Stems break. Cause: Stalk borers. These long, thin, striped caterpillars eat their way through cosmos stalk centers. Small, round stem holes may betray their presence. Once plants are visibly affected, there is no cure. If borers are a regular problem, spray plants with BTK early in the season as borers enter plants.

Leaves, stems, and buds distorted. Cause: Aphids. See "Leaves, stems, and buds distorted" on page 20 for controls.

Leaves greenish yellow; growth poor. Cause: Aster yellows. For more information and control measures, see "Leaves greenish yellow; growth poor" on page 19.

Plant wilts suddenly. Cause: Bacterial wilt. This disease causes a soft rot at the base of the stem, killing plants quickly. Remove and destroy infected plants. Do not replant cosmos in that area.

Cotoneaster

Cotoneaster. Shrubs.

Cotoneasters are alternate-leaved evergreen or deciduous shrubs with persistent red or black fruit. They are useful as groundcovers, in the rock garden, in foundation plantings, or overhanging a wall.

Plant in spring as balled-and-burlapped or container-grown plants. Although not particular as to soils, avoid poorly drained sites. Cotoneasters do best in full sun. Cotoneasters can withstand wind, dry soil, and even salt spray.

Problems

Leaves, flowers, and branches blackened. Cause: Fire blight. This bacterial disease causes new shoots to wilt suddenly, turn dark, and die back. The disease eventually spreads, killing the whole plant. Lush new growth is particularly susceptible, so avoid overfertilizing. Prune out diseased tissue, cutting back at least 6″ beyond the discolored area; destroy or dispose of prunings. Disinfect pruners after each cut in a 10 percent bleach solution (1 part bleach to 9 parts water) to avoid spreading the disease to healthy wood. Spray copper or bordeaux mix the following spring during flowering.

Leaves wrinkled and discolored. Cause: Aphids. See "Leaves wrinkled and discolored" on page 235 for controls.

Leaves pale and mottled. Cause: Lace bugs. These 1/8″–1/4″, dark-colored bugs have lacy wings. They feed on the undersides of leaves and deposit small black spots of excrement. Severe infestations can cause early leaf drop. Control by spraying leaves with superior oil or insecticidal soap.

Leaves yellow; stems and leaves covered with small bumps. Cause: Scales. See "Leaves yellow; stems and leaves covered with small bumps" on page 237 for more information.

Leaves skeletonized; branches webbed. Cause: Cotoneaster webworms. These 1/2″, yellowish green to brown caterpillars feed on leaves and spin silken webs around leaves and stems. Young plants may die; older ones can be seriously weakened. Break up and remove the webs; spray plants with BTK.

Leaves stippled with yellow; foliage webbed. Cause: Spider mites. See "Leaves stippled with yellow; foliage webbed" on page 236 for controls.

Leaves with brownish blisters underneath. Cause: Pear leaf blister mites. These microscopic mites live and feed on tissue inside leaves. They overwinter on buds and infest new leaves in spring. Spray plants in late winter with dormant oil or lime-sulfur.

Trunk or branches with small holes; limbs die or break off. Cause: Borers. See "Trunk or branches with small holes; limbs die or break off" on page 238 for controls.

Crataegus

Hawthorn. Trees.

Hawthorns are small, thorny, deciduous trees with alternate leaves and showy, white, pink, or red flowers. Hawthorns make attractive specimen trees, barriers, or hedges.

Set out balled-and-burlapped plants in full sun in early spring. Hawthorns are undemanding as to soil, but avoid poorly drained areas. Choose planting sites with good air circulation to minimize disease problems. To avoid rust diseases, don't plant hawthorns where Eastern red cedar (*Juniperus virginiana*) is common.

Hawthorns share many insect and disease problems with apples. For symptoms and controls, see the Apple entry beginning on page 21.

Crocus

Crocus. Corms.

Much like the first robins, dainty crocuses signal spring's arrival. These low-growing flowers bear blooms of yellow, white, pink, and purple amid dark green, grasslike foliage in late winter and early spring. Fall-blooming species include showy crocus (*Crocus speciosus*), a blue- or white-flowered plant that blooms before leaves extend fully. Most species grow only 3″-6″ tall.

Plant both spring- and fall-flowering species as soon as corms are available. Corms are occasionally infected with dry rot; examine corms carefully when purchasing (see the illustration on page 49). Full sun and well-drained soil are crocuses' main requirements; corms will grow and multiply undisturbed for years. Set 3″-4″ deep and 2″-6″ apart with pointed growing tips right side up. Avoid cutting foliage back until it fades naturally; early removal of leaves reduces flowering in subsequent years. Divide large clumps, as needed, after leaves die back; replant full-size corms and cormels (small corms growing from base of parent corm) immediately.

Problems

Leaves, stems, and buds distorted, sticky; clusters of small insects. Cause: Aphids. Aphids may feed on foliage; see "Leaves, stems, and buds distorted, sticky; clusters of small insects" on page 51 for control information.

Plant disappears; corms missing. Cause: Animal pests. Rodents such as mice, voles, squirrels, and gophers all will eat crocus corms and can quickly make a planting disappear. The presence of a pet dog or cat often deters rodents; if pets aren't an option, try other techniques (see "Stopping Animal Pests" on page 408). Line planting beds with screen or hardware cloth to keep pests from burrowing around corms. Lay mesh over the top, too; crocus shoots can still squeeze through.

Leaves yellow, distorted; flowers absent; corms rotted. Cause: Bulb mites. If you suspect these pests, dig the corms and use a magnifying glass to look for whitish, 1/50″-1/25″ mites. Bulb mites favor rotting bulbs, but move from there into healthy bulbs, carrying harmful bacteria and fungi with them. Their feeding causes corky, brown spots that become powdery. Inspect corms carefully for damage or signs of infestation. Destroy badly infested bulbs and discard surrounding soil. Solarize the soil before planting corms in previously infested ground.

Cucumber

Cucumis sativus (Cucurbitaceae)

Cucumbers are annual vegetables grown for their crisp and crunchy fruit. They add life to summer salads and make tasty pickles.

Culture

Cucumber seeds need 60°F soil to germinate, so wait until weather is warm to plant. Make a second planting 4–5 weeks after the first so you will have fruit all season. Cover plants with floating row cover to protect them from insects and late cold snaps. Remove row cover when plants begin to flower so insects can pollinate the blossoms, or you will not get any fruit.

Cucumbers do best in well-drained, loose-textured soils with lots of organic matter. They will grow in soils with a pH between 5.5 and 6.8, but prefer a pH above 6.0. Plants need lots of water, but don't let soil become saturated. Prevent disease problems by keeping leaves dry. Mulch cucumbers to help conserve water; black plastic is a good choice for central and northern areas, but in extremely warm areas it can warm the soil too much. Organic mulches are good, too, but may provide shelter for pests like squash bugs. Foil mulches help prevent aphid problems. If rotting fruit is a problem, raise fruit off the ground by placing scraps of wood under them.

Rotate crops so that no member of the cucurbit family (squash, melon, and cucumber) is grown in the same place more often than every 4 years.

Caution: Cucumber leaves are easily burned by insecticidal soap and copper sprays. Use the most dilute spray recommended and use sparingly. Do not spray plants in direct sun or if temperatures are above 80°F, and don't spray drought-stressed plants.

Leaf and Vine Problems

Leaves with chewed holes. Cause: Cucumber beetles. Adults are 1/4" long, greenish yellow beetles with black stripes or spots. See page 214 for an illustration of the insects and damage. They attack young leaves and should be controlled immediately, as they can spread bacterial wilt or viruses. Spray plants with sabadilla or a commercial pyrethrin/rotenone mix. Prevent problems by planting cultivars, such as 'Liberty' and 'Wisconsin SMR58', that are tolerant of cucumber beetles.

Leaves with pale green patches; afflicted leaves wilt and blacken. Cause: Squash bugs. Adults are brownish black, 1/2" long bugs. Immature bugs are whitish green or gray with dark heads and legs. Both emit a strong, sharp smell when crushed. Eggs are bright orange and laid on undersides of leaves. See page 214 for an illustration of the insect and its damage. Handpick adults and eggs. Trap bugs by laying a board near plants. Squash bugs will hide underneath it and can be destroyed each morning.

Leaves with yellow patches; older leaves mottled and distorted. Cause: Mosaic. For an illustration of this disease, see page 214. Remove and destroy diseased plants. Control aphids and cucumber beetles because they spread the virus. Reduce problems by planting cultivars, such as 'Comet', 'Fancipak', 'Liberty', 'Monarch', 'Score', 'Slicemaster', 'Striker', and 'Wisconsin SMR58', that are resistant to mosaic diseases.

Leaves yellow, curled, and wilted. Cause: Aphids. Look for small, green, pink, gray, black, or white fluffy-coated, soft-bodied insects feeding on plants. Aphids can also transmit viruses. For mild infestations, knock pests off plants with a blast of water. Control with a weak insecticidal soap spray; see the caution on page 213 before spraying. Prevent problems by using a foil mulch, which keeps aphids from finding plants.

Leaves mottled yellow between veins; leaf undersides have purple spots. Cause: Downy mildew. As the disease progresses, older leaves turn brown and die and younger leaves become infected. Treat plants with a dilute solution of copper spray to control; see the caution on page 213 before spraying. Prevent problems by planting tolerant cultivars such as 'Fancipak', 'Liberty', 'Poinsett 76', and 'Slicemaster'.

Leaves with spots, blotches, or brown areas. Causes: Powdery mildew; angular leaf spot; scab anthracnose. Various diseases attack cucumbers. Reduce problems by keeping foliage dry when watering and by not touching plants when wet. Spray infected plants with a dilute solution of copper spray to control outbreaks; see the caution on page 213 before spraying.

Powdery white spots on leaves, especially on upper surfaces are caused by powdery mildew. Prevent problems by planting resistant cultivars such as 'Fancipak', 'Liberty', 'Poinsett 76', and 'Slicemaster'.

Water-soaked spots that turn gray, die, and drop out leaving shotholes are caused by angular leaf spot or by scab. Angular leaf spot causes small, brown, angular spots on fruit. Scab-damaged fruit develops sunken, brown spots with a sticky ooze. Scab is worse in cool, moist weather. 'Fancipak', 'Score', and 'Slicemaster' are resistant to scab.

Yellow spots that turn brown are caused by anthracnose. Infected leaves eventually die. Infected fruit has circular, black, sunken cankers. Prevent problems by planting cultivars, such as 'Fancipak', 'Liberty', 'Poinsett 76', 'Score', and 'Slicemaster', that are resistant to anthracnose.

Vines wilt suddenly. Cause: Squash vine borers. These fat, white, 1″ long larvae burrow into stems and exude masses of yellow-green, sawdustlike excrement. See the illustration of this insect and its damage on page 214. Slit stems lengthwise above each injury with a sharp knife and kill the larvae. Cover cut stems with moist soil so stems will form new roots. Injecting stems with BTK or parasitic nematodes may also help control borers. To reduce problems, plant the cultivar 'Sweet Mama Hybrid', which is resistant to vine borers. Or spray the base of stems with BTK once a week in late spring and early summer.

Vines wilt at midday, starting with younger leaves; leaves remain green. Cause: Bacterial wilt. See the illustration of disease damage on page 148. As the disease progresses, leaves fail to recover, and die. Cut wilted stems and press out drops of sap. If it is milky, sticky, and astringent, your plant is infected. Destroy infected plants immediately. Prevent problems by controlling cucumber beetles, since they spread the disease.

Fruit Problems

Fruit shrivels. Cause: Bacterial wilt. See "Vines wilt at midday, starting with younger leaves; leaves remain green" above for controls.

Fruit misshapen, one end not filled out. Causes: Diseases; poor pollination; nutrient deficiency. Many diseases cause misshapen fruit; use leaf symptoms to determine the disease causing the problem. If leaves are healthy, high temperatures may have damaged pollen, or bees may not have been active. Early in the season, wait for better conditions, or pollinate the flowers yourself by dusting pollen from male flowers onto female flowers (see page 214 for an illustration of male and female flowers). Later in the season, pull plants and wait for second planting to bear. If plants appear to be healthy and temperatures are not extreme, try spraying plants with compost tea to correct possible nutrient deficiency.

Fruit with spots; flesh may rot. Causes: Angular leaf spot; Alternaria blight; black rot; scab; other fungal or bacterial diseases. See "Leaves with spots, blotches, or brown areas" on page 85 for controls.

Fruit dull bronze. Cause: Phosphorus deficiency. Spray plants with compost tea to correct possible nutrient deficiency. Check with a soil test and amend soil as needed.

Currant

Ribes spp. (Saxifragaceae)

Currants are upright or spreading, deciduous bushes growing 3′-5′ high. European black currants (*Ribes nigrum*)bear fruit mostly on the previous season's growth; red and white currants (*R. petraeum, R. rubrum,* and *R. sativum*) bear on the previous season's growth and on spurs—short branches that elongate only a fraction of an inch per year—on older stems. Currants are hardy in Zones 3-8.

Plant currants in full sun or partial shade and apply a thick, organic mulch. To winter-prune red and white currants, cut all but 6 of the previous season's shoots to the ground. Remove all shoots more than 3 years old. To winter-prune black currants, cut 2-5 of the oldest branches to the ground and shorten tall, old branches to vigorous young sideshoots. Most currants are self-pollinating.

Problems

Leaves blistered and reddened. Cause: Aphids. You'll find these tiny insects clustered on leaf undersides. If damage is not severe, ignore it; otherwise spray with insecticidal soap or rotenone, making sure to get the undersides of the leaves. Dormant oil applied in the winter helps prevent aphids.

Leaves yellow; stems die back. Cause: Currant borers. As spring growth begins, you can easily spot borer-infested canes because the leaves look weak and sickly. Currant borers spend the winter in canes, pupate in the spring, and emerge as moths to reinfect currant bushes. The easiest control is to cut out and destroy infested canes as soon as you notice them.

Leaves spotted, yellow, and drop early. Causes: Septoria leaf spot; anthracnose. Since both of these fungal diseases overwinter on old leaf litter, rake up the leaves or bury them under a thick mulch in autumn. Copper sprays also control both diseases. If defoliation occurs late in the season, it does plants little harm.

Leaves with rust-colored pustules on undersides. Cause: White pine blister rust. This fungal disease is passed back and forth between white pines (or other 5-needled pines) and various species of *Ribes.* The disease is much more devastating to pines than to *Ribes,* so *Ribes* plantings are sometimes banned where pines are economically important. European black currants are most susceptible. Red and white currants generally are not much affected. Avoid planting susceptible currants near white pines.

Leaves with white powdery coating. Cause: Powdery mildew. Usually this fungal disease is harmless, but if it becomes severe, spray plants with sulfur, lime-sulfur, or a solution of sodium carbonate (washing soda) or sodium bicarbonate (baking soda).

Fruit colors and drops early. Cause: Currant fruit fly larvae. For more information, see "Fruit colors and drops early" on page 107.

Foliage stripped from bushes just as leaves expand. Cause: Imported currantworms. For controls, see "Foliage stripped from bushes just as leaves expand" on page 108.

Dahlia

Dahlia. Tuberous roots.

Descended from species native to Central and South America, modern dahlia hybrids are most often grown as annuals in temperate climates and are dug, overwintered indoors, and replanted the following spring. Thousands of cultivars offer a huge array of flower shapes, colors, and sizes to gardeners willing to meet dahlias' rigorous growth requirements. Thick-stemmed, rather coarse-textured plants range from 1′ to 8′ tall; midsummer to fall blossoms may be 1″-15″ in diameter.

Select a planting site in full sun and prepare the soil to a depth of 1′, incorporating plenty of compost or other organic matter. Dahlias require good drainage; raised beds help meet this need. Plant tuberous roots after all danger of frost has passed; placing them horizontally in the soil with the buds upward and fleshy roots spread out in the planting hole, 6″-8″ below the soil surface. Cover the buds with about 3″ of soil at planting, then gradually fill in as the shoots grow. All but the shortest cultivars need support for heavy flower heads. Set stakes at planting time.

Pinching shoots back during the growing season promotes bushier growth and increased flowering; removal of some buds increases the size of the remaining flowers. These heavy feeders benefit from side-dressings of compost or general-purpose fertilizer throughout the summer. Avoid high-nitrogen fertilizers, which encourage weak growth and pest and disease problems. Dahlias need at least 1″ of water weekly; soil should be neither soggy nor completely dry. Mulch to retain moisture.

Dahlias survive winter outdoors in Zones 9-10, given good drainage; otherwise they must be dug and stored. Cut plants back to about 1′ and dig the roots in fall before the first frost. Shake away loose soil and lay roots in the sun to dry for several hours. After drying, remove the tops and store the roots in a cool, dark, dry place. Check roots at regular intervals to make sure they're not so dry that they shrivel up.

Problems

Leaves wilt; stems collapse. Cause: Borers. Both European corn borer and stalk borer caterpillars are long, thin, and striped and tunnel within dahlia stems and flower stalks, causing collapse. Inspection of damaged plants reveals small, round holes in stems. Remove and destroy infested stems; clean up debris where pests overwinter. If borers have been a problem in past years, spray BTK weekly in early summer before borers enter stems.

Leaves mottled white or yellow. Cause: Leafhoppers. Feeding by wedge-shaped leafhoppers gives leaves a speckled appearance, followed by dry, brown blotches; foliage becomes distorted and falls off. In addition to feeding injury, leafhoppers can spread aster yellows and other diseases. Encourage natural predators such as bigeyed bugs and parasitic wasps. Spray leafhopper infestations with insecticidal soap first; if control is unsatisfactory, treat with a commercial pyrethrin/rotenone mix.

Leaves yellow-green or deformed; shoots spindly, stunted. Cause: Viral diseases. Several viruses infect dahlias, causing stunted growth and yellowed leaves with ring spots, mottling, and pale or dead areas. Aphids and leafhoppers spread viruses; control pests to limit infection. Dig and destroy dahlias showing symptoms.

Flowers and foliage with gray mold. Cause: Botrytis blight. During wet, cloudy weather, *Botrytis* fungi can cause flowers to turn brown and rot. Pick and destroy infected plant parts. If blight is a persistent problem, apply sprays of bordeaux mix.

Daphne

Daphne. Shrubs.

Daphnes are small evergreen or deciduous shrubs; most produce fragrant flowers. They are excellent foundation shrubs.

Daphnes are not the easiest plants to grow; there is considerable controversy as to their soil preferences. Furthermore, for no known reason, plants will occasionally die. Plant them in organically enriched soils in sun or light shade and be sparing of fertilizer and water. Spring planting as container-grown or balled-and-burlapped plants suits them best.

Daphnes are little troubled by insects. In warmer climates scales may be a problem; see "Leaves yellow; stems and leaves covered with small bumps" on page 237.

Delphinium

Delphinium, larkspur. Perennials.

Dense flower spikes of lavender, true blue, pink, and white make delphiniums summer garden showstoppers. Towering flower stalks range in height from 1½′ to 8′, depending on the cultivar. Large, lobed, and deeply toothed leaves cluster near the ground.

Soil for delphiniums must be high in organic matter, slightly alkaline, well-drained, and moist. Select a site protected from damaging winds, and provide stakes for brittle, hollow flower stems. These plants prefer full sun, but benefit from afternoon shade where summers are long and hot. Delphiniums need cool summers and may die during extended hot, humid periods. Even under favorable conditions, most plantings lose vigor and need renovation every 3-5 years.

Problems

Leaves, stems, and buds distorted, sticky; clusters of small insects. Cause: Aphids. See "Leaves, stems, and buds distorted, sticky; clusters of small insects" on page 177 for controls.

Leaves with tan or brown blotches or serpentine tunnels. Cause: Leafminers. For recommended controls, see "Leaves with tan or brown blotches or serpentine tunnels" on page 27.

Leaves with large, ragged holes. Cause: Slugs and snails. See "Leaves with large, ragged holes" on page 177 for control information.

Leaves covered with white powder. Cause: Powdery mildew. See "Leaves covered with white powder" on page 177 for controls.

Stems exude sawdustlike material and break; leaves wilt. Cause: Borers. Long, thin, striped stalk and burdock borers make small holes in the base of delphinium stems as they enter to feed. Sawdustlike castings around the holes mark the presence of these moth larvae. Prevent borers by removing weeds and plant debris where eggs might overwinter; applications of BTK give control only if applied just as borers are entering plants. You can also control borers by slitting affected stalks lengthwise, removing the borers, and binding the stems together—a long task in a large planting.

Leaves stippled, reddish to yellow, with fine webbing. Cause: Spider mites. These 1/50″ arachnids quickly disfigure ornamentals. Their feeding causes leaves to yellow, dry up, and die. Excessive feeding turns foliage almost white; fine webs cover leaves and growing tips. Mites target water-stressed plants in hot, dry weather; adequate watering guards against them. Hose down infested plants; a strong stream knocks mites off plants, but may damage brittle flower stalks. Introduce predatory mites (1,000 per 500 square feet); spray severe infestations with avermectins, citrus oils, insecticidal soap, neem, pyrethrin, or rotenone.

Stems blacken at base; leaves wilt; plant falls over. Cause: Crown or root rots. Both fungi and bacteria cause crown and root rot of delphiniums. Plants may wilt suddenly or yellow and wilt slowly. Stems and roots may turn black; mold may appear. Poorly drained, overly wet soil promotes rot; choose well-drained sites and add organic matter to improve drainage. Avoid injury to crowns when digging near plants; keep winter mulch away from crowns. Remove and destroy infected plants; wash tools used around diseased plants. Solarize the soil before replanting.

Leaves with yellow, brown, or black spots. Cause: Fungi. Several fungi cause spots on leaves. As spots enlarge, entire leaves may wither. Remove and destroy infected foliage; space plants to encourage good air circulation; avoid wetting foliage when watering. Dig and destroy seriously infected plants. Keep garden free of debris, and cut plants to the ground at the end of the growing season. Apply preventive sulfur sprays if leaf spot is a serious problem.

Deutzia

Deutzia. Shrubs.

Deutzias are a group of low-growing, deciduous, spring-blooming shrubs of rounded form. They are excellent in the shrub border or massed as an informal hedge.

Deutzias tolerate virtually any soil in sun or very light shade. Set out in spring as bareroot, balled-and-burlapped, or container-grown plants. They bloom on the previous year's wood, so prune after blooming, cutting back the branches that have flowered.

While deutzias are relatively free of major insect problems, aphids may feed on the leaves. For controls, see "Leaves wrinkled and discolored" on page 235. Leafminers may occasionally damage the leaves; see "Leaves tunneled" on page 237.

Dianthus

Pink, carnation. Perennials.

Sprightly pinks bear fringed flowers in shades of pink, white, maroon, and red over tidy, gray-green grasslike leaves. *Dianthus* spp. range in height from 4″ to 18″ with spreads of

1′-2′; most bloom profusely from late spring through summer. Popular species include sweet William (*D. barbatus*), a short-lived perennial often grown as an annual, and mat-forming maiden pinks (*D. deltoides*).

Well-drained, average soil with neutral to slightly alkaline pH is best for pinks. Most prefer full sun, but tolerate light shade, especially where summer temperatures are high. Pinks may decline in the heat of summer; select heat-resistant cultivars for southern gardens. Avoid overwatering but don't let plants wilt from drought stress. Cut plants back at the end of the blooming season to encourage new growth. Rejuvenate plantings by dividing them every 2-3 years.

Problems

Leaves stippled, reddish to yellow, with fine webbing. Cause: Spider mites. These 1/50″ arachnids quickly disfigure ornamentals. Their feeding on the undersides of foliage causes it to yellow, dry up, and die; excessive feeding turns foliage almost white and fine webs cover leaves and growing tips. Mites seek out water-stressed plants in hot, dry weather; adequate watering helps prevent infestations. Hose down plants when symptoms first appear; a strong stream knocks mites off plants and may give sufficient control. Introduce predatory mites (1,000 per 500 square feet); spray severe infestations with avermectins, citrus oils, insecticidal soap, neem, pyrethrin, or rotenone.

Leaves, stems, and buds distorted, sticky; clusters of small insects. Cause: Aphids. Also known as plant lice, tiny aphids damage plants by sucking sap from leaves, stems, and buds. In addition, aphids transmit viruses that can injure plants long after the pests are vanquished. Use a strong spray of water to knock aphids off plants, or apply insecticidal soap sprays. Encourage natural predators such as lady beetles and lacewings. Treat severe problems with pyrethrin or a commercial pyrethrin/rotenone spray or dust.

Leaves and flowers greenish yellow, distorted; new growth spindly. Cause: Viruses. Several viruses may infect pinks, stunting growth and mottling and spotting foliage. Control aphids and leafhoppers that spread viruses; there is no cure for infected plants. Remove and destroy plants with viral symptoms; don't compost diseased materials.

Leaf surfaces pale; powdery orange spots beneath. Cause: Rust. This fungus sometimes bothers sweet Williams, deforming leaves and stems and causing early leaf drop. Its orange spots release fungal spores that spread via wind and rain. Avoid wetting foliage when watering and encourage good air circulation. In problem areas, apply wettable sulfur several weeks before rust normally appears. Remove infected plant parts; clear debris from the garden in the fall.

Plant wilts; stems rot at soil level. Cause: Fungal or bacterial rots. Pinks are susceptible to various diseases that cause plants to wilt suddenly or to rot at or below the soil surface. This is especially true of plants growing in wet or poorly drained soil or in crowded conditions. Prevent problems by planting in well-drained sites, and avoid overwatering and overcrowding; keep mulch away from stem bases. Remove and destroy infected plants; do not compost diseased materials. If the problem is serious, solarize the soil before replanting a site.

Dicentra

Bleeding heart. Perennials.

These graceful plants carry prolific stems of dainty, heart-shaped flowers. Common bleeding heart (*Dicentra spectabilis*) is a Japanese native with dangling, pink and white blooms in late spring to early summer; native fringed bleeding heart (*D. eximia*) bears pink or white blossoms on 1′ spikes throughout the summer. Plants range from 1′ to 3′ in height and spread. Medium green leaves are fernlike.

Plant bleeding hearts in moist, rich, and well-drained soil under light shade. Plants tolerate full sun only where summers are cool. Avoid sites with standing water, an invitation to stem rot. Mark the spot where you plant bleeding heart's fleshy roots—common bleeding heart foliage may die back after blooming and all species disappear in winter. Plant bleeding heart roots 2′ apart and don't disturb them; roots are brittle and do not transplant or divide easily. Few pests or diseases pose serious trouble for well-grown bleeding hearts.

Problems

Stems blacken at base; leaves wilt; plant falls over. Causes: Wilts; stem rots. Many destructive fungi live in soil and will attack plants at soil level. Plants may wilt suddenly or yellow and wilt slowly. Stems and roots may turn black; mold may appear. Discolored, water-soaked lesions girdle stems; plants gradually die. Poorly drained, overly wet soil promotes wilt and rot; choose well-drained sites and add organic matter to improve drainage. Avoid overwatering and overcrowding; keep mulch away from the base of stems. Remove and destroy infected plants; wash tools used around diseased plants.

Digitalis

Foxglove. Perennials.

Common foxgloves (*Digitalis purpurea*) are actually biennials that self-sow so well that they function as perennials. Bell-like flowers in shades of purple, pink, and white cover 2′-5′ spires in late spring. Large, coarse, dark green leaves cluster near the plant's base. True perennial species are available, but their blooms are fewer and less showy than common foxglove's.

Foxgloves enjoy a shady site with rich, moist, well-drained, slightly acidic soil. Plants tolerate full sun, but favor cool weather and need shade in extreme summer heat. Water regularly to maintain soil moisture. Stake tall cultivars to support flower spikes; plant in a sheltered site. Remove foxgloves after they have set seed and self-sown next year's crop. Cutting flowers before they set seed promotes second-year blooms, but flowering is reduced the second year.

Problems

Leaves and/or flowers with holes. Cause: Japanese beetles. These pests can seriously damage foxglove foliage; see "Leaves and/or flowers with holes" on page 176 for control information.

Leaves covered with white powder. Cause: Powdery mildew. For information on controlling this common and troublesome fungal disease, see "Leaves covered with white powder" on page 177.

Leaves with large, ragged holes. Cause: Slugs and snails. Slugs and snails may hide below foxgloves' low, broad foliage. Limit them by removing faded leaves and plant debris. See "Leaves with large, ragged holes" on page 177 for more control information.

Leaves, stems, and buds distorted, sticky; clusters of small insects. Cause: Foxglove aphids. These pests feed on other plants but overwinter as eggs on foxgloves. Good gardening practices will slow the spread of foxglove aphids. Remove spent plants in fall to reduce aphid activity in following years. See "Leaves, stems, and buds distorted, sticky; clusters of small insects" on page 177 for controls.

Dill

Anethum graveolens (Umbelliferae)

Dill is an annual herb grown for its tangy leaves and seeds. Sow seed outdoors where the plants are to grow after the soil is warm. Dill prefers a site in full sun with rich, well-drained, moist soil and a pH between 5.0 and 8.0.

Dill is rarely troubled by pests or diseases. Virus-infected leaves are dwarfed and deformed and abnormally colored. Destroy infected plants and control leafhoppers as they spread viruses. Leaf blight causes dark spots with yellow edges. To control this fungal disease, spray plants with fish emulsion or in severe cases, with sulfur. See the Herbs entry beginning on page 116 for other possible problems.

Echinacea

Purple coneflower. Perennials.

Purple, daisylike petals surround the prickly, brown centers of purple coneflower's (*Echinacea purpurea*) 1″-4″ wide blossoms. Plants grow 3′-4′ tall and bloom from early summer until frost. The cone-shaped centers persist throughout the winter, providing interest after the petals have fallen. Pale coneflower (*E. pallida*) has creamy white petals.

These durable prairie natives like full sun, but endure poor soils and tolerate some drought once established. Shaded sites cause tall, spindly growth, as does excess fertilization. Pinch in spring to encourage sturdy, branching growth. Coneflowers are somewhat short-lived; divide clumps every 3 years or so to rejuvenate plantings, placing divisions 20″ apart. Plants also start easily from seeds, which need light to germinate.

Japanese beetles can quickly skeletonize the foliage of purple coneflowers. See "Leaves and/or flowers with holes" on page 176 for controls.

Eggplant

Solanum melongena var. *esculentum* (Solanaceae)

Eggplants are tender perennials that are grown as annuals in temperate regions for the firm-fleshed fruit. Fruit shapes range from long and thin to short and blocky and may be white, yellow, or dark purple. Harvest fruit at any size, preferably before seeds turn brown and harden.

Culture

Eggplants do best in full sun and well-drained, fertile soil with lots of organic matter. They prefer a pH between 6.0 and 6.8, but will tolerate a pH as low as 5.5. Eggplants need a high level of nitrogen and moderate levels of phosphorus and potassium. Have the soil tested and correct any deficiencies. They grow best at temperatures between 70° and 85°F, and poorly above 95° or below 65°F.

Eggplants need lots of water. Keep the soil evenly moist, and never let it dry out. They do well in mulched, raised beds with drip irrigation. Black plastic mulch is a good choice because it warms the soil.

Purchase stocky, insect- and disease-free plants, or start your own from seed indoors. Soaking seed in 122°F water for 30 minutes before planting can help reduce seed-borne diseases. (Be aware that this treatment can damage seed viability; for complete instructions, see page 422.) Eggplant seeds germinate best between 80° and 90°F. Once seedlings are up, they grow best at 70°F. Do not plant out before average daily temperatures have reached 65°-70°F. Protect transplants from wind, and water new transplants well with seaweed extract or compost tea to give them a good start. Spray plants with seaweed extract with 1 teaspoon of Epsom salts added per gallon when the first flowers open to improve fruit set.

Eggplants are susceptible to many of the same problems, pests, and diseases as tomatoes, including flower drop or misshapen fruit due to extreme temperatures, flea beetles, Colorado potato beetles, aphids, hornworms, mites, Verticillium and Fusarium wilts, tobacco mosaic virus, and anthracnose fruit rot. See the Tomato entry beginning on page 227 for symptoms, causes, and controls.

Prevent problems by planting tobacco mosaic virus-tolerant cultivars such as 'Blacknite', 'Classy Chassis', 'Dusky Hybrid', 'Epic', and 'Vernal'.

Problems

Leaves turn yellow, then brown. Cause: Lace bugs. Leaves eventually die. These flat, gray to brown, 1/10" long insects have lacelike wings and feed on the undersides of leaves. Spray plants with a commercial pyrethrin/rotenone mix if damage is severe. Prevent problems by covering plants with row cover until they flower.

Leaves with large holes. Cause: Blister beetles. These 3/4" long, elongated beetles have yellow and black stripes. Wear gloves to handpick, since these beetles secrete a substance that may cause blisters. Blister beetle larvae help control grasshoppers, so think twice before spraying to control them. Spray plants with a commercial pyrethrin/rotenone mix if damage is severe.

Leaves with light-centered, gray to brown spots; fruit with pale, sunken spots. Cause: Phomopsis blight. Stems may also develop dark areas. Fruit spots enlarge and run together, affected flesh is discolored and may rot and shrivel. Spray plants with copper if symptoms are present and if weather is wet or humid. Prevent problems by planting resistant cultivars like 'Florida Beauty' and 'Florida Market'.

Fruit with dry, brown chew marks. Cause: Colorado potato beetles. See "Leaves with large, ragged holes or leaves missing" on page 190 for description and controls.

Epimedium

Epimedium. Perennials.

These 6″-12″ tall, evergreen to semi-evergreen plants bear clusters of ¾″, yellow, white, or pink flowers in early spring. Compound leaves consist of 2″-3″ long, heart-shaped leaflets, tinged pink or red in spring and turning yellow or bronze in fall.

Epimediums are slow to establish, but can form a hardy groundcover for moist, partially shaded locations and will grow in sites where low soil fertility limits other perennials. Cut foliage low to the ground in early spring to keep it from hiding the flowers.

Like most groundcovers, epimediums offer a haven for slugs and snails. See "Leaves with large, ragged holes" on page 177 for controls.

Euonymus

Euonymus, spindle tree. Trees, shrubs, and vines.

Euonymus are evergreen or deciduous trees, shrubs, or vines, always with opposite leaves and generally with toothed leaves. Shrubby species are good for hedges or specimen plants; vining species are great as groundcovers or on walls.

Set euonymus out in spring or fall as balled-and-burlapped or container-grown plants. The plants are tolerant of sun or shade, but heavy shade may decrease fruiting.

Problems

Leaves yellow; stems and leaves covered with small bumps. Cause: Scales. Scale insects are the most common insect pests on euonymus plants. See the opposite page for an illustration of euonymus scale and its damage. For control measures, see "Leaves yellow; stems and leaves covered with small bumps" on page 237.

Trunk or roots with swollen, wartlike growths. Cause: Crown gall. For an illustration of this bacterial disease, see the opposite page. For controls, see "Trunk or roots with swollen, wartlike growths" on page 238.

Leaves wrinkled and discolored. Cause: Aphids. See "Leaves wrinkled and discolored" on page 235 for controls.

Leaves and shoots blackened; leaves with moist or brown sunken spots. Cause: Anthracnose. See the opposite page for an illustration of this disease. For control measures, see "Leaves and shoots blackened; leaves with moist or brown sunken spots" on page 238.

Leaves with powdery white coating. Cause: Powdery mildew. See "Leaves with powdery white coating" on page 237 for controls.

Euphorbia

Spurge. Perennials.

The species of perennial *Euphorbia* range in size from prostrate myrtle euphorbia (*E. myrsinites*) to 1′-1½′, clump-forming cushion

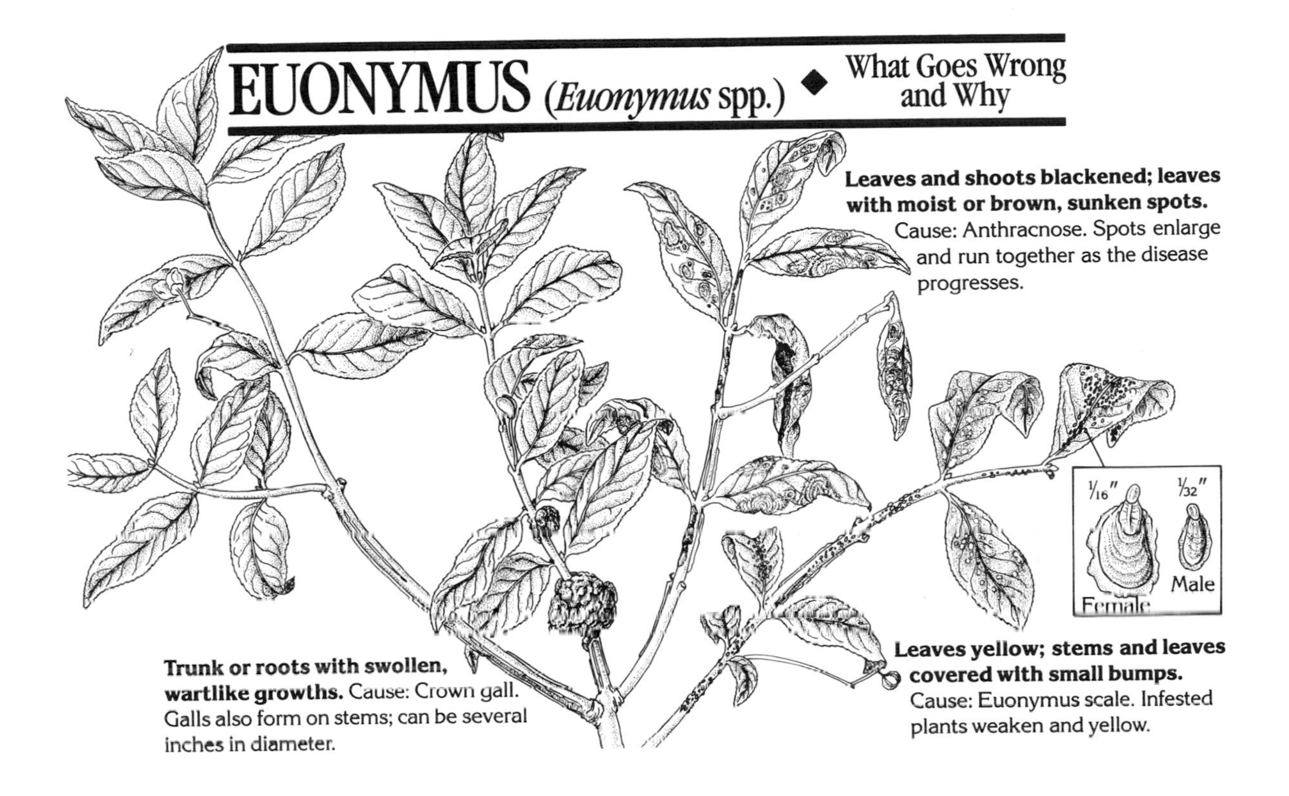

spurge (*E. epithymoides*) to 2′-3′ tall Griffith's spurge (*E. griffithii*). Brightly colored flowers, actually showy bracts, bring shades of yellow and red-orange to the garden from early spring through summer, depending on the species. Plants remain attractive after bracts have faded; some offer red fall color. Also characteristic of this genus is milky sap that can cause skin irritation and burning; wear gloves when handling spurges and avoid contact with the sap.

Spurges prefer full sun and well-drained, moist soil, but may become invasive under these ideal conditions. In southern zones, they need partial shade during the heat of summer to prevent leggy, open growth. Spurges don't mind average to poor soil and most tolerate some drought. Divide vigorously growing plants every 2-3 years to check their spread, being sure to leave solid roots on each division.

Problems

Plant wilting. Cause: Lettuce root aphids. Usually sturdy and resistant to most pests, spurge is occasionally infested by tiny white to yellow insects that feed underground on its roots as well as those of lettuce. Remove weeds to prevent aphids from multiplying in early spring. Add compost to soil to encourage beneficial nematodes, ground beetles, and other soil predators. Improve watering to help plants withstand root damage. In severe cases, drench the soil around spurges with nicotine tea.

Fagus

Beech. Trees.

Beeches are deciduous trees with gray bark and alternate leaves. They are one of the finest specimen trees for large landscapes. They do not, however, thrive on city conditions, and their spreading growth and extensive surface roots make it virtually impossible to grow other plants beneath them.

Set out balled-and-burlapped plants in spring in well-drained, evenly moist soil. Full sun is best but they'll tolerate part shade. Beeches are very sensitive to root zone disturbances—cuts, fills, compaction—so keep them safe from such abuse. A thick layer of organic mulch will protect the shallow roots and help keep them cool and moist.

Problems

Leaves skeletonized. Causes: Cankerworms; other caterpillars. Cankerworms are ½″-1″, yellow, green, or brown caterpillars that feed on beech leaves, often defoliating the tree. To avoid this problem, apply a band of sticky coating, such as Tanglefoot, around the base of the tree in spring to keep females from climbing up and laying their eggs. Apply fresh coats in September and February to trap both spring and fall cankerworms. If their feeding becomes a problem, spray leaves with BTK as soon as you notice the pests; continue every 2 weeks until they are gone.

Several other caterpillars, including gypsy moths, loopers, and tent caterpillars, feed on beeches. See "Leaves skeletonized or with large holes; branches may be webbed" and "Plant defoliated" on page 236 for more information and controls.

Trunk or branches with small holes; limbs die or break off. Cause: Borers. See "Trunk or branches with small holes; limbs die or break off" on page 238 for controls.

Leaves yellow; stems and leaves covered with small bumps. Cause: Scales. A number of scale insects attack beeches, including the white-colored beech scale. Note that beeches may be damaged by applications of dormant oil; instead, spray lime-sulfur on the trunk and branches in late winter. For other controls, see "Leaves yellow; stems and leaves covered with small bumps" on page 237.

Trunk or branches with oozing lesions; branch tips die back. Cause: Canker. Bleeding canker is spread by the beech scale, which feeds on the bark, causing wounds that permit entry of the fungus. Control the insect and you'll prevent the disease. For more details on dealing with canker, see "Trunk or branches with oozing lesions; branch tips die back" on page 238.

Leaves wrinkled and discolored. Cause: Aphids. See "Leaves wrinkled and discolored" on page 235 for controls.

Leaves with powdery white coating. Cause: Powdery mildew. See "Leaves with powdery white coating" on page 237 for controls.

Leaves with spots. Cause: Leaf spots. See "Leaves with spots" on page 237 for controls.

Fig

Ficus carica (Moraceae)

Figs are deciduous, subtropical trees or bushes that grow 10′–30′ tall and bear fruit on 1- and 2-year-old wood. They are hardy in Zones 8–10.

Fig trees need abundant sunlight but tolerate of a wide range of soil types. In colder areas, plant figs against a wall or wrap them for the winter to minimize dieback from cold. Each winter, head back long branches to maintain tree shape and induce vigorous new growth. Thin out crowded, weak, diseased, or crossed branches. Cultivars such as 'Celeste' and 'Mission', which fruit best on last year's wood, should be pruned more lightly than cultivars such as 'Magnolia' and 'Kadota', which fruit heavily on new growth. Some figs need cross-pollination; others develop fruit without pollination. Check pollination needs before buying plants.

Fruit Problems

Fruit soured, mushy, or smutty. Cause: Sour bugs. These small black beetles crawl into the eye of fruit and lay eggs that hatch into white grubs. The grubs carry bacteria that cause figs to turn sour, mushy, or smutty. Soured fruit may have a drop of pink, sticky fluid at its eye and a fermented odor. Smut produces dusty black spores. To prevent sour bug damage, grow closed-eye cultivars such as 'Brown Turkey', 'Celeste', 'Green Ischia', and 'Mission'. Clean up fallen fruit, which attracts beetles, and trap beetles in containers of fermenting fruit.

Fruit with ants inside. Cause: Insect feeding. Ants enter the fruit eye to feed on ripe fruit. To thwart ants, band stems with plastic or heavy paper coated with sticky material. Or place ant bait stations containing boric acid around trees. Wood ashes around the base of trees also helps prevent ants from climbing up to fruit.

Leaf and Branch Problems

Leaves with raised, rusty spots on undersides. Cause: Rust. Leaves may yellow and die. To control this fungal disease, rake up and destroy fallen leaves. Where the problem is severe, spray with copper.

Leaves spotted. Cause: Leaf blight. Various fungal diseases attack fig leaves and twigs. Control all these conditions by raking up and destroying fallen leaves and fruit. Also prune plants to remove infected twigs and allow sunlight to reach inner branches.

Whole Plant Problems

Whole plant lacking vigor; bumps on twigs and leaves. Causes: Scales; nematodes. Scales look like yellow, gray, white, or reddish or purplish brown bumps that you can scrape off stems or leaves with your fingernail. Many natural enemies, including lady beetles and chalcid wasps, help keep scales in check. Prune off heavily infested branches and destroy them. For serious infestations, spray before growth begins with dormant oil.

Nematodes are soil-dwelling, microscopic worms. Some nematodes perform helpful tasks, such as breaking down organic matter; others attack plant roots and cause diseaselike symptoms. Figs grown in sandy soils are especially susceptible to nematode infestation. If you have sandy soil, plant your fig tree near a building and apply a thick, organic mulch. Both of these strategies seem to deter nematodes. If nematodes are common in your area, solarize the soil for 1 or 2 months during the summer before planting figs. Planting a cover crop of French dwarf marigolds or African marigolds or adding chitin to the soil also helps control nematodes. 'Celeste' and 'Hunt' are 2 nematode-resistant cultivars.

Filbert

Corylus avellana, C. americana
(Betulaceae)

Filberts grow as deciduous trees or shrubs reaching about 20′. They are hardy in Zones 4-8. To set nuts, most filbert cultivars need cross-pollination with a second cultivar.

Filberts blossom very early in the spring, so they need a site not prone to spring frosts. Plant in full sun. Since plants blossom on last year's wood, annual pruning is needed to stimulate current season's growth, which will bear next year's nuts.

Problems

Nut kernels blackened; nut shell has small hole. Cause: Filbertworms. These are the larvae of a brown moth. The easiest control is to keep trees free of ground debris, such as leaves, fallen nuts, and shells, because the larvae pupate on the ground throughout the winter.

New leaves twisted or curled and covered with a sticky coating. Cause: Aphids. Look on leaf undersides for clusters of these tiny black insects. Leaves also may be covered with a black fungus, called sooty mold, which feeds on honeydew secreted by these pests. For information on controlling aphids, see "New leaves twisted or curled and covered with a sticky coating" on page 67.

Flower buds swollen. Cause: Bud mites. Infected flowers don't produce nuts. Plant resistant cultivars, including 'Barcelona', 'Cosford', 'Italian Red', and 'Purple Aveline'. In severe cases, spray trees with superior oil around May when the mites crawl out of the swollen buds to attack healthy buds.

Branches blighted. Causes: Eastern filbert blight; western filbert blight. Eastern blight has no cure. It is present in the East and carried on American filberts. Young branches are destroyed first, then older branches and the trunk, without killing the roots. American filberts have some tolerance for the fungus, so they have been used in breeding with other filbert species in an attempt to create disease-resistant plants bearing high-quality nuts.

In the Pacific Northwest, blight is caused by western filbert blight, a bacterial disease. Small, angular spots appear on leaves, water-soaked at first, then turning reddish brown. Cankers form on branches, which then may die. Control this blight by preventing sunburn and winter injury, especially on young plants, which are most susceptible. On older plants, only smaller twigs die. Spray with copper and prune out infected twigs in winter.

Forsythia

Forsythia. Shrubs.

Forsythia is the most well-known spring-flowering, deciduous shrub. It is effective as a hedge, grouped for landscape accent, or in a shrub border.

Set out in spring or fall as bare-root or balled-and-burlapped plants. They require at least partial sun but flourish in almost any soil. Right after bloom, cut out some of the oldest canes to maintain strong flowering.

Forsythias are troubled little by insects. Crown gall may be a problem; see "Trunk or roots with swollen, wartlike growths" on page 238.

Fraxinus

Ash. Trees.

Ashes are deciduous trees with opposite, compound leaves that turn yellow to purple in the fall. Male and female flowers may appear on the same or different trees; all-male trees are preferred for landscaping since they don't produce undesirable seedlings. Ashes are used as street trees or specimens for large properties.

Set out trees in spring or fall. Full sun is best; the soil should be deep and moist but well-drained. Although often found growing along streams, ashes tolerate dry conditions and a range of soil pH. Vigorously growing ashes are fairly trouble-free.

Problems

Leaves yellow; stems and leaves covered with small bumps. Cause: Scales. Probably the most serious pests of ash are scale insects, with oystershell scale as the most important, especially in the Midwest. Putnam scale, San Jose scale, and soft scale (in warmer climates) also occur. For control information, see "Leaves yellow; stems and leaves covered with small bumps" on page 237.

Trunk or branches with small holes; limbs die or break off. Cause: Borers. See "Trunk or branches with small holes; limbs die or break off" on page 238 for controls.

Leaves skeletonized or with large holes; branches may be webbed. Cause: Caterpillars. For control information, see "Leaves skeletonized or with large holes; branches may be webbed" on page 236.

Branches with small masses of galls. Cause: Ash flower gall mites. These tiny mites attack male flower clusters on white ash (*Fraxinus americana*). Infested flowers develop abnormally, producing galls that dry and remain on the tree. Spray branches and galls with dormant oil in late winter for control.

Leaves with spots. Cause: Leaf spots. See "Leaves with spots" on page 237 for details.

Trunk or branches with oozing lesions; branch tips die back. Cause: Canker. See "Trunk or branches with oozing lesions; branch tips die back" on page 238 for controls.

Leaves distorted, orange-yellow spots on leaf undersides. Cause: Rust. This fungus produces orange pustules, distorting the leaves and making branch tips swell. Rust usually doesn't harm the tree much, except for detracting from the appearance. Clean up and destroy fallen leaves and branches in autumn. To prevent the spread of mild infections, spray leaves with sulfur early in the season.

Leaves skeletonized. Cause: Sawflies. See "Leaves skeletonized" on page 236 for controls.

Fritillaria

Fritillary. Bulbs.

Red, orange, or yellow, bell-shaped blooms hang downward in groups of 8–10 from crown imperial's (*Fritillaria imperialis*) 3′ tall flower stalks. Smaller checkered lilies (*F. meleagris*) bear individual purplish flowers, marked with a checkerboard of dark purple on 1′ stems. Both the bulbs and the lance-shaped, dark green foliage have a musky odor that is reputed to discourage rodents.

Plant bulbs in early autumn as soon as they are available. Handle carefully to avoid bruising, and don't let them dry out before planting. Set bulbs 6″ deep in humus-rich, moist, very well drained soil under partial shade. Plants bloom in early spring; choose a site that offers protection from late frosts. Water during bloom. Well-grown fritillaries are troubled by few pests or diseases.

Fruit

Unlike vegetable crops, which share many of the same pests, fruit crops are affected by a wide variety of insects and diseases. Because fruit crops are borne on so many types of plants—including fruit trees, berry bushes, vines, and herbaceous plants such as strawberries—they have a wide range of cultural requirements as well. However, regardless of the crop you are growing, there are basic steps you can take to help prevent and control insects and diseases.

Site Selection

Full sun is a must for nearly all fruit crops: Even 1 or 2 hours of shade a day may result in smaller crops and less-flavorful fruit. Well-drained soil is also essential. In sites where drainage is a problem, plant in raised beds.

A location near the top of a gentle slope is ideal for fruit growing. A north-facing slope will help delay spring flowering, which is a plus in areas where frost damage is common. Planting about 15′ from the north side of a building also helps delay flowering. To help protect winter-tender trees, look for a sheltered site on the south side of a building.

Avoid planting in frost pockets where late spring frosts may damage blossoms and early fall freezes may shorten the harvest season. Frost pockets develop because cold air tends to sink and collects in depressions. Frost pockets occur at the bottom of valleys and on the uphill side of woods or buildings on slopes. If your whole property is a frost pocket, try training trees taller so they can blossom above the frost pocket, and try planting hardier cultivars.

While good air circulation helps reduce disease problems, blustery winds in open areas or on hilltops can make training difficult, knock fruit off trees early, or topple trees altogether. To encourage air movement, space plants far enough apart so they won't grow into each other or nearby plants.

Fruit Selection

Fruit trees and berry bushes are a long-term investment. Before you plant, learn about the types of fruit suitable to the local climate and soil type. Look for cultivars that will tolerate local conditions. Northern gardeners should choose cultivars that will survive winter cold, blossom late enough to escape late spring frosts, yet still set and mature fruit before the end of the growing season. Southern gardeners need cultivars that will tolerate intense summer heat and humidity and don't require much of a winter chilling period. If available, disease-resistant cultivars are desirable.

It's a good idea to consult nearby nurseries, botanical gardens, and fruit hobbyists—as well as local orchardists and extension agents—for information on growing fruit in your area. The more you know about local problems affecting fruits, the better your chances for selecting the right cultivars.

Almost all fruit trees and many grape vines are grafted onto a rootstock selected for strong rooting characteristics, disease resistance, or dwarfing effect. Many nurseries only offer 1 or 2 rootstocks that grow moderately well in a wide range of conditions. Some mail-order nurseries, however, offer a selection of rootstocks for the same fruit cultivars. Once you have identified a specific rootstock that should do well in your conditions, you may want to look for a source that carries it.

Some fruit trees come grafted on rootstocks that produce dwarf, semi-dwarf, and standard-size trees. Most home gardeners prefer dwarf or semi-dwarf trees, which fruit at a younger age than standard trees and are easier to tend.

Setting Fruit

When selecting fruit trees, it's important to understand their pollination requirements. Many fruit tree cultivars will only set fruit if the flowers receive pollen from a different but compatible cultivar. For example, 'Red Delicious' apples are self-unfruitful, meaning 'Red Delicious' trees can't pollinate one another, although they are good pollinators for other cultivars. On the other hand, 'Winesap' apples are not only self-unfruitful, they're also unable to pollinate other cultivars. To get fruit from a self-unfruitful plant, you need to plant another suitable cultivar for cross pollination. 'Golden Delicious' is both self-pollinating (self-fruitful) and a good pollinator for other apples—including 'Winesap' and 'Red Delicious'.

Reputable nurseries will describe pollination requirements and suggest suitable pollinators. Most apples, pears, sweet cherries, and Japanese plums require cross-pollination to set a good crop. Most peaches, apricots, sour cherries, and some European plums are self-fruitful, although they often bear more heavily if cross-pollinated. Not all cultivars are compatible, so check with your supplier.

Planting

Proper planting will encourage healthy, vigorous plants that resist attack by insects and diseases. Choose your location and prepare the planting site before the plants arrive. Test the soil and amend as needed. Work lots of organic matter into the top 6″-8″ of soil, and remove all perennial weeds.

Most fruit plants are shipped bare-root and dormant. As soon as they arrive, inspect them for damage like broken roots or branches. Keep plants cool and the roots moist, but not wet, until you can plant them.

Dig a hole large enough in which to spread out the roots without cramping them. Cut back any damaged roots and shorten unusually long ones, if necessary. Most plants should be planted the same level at which they were growing in the nursery. Plant grafted trees with the graft union (which looks like an angle or an enlarged section near the base of the trunk) at least 1″ above the soil line. Exceptions to this rule may be noted in special instructions supplied with the plant.

Refill the hole with the soil you had removed, and water the plant well to give the plant a good start. Stake or trellis as needed to prevent damage, and reduce disease by encouraging good air circulation around the branches and between plants. Mulch to keep weeds under control, supply a slow release of organic matter, and conserve soil moisture.

Protect your new trees by painting the trunks with white latex paint diluted with an equal amount of water to prevent sunscald. Frustrate gnawing rodents by placing a 1′ high, ¼″ hardware cloth guard around the trunk, sunk into the soil 1″.

Pruning and Training

Fruit trees are pruned to develop strong branches, to remove dead or damaged growth, and to allow more air and sunlight into the

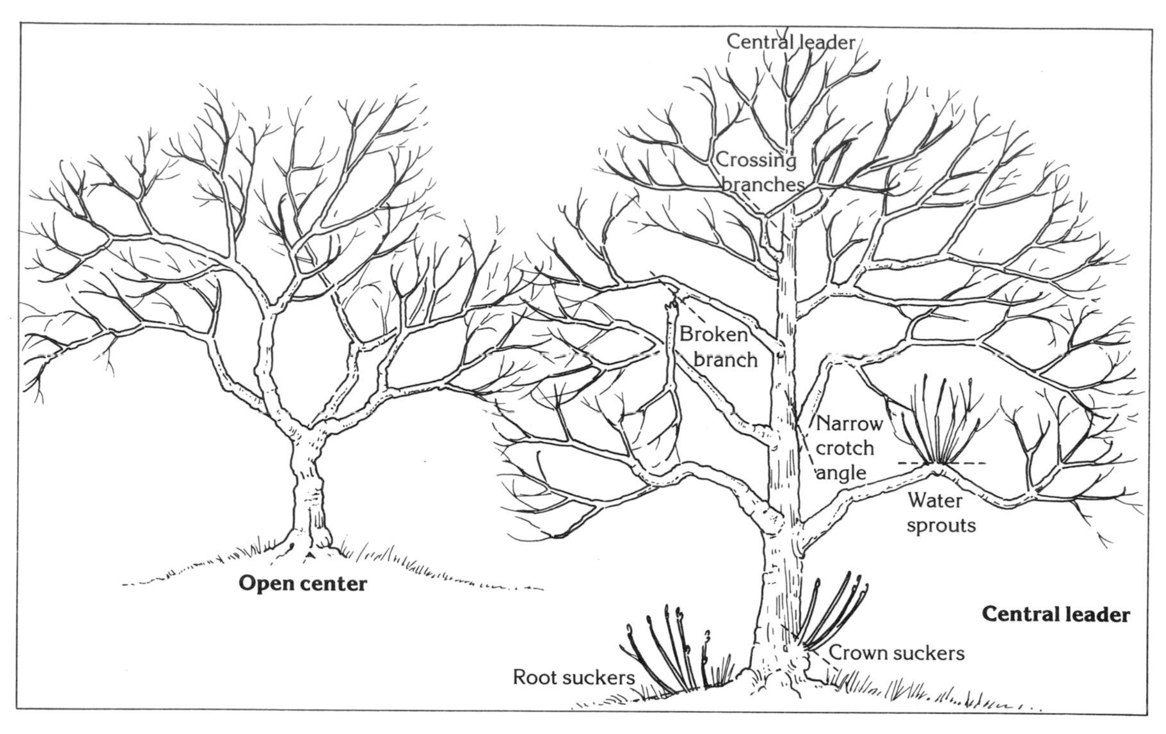

Open center (left) *and central leader* (right) *are 2 ways to train fruit trees to obtain maximum yield and exposure to sunlight. The best method depends on the growth habit of your fruit tree. In general, to improve tree health and performance, also remove any crossing branches, water sprouts, crown or root suckers, diseased or broken branches, or limbs that emerge at a narrow crotch angle.*

center of the plant to help prevent disease and insect problems. Pruning and training also help control size and encourage sturdy, well-spaced branches with wide crotch angles.

Start training trees as soon as they are planted. Prune off any damaged branches and any that are growing at a narrow angle to the trunk. Let the rest of the branches grow for a full year before starting to select your main branches.

It's important to establish sturdy, well-placed main branches while trees are young. Training can reduce the upward growth (affecting the amount of pruning you need to do) and will stimulate early fruit bearing. Train young branches to grow at a wide angle (about 45 degrees) by clamping a clip clothespin to the trunk just above each branch while they are still green and malleable. Older branches can also be spread, but with more difficulty. Use 1′-2′ long boards with a notch in each end and brace them between the trunk and the branch, or tie the branch to a weight (such as a rock) or the trunk. Gradually spread the branch during the season by moving the brace or tightening the string.

Prune apples, pears, grapes, and berries in late winter or early spring before they break dormancy. Prune peaches, plums, and cherries just after the buds burst in the spring, when they are less prone to canker infection. Remove dead and diseased branches on all kinds of fruit-bearing plants if they appear during the growing season. Summer pruning is also useful for controlling the growth of overly vigorous trees, as it stimulates less regrowth than dormant pruning does. Do not prune after August; fall pruning can increase the risk of winter injury.

There are 2 general types of pruning cuts: heading and thinning. Cutting a branch back partway to shorten it is a heading cut. Heading back a branch tends to encourage a flush of new shoots to sprout near the cut end. This is a useful technique for stimulating new branches when necessary, but too many heading cuts can cause overgrown, overly bushy trees. A thinning cut is used to remove a branch altogether by cutting it back to a main or side branch. Thinning encourages the remaining branch to grow, and unlike a heading cut, does not stimulate new ones to sprout. For this reason, thinning cuts are the best choice for developing an open framework and a sturdy, compact tree.

Fruit trees are usually trained and pruned into one of two general shapes: open center or central leader, which are illustrated on the opposite page. Modified central leader is intermediate between the two. The individual fruit tree entries include training recommendations. Grapes are usually trained on a trellis or arbor. Berry plants are thinned out in the appropriate season. There are a number of good references on page 493 that will give you complete, easy-to-follow instructions for pruning and training whatever type of fruit-bearing plants you choose.

Thinning

Fruit trees often set more fruit than they can ripen. Resulting fruit will be small and the excess weight can damage branches. Crowded fruit is also more susceptible to fungal diseases. Thin early in the season when fruit is as small as possible. First, clip or twist off all insect-damaged or deformed fruit. Then, remove the smaller fruit, leaving the biggest and best.

Space fruit so it will not touch each other as it grows. If you can't reach the upper limbs, tap them with a padded pole to shake loose some of the extras.

Preventive Care

Many disease and insect problems can be controlled or greatly reduced by simple sanitation. Clean up and dispose of branches after you prune, especially when removing diseased wood. Many insects pupate in the soil or in loose bark or dead plant debris. For this reason, rake and remove fallen leaves in fall. Also remove dropped fruit during the season and dry, diseased fruit (known as mummies) clinging to the branches after harvest. Inspecting bark in the winter and removing and destroying egg masses is also effective.

It's a good idea to find out what insects and diseases may attack your plants. Inspect plants regularly and control problems before they get out of hand. When problems do arise, start with the most environmentally gentle controls, such as handpicking, pruning, releasing predatory insects, or using mating disruption. Then move on to superior oil, insecticidal soap, BT, or sulfur. Use botanical insecticides, such as pyrethrin or rotenone, or copper compounds, only as a last resort after you have tried other less-toxic methods.

Garlic

Allium sativum and *A. ophioscorodon* (Liliaceae)

Garlic is a perennial herb grown for its pungent bulbs and greens. It is easy to grow healthy, trouble-free garlic. Plant cloves in the fall and cover with mulch after the weather turns cold. Garlic prefers full sun and rich, deep, moist, but well-drained soil with lots of organic matter and a pH between 6.0 and 7.0. Plants need extra water during summer while bulbs are forming. Water deeply when soil is dry. Once stalks begin to droop and brown at the end of summer, withhold water.

Garlic is closely related to onions and prone to the same problems, especially onion maggots. See the Onion entry beginning on page 155 for more information.

Geranium

Cranesbill. Perennials.

The cranesbills, *Geranium* spp., are hardy, easy-to-grow perennials that are often confused with the more tender zonal or bedding geraniums, which actually belong to the genus *Pelargonium.* Although related to the better-known zonal geraniums, cranesbills (also called hardy or true geraniums) are distinctly different from their tender relatives. Most have a mounding growth habit with spread roughly equal to height. Plants range from 4″ to 4′ tall with lobed to deeply divided foliage. Flowers bloom in spring and early summer in shades of pink, purple, red, and blue; many blossoms feature darker contrasting veins in petals or dark eyes. Some geraniums have a spreading growth habit; these are ideal for use as groundcovers.

Grow cranesbills in full sun to partial shade. Some species will grow in heavy shade, but, in general, too much shade reduces flowering and causes leggy growth. Plants tolerate some drought, but perform best in moist, well-drained soil of average fertility. Rich, moist soils encourage invasive growth in some species. Divide every 3-4 years in spring or fall. Some cranesbills form taproots that resist division; increase these types through root cuttings, stem cuttings, or seeds.

Hardy and fairly pest-free, cranesbills suffer few problems beyond those common to most perennials. Geranium aphids and four-lined plant bugs may attack them; see "Leaves, stems, and buds distorted, sticky; clusters of small insects" on page 177 and "Buds and leaves deformed or dwarfed" on page 178 for information on controls.

Gladiolus

Gladiolus, glad. Corms.

Showy, 2′-4′ tall glads brighten summer with tall spikes of trumpet-shaped, often bicolored flowers. The 6-petaled blooms range from 2″ to 5″ wide and open from the bottom of the spike upward. Glads are upright plants with narrow, sword-shaped, 1′-1½′ long leaves. Hundreds of cultivars of common gladiolus (*Gladiolus* × *hortulanus*) offer gardeners a spectrum of flower colors and forms from which to choose.

Select and plant gladiolus corms in spring. Do not buy or use corms that are lightweight, spongy, or showing signs of decay. See the illustration on page 49 for more information. Good corms are about 1½″ wide with high tops. Large, healthy corms flower more quickly than smaller ones. Plant corms after danger of frost has passed. To create a succession of bloom, plant at 2-week intervals or plant different-size corms all at one time. Allow at least 3 months between planting and the first frost.

Prepare soil to 1′ deep in a sunny location. Add plenty of organic matter to ensure good drainage. Plant corms from 3″ to 8″ deep—about 4 times their thickness; space about 6″ apart. Make sure pointed growing tips face upward. Water regularly; glads need about 1″ of water per week. Mulch to help retain soil moisture. Side-dress with compost or slow-release, general-purpose fertilizer when plants emerge and when flowers show color. Plan to stake up the tall, flower-laden spikes; nearly all cultivars require some form of support.

In the South, gladiolus corms overwinter safely in the ground, but in most of the country they must be dug and stored indoors before hard frost. Allow leaves to die back naturally for about 6 weeks after flowering stops. After leaves and stalks turn brown, dig corms and shake off loose soil; cut away stems just above the top of the corms. Let corms dry out of the sun for a few days. Do not remove husks, as they help retain moisture. Discard withered old corms. Separate cormels, the tiny offspring surrounding a mature bulb, and store for planting—these bloom in 1-2 years. Check all corms for decay, spotting, or other disease symptoms. Dust with sulfur or copper-based fungicide to guard against disease problems; store in a cool (40°F), dry, well-ventilated place.

Problems

Flowers deformed; leaves and petals with white flecks. Cause: Thrips. Gladiolus thrips are a very destructive common pest. These 1/25″, yellow to black, flying insects feed by rasping petals and leaf surfaces, leaving silvery spots and streaks. They hide under leaf sheaths and inside flowers. Other symptoms are partial bloom, failure to bloom, and shriveling. Thrips' wastes may appear as black spots on the undersides of leaves. Infested corms are dark, sticky, and rough.

Cut away and destroy severely infested plant parts. Use blue sticky traps to monitor and trap pests. Encourage native predators such as pirate bugs, lacewings, and lady beetles. Insecticidal soap sprays give some control of thrips populations, but may affect beneficials as well. To limit thrips infestation, dig corms early in fall and cut off tops before thrips move down into corms. Bag and discard debris. Dust corms with pyrethrin to control thrips in storage. Or soak them in a Lysol solution (1½ tablespoons in 1 gallon water) for several hours before planting.

Leaves, stems, and buds distorted, sticky; clusters of small insects. Cause: Aphids. Plants infested with aphids have growth that may be curled, puckered, or stunted. Leaves may turn yellow or brown; feeding can seriously damage flower buds or blooms. As aphids feed, they excrete sticky honeydew; black, sooty mold fungus often forms on honeydew, further disfiguring plants.

Use a strong spray of water in the morning to knock aphids off glads, paying particular attention to the undersides of leaves. Encourage predators such as lacewings and lady beetles. If control is not satisfactory, spray with insecticidal soap. Destroy seriously infested plants. Treat severe problems with pyrethrin or a commercial pyrethrin/rotenone mix.

Leaves yellow, distorted; flowers absent; corms decayed. Cause: Bulb mites. These 1/50", whitish pests favor damaged, rotting corms, but travel from those to healthy ones. Feeding causes corky, brown spots on corms, which eventually turn dry and crumbly. Dig and destroy seriously infested corms and the surrounding soil. Don't replant corms in infested soil. Kill mites by dipping corms in hot (120°F) water for a few minutes. Examine corms carefully when purchasing. For hints on choosing healthy bulbs, see the illustration on page 49.

Leaves yellow; plant dies early. Cause: Dry rot. Other symptoms of this fungal disease include dry, brown or black, corky spots on corms and husk coverings. You may see black fungal growth spots on decayed leaf bases. Plants may turn yellow and die prematurely. Choose corms carefully to avoid infected specimens. Destroy infected plants. Do not replant in same area. Replant only in well-drained soil.

Leaves mottled yellow and distorted; flowers small and faded. Cause: Fusarium yellows. This soilborne fungal disease is first seen as bending and curling of leaves and stems; foliage yellows and dies, starting with the oldest leaves. On corms small reddish brown lesions enlarge and darken; entire corms may become hard, dry, and mummified. Immediately destroy infected corms. Do not replant corms in same area. Some cultivars are more resistant than others; check with a reputable nursery or call your local extension agent for area-specific recommendations.

Leaves stippled and pale; growth poor. Cause: Spider mites. Mites feeding on leaves remove plant nutrients; fine webbing may appear on foliage. Use strong sprays of water when first symptoms appear. Mites multiply quickly and prefer drought-stressed, dusty plants. Keep gladiolus well-watered and healthy. Remove and destroy badly infested plants; treat severe infestations with avermectins, citrus oils, insecticidal soap, neem, pyrethrin, or rotenone.

Stems eaten through at base. Cause: Wireworms. These click beetle larvae are common in soils formerly covered by sod or with high organic content. These slender, hard-bodied, yellowish worms drill holes in corms as they feed, damaging them and hastening decay. Set out traps made from buried pieces of raw potato or carrot; check every 1-2 days and destroy wireworms; replace traps as necessary. Delay planting corms until the soil is very warm; apply parasitic nematodes to soil to reduce wireworm populations.

Flowers, leaves, and stalks spotted; corms rotted. Cause: Botrytis blight. This fungal disease also causes slimy, collapsed leaves and flowers. It is most prevalent in cool, damp weather. Prevention includes planting gladiolus in areas with good air circulation. Avoid low or shaded areas. Water so leaves dry before sunset. Remove all debris at the end of the season and destroy.

Leaves with reddish brown spots; corms with pale to brown spots. Cause: Scab. This bacterial disease also causes brown spots on blossoms; plants rot at the base and fall over. Spots on corms turn dark brown with sunken centers and scabby margins. Dig and destroy infested plants; do not replant corms in infested soil.

Gleditsia

Honey locust. Trees.

Honey locusts are deciduous trees with alternate, fine-textured leaves. They cast dappled shade and, except for their thorniness and messy pods, are excellent as specimen or street trees. Thornless, nonfruiting cultivars are preferable. *Gleditsia triacanthos* var. *inermis* is thornless honey locust, from which most cultivars have been selected for landscape use.

Set out in spring or fall as bare-root or balled and burlapped plants. Honey locusts are undemanding and are well-suited to city conditions. While they prefer moist, fertile, slightly alkaline soil, they tolerate drought and salt.

Problems

Leaves skeletonized or with large holes; branches may be webbed. Cause: Caterpillars. Webworms can be a serious problem on honey locust. Some cultivars, such as 'Moraine', show some resistance to webworm damage. For other control information, see "Leaves skeletonized or with large holes; branches may be webbed" on page 236. Gypsy moths also feed on leaves, but they don't form webs. See "Plant defoliated" on page 236 for controls.

Plant defoliated; branches bear cocoonlike bags. Cause: Bagworms. See "Plant defoliated; branches bear cocoonlike bags" on page 236 for control suggestions.

Trunk or branches with small holes; limbs die or break off. Cause: Borers. See "Trunk or branches with small holes; limbs die or break off" on page 238 for controls.

Trunk or branches with oozing lesions; branch tips die back. Cause: Canker. For details, see "Trunk or branches with oozing lesions; branch tips die back" on page 238.

Gooseberry

Ribes hirtellum, R. uva-crispa
(Saxifragaceae)

Gooseberries are thorny bushes with arched branches growing 3′-5′. The inconspicuous flowers are borne laterally on 1-year-old wood and on short spurs—short branches that elongate only a fraction of an inch per year—on older wood. Gooseberries are hardy in Zones 3-8.

In cool climates plant gooseberries in full sun. Choose partial shade in hot climates. Maintain a thick, organic mulch beneath plants to keep soil cool and moist. Prune mature plants every winter, cutting away all but 6 of the shoots that grew the previous season at ground level. Also remove all wood more than 3 years old. Gooseberries are self-pollinating.

Fruit Problems

Fruit with powdery white or gray coating. Cause: American gooseberry mildew. This fungal disease spreads most rapidly during periods of warm days followed by cool nights. Under these conditions the crop can be ruined overnight. For more information, see "Leaves with powdery white patches; leaves stunted and deformed" below.

Fruit with holes when nearly ripe; fruit and leaves covered with webs. Cause: Gooseberry fruitworms. These are the yellow-green larvae of a moth that infests both gooseberries and currants. Damaged fruit may be hollowed out and may change color prematurely. Clean up and destroy damaged fruit and spray rotenone as soon as you notice the webbing.

Fruit colors and drops early. Cause: Currant fruit fly larvae. The currant fruit fly lays eggs on gooseberries and currants in the spring. The larvae that hatch from these eggs feed on berries and cause them to drop. Infested berries have a dark spot surrounded by a red halo. Early-bearing cultivars, such as 'Oregon Champion' and 'Welcome', may escape damage. Control this pest by destroying all infested berries as soon as you find them.

Leaf Problems

Leaves with powdery white patches; leaves stunted and deformed. Cause: American gooseberry mildew. To prevent this fungal disease, avoid overfertilizing plants. Lush, sappy growth is more susceptible to infection. Plants well-supplied with potassium are best able to resist mildew. For serious infection, spray plants with lime-sulfur. Or spray a solution made from 1 pound of washing soda (sodium carbonate) plus 1/4 pound of soap in 1 gallon of water. Don't spray sulfur on the cultivars 'Bedford Yellow', 'Langley Gage', and 'Leveller'; these are easily damaged by sulfur. Cultivars resistant to gooseberry mildew include 'Hinnomaki Yellow', 'Lepaa Red', 'Poorman', and 'Welcome'.

Leaves with spots; leaves yellow and drop. Causes: Septoria leaf spot; anthracnose. Since both these fungal diseases overwinter on leaf litter, rake up leaves or bury them under a thick mulch. Copper sprays also control both diseases. If defoliation occurs late in the season, it does little harm.

Foliage stripped from bushes just as leaves expand. Cause: Imported currantworms. These are the larvae of a sawfly that lays eggs on gooseberry bushes. The larvae devour the leaves, then drop to the ground to pupate. Control with a commercial pyrethrin/rotenone spray. Make sure to spray into the center of the bush. Controlling of the first brood may eliminate the need for another spray.

Leaves blistered and reddened. Cause: Aphids. These tiny insects cluster on leaf undersides. If damage is severe, spray with insecticidal soap or rotenone, making sure to get the undersides of the leaves.

Grape

Vitis spp. and hybrids (Vitaceae)

Grapes, one of the most ancient cultivated crops, are perennial, woody vines that cling to support by means of tendrils. Under ideal conditions vines may live for over 100 years. Fruit is produced on shoots that grow off 1-year-old wood.

There are 4 main types of grapes grown in North America: European, or wine grapes (*Vitis vinifera*); American (*V. labrusca*), such as 'Concord'; hybrids between European and American; and muscadine (*V. rotundifolia*). European grapes generally are most susceptible to diseases. American grapes are the most cold-hardy. Their hardiness (Zones 4–10) depends on the species and cultivar.

Culture

Plant in a sunny site with deep, well-drained, moderately fertile soil and good air circulation to promote disease resistance. Train vines to a fence or trellis. Prune in late winter to keep the fruit within reach, to increase cluster size, and to allow air and sunlight to penetrate the branches. Enclose fruit clusters in paper bags to keep birds away from berries and to control some insects and diseases. Many muscadines need cross-pollination, but the other types are self-pollinating.

Fruit Problems

Young fruit covered with a white coating; fruit ripens unevenly or covered with blotches. Causes: Powdery mildew; downy mildew. Early in the season, fruit infected with powdery mildew has a white, dustlike coating on the berries. Infected plants may set few berries. Later on, the infection can halt the growth of

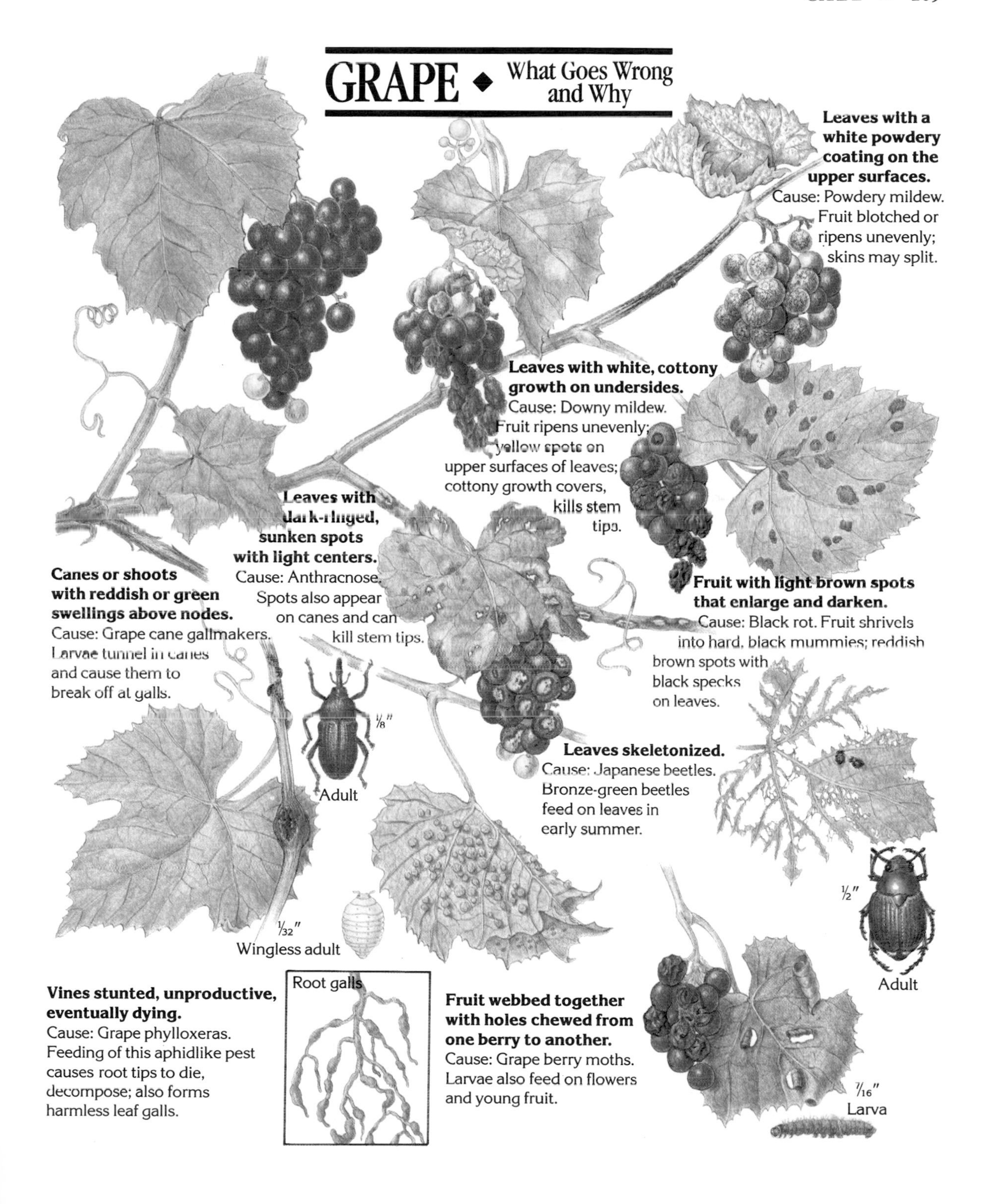
GRAPE ◆ What Goes Wrong and Why
Leaves with a white powdery coating on the upper surfaces.
Cause: Powdery mildew. Fruit blotched or ripens unevenly; skins may split.
Leaves with white, cottony growth on undersides.
Cause: Downy mildew. Fruit ripens unevenly; yellow spots on upper surfaces of leaves; cottony growth covers, kills stem tips.
Leaves with dark-ringed, sunken spots with light centers.
Cause: Anthracnose. Spots also appear on canes and can kill stem tips.
Canes or shoots with reddish or green swellings above nodes.
Cause: Grape cane gallmakers. Larvae tunnel in canes and cause them to break off at galls.
Fruit with light brown spots that enlarge and darken.
Cause: Black rot. Fruit shrivels into hard, black mummies; reddish brown spots with black specks on leaves.
1/8″
Adult
Leaves skeletonized.
Cause: Japanese beetles. Bronze-green beetles feed on leaves in early summer.
1/2″
Adult
1/32″
Wingless adult
Root galls
Vines stunted, unproductive, eventually dying.
Cause: Grape phylloxeras. Feeding of this aphidlike pest causes root tips to die, decompose; also forms harmless leaf galls.
Fruit webbed together with holes chewed from one berry to another.
Cause: Grape berry moths. Larvae also feed on flowers and young fruit.
7/16″
Larva

the berry skins, causing them to split while green. Ripened clusters have blotchy, poor-tasting fruit, but have no white growth.

Ripening fruit clusters infected with downy mildew may have a mix of hard, reddish green, infected berries and soft, juicy, healthy berries. The fuzzy coating on fruit apparent early in the season may not be present. For information on controls for these diseases, see "Leaves with a white powdery coating on the upper surfaces" below for powdery mildew symptoms; for downy mildew, see "Leaves with white, cottony growth on undersides" below.

Fruit with light brown spots that enlarge and darken. Cause: Black rot. Common east of the Rockies, especially in hot, humid weather, black rot causes fruit to shrivel into hard, black shriveled berries (known as mummies) that remain on the cluster. Overwintering mummies and infected canes or shoots carry the disease from one growing season to the next. For control, remove and destroy all mummies. For persistent infection, apply copper sprays. Cultivars moderately resistant to black rot include 'Beta', 'Campbell's Early', 'Cascade', 'Chancellor', 'De Chaunac', 'Delaware', 'Elvira', 'Fredonia', 'Hunt', 'Ives', 'James', 'Scuppernong', 'Sheridan', and 'Worden'. See page 109 for an illustration of this disease.

Fruit with small, dark-ringed, sunken spots with light centers. Cause: Anthracnose. For more information on this disease, also called bird's-eye rot, see "Leaves with dark-ringed, sunken spots with light centers" below.

Fruit clusters enveloped with a fluffy, gray-brown coating. Cause: Botrytis bunch rot. This fungal disease appears in tight fruit clusters and in vines with poor air circulation. For prevention, thin berries within clusters and remove some leaves around fruit. Clean up prunings, cluster stems, and mummified fruit by early spring. Plant resistant cultivars, including 'Baco No.1', 'Cascade', 'Catawba', 'Concord', 'Delaware', 'De Chaunac', 'Fredonia', 'Ives', and 'Niagara'.

When the timing of this infection is perfect, it imparts a special taste to European wine grapes. In this case it is called "noble rot," for the taste it imparts to wine.

Fruit webbed together with holes chewed from one berry to another. Cause: Grape berry moths. This pest, common east of the Rockies, is a green or brown, 7/16" caterpillar that moves from fruit to fruit, feeding on pulp and seeds. Pick and destroy infested berries, which shrivel and color prematurely. Larvae also feed on flowers and newly set fruit. Larvae cut flaps in the edges of leaves and roll them over with webbing to pupate. At the end of the season, collect and destroy or bury fallen leaves, which may harbor overwintering pupae. For heavy infestations, spray BTK to kill the caterpillars. See page 109 for an illustration of this pest.

Fruit covered with a sticky coating and black mold. Causes: Grape whiteflies; grape mealybugs. Whiteflies and mealybugs secrete a sticky coating, called honeydew; sooty mold, a black superficial fungus, feeds on the honeydew. Both whiteflies and mealybugs are serious grape pests in California.

To control whiteflies, remove nearby buckthorns (*Rhamnus* spp.), where the pest overwinters. Or spray nearby buckthorns in the winter with dormant oil spray.

Natural enemies usually keep mealybugs in check, but for heavy infestations, introduce mealybug destroyers (*Cryptolaemus* spp.), also called Australian lady beetles. Buy these natural predators from commercial insectaries or natural farm and garden suppliers. Also, ants protect mealybugs from predators and feed on their honeydew. To make mealybugs more susceptible to predators, control ants by placing bait stations containing boric acid on the ground around vines.

Fruit with holes. Cause: Birds. Birds peck

holes in ripening berries and are especially fond of red or blue cultivars. Damaged fruit attracts bees and wasps for further feeding. Drape a net over the whole vine or bag individual clusters in paper bags.

Leaf and Branch Problems

Leaves with a white powdery coating on the upper surfaces. Cause: Powdery mildew. Infected leaves eventually become distorted, turn brown, and fall. Dark patches appear on canes. This fungal disease weakens vines, diminishes yield, and makes plants more susceptible to winter injury. More common in the West, it spreads fastest when days are dry and warm and nights are cool. For susceptible plantings, apply sulfur spray in the spring. Plant resistant cultivars, including 'Canadice', 'Cayuga White', 'Ives', and 'Steuben'. See page 109 for an illustration of this disease.

Leaves with white, cottony growth on undersides. Cause: Downy mildew. Other symptoms of this fungal disease include small yellow leaf spots, distorted or brown leaves, and early leaf drop. Older leaves are affected first, but the disease also attacks shoots and tendrils. To control downy mildew, remove and destroy all diseased leaves and tendrils in the fall. Since the fungus overwinters on bud scales and shoots, you may have to spray vines with copper several times throughout the growing season to control the disease. Do not spray copper during flowering. Plant resistant cultivars, including 'Aurora', 'Baco No.1', 'Canadice', 'Cascade', 'Concord', 'Foch', 'Himrod', and 'Steuben'. See page 109 for an illustration of this disease.

Leaves with dark-ringed, sunken spots with light centers. Cause: Anthracnose. Also called bird's-eye rot, this fungal disease weakens but usually doesn't kill vines. For control, remove and destroy diseased portions, including infected berries, and spray in the spring with lime-sulfur. Muscadine grapes and the American cultivars 'Concord', 'Delaware', 'Moore Early', and 'Niagara' are anthracnose-resistant. See page 109 for an illustration of this disease.

Leaves with reddish brown spots with black specks. Cause: Black rot. Leaves may wilt and shoots may show large, black, elliptical lesions. For more information, see "Fruit with light brown spots that enlarge and darken" above.

Leaves or leaf petioles with reddish swellings. Cause: Grapevine tomato gall. The ½″ swellings are galls made by a very small fly called a midge. If you slit one of the galls open, you may find small pinkish orange maggots inside. These galls are harmless. Just prune them off.

Leaves with green, pealike swellings on undersides. Cause: Grape phylloxeras. Phylloxeras are aphids that make leaf galls to hold eggs and young. Inside 1 gall, you may find hundreds of yellowish nymphs. Grape phylloxeras are native to the United States and are tolerated by American grapes, but they're serious pests of European grapes. In fact, when phylloxeras were accidentally transported to France in the late 1800s, they spread rapidly through vineyards and killed almost a third of the grapevines there. To avoid problems with phylloxeras, plant American or French hybrid grapes, both of which are resistant to this pest. Phylloxeras also attack roots of susceptible species, causing death of the whole vine. See page 109 for an illustration of this pest.

Leaves with pale stipples along leaf veins. Cause: Leafhoppers. These green or brown, $1/10''$–$1/2''$ insects suck juices from leaf undersides, causing foliage to be stippled with tiny white spots. Heavily infested leaves may turn yellow or brown and drop from the vine. Though they are quite common, leafhoppers rarely cause serious damage to backyard grapes. Many natural enemies keep them in check. Where further control is needed, spray leaves with insecticidal soap.

Leaves skeletonized. Cause: Japanese beetles. These 1/2″, metallic blue-green insects with bronze wing covers feed on leaves in early summer. For light infestations, visit plants in the morning while beetles are sluggish and knock them off leaves into jars filled with soapy water. For more information, see "Leaves skeletonized" on page 46. See page 109 for an illustration of this pest.

Canes or shoots break easily. Cause: Grape cane girdlers. This 1/8″ beetle punctures a cane, lays a single egg inside, and then encircles the cane with 2 rows of punctures. Damaged canes or shoots then break off easily. Damage is usually minor, but to control this pest, remove and destroy the injured cane a few inches below the area with puncture marks.

Canes or shoots with reddish or green swellings above nodes. Cause: Grape cane gallmakers. This 1/8″ beetle deposits a single egg in a cane and then makes additional punctures in a vertical row above the original egg-laying site. The area where the egg is deposited enlarges to a red or green gall about twice the diameter of the cane. Larvae tunnel in canes, and canes often break off at the galls. Damage is usually minor, but to control this pest, remove and destroy the injured cane a few inches below the gall. See page 109 for an illustration of this pest.

Whole Plant Problems

Vines stunted, unproductive, eventually dying. Causes: Grape phylloxeras; grape scale; Pierce's disease. Grape phylloxeras are aphids that infest both leaves and roots. Leaf infestation causes harmless, pealike leaf galls. Root infestation causes knotlike root galls that prevent nutrient uptake and cause stunting and death of the vine.

Phylloxeras are difficult to eradicate. Where they are a problem, plant American grapes or European grapes grafted onto American grape rootstock, both of which are resistant to this pest. For more information on grape phylloxeras, see "Leaves with green, pealike swellings on undersides" on page 111. Also see page 109 for an illustration of this pest.

Grape scales are tiny, round, immobile insects resembling light gray bumps. Scales usually hide under the loose bark of older canes or trunks, where they suck sap and cause the vine to slowly decline. For control, spray with dormant oil in late winter and prune old growth severely.

Pierce's disease, a bacterial disease common in the South, is spread by leafhoppers. Infected vines typically show scorched, dried leaves in midsummer, wilted dried fruit, and eventual death of the vine. There is no cure for this disease. Dig up infected plants and replant with disease-free stock. Muscadine grapes are resistant to Pierce's disease, as are the American grape cultivars 'Champanel', 'Herbemont', and 'Lenoir'.

Gypsophila

Baby's-breath, gypsophila. Perennials.

The delicate white or pink flower sprays of baby's-breath (*Gypsophila paniculata*) are a mainstay of floral arrangements. Sprays may have more than 1,000 single or double flowers, each just 1/16″ wide. This mounding plant is equally delightful in the garden where it grows 2′-3′ tall and 2′ wide, bearing copious blooms over slender, gray-green leaves. Creeping baby's-breath (*G. repens*) forms a blossom-covered mat only 4″-8″ tall and is useful for edging or rock gardens.

Plant baby's-breaths in sites where they won't be disturbed or moved; their large fleshy roots do not transplant well. Given full sun, alkaline soil, and good drainage, plants thrive with little care. Remove spent flowers to prolong bloom. Protect plants with winter mulch; don't cover crowns until after the ground is frozen to avoid rotting.

Problems

Leaves stippled; growth poor. Cause: Leafhoppers. Slender, wedge-shaped, greenish yellow pests only 1/8″ long, aster leafhoppers suck sap from the undersides of leaves as they feed, discoloring and distorting the foliage. Damaged leaves shrivel and drop. Encourage predators such as bigeyed bugs and parasitic wasps, or spray with insecticidal soap. In severe cases treat plants with a commercial pyrethrin/rotenone mix.

Leaves and flowers greenish yellow, distorted; new growth spindly. Cause: Aster yellows. Feeding leafhoppers may transmit this disease. There is no control for aster yellows. Remove and destroy infected plants; do not compost them. Prevent the disease spread by controlling sucking pests such as leafhoppers and aphids.

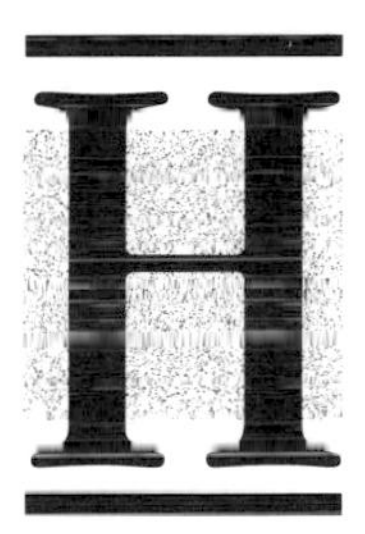

Hamamelis

Witch hazel. Shrubs and small trees.

Witch hazels are alternate-leaved shrubs with narrow-petaled, twisted, yellow or coppery flowers appearing either in late fall or late winter. They flourish in well-drained but moisture-retentive soils, enriched with lots of organic matter.

Witch hazels are relatively free of serious insect pests. Caterpillars and Japanese beetles may cause some damage; see "Leaves skeletonized or with large holes; branches may be webbed" and "Leaves skeletonized" on page 236 for controls. Witch hazel cone gall, which appears as a conical gall on the upper leaf surfaces, is caused by a kind of aphid, as are elliptical galls on the flower buds. Neither is usually a major problem; control both by repeated applications of insecticidal soap in late spring, early summer, and autumn.

Hedera

Ivy. Vines.

Ivies are vigorous evergreen vines, climbing by means of rootlike appendages. They are widely planted as groundcovers, as vines on walls, or as houseplants.

Set out bare-root plants in spring or fall. Almost any moisture-retentive soil will suit ivy. In the North, strong winter sun can scorch leaves; it's best to choose a site where some shade is available. Cut back in late winter to promote denser growth.

Problems

Leaves wrinkled and discolored. Cause: Aphids. For controls, see "Leaves wrinkled and discolored" on page 235.

Leaves yellow; stems and leaves covered with small bumps. Cause: Scales. See "Leaves yellow; stems and leaves covered with small bumps" on page 237 for control information.

Leaves stippled with yellow; foliage webbed. Cause: Spider mites. See "Leaves stippled with yellow; foliage webbed" on page 236 for controls.

Leaves with circular brown spots. Cause: Bacterial leaf spot. This bacteria can cause water-soaked spots on lower leaves. These spots turn brownish black and may spread to form large patches of dead tissue. To prevent the spread of this disease, avoid working around wet plants. Remove and destroy affected parts; spray bordeaux mix on remaining plants.

Trunk or branches with oozing lesions; branch tips die back. Cause: Canker. See "Trunk or branches with oozing lesions; branch tips die back" on page 238 for controls.

Leaves with powdery white coating. Cause: Powdery mildew. For controls, see "Leaves with powdery white coating" on page 237.

Helianthus

Sunflower. Annuals.

Rapid growth, showy flowers, and tasty seeds make sunflowers fun to grow as well as nutritious. The plants can grow to 10′ tall. Individual flowers may reach 14″ across, with golden petals and a brown center that becomes a mass of seeds by late summer.

Sunflowers need warmth to germinate. Direct-sow seeds ½″ deep in average soil in early May. They will germinate within 10 days. Space tall-growing cultivars 2′ apart, or they will crowd each other. Sunflowers perform well in average soil and full sun, although they will tolerate light shade. Apply a general-purpose fertilizer once, about midway through the growing season.

Problems

Leaves, stems, and buds distorted. Cause: Aphids. For control measures, see "Leaves, stems, and buds distorted" on page 20.

Leaves with ragged holes. Cause: Caterpillars. Caterpillars of all kinds find large sunflower leaves a tasty food supply. A healthy plant can generally withstand an attack. Handpick pests, or spray leaves with BTK.

Leaves wilt. Causes: Lack of water; fungal wilt. If a good soaking does not perk up your plant by the next day, suspect a fungal wilt. These fungi live in the soil and move upward through the plant. Leaves may be mottled with green and yellow, and dark brown areas may appear on and within sunflower stems. There is no control; remove and destroy infected plants. Avoid replanting sunflowers in that area.

Leaves with powdery white patches. Cause: Powdery mildew. For controls, see "Leaves with powdery white patches" on page 19.

Helichrysum

Strawflower. Annuals.

These 1½′-3′ Australian natives are best known as an integral part of dried flower arrangements. If picked when partially opened, strawflowers hold their color well and last for several years indoors. Colors include white, yellow, salmon, red, and pink.

Sow seeds indoors 6-8 weeks before the last outdoor frost. Do not cover seeds; they need light during the 7-10 day germination period. Transplant seedlings 9″-12″ apart. Where summers are long, direct-sow strawflowers in late spring or early summer. Plant in full sun in well-drained, moderately fertile, slightly moist soil. They are fairly drought-tolerant once established. Taller cultivars may need staking. Little fertilizer is needed; too much fertilizer results in many leaves and few flowers. Pests and diseases are rarely a problem.

Helleborus

Hellebore. Perennials.

Hellebores flower in winter and early spring, earning names such as Christmas rose (*Helleborus niger*) or Lenten rose (*H. orientalis*). Most have green, creamy white, or dull purple blooms, carried over low evergreen foliage.

Plant in partial shade, such as that found under deciduous trees, in moist, well-drained, neutral soil rich in organic matter. Maintain even soil moisture with a summer mulch. The brittle rhizomes make this plant difficult to divide or transplant; if you must move them, do so with care. Dark brown to black leaf spots, yellowing leaves, and stem cankers indicate fungal infection; to prevent, spray with bordeaux mix or sulfur.

Hemerocallis

Daylily. Perennials.

Favored for their ability to thrive under tough conditions, daylilies grace many urban plantings. There are thousands of hybrids from which to choose. Plants range in height from 1½′ to 3′ and offer blossoms in every color but true white and blue. A mature planting provides hundreds of 2″-6″, trumpet-shaped flowers and graceful, swordlike, bright green leaves. Each flower lasts a day, then closes at night, but numerous buds provide extended periods of bloom.

Daylilies are easy-to-grow, nearly problem-free plants that perform best in full sun. Most also tolerate partial shade; in fact, plants may grow better in some shade in southern zones. Plant in well-drained, average soil and water regularly to get the plants established. After that, limited watering is fine, but don't let them dry out while blooming. Excess fertility encourages lush foliage and fewer flowers. Plants grow vigorously if undisturbed; clumps expand over time but are usually not invasive. Remove spent flower stalks to prolong bloom. In fall, cut all stems to the ground and remove dead foliage. Divide large clumps every 3-5 years. Roots are tough, heavy, and tuberous, so this is not an easy job. Leave a crown bud and several solid roots on each division.

Problems

Flower buds die; stems have corky lesions. Cause: Flower thrips. These 1/25″ insects occasionally infest daylilies, feeding on buds, stem tips, and flowers. Infested plants have distorted blooms, and in severe cases, flower stalks fail to develop. Stems may develop corky lesions that extend several inches. Thrips are difficult to control because they burrow into plant tissue. Use blue sticky traps to monitor populations and trap pests. Pick off and destroy infested plant parts, and encourage natural predators such as pirate bugs, lacewings, and lady beetles. Regular sprays of insecticidal soap may give some control once pests are spotted on traps.

Leaves stippled, reddish to yellow, with fine webbing. Cause: Spider mites. These 1/50" arachnids feed on the undersides of leaves, causing them to yellow, dry up, and die; excessive feeding turns foliage almost white and fine webs cover leaves and growing tips. Mites seek out water-stressed plants in hot, dry weather; adequate watering guards against them. Hose off plants when symptoms first appear; a strong stream knocks mites off plants and may give control. Introduce predatory mites; spray severe infestations with avermectins, citrus oils, insecticidal soap, neem, pyrethrin, or rotenone.

Herbs

Herbs are grown for their savory, aromatic, or medicinal parts. They include annuals, biennials, and perennials. In general, herbs are little troubled by pests and diseases if given good growing conditions.

To grow healthy, trouble-free plants, choose a well-drained site with as much sun as possible. Most do not need rich soil, and some actually prefer dry, poor soil; check individual entries for specific preferences.

Whole Plant Problems

Seedlings fall over; stems girdled or rotted at soil line. Cause: Damping-off. This disease is caused by soil-dwelling fungi. To prevent it, disinfect reused pots and flats before filling them with fresh seed-starting mix. To disinfect, dip them in a 10 percent bleach solution (1 part bleach to 9 parts water) and let them air-dry. Sow seed thinly to allow for air movement around seedlings. Cover seed with a thin layer of soilless mix or vermiculite. Water only enough to keep soil moist, not soggy.

Seedlings clipped off at soil line. Cause: Cutworms. Check for fat, 1" long, brown or gray caterpillars in the soil near the base of plants. Once they chew a seedling in two, there is nothing you can do except protect the remaining seedlings from their nocturnal attacks. Place cutworm collars around transplants, sprinkle moist bran mixed with BTK on the soil surface in the evening, or add parasitic nematodes to the soil to control.

Plant yellow and stunted, wilts during bright, hot days; roots have swollen galls. Cause: Root knot nematodes. These microscopic, wormlike creatures invade and feed on plant roots. Pull and destroy infected plants. Control pest nematodes by adding chitin or parasitic nematodes to the soil. Solarizing problem areas also reduces future problems.

Plant wilts; cut-open roots are discolored. Cause: Verticillium or Fusarium wilt. There is no cure for these diseases. They are caused by a number of fungi, each of which attack only specific plants or plant families. Pull and destroy infected plants. Avoid planting susceptible members of the same plant family for 3 or more years in soil where disease symptoms have developed.

Leaf Problems

Leaves turn yellow, older ones first; plant stunted. Cause: Nitrogen deficiency. This may be caused by a low level of nitrogen in the soil or by waterlogged soil. Waterlogged soil damages roots and prevents them from using nutrients available in the soil. Check soil moisture, and reduce watering if soil is soggy. Spray foliage with compost tea or fish-meal tea, or side-dress plants with compost to alleviate symptoms. Prevent problems by choosing well-drained sites, adding organic matter to the soil to add nitrogen and improve drainage, and by planting in raised beds.

Leaves yellow and curled; new growth distorted. Cause: Aphids. These small, soft-bodied, green, black, gray, pink, or white fluffy-coated insects suck plant juices. They leave a sticky honeydew on leaves and fruit that in turn supports the growth of black, sooty molds. Spray with a strong jet of water or insecticidal soap if populations are low. If populations are high, spray with a commercial pyrethrin/rotenone mix.

Leaves yellow; plant wilts; stems, crowns, or roots water-soaked and rotted. Cause: Stem, crown, or root rot. These diseases are often due to poor drainage or overwatering. Destroy infected plants or plant parts. Thin plants to increase air movement and reduce moisture around plants. Spray with a copper fungicide if weather is warm and wet and disease is severe. Plant in well-drained soil or in raised beds.

Texas root rot occurs in warm climates and is favored by high temperatures and a pH above 7.0. If pH is high, add sulfur to the soil to lower it. Use a 3-year rotation to starve the fungus out of the soil.

Leaves with dark, yellow-bordered spots. Cause: Leaf blight. This disease is caused by various fungi. Leaves may turn yellow or drop off as the disease progresses. Spray foliage with fish emulsion or with sulfur to prevent the spread of mild infections.

Lower leaves covered with tan, fuzzy growth. Cause: Gray mold. This fungal disease attacks a wide range of edible and ornamental plants. Pick off and destroy moldy leaves. Thin plants to increase air movement and reduce moisture around leaves, since the mold thrives in damp conditions. Spray foliage with compost tea to control mold. Spray with sulfur if the weather is wet and cool and the disease is severe.

Leaves with powdery white growth on upper surface. Cause: Powdery mildew. This symptom is caused by a number of fungi, each of which attacks only specific plants. Thin plants to increase air movement and reduce moisture around leaves. Spray plants with compost tea, sulfur, or bordeaux mix to prevent the spread of mild infections.

Heuchera

Alumroot, coral bells. Perennials.

Although no other species in this genus compares with the dainty blossoms of coral bells (*Heuchera sanguinea*), several are worthy additions to the perennial border. Low-growing clumps of rounded, lobed, heart-shaped, or maplelike foliage are usually evergreen and interesting throughout most of the year. Cultivars of American alumroot (*H. americana*) and small-flowered alumroot (*H. micrantha*), both grown more for foliage than flowers, offer leaves with purple coloring.

Plant alumroots in well-drained, richly organic soil in a sunny or lightly shaded spot. Partial shade is best for foliage and for plants in the South. Shallow, fleshy roots need regular watering and benefit from summer mulching; however, excess moisture, combined with poor drainage, promotes rots.

Problems

Leaves covered with white powder. Cause: Powdery mildew. Powdery mildew may mar foliage and is best limited through sanitation and air circulation. See "Leaves covered with white powder" on page 177 for controls.

Crowns blacken and die; leaves notched. Cause: Strawberry root weevils. The grubs of 1/4", black snout beetles feed on alumroots' crowns starting in early spring. Adults feed at night on foliage. Remove and destroy infested plants; clean up debris where pests overwinter. Drench soil around roots with parasitic nematodes; spray pyrethrin or rotenone on leaves at night to kill adults.

Hibiscus

Hibiscus, rose-of-Sharon.
Shrubs and trees.

These old-time favorites have bushy but upright habits and bloom in mid- to late summer. They are easy to grow, even withstanding seashore conditions. Avoid unwanted volunteer seedlings by planting sterile cultivars.

Set out in spring or fall as bare-root or container-grown plants. They prefer full sun and average soil with added organic matter.

Hibiscus are fairly problem-free. Aphids or Japanese beetles may feed on foliage; see "Leaves wrinkled and discolored" on page 235 and "Leaves skeletonized" on page 236.

Hosta

Hosta, plantain lily. Perennials.

Grown primarily for their attractive, 5"-10", broadly lance-shaped leaves and durability as groundcovers, hostas are unparalleled perennials for shady areas. Handsome ribbed leaves may be short and narrow or broad and up to 1½' long; leaf colors range from light to dark green or blue-green and may be variegated with yellow, light green, cream, or white. Spikes of tubular, 2", white to purple flowers show nicely above foliage from mid- to late summer.

Hostas look their best in light to deep shade. Sunny sites encourage flowering but also cause foliage color to bleach and brown and sunburned spots to appear on leaves. Set plants 1'-3' apart, depending on leaf size, in moist but well-drained, organic soil. Give plants ½"-1" of water weekly; mulch to avoid splattering water onto leaves.

Hostas may be increased by division, although plantings will prosper undisturbed for many years. If desired, divide plants every 5 years or so. Remove faded flower stalks to improve appearance, and cut back foliage at the end of the growing season.

All hostas go dormant in winter. New leaves emerge in spring. Try not to select heavily puckered leaf forms for use under trees. The puckers or wrinkle designs catch leaf drippings, including gum and sap, which detract from the plants' appearance.

Problems

Leaves small, light brown, papery, scorched, or bleached. Causes: Lack of water; too much sun. An easy way to ruin hostas is to let them dry out, even for a little while. Even if they continue to grow, plants will be stunted. Sunburned edges or spots reduce hostas ornamental value significantly and are best avoided through proper selection of planting site.

Leaves with large, ragged holes. Cause: Slugs and snails. Hostas' broad, low-growing foliage welcomes these slimy pests. See "Leaves with large, ragged holes" on page 177 for controls.

Stems blacken at base; leaves yellow and wilt; plant collapses. Cause: Crown rot. Hostas are susceptible to fungal and bacterial attack when grown in poorly drained soil. Crown injury and excess moisture during dormancy also increase the chances of infection. Choose well-drained sites; add organic matter to improve soil drainage; keep winter mulch away from crowns. Remove and discard infected plants and the surrounding soil.

Houseplants

The world of houseplants covers a vast array of plant species from all over the world. And any enthusiast's collection may include species native to a wide variety of climates, from tropical jungle-dwellers, such as some orchids or philodendrons, to desert-dwelling cacti. Despite the diverse cultural requirements different houseplants prefer, there are basic techniques that will ensure healthy plants that resist pests and diseases.

Culture

Since houseplants live in an artificial environment, maintaining pest- and disease-free plants requires attention to 4 main areas of care: light, temperature, humidity, and water. As with plants grown outdoors, healthy houseplants are less susceptible to attack by pests and diseases. Plants stressed by light exposure, temperature extremes, improper watering, or low humidity are more prone to problems than those provided with conditions that meet their basic requirements.

Unless you own an artificial lighting system especially for houseplants, choose plants with light needs similar to the natural light in your home. Too little light on a sun-loving plant causes pale leaves and leggy growth. Too much sun on a plant that prefers low light causes leaf scorch, yellowing, and dehydration. North-facing windows provide indirect sun and the lowest light level; they're also often the coolest exposures. East- and west-facing windows provide moderate amounts of bright light; west-facing windows, which receive hot afternoon sun, tend to be warmer than east-facing ones. South-facing windows provide the brightest and longest-lasting light and also the most heat. Keep in mind that overhanging awnings, drapes, or nearby trees affect the amount of light plants receive.

As a general rule, plants grow best at daytime temperatures of 60°-80°F and nighttime temperatures of 58°-65°F. Avoid wide temperature swings; keep plants away from heat vents, radiators, air-intake ducts, and air-conditioning vents.

Low humidity is a familiar problem to indoor gardeners. During the heating season, indoor humidity may only be 10 percent. Most tropical species require 40-60 percent humidity to thrive. The easiest way to increase humidity is to run a whole-house or portable humidifier. Not quite as efficient, but still worthwhile, is the practice of keeping a pan of water near a heat source. Grouping plants together also improves indoor humidity because leaf transpiration from a cluster of plants increases the air moisture in the immediate area. To further humidify a plant grouping, set plants on trays filled with 1″ of pebbles and ½″ of water.

Most houseplants should be watered only when the soil is dry. Test soil dryness by digging your finger an inch into the soil. Water only if this soil is dry to the touch. Don't allow excess water to stand in saucers, and avoid wetting foliage. Improper watering may cause wilting that can be mistaken for diseases.

Plants growing in a commercial potting mix need little additional fertilizer if repotted regularly. During active growth in spring and summer, most plants benefit from regular doses of a liquid organic fertilizer such as fish emulsion; mix and apply according to label directions. Fertilize less during the winter, when lower temperatures and light levels slow the growth of most plants.

To catch problems before they become serious, it's a good idea to inspect plants regularly for signs of insects or diseases. Rinsing the plant occasionally with room-temperature water in the sink or bathtub is also a good idea: It rinses away insects along with dust that can inhibit photosynthesis.

Problems

Leaves spotted, yellowed, browned, or wilted; roots, stems, or leaves with rotted spots. Cause: Cultural problems. Too much or too little water, persistent hot or cold drafts, improper light exposure, potbound roots, or nutrient starvation can all cause symptoms that can be mistaken for pests or diseases.

To diagnose cultural problems, review the conditions the plant prefers and examine the plant. Also consider the amount of light the plant receives, review watering patterns, and look for drafts or other problems that may not be immediately apparent. Once you've decided on a cause, correct the situation. See "Culture" on page 119 for more information.

Leaves and stems with white fuzzy growth that later turns brown. Cause: Powdery mildew. Cut out all diseased parts; place the plant in an area where it will receive excellent air circulation. If necessary, use a small fan to move air around the plant.

Leaves and stems with white cottony patches. Cause: Mealybugs. Isolate affected plants; dip a cotton swab in 70 percent isopropyl alcohol and wipe each insect. A few days later, rinse the plant in the shower, or outdoors (in shade) with a fine mist of water from a hose. Check for mealybugs again after a month and re-treat if necessary.

Leaves pale and discolored; tiny brown bumps on undersides and stems. Cause: Scales. These immobile insects suck fluids from plant cells. They may excrete sticky honeydew onto nearby leaves. Isolate affected plants. Remove heavily infested leaves. Gently scrape insects from stems with your fingernail; spray plants with insecticidal soap. (Don't use soap on ferns.) After a week, rinse plants in water. If the problem persists, wipe leaf undersides with a solution of 1 tablespoon vegetable oil in ¾ cup warm water.

Leaves pale or stippled with fine webbing on undersides. Cause: Spider mites. These spiderlike creatures, no larger than a speck of dust, suck juices from plant leaves. They spread most rapidly in hot, dry conditions. Isolate affected plants; ideally, move them to a more humid location. Remove heavily infested leaves and stems. Spray thoroughly with insecticidal soap (do not use soap on ferns), then mist with water daily for a week.

Leaves crinkled or curled with a sticky coating on undersides. Cause: Aphids. These tiny, sucking insects secrete a sugary honeydew that may attract ants. Isolate affected plants. Remove heavily infested leaves and stems. Wipe insects off stems with a cotton swab dipped in warm soapy water. Spray with insecticidal soap. (Do not use soap on ferns.)

Leaves with dark, sooty blotches. Cause: Sooty mold. This fungus lives on the honeydew excreted by aphids, mealybugs, and scales. Treat the plant for the insect infestation. Then clean the foliage with warm soapy water and rinse. See "Leaves and stems with white cottony patches," "Leaves pale and discolored; tiny brown bumps on undersides and stems," and "Leaves crinkled or curled with a sticky coating on undersides" above for information on controlling these insects.

Leaves, stems, or flowers with gray, moldy growth. Cause: Botrytis blight. This fungal disease thrives on plants that have been overwatered, overfed with nitrogen, or overcrowded. Remove and discard affected plant parts. If the plant has tender growth due to excess fertilizer, withhold feeding until current growth hardens off. Divide overgrown or crowded plants, improve air circulation, and withhold water until soil is nearly dry.

Hyacinthus

Hyacinth. Bulbs.

Cultivars of *Hyacinthus orientalis,* the only species in this genus, are prized for the heady aroma they bring to the spring garden. Tubular or bell-shaped blossoms cover dense, blunt, 10″-12″ tall, upright spikes; each bulb produces 1 spike of white, pink, blue, or yellow flowers. Narrow, straplike leaves surround the base of the stalk.

Plant bulbs in fall at least 1 month before the ground freezes. Select a site in full sun with humus-rich, well-drained soil; sheltering from wind reduces injury to rigid flower stalks. Set bulbs 4″-5″ deep with pointed growing tips facing upward. Hyacinths multiply less freely than other bulbs and flowering tends to diminish with age. Plant new bulbs every 2-3 years for optimal floral display. Remove spent flowers to prevent seed formation; let foliage die back naturally.

Problems

Leaves yellow or distorted; bulbs decayed. Cause: Bulb mites. Like most bulbs, hyacinths may be infested by bulb mites. See "Leaves yellow or distorted; bulbs decayed" on page 50 for controls.

Leaves and stems with spots, gray mold, or yellow slime; bulbs decayed; growth poor. Cause: Bacterial and fungal rots. Hyacinths are subject to a number of rot diseases. Rots are prevalent in poorly drained soil; control through careful bulb selection and culture. For tips on buying healthy bulbs, see the illustration on page 49.

Leaves with swollen spots; plant fails to grow. Cause: Bulb and stem nematodes. Tiny roundworms feed on bulbs; plants fail to grow in spring or fail to bloom. Swollen, yellow-green spots appear on leaves; an infested bulb, cut in half, reveals dark blotches or rings. Wet soil encourages nematodes. Dig and destroy infested bulbs; check new bulbs carefully for pests. Plant in well-drained soil.

Hydrangea

Hydrangea. Shrubs and vines.

Hydrangeas are opposite-leaved, deciduous shrubs and vines. The flowers are small but borne in large clusters. Hydrangeas are ideal for shrub borders or grouped for landscape interest.

Hydrangeas are among the few flowering shrubs that bloom well under shaded conditions, although they flourish in full sun in cooler climates. They prefer acid soils. Set out in spring or fall as balled-and-burlapped or container-grown plants.

Problems

Leaves wrinkled and discolored. Cause: Aphids. For control information, see "Leaves wrinkled and discolored" on page 235.

Leaves tied together over flower buds. Cause: Hydrangea leaftiers. The ½″, green-bodied, brown-headed caterpillar causes the foliage to become ragged, turn brown, and die. If the infestation is light, break open the "envelopes" and pick off the caterpillars; for heavier infestations, spray plants with BTK.

Leaves deformed; shoot tips blackened. Cause: Tarnished plant bugs. This pest is an active, ¼″, light green to brown bug mottled with white, yellow, red, and black. When it feeds on the leaves, it produces a toxin that deforms the leaves and blackens the terminal shoots and flowers. Tarnished plant bug overwinters on weeds and under leaves; removing these havens will greatly reduce the likelihood of subsequent infestation. Control severe infestations by dusting leaves with rotenone.

Leaves with powdery white coating. Cause: Powdery mildew. See "Leaves with powdery white coating" on page 237 for controls.

Flowers rot. Cause: Botrytis blight. In wet seasons, *Botrytis* fungi can cause a blight on hydrangeas that will spoil the flowers. Pick off and destroy affected blooms.

Leaves stippled with yellow; foliage webbed. Cause: Spider mites. See "Leaves stippled with yellow; foliage webbed" on page 236 for suggested controls.

Leaves and flowers skeletonized. Cause: Rose chafers. This 1/3″, reddish brown beetle with thick, yellowish hairs on its wing covers appears in late spring and damages both leaves and flowers. Handpicking is the best control. Reduce populations of larvae by treating lawn areas with milky disease spores. Spray plants with pyrethrin or rotenone as a last resort.

Leaves and flowers browned and wilted. Cause: Bacterial wilt. This disease can spread quickly, killing the plant in hot weather. Remove infected parts as soon as you see them; destroy seriously infected plants.

Leaves with small brown spots. Cause: Rust. Infected leaves have bright orange spots underneath and pale yellow spots on the upper surfaces; they eventually wilt and drop. Remove and destroy leaves infected with this fungal disease. As a last resort, spray or dust with sulfur, repeating as needed.

Iberis

Candytuft. Perennials.

These very hardy evergreen border plants brighten spring with clusters of tiny blossoms of pure or pink-tinged white. Evergreen candytuft (*Iberis sempervirens*) grows to about 1′ tall with a spread of 1½′-2′, while rock candytuft (*I. saxatilis*) is even more compact, reaching heights of 3″-6″. The 1½″-2″ inflorescences last several weeks.

Plant candytufts in full sun in well-drained soil. Light shade is tolerated but reduces flowering, as does drought. Cut plants back after flowering to encourage compact growth. Foliage may brown in cold, windy conditions; mulch lightly to reduce injury. Few pests or diseases trouble candytufts.

Ilex

Holly. Trees and shrubs.

Hollies are deciduous or evergreen trees or shrubs with alternate leaves. The male and female flowers are borne on different plants; female plants produce striking red or black fruit. Hollies are useful as foundation shrubs, accent plants, informal hedges, and specimen plants.

Hollies prefer a moist, acid soil well-enriched with organic matter. Plant in sun or shade. Set out as balled-and-burlapped or container-grown plants in spring or fall; make sure that the top of the root ball is the same level at which the plant grew in the nursery. Their roots grow close to the surface, so don't cultivate around them; a deep organic mulch will keep down weeds and provide a cool, moist root zone.

Protect evergreen hollies from winter sun and drying winds with a burlap screen or an antitranspirant spray. Remember to plant both male and female plants, so you'll get the pollination necessary for berries.

Problems

Leaves tunneled. Cause: Leafminers. This tiny insect is by far the most serious pest of hollies. For control information, see "Leaves tunneled" on page 237.

Leaves webbed together. Cause: Bud moths. These 3/8", greenish white caterpillars web together and feed on the tips of new holly shoots in May. Fully grown larvae usually drop to the ground and overwinter in plant debris. Remove fallen leaves around plants. Destroy webs and handpick caterpillars. Spray leaves with BTK in April and May.

Leaves yellow; stems and leaves covered with small bumps. Cause: Scales. See "Leaves yellow; stems and leaves covered with small bumps" on page 237 for control information.

Leaves stippled with yellow; foliage webbed. Cause: Spider mites. For controls, see "Leaves stippled with yellow; foliage webbed" on page 236.

Leaves with notched margins. Cause: Japanese weevils. The 1/4", brown adults attack foliage, and the legless white grubs feed on roots. Apply a sticky substance to the lower stems to prevent adults from climbing up the plants. Drench the soil around the base of the plant with a solution of parasitic nematodes to control weevil larvae. Spray leaves several times with pyrethrin for major infestations of adult weevils.

Leaves yellowed. Cause: Nitrogen deficiency. This is usually most noticeable on older leaves. Regular fertilization helps keep hollies green and vigorous. Apply cottonseed meal or a generous mulch of compost in early spring.

Leaves and shoots blackened; leaves with moist or brown sunken spots. Cause: Anthracnose. See "Leaves and shoots blackened; leaves with moist or brown sunken spots" on page 238 for more details.

Leaves with spots. Cause: Leaf spots. See "Leaves with spots" on page 237 for controls.

Leaves with powdery white coating. Cause: Powdery mildew. For control information, see "Leaves with powdery white coating" on page 237.

Trunk or branches with oozing lesions; branch tips die back. Cause: Canker. See "Trunk or branches with oozing lesions; branch tips die back" on page 238.

Leaves mottled yellow. Cause: Whiteflies. These tiny, white, mothlike insects and their even smaller larvae feed on the undersides of leaves, weakening the plant. Spray leaves (make sure you get the undersides, too!) with insecticidal soap, or pyrethrin as a last resort; repeat weekly until pests are under control.

Impatiens

Impatiens, balsam.
Perennials grown as annuals.

In American gardens, the 2 most widely grown species of the genus *Impatiens* are garden balsam (*I. balsamina*) and impatiens (*I. wallerana*). Garden balsam grows upright to 2′ with a 1½′ spread. The clustered flowers resemble small double camellias. Impatiens reach 6″-24″ high and about 1′ wide, with open, single flowers. Flower colors include white, red, rose, and orange.

While many gardeners prefer to buy nursery-grown plants, the best color choices usually come in seed packets. Sow seed indoors 8 weeks before last frost. Do not cover the small seeds with soil; they need light to germinate. Germination time is 5 days for balsam, 2 weeks for impatiens. Transplant outdoors 2 weeks after last frost.

Balsam can grow in full sun, while impatiens prefer some shade, especially in warm weather. These otherwise rather care-free plants do not tolerate cold, wet weather. They do require slightly moist soil with lots of organic matter. Water regularly to keep the soil moist but not soggy. Feed balsam once a month to maintain the dark green leaf color. Impatiens don't need extra feeding; too much fertilizer gives ample foliage but few flowers. Pinch back young plants for denser growth. Plants often self-seed.

Problems

Leaves, stems, and buds distorted. Cause: Aphids. For controls, see "Leaves, stems, and buds distorted" on page 20.

Leaves with large, ragged holes. Cause: Slugs and snails. See "Leaves with large, ragged holes" on page 18 for controls.

Seedlings die. Cause: Damping-off. See "Seedlings die" on page 20 for controls.

Ipomoea

Morning glory. Annuals.

These quick-growing tropical vines have white, pink, or blue, trumpet-shaped flowers that may reach 3″ wide. Older types close by midday, but new cultivars stay open longer. Vines may reach 10′ high within 2 months of seeding and flower profusely from July until November. Use morning glories to beautify a fence or wall, or as a temporary groundcover.

Direct-seed outdoors in early April. Soak the hard-coated seed in tepid water for 8 hours, or notch them with a file to speed germination. Place in loose, well-prepared, average soil. Or start seeds indoors in 4″ pots and transplant outdoors after frost.

Full sun is best. Plants require little water or fertilizer once established. Insects and diseases rarely attack plants.

Iris

Iris, flag. Bulbs or rhizomes.

Irises offer a huge range of colors and patterns, heights, and bloom times, with variations on a common flower shape and plant form. The basic iris flower consists of 3 inner (often erect) petals, called standards, surrounded by 3 outer petals (usually arching out), called the falls. Long, flat leaves resemble swords or grass; they grow in rather open to quite dense upright or arching clumps from bulbs or creeping rhizomes.

By far the most popular group is the large collection of hybrids termed the bearded irises, named for the hairy, caterpillar-like feature creeping out of the center of each fall. Flowers range from 2″ to 7″ wide in one of the widest color ranges of any plant group, lacking only pure red. They bloom in early summer, from 2″ to nearly 5′ above stiff, swordlike leaves.

In place of a beard, "beardless" irises flaunt a colorful spot, called a signal, or an intricate pattern of lines. Blooms on Siberian iris (*Iris sibirica*) rarely exceed 3″ wide; they occur in shades of white, red-violet, blue, and purple (occasionally pinks and yellows) in upright, grassy clumps averaging 3′ tall. They bloom as bearded irises stop flowering. Japanese irises (*I. ensata,* formerly *I. kaempferi*) bear 4″-10″ flattish or double flowers in shades of white, pinkish lavender, red-violet, blue, and violet, often edged, lined, or speckled. Most grow to about 3′ and bloom a few weeks later than Siberian irises.

Very early spring-blooming, bulbous reticulated iris (*I. reticulata*) hybrids have fragrant, narrow-petaled, 3″ blooms, mostly in blues and purples with orange or yellow signals, amid sparse, 4-sided leaves that grow to 1½′ after bloom.

Most bearded irises are easy to grow, but they do have specialized needs. Plant and divide (every 3-4 years) in summer or early fall, splitting them into individual "fans" with the rhizome attached, or into divisions with a few

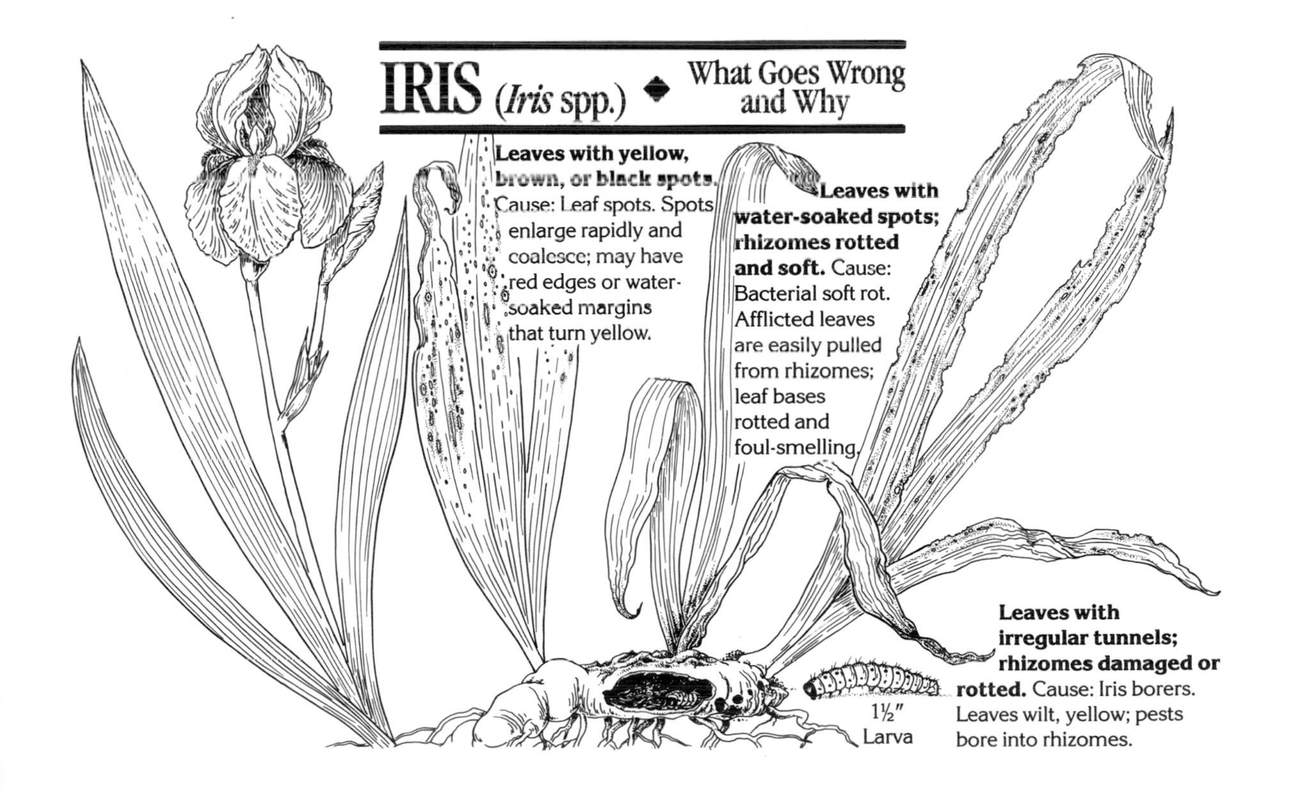

fans. Trim leaves back before planting to make up for root loss. Plant in full sun or very light shade and average to rich, well-drained soil. Barely cover the rhizome and point the leafy end in the direction you want it to grow.

Bearded irises tolerate drought very well when dormant (usually beginning about 6 weeks after bloom), but water them well until dormancy sets in and after division. Fertilize routinely in spring and early fall, keep weeds and other plants away from rhizomes, mulch loosely the first winter after division, and plan to stake tall-growing cultivars when in bloom.

Siberian irises enjoy conditions similar to those favored by bearded irises, but tolerate wetter soil and need less-frequent division in spring or fall. Replant as soon as possible after dividing.

Grow Japanese irises in much the same way, providing shade from the hottest sun. Water well before and during bloom. They need acid soil and benefit from a few inches of mulch in summer.

Plant reticulated irises in fall, about 3″ deep and a few inches apart in average to more fertile, very well drained soil. Grow with annuals and perennials to fill gaps left by their leaves, which wither by summer.

Problems

Leaves with irregular tunnels; rhizomes damaged or rotted. Cause: Iris borers. The most destructive pest of irises, these moth larvae favor bearded irises, but may feed on all species. They seldom bother Siberian irises. Borer eggs hatch in spring, producing up to 2″ long, fat, pinkish larvae. The larvae enter a fan at the top and tunnel down toward the rhizome, where they may eat the whole interior without being noticed. Borers often introduce soft rot bacteria into rhizomes as they feed.

In fall, remove dead, dry leaves, which often carry borer eggs, and destroy badly infested fans in spring. You can also crush borers in the leaves by pinching toward the base of the telltale ragged-edged leaves or by running your thumb between the leaves and squashing any borers you find. Check rhizomes when you divide the clumps for this pest. If you find a few borers, try cutting them out; destroy badly infested rhizomes. In spring, dust the base of plants with pyrethrin to kill emerging larvae.

Leaves with large, ragged holes. Cause: Slugs and snails. These slimy pests live and feed amid dense iris foliage. See "Leaves with large, ragged holes" on page 50 for controls.

Flower buds die; petals distorted; growth stunted. Cause: Thrips. Several species of tiny thrips infest irises; Japanese irises are especially susceptible. Thrips feed on inner folds of leaves, causing stunted growth and russet or sooty areas on leaves. Tops of plants eventually turn brown and die. Flowers may appear discolored, flecked with white, or deformed. These pests are difficult to control, for they burrow into plant tissue. Don't buy sickly looking irises that may be infested. Remove and destroy severely infested plant parts. Use blue sticky traps to monitor and trap pests. Applications of insecticidal soap may provide some control once pests are spotted in traps.

Leaves with yellow, brown, or black spots. Cause: Leaf spots. Irises may develop leaf spot, caused by several different species of fungi, especially in wet weather. Remove infected plant parts; clean up debris in fall to remove disease spores.

Leaves with water-soaked spots; rhizomes rotted and soft. Cause: Bacterial soft rot. Soft rot attacks during wet seasons in poorly drained soil, entering through wounds in the rhizome made by premature leaf removal or cultivation or carried on the bodies of iris borers. Crowded plants in shady locations are more susceptible to this disease. Infected rhizomes are dry on the outside, but wet, smelly, and slimy inside. This rot may start in the leaves following borer attack. Water-soaked streaks appear on leaves, which then turn yellow and wilt, starting from the tips. Eventually the entire leaf cluster may fall to the ground. Bulbous irises are not infected by this particular rot.

Control borers to reduce soft rot infection; see "Leaves with irregular tunnels; rhizomes damaged or rotted" above. Remove and destroy rotting rhizomes. Wash tools when cultivating or dividing irises to avoid transmitting the infection. Choose rhizomes carefully, inspecting them for signs of infestation; plant in well-drained soil with adequate sunlight.

Leaves mottled or streaked. Cause: Mosaic virus. Bulbous irises that are stunted and streaked with yellow may carry this virus. Flowers may be mottled and smaller than normal. There is no cure for infected plants; remove and destroy irises showing symptoms. Mosaic virus is spread by sucking insects such as aphids and leafhoppers. Control pests to reduce risk of infection. See "Leaves, stems, and buds distorted, sticky; clusters of small insects" on page 51 for controls.

Stems rot at base. Cause: Crown rot. Crowded plants are most susceptible to this fungal disease. Leaves and stems turn brown at the base, foliage turns yellow, and black spores may appear on stems. White or brown mold may be present. Rhizomes may also rot. Dig and divide iris clumps every few years to avoid overcrowding. Plant in well-drained soil. Avoid damaging crowns when cultivating; keep winter mulch away from crowns.

Juniperus

Juniper, red cedar. Trees and shrubs.

Junipers are a large group of evergreen trees and shrubs. Foliage in young plants is awl-shaped, in mature plants, flat and scalelike. In some cases, both types of foliage can exist simultaneously on the same plant. Plants are either male or female; the females bear small, berrylike cones. Many species and cultivars are valuable in the landscape. Depending upon their form, junipers are valuable for foundation plantings, as groundcovers, or even as specimen trees. Because they respond well to pruning for shape, they also make effective evergreen hedges.

Junipers are widely tolerant of all but

poorly drained soils. They will withstand drought and grow in low-fertility soils—even seaside conditions. Set them out as balled-and-burlapped or container-grown plants in spring or fall. Full sun suits them best; grown in too much shade, they become spindly and unattractive.

Problems

Branches with large galls. Cause: Rust. Cedar-apple rust, cedar-hawthorn rust, and cedar-quince rust are fungal diseases that spend part of their life cycles on junipers, the rest on alternative hosts, such as apples and crab apples. Large galls form and eventually swell to release the spores, especially in warm, moist weather. For an illustration of rust galls on junipers, see below.

Other than producing the galls, rust does little harm to junipers. Cut off and destroy or dispose of galls before early spring. Avoid planting Eastern red cedar (*Juniperus virginiana*) and its cultivars, which are very susceptible to rusts. See "Leaves with pale yellow spots that enlarge and turn orange" on page 25 for information about this disease on apples.

Plant defoliated; branches bear cocoonlike bags. Cause: Bagworms. See below for an illustration of this pest. See "Plant defoliated; branches bear cocoonlike bags" on page 236 for control measures.

Leaves wrinkled and discolored. Cause: Aphids. Several kinds of aphids feed on junipers. For control information, see "Leaves wrinkled and discolored" on page 235.

Leaves yellow; stems and leaves covered with small bumps. Cause: Scales. For control measures, see "Leaves yellow; stems and leaves

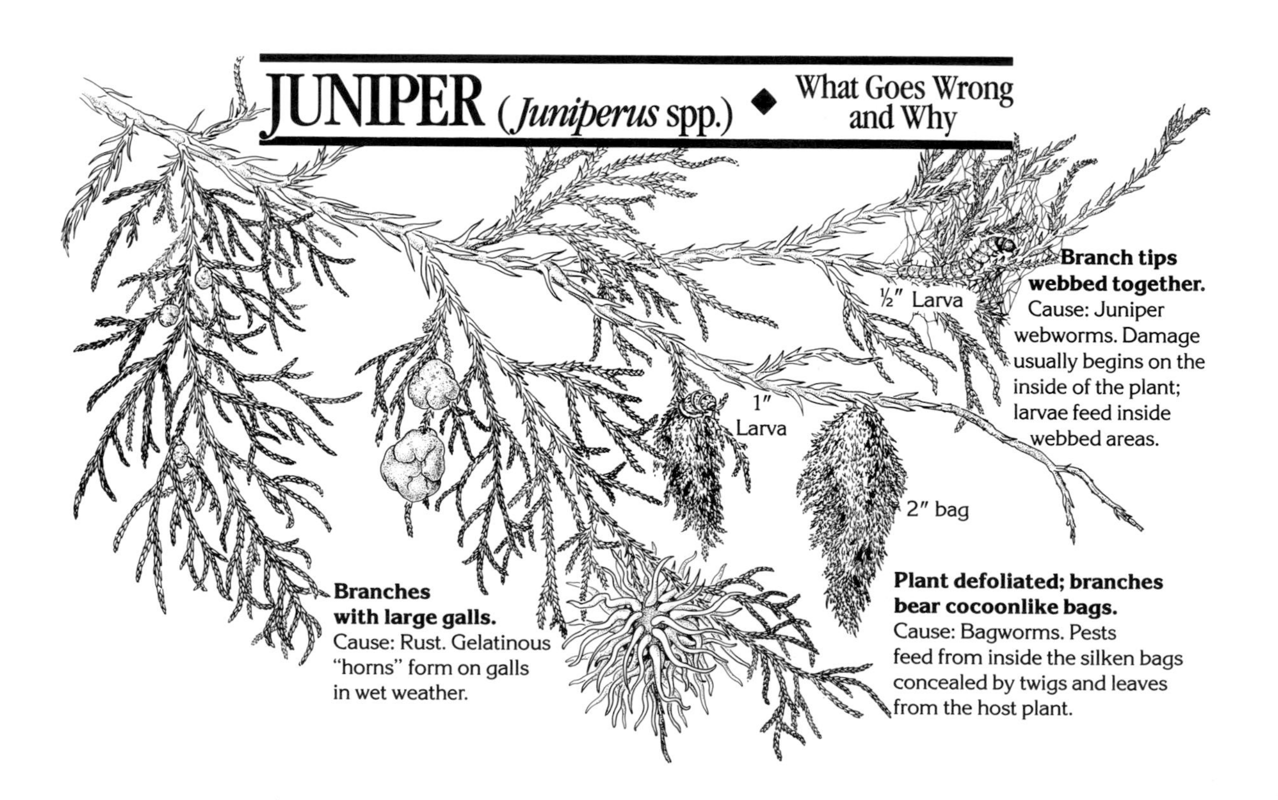

covered with small bumps" on page 237.

Leaves yellow; whole plant weakened. Cause: Juniper mealybugs. The tiny, powdery white adults generally congregate on the trunks and interior branches, making them difficult to see until the plant starts to weaken. Control by spraying the plant (especially the bark) with superior oil or repeated applications of insecticidal soap.

Leaves stippled with yellow; foliage webbed. Cause: Mites. For control measures, see "Leaves stippled with yellow; foliage webbed" on page 236.

Branch tips webbed together. Cause: Juniper webworms. If left unchecked, this pest can kill the twigs. Remove and destroy the nests, and spray with BTK or pyrethrin as a last resort. See the opposite page for an illustration of this pest.

Needle bases blistered; shoot tips die. Cause: Juniper midges. The tiny yellow larvae of this pest feed on the bases of needles, damaging shoot tips. Prune off infested parts, then spray the whole plant with insecticidal soap. Clean up debris around the base of the tree to reduce the pest population.

Branch tips browned. Causes: Twig blights; cedar bark beetles. Fungal blights may attack junipers in spring or summer. These diseases spread progressively, killing branches or entire plants. Control by pruning and destroying affected branches on a dry day. Avoid by using resistant cultivars, such as *J. chinensis* var. *sargentii* 'Glauca'; ask your local nursery owner or extension agent for others recommended for your area.

Similar symptoms on recently transplanted junipers, or those weakened by drought or by the onslaughts of other insects, may be due to cedar bark beetles. The white larvae of this tiny black beetle tunnel under the bark, sometimes killing weakened trees. The adult beetles chew holes in twig crotches, often causing shoot tips to dangle from the branch. The defense is to keep trees in healthy condition, by watering during dry spells and fertilizing regularly. If damage occurs, prune off damaged tips. Spray with pyrethrin, making 2 applications 3-4 days apart.

Kale

Brassica oleracea
Acephala group (Cruciferae)

Kale is a cool-season vegetable grown for its crinkly, blue-green leaves. It is a very hardy member of the cabbage family, tolerating temperatures below 40°F. In fact, frost improves its taste. It is less tolerant of heat than other members of the cabbage family. In warmer climates grow kale in soil with a high clay content if you have a choice, and shade soil around roots.

Grow kale as you would cabbage. It is occasionally bothered by the same diseases and insects as cabbage. See the Cabbage entry beginning on page 52 for culture and information on problems.

Kalmia

Mountain laurel. Shrubs.

Mountain laurels are broad-leaved evergreen shrubs with clusters of white, pink, or red flowers in spring. They are splendid for woodland plantings, shrub borders, or foundation plantings.

In the spring, set out balled-and-burlapped or container-grown plants. Select a partially shaded location (the further south, the more shade) with moist but well-drained acid soil rich in organic matter. Don't cultivate around these shallow-rooted plants; instead, use a deep organic mulch, which will keep weeds down and provide the cool, moist root zone the plants need.

For information on pests and diseases, see the Rhododendron entry beginning on page 200.

Kiwi

Actinidia arguta, A. deliciosa,
and *A. kolomikta* (Actinidiaceae)

Kiwis are vigorous, twining vines. Fruit is borne toward the base of new shoots that grow off last year's canes. The kiwi found in produce markets, *Actinidia deliciosa,* grows in Zones 7-9. *A. arguta* is hardy in Zones 4-7; *A. kolomikta* is hardy in Zones 3-7.

Grow kiwis in full sun or partial shade in perfectly drained soil. Allow about 200 square feet of trellis or arbor for each plant. Male and female flowers appear on separate plants; therefore, a plant of each sex is required for fruit set. A few cultivars are self-pollinating. Check before you buy.

Problems

Leaves skeletonized. Cause: Japanese beetles. These ½″, metallic blue-green insects with bronze wing covers like to feed in the sun. Check plants early in the morning while beetles are sluggish and knock them into jars filled with soapy water. Scent-baited traps may be effective if placed at some distance from plants. Some gardeners find traps attract extra beetles into the garden, so be sure to place them well away from the plants you wish to protect. Populations of this pest naturally decline by midsummer, but for very heavy infestations, spray with rotenone. For more information, see "Leaves skeletonized" on page 46.

Vines chewed, bedraggled. Cause: Cats. Strange as it may seem, cats like kiwis as much as they like catnip. Protect plants with chicken wire, if necessary.

Plant stunted; leaves yellow. Cause: Crown rot. Kiwis require perfectly drained soil to avoid crown rot. Avoid planting in poorly drained sites and modify soil enough to assure good drainage.

Kohlrabi

Brassica oleracea
Gongylodes group (Cruciferae)

Kohlrabi is a cool-season vegetable grown for its crisp, bulbous stems. It looks like an aboveground turnip. Kohlrabi can tolerate temperatures below 40°F and a pH between 5.5 and 6.8.

The trick to growing tender kohlrabi is encouraging rapid growth. Keep young plants well-watered. Harvest bulbs when they are 2″-3″ in diameter; they may become fibrous and bitter if allowed to grow larger.

Kohlrabi and cabbage require similar culture and are attacked by the same diseases and insects. See the Cabbage entry beginning on page 52 for culture and information on problems.

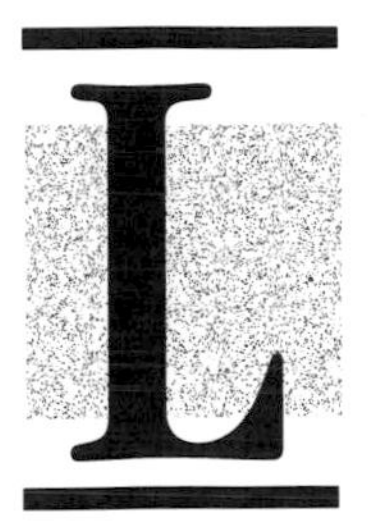

Lagerstroemia

Lagerstroemia, crape myrtle. Shrubs and trees.

Lagerstroemias, also known as crape myrtles, are popular summer-flowering shrubs and small trees throughout the warmer portions of the United States.

Lagerstroemias require full sun and a moist but well-drained soil enriched with organic matter. They will withstand severe pruning in early spring and still bloom because they produce flowers on the current season's wood.

Powdery mildew can be a serious problem; choose resistant cultivars. For more information, see "Leaves with powdery white coating" on page 237. Aphids may feed on leaves; see "Leaves wrinkled and discolored" on page 235 for controls.

Lantana

Lantana, shrub verbena.
Tender perennials grown as annuals.

Lantana brings masses of color to containers, rock gardens, and hanging baskets. The quick-growing plants can grow to 4′ high by 8′ wide. They are covered with 1″–2″ clusters of small flowers in shades of pink, yellow, orange, red, and bicolors. The plants are frost-sensitive, although they may bloom throughout the year in mild areas. Birds, bees, and butterflies find the plants quite attractive.

Take softwood cuttings from existing plants, or start seeds indoors in midwinter (they germinate in about 8 weeks). In warm climates direct-sowing is effective. Lantanas need full sun and average, well-drained soil. Plant 1½′ apart. Water deeply but let soil dry out between waterings. Fertilize lightly; over-fertilization or excess water decreases bloom.

Problems

Leaves yellow; plant weakened. Cause: Whiteflies. For controls, see "Leaves yellow; plant weakened" on page 18.

Leaves and stems with white, cottony clusters. Cause: Mealybugs. See "Leaves and stems with white, cottony clusters" on page 20 for control measures.

Leaves mottled, with shiny black flecks underneath. Cause: Lace bugs. Adult lace bugs are 1/10″, flattened, dark insects with lacy-patterned wings. Both adults and the tiny nymphs feed on the undersides of leaves and produce tiny dots of brown excrement. At the first sign of damage, spray with insecticidal soap, paying particular attention to the undersides of the leaves. Use pyrethrin as a last resort.

Larix

Larch. Trees.

Larches comprise a unique genus—a group of deciduous, needle-leaved conifers. They are trees for cold climates.

Set out in spring or fall as balled-and-burlapped plants. Plant in full sun in moist, acid soil. Larches do not tolerate shade, dry soils, or urban conditions.

Problems

Leaves mined, yellowed, and shriveled. Cause: Larch casebearers. The tiny reddish brown larvae of this moth feed and overwinter inside the leaves. Adult moths emerge in May or June. Natural parasites usually keep this pest under control. For severe infestations, apply a dormant spray of lime-sulfur to branches.

Branches defoliated. Cause: Larch sawflies. In June, the small, wasplike adults lay eggs in the side of larch shoots, causing shoots to twist. These ½″-1″, grayish green, caterpillar-like larvae emerge about a week later. They feed on the leaves, starting with lower branches. Remove debris, which provides an overwintering site for the pest, from around trees. Spray shoots with superior oil or rotenone to control heavy populations.

Trunk or branches with oozing lesions. Cause: Larch branch canker. This fungus produces sunken areas on the bark that are surrounded with drops of resin. Cut out and destroy affected parts as soon as detected; once established, there is no control. European larch (*Larix decidua*) is prone to serious infections; Japanese larch (*L. kaempferi*) is less susceptible to the disease.

Lathyrus

Sweet pea. Annuals.

These fragrant, climbing annuals make excellent cut flowers. Clusters of lavender, pink, or white blossoms appear in spring. Plants can grow 1′-6′ tall. Hot weather kills cultivars that aren't heat-resistant.

Soak seeds for 24 hours in warm water to speed germination. Direct-sow in early spring as soon as soil can be worked. Indoors, start in individual peat pots about 7 weeks before last frost. Seeds germinate in 2 weeks.

Plant sweet peas in full sun in a deeply worked bed amended with lots of organic matter. Water deeply and frequently. Remove spent flowers to prolong blooming.

Powdery mildew can occur if air circulation is poor; see "Leaves with powdery white patches" on page 19. If aphids are a problem, see "Leaves, stems, and buds distorted" on page 20.

Lawns

Good organic lawn maintenance is the secret to having an attractive lawn without using synthetic chemicals. If you select the right mix of grass species for your area and maintain soil fertility and organic matter content, you'll have few problems with insects and diseases.

If your lawn is out of condition, or if you've been relying on synthetic fertilizers and pesticides to keep it growing, here are some steps you may need to take to develop a lush, organic lawn.

■ Mow regularly to the recommended height for the grasses in your lawn, and leave the clippings on the lawn. Grass clippings are one of the best organic fertilizers for lawns.

■ Check soil pH, and adjust if necessary. Most grasses do best at pH 6.0-7.0.

■ Top-dress your lawn with a ¼″-½″ layer of finished compost, dehydrated cow manure, or other fine organic matter.

■ Fertilize once a year—in spring and early fall for northern lawns, in spring or summer for southern lawns.

■ Water deeply and infrequently to encourage grasses to develop deep roots so they'll be more drought-tolerant.

■ If your lawn gets lots of foot or vehicle traffic or is growing in heavy soil, aerate the soil to loosen it and encourage deeper root growth. You can rent a motorized or power-driven aerator, or aerate manually with a spading fork. At 1′ intervals all over your lawn, insert a spading fork into the turf at a 45-degree angle. Push the tines in to a depth of 4″. Press down slightly on the fork handle to loosen the soil, and then pull out the fork.

■ If your lawn has a buildup of thatch—undecomposed plant debris at the soil surface—thin it out by raking the lawn with a special

LAWNS ◆ What Goes Wrong and Why

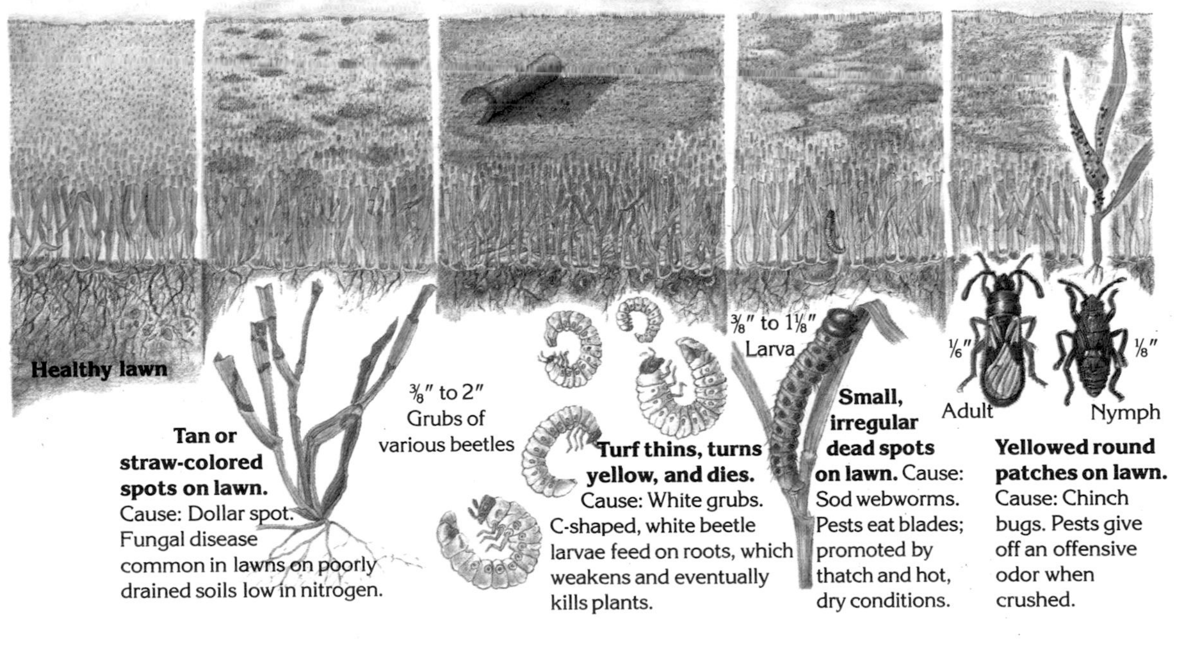

thatching rake, available from nursery suppliers and mail-order tool companies. For very large lawns, you will need to rent a machine called a dethatcher.

■ If your lawn is in bad condition, or suffering from severe insect or disease problems, you may decide to start from scratch. If so, be sure to choose the best grass or mixture of grasses for your area. Cool-season grasses, which grow the most in spring and fall and are dormant in midsummer, are best for the northern half of the country. Warm-season grasses, which are dormant in winter and begin growing in early summer, are best for the desert southwest and the Sun Belt. There are many disease- and insect-resistant cultivars available. One group, the endophyte-containing grasses, hosts fungi that produce a substance that deters feeding by some insect pests and is actually toxic to other pests. Check with your local extension office or garden center for the best new cultivars for your areas.

Problems

Turf has bare or ragged patches. Cause: Armyworms. These greenish brown caterpillars with white stripes and black heads chew grass blades down to the crowns, leaving bare areas or ragged patches of grass. They are a common pest on bermudagrass during cool, wet periods. In the South as many as 6 generations may occur in 1 year. Spray affected areas with parasitic nematodes while larvae are still feeding. Spray feeding larvae with BTK. Remove dead areas of turf. Reseed and overseed with resistant grasses such as the endophyte-containing cultivars.

Grass blades yellowed; thinned, brown turf. Cause: Mites. These tiny, 8-legged creatures suck sap from grass blades, causing them to turn yellow or straw-colored. Heavy infestations can kill plants, leaving turf looking brown and sparse. They thrive on poorly fed lawns and during dry conditions. Improve fertility and keep lawns well-watered during dry spells. For serious infestations, spray with insecticidal soap.

Turf with yellow or brown patches. Cause: Billbugs. Billbug larvae are white grubs with yellow-brown heads that feed on grass stems, causing shoots to turn brown and die. In warm weather the grubs tunnel into the soil and feed on roots and rhizomes. Billbugs are brown or nearly black, ¼″-½″ weevils. Control grubs by aerating the lawn, watering deeply, removing thatch, and adding organic matter. Reseed or overseed with resistant cultivars.

Yellowed round patches on lawn. Cause: Chinch bugs. These bugs suck plant sap, causing grass to turn yellow and die off in patches. Adult chinch bugs have ⅙″, dark bodies with a black triangular pad between white, folded wings. Nymphs are tiny bright red insects with a white band across their backs. Both adults and nymphs cause damage, especially during dry conditions. See page 133 for an illustration of the pest and its damage.

The pests congregate in open, sunny parts of the lawn. They smell bad, especially when crushed. You may be able to detect the odor simply by walking across a severely infested lawn. To check for chinch bugs: Remove both ends from a large coffee can. Push one end 2″ into the sod, using a knife to cut the ground so the can inserts easily or a board to hammer it in place. Fill the can with water. Chinch bugs will float to the surface within 10 minutes. If there are 20 or more bugs in the can, monitor the population weekly, and take steps to control, if necessary.

Encourage native predators, including bigeyed bugs, minute pirate bugs, lacewings, lady beetles, and birds. Drive out chinch bugs by keeping the soil very moist. Wet it to a depth of 6″, and maintain that moist condition for 3-4 weeks. To control chinch bugs in small lawns, soak sod with soapy water (1 ounce liquid dish soap to 2 gallons water), then lay a flannel sheet over the grass to snare the bugs as they are driven out by the soapy solution. Pick up the sheet and rinse it out in a bucket of soapy water to kill the chinch bugs. Rinse the soap off the treated grass by watering well. Seed and reseed treated areas with resistant cultivars.

Grass chewed. Cause: Grasshoppers. These common insects chew on grass, but usually aren't a serious threat to lawns. Broadcast *Nosema locustae* spores mixed in a bran bait (available commercially) as soon as grasshoppers emerge in spring: Effects of this measure usually aren't realized until the second summer after application.

Turf has irregular streaks of brown grass. Cause: Mole crickets. These 1½″, light brown insects have short forelegs and shovel-like feet. They are serious lawn pests in the South. The crickets tunnel under the lawn and feed on grass roots. Infested grass wilts and turns brown, so irregular dark streaks appear on lawns. Apply parasitic nematodes and water the soil well before and after application, or apply milky disease spores.

Small, irregular dead spots on lawn. Cause: Sod webworms. Webworms sever grass blades just above the thatch line and pull the blades into a silken tunnel in the ground to eat. As they feed and tunnel, irregular dead patches appear in the lawn. They are most often a problem on bluegrass, hybrid bermudagrasses, and bentgrasses in the South. Webworms are the larvae of the buff-colored lawn moth, which fly over lawns in late spring. Hot, dry conditions and thatch buildup encourage the pests.

Control webworms by saturating infested areas with a soap drench (2 tablespoons liquid dish soap to 1 gallon water) to drive the larvae to the surface. Rake the pests into a pile and dump them into a bucket of soapy water. Apply BTK or drench the soil with parasitic nematodes when pests are in their larval stage (usually about 2 weeks after moths appear). Spray severely infested areas with pyrethrin if all other attempts at control fail.

Turf thins, turns yellow, and dies. Cause: White grubs. These curved, fat, whitish larvae of Japanese beetles and other beetle species chew on grass roots, leaving sections of lawn that appear burned. Damaged turf can be easily lifted from the lawn. Ten or more grubs per square foot is a serious infestation. Japanese beetle grubs are most common on cool-season grasses. Grubs of other species such as scarab beetles, European chafers, and oriental beetles feed on bentgrasses, Kentucky and annual bluegrasses, and fescues.

Walk the turf in spiked sandals (available through mail-order catalogs) that will pierce and kill the grubs. For a large lawn, apply milky disease spores: This treatment will eliminate Japanese beetle grubs over a few seasons. Apply the material in late spring or fall when soil temperature is at least 70°F, but while grubs are still present. Or apply predatory nematodes to control both Japanese beetle grubs and other white grubs. Water the soil well before and after application to improve results.

Turf has brown circular patches. Cause: Brown patch. This fungus causes circular areas of grass up to 2′ in diameter to turn brown and die. It tends to attack St. Augustine grass, bentgrass, bermudagrass, zoysia grass, tall fescues, and ryegrasses, especially during hot, humid weather. Close cutting, poor drainage, overwatering, excessive nitrogen, and low pH all contribute to brown patch. Control the disease by reducing nitrogen fertilization, mowing less frequently, aerating, and dethatching. Top-dress with humus-building material. Water less frequently and only during the day so grass dries off quickly. Rake out dead grass and replant bare spots with resistant grasses.

Tan or straw-colored spots on lawn. Cause: Dollar spot. This fungus causes tan or straw-colored spots the size of silver dollars to appear on the lawn. The fungus occurs widely on golf greens, but may also be a severe problem on low-nitrogen, poorly drained lawns. Aerate the soil and improve drainage by top-dressing with organic matter. In the meantime, keep soil well-watered. Apply a high-nitrogen fertilizer; applying seaweed extract is also helpful. Mow less frequently, if possible. Overseed in fall with resistant cultivars.

Green spots outlined with brown on lawn. Cause: Fairy ring. If your lawn has bright green circular areas that seem to grow more rapidly than the rest of the lawn, the grass is probably infected by fairy ring fungus. A ring of grass around the green spots turns brown; the green areas eventually will brown out also. A circle of mushrooms usually develops around the edge of the infected area. Rake and discard the mushrooms as they appear. Spike the area with a spading fork every day. Water well. Encourage beneficial soil microbes by top-dressing with finished compost or an organic lawn fertilizer. Eradicate the fungus by digging out the soil in the area of the ring. Dig down 2′ and extend the hole outward at least 1′ out from the ring. Remove the soil carefully, being sure not to spill any infected soil on the healthy lawn. Fill in the hole with humusy topsoil and finished compost and reseed.

Reddish brown, tan, or yellow patches develop on lawn. Cause: Fusarium blight. This disease is common on Kentucky bluegrass during periods of hot, humid weather. Infected lawns develop spots of reddish brown grass 2″–6″ in diameter. The spots later turn tan and finally yellow. Roots will rot and may be covered with pink mold. Dethatch and aerate the lawn. Apply 1″–2″ of water each week. Raise the mowing height in summer. Don't fertilize in late spring or early summer. Rake out dead grass and replant with Fusarium-resistant cultivars.

Grass blades have dark spots. Causes: Leaf spots; leaf blights. Several fungi cause grass to develop reddish brown to black spots on the leaf blades. Grass shrivels and roots rot. Hot, humid conditions favor fungal growth. Fight leaf spots and blights by building soil fertility. Set mowing height as high as recommended for your grass mix. Don't mow during disease outbreaks to avoid spreading the fungi. Keep soil evenly moist. Restore diseased areas by raking out diseased grass and top-dressing with finished compost. Reseed or overseed with resistant cultivars.

Grass water-soaked, blackened. Cause: Cottony blight. Infection by *Pythium* fungi causes patches of grass to turn black and look water-soaked. In humid conditions a cottony mold may appear. The disease usually occurs in wet, poorly drained areas that have been overfertilized. It spreads rapidly once established. Alkaline soils and calcium deficiency encourage the disease. Aerate and dethatch lawn. Have the soil tested and correct calcium deficiency, if necessary. Reduce nitrogen fertilization, particularly in fall. Keep soil evenly moist. Maintain slightly acid soil. In severely infected areas enrich soil with organic matter to improve drainage, and replant.

Circular, scorched patches or pink, gelatinous masses on lawn. Causes: Red thread; pink patch. Lawns infected by red thread fungus have circular patches of dried grass that have red or rusty threads on the blades. Lawns suffering from pink patch will develop pink, gelatinous masses on leaf blades. These related fungal diseases are common on bentgrass, bluegrass, fescue, and ryegrass, especially in cool, humid regions. Apply an organic fertilizer with nitrogen in readily available form, such as seaweed extract. Mow regularly to remove infected leaf tips. Water regularly and thoroughly.

Grass blades develop yellow to red, powdery blisters. Cause: Rust. Infection by this fungus causes yellow to rusty red, powdery spore blisters to appear on leaf blades. Seriously infected lawns turn yellow and wither. Rust typically occurs in late summer on dry lawns lacking nitrogen. To avoid problems, water the lawn well in the early morning, so the grass can dry off quickly. Fertilize using seaweed extract or other nitrogen-rich fertilizer. Mow regularly. Rake out dead grass and overseed with resistant cultivars.

Lettuce

Lactuca sativa (Compositae)

Lettuce is a hardy annual grown for its tender leaves. The 3 most common types of lettuce are: leaf or bunching, head (including butterhead and crisphead), and cos or romaine.

Culture

Lettuce grows best in rich, loose soil with a pH between 6.0 and 6.8. It likes full sun, but in hot weather does better with light shade in the heat of the day. Lettuce needs to grow rapidly and without interruption. Provide plenty of nitrogen in both quicker-release forms, such as blood meal or soybean meal, and slower-release forms, such as compost or alfalfa meal. Spray the plants with compost tea and/or seaweed extract every other week to give them an extra boost. Spraying with compost tea may also help prevent some fungal disease problems.

Lettuce grows best at temperatures between 60° and 65°F. Most lettuce cultivars grow poorly above 75°F, but will tolerate temperatures as low as 45°F. Plants exposed to high temperatures will bolt. Prevent bolting by providing plants with partial shade in the heat of the summer, harvesting promptly, and planting bolt-resistant cultivars.

To grow tender, trouble-free lettuce, keep the soil moist, but not soggy, and do not allow it to dry out. Unlike most vegetables, lettuce responds well to having its foliage sprinkled with water. Plant in raised beds to improve drainage. To help prevent disease problems, do not plant lettuce in soil where it has been grown within the last 3 years.

Lettuce seeds will not germinate if soil temperatures are above 80°F. In the heat of the summer, start seedlings in a cool, shaded location and then transplant them into the garden.

Various nutrient deficiencies can affect lettuce. Boron or phosphorus deficiency causes malformed plants. Calcium deficiency causes browning of young leaves. Poor heart formation is a symptom of molybdenum deficiency. Copper deficiency prevents heads from forming. Spray seedlings with seaweed extract to help prevent nutrient deficiencies. Continue to spray plants with seaweed extract or compost tea every 2 weeks to boost plant health. Do a soil test to confirm the deficiency, and amend soil accordingly.

Problems

Seedlings rot near soil line and fall over; seeds do not germinate. Cause: Damping-off. Keep soil moist, but not soggy. Thin seedlings and spray with compost tea as soon as first true leaves open to help prevent problems.

Seedlings clipped off at soil line. Cause: Cutworms. Once a seedling is clipped off, there is nothing to do but protect the remaining seedlings from nocturnal cutworm attacks. Check for fat, 1″-2″ long, brown or gray caterpillars in the soil near the base of plants. Place cutworm collars around transplants, sprinkle moist bran mixed with BTK on the soil surface in the evening, or add parasitic nematodes to the soil at least 1 week before planting to control them.

Leaves yellow. Causes: Excessive heat; nitrogen deficiency; waterlogged soil. Plants are stunted and leaves are tough and bitter. Temperatures above 80°F will produce these symptoms. Prevent problems by planting heat-tolerant cultivars and providing partial shade for plants in the heat of the summer.

If weather is not extremely hot, the problem is probably nitrogen deficiency. Spray plants and drench roots with fish emulsion or fishmeal tea to alleviate symptoms. Waterlogged soil produces the same symptoms by damaging roots. Keep soil moist, but not soggy. Plant in raised beds if drainage is a problem.

Plant yellow and stunted; plant wilts during bright, hot days and recovers at night. Causes: Wireworms; nematodes. Wireworms are yellow to reddish brown, slender, tough-bodied, segmented worms up to 1½" long, with brown heads. Adults are dark-colored, elongated click beetles. Apply parasitic nematodes to the soil to control them, and avoid planting where sod was the previous year.

Root knot nematodes cause enlargements or galls on roots. Control pest nematodes by applying chitin or parasitic nematodes to the soil.

Young leaves yellow and distorted. Causes: Tarnished plant bugs; aphids; big vein. Tarnished plant bugs are oval, light green to brown, ¼" long insects with triangles on their backs. Nymphs are smaller and yellow-green. When they feed, they inject a toxin into plants that causes distorted growth. Trap them with white sticky traps. Spray plants with a commercial pyrethrin/rotenone mix in the evening if infestation is heavy. Till soil after harvest to reduce overwintering pests.

Aphids are soft-bodied, small, green, black, gray, pink, or white fluffy-coated, sucking insects that can spread diseases. For light infestations, knock pests off plants with a blast of water. Spray plants with insecticidal soap to control, or with a commercial pyrethrin/rotenone mix if infestation is heavy. Cover plants with row cover when they come up to prevent problems.

Big vein is a disease that causes light green or yellow, crinkled leaves with lighter, enlarged veins. Infected plants are stunted. The disease is carried by a fungus and may be caused by a viruslike organism. Destroy infected plants. Prevent problems by not planting in cold, wet soils or where big vein has been a problem in the past.

Leaves pale; plant stunted. Cause: Leafhoppers. Leaves may appear stippled. These green or brown, wedge-shaped, 1/10"–½" long insects feed on plant sap and can spread diseases. Spray infested plants with insecticidal soap in the evening or with a commercial pyrethrin/rotenone mix if infestation is severe. Prevent problems by covering plants with row cover when they germinate.

Leaves dull gray-green or silvery. Cause: Thrips. Leaves may turn brown and papery. These tiny, spiderlike insects can barely be seen with the naked eye. To check for thrips, tap a leaf over a white paper and look for moving specks. Trap thrips with sticky traps or spray plants with insecticidal soap to control them. Try blue, yellow, and white sticky traps placed just above plant height to see which color works best.

Young leaves are dwarfed, curled, or twisted. Cause: Aster yellows. There is no cure for infected plants; destroy them. Control leafhoppers because they spread this disease. See "Leaves pale; plant stunted" above for leafhopper controls.

Leaves mottled and ruffled. Cause: Mosaic virus. Plants are stunted. Destroy infected plants as soon as possible. Control aphids that spread the disease. See "Young leaves yellow and distorted" above for controls. Prevent problems by planting resistant or tolerant cultivars such as 'Don Juan', 'Greenfield', 'Montello', 'Nancy', 'Paris Island Cos', 'Salad Bibb', and 'Salad Crisp'.

Leaves covered with white powdery coating. Cause: Powdery mildew. Older leaves are usually affected by this fungal disease. Infected leaves curl, turn yellow, and eventually turn brown and die. Spray plants with a baking soda solution (1 teaspoon per quart of water) or sulfur as soon as disease appears to prevent further infection, and thin plants to increase air movement.

Leaves with yellow or light green spots. Cause: Downy mildew. Undersides of spots develop a fluffy white growth. Affected areas eventually turn brown. In severe cases plants become brown and stunted. Downy mildew is a fungal disease common in warm, damp, foggy weather. Spray plants with sulfur in the evening to prevent further infection. To prevent problems, plant cultivars, such as 'Alpha DMR', 'Cal K-60', 'Don Juan', 'El Toro', 'Erthel', 'Morangold', 'Salad Bibb', 'Salinas', and 'Tania', that are tolerant of downy mildew.

Leaves with dark or water-soaked spots. Cause: Various bacterial or fungal leaf spot diseases. Destroy badly spotted plants or leaves. Thin plants to increase air movement, and avoid wetting leaves when watering. Prevent problems by keeping plants well-fed.

Leaf margins brown and dried. Causes: Tipburn; freezing injury. Tipburn also causes dark spots on veins. This condition is prevalent in hot weather and is related to uneven soil moisture and a shortage of calcium in the leaves. To prevent problems, keep the soil moist, but not soggy, and do not allow it to dry out. Plant cultivars that are somewhat resistant to tipburn, such as 'Cal K-60', 'Canasta', 'Don Juan', 'Empire', 'Grand Rapids TBR', 'Green Lake', 'Mesa', 'Montello', 'Paris Island Cos', 'Salinas', 'Slobolt', and 'Waldmann's Green'.

Temperatures below 35°F cause outer leaf margins to turn tan and leaves to blister. Protect plants with row cover if low temperatures are predicted.

Leaves with discolored midribs. Cause: Excessive heat. Midribs remain firm. Temperatures above 80°F can cause midrib discoloration. Cover plants with shade cloth when temperatures soar, or plant in partial shade. Certain cultivars, such as 'Ithaca', are less susceptible to this condition.

Whole plant collapses or rots. Causes: Bottom rot; lettuce drop; gray mold. Bottom rot first infects the lower leaves that touch the ground. Dark, sunken spots develop on midribs, then entire leaves turn brown and slimy. Entire plant may be affected in severe cases. The same fungus causes diseases on many different vegetables, so preventive rotation is difficult. Prevent bottom rot by planting resistant cultivars such as 'Canasta'. Lettuce drop fungus causes plants to wilt and collapse, outer leaves are affected first. Gray mold fungus also starts at the bottom of the plant and turns it into a slimy, brown mess. The center stem and heart may rot out before the outer leaves are affected. 'Montello' is tolerant of gray mold.

Destroy infected plants or leaves and thin remaining plants to increase air movement. Prevent problems by working lots of compost into the soil before planting, and plant lettuce in raised beds to improve soil drainage. Spray plants with compost tea every 2 weeks to help suppress diseases.

Leaves with wandering, white or translucent tunnels. Cause: Leafminers. Larvae are white and maggotlike and burrow through leaves, leaving empty tunnels. Adults are tiny black-and-yellow insects. Destroy infested leaves. Catch adults with sticky traps; try blue, yellow, and white to see which color works best. Control heavy infestations of adults by spraying plants with a commercial pyrethrin/rotenone mix. Once the larvae are inside leaves, spraying is useless. Certain parasitic nematodes will attack leafminer larvae in the leaf tunnels. Prevent problems by covering plants with row cover when they germinate.

Leaves with small holes. Cause: Flea beetles. These small, shiny, black beetles hop when disturbed. Spray plants with a commercial pyrethrin/rotenone mix to control severe infestations. Protect plants with row cover as soon as they germinate.

Leaves with large holes. Causes: Caterpillars; slugs and snails. If there are green droppings on leaves and below plants, look for caterpillars feeding on leaves. Spray plants with BTK to control them.

If there are shiny slime trails on the remaining leaves, slugs and snails are at work. Sprinkle wood ashes or diatomaceous earth around plants, or trap the pests in shallow pans filled with stale beer and empty traps daily.

Leaves cut back or missing. Cause: Animal pests. Lettuce is a favorite snack for rabbits and groundhogs. See "Stopping Animal Pests" on page 408 for control methods.

Liatris

Gayfeather, blazing-star. Perennials.

These striking upright plants provide bottle-brush-type flower spikes of pink to purple with stemless, grasslike leaves that become smaller as they progress up the spikes. Gayfeathers range in height from 3′ to 6′; flowers open from the tops of spikes downward and bloom in summer and into fall.

Plant in full sun. These tough natives of the Great Plains tolerate poor to average soils, heat, cold, and some drought, but are at their best in moist, well-drained soil. Gayfeathers may self-sow under optimum conditions but are rarely invasive. Cut spent spikes back by a third, leaving the rest to produce food for continued growth. Tall species and cultivars may need staking. Divide every 4 years if plants seem crowded.

Problems

Plant fails to thrive. Cause: Root knot nematodes. Gayfeathers have no significant pests or diseases, but in some states the southern root knot nematode is a problem. Microscopic in size, soil-dwelling nematodes are roundworms that feed on roots. Their feeding impairs uptake of nutrients and water. Plants fail to thrive, but otherwise show few aboveground symptoms beyond midday wilting in summer drought. Root knot nematode feeding causes numerous knobby root swellings that don't rub off. Control is difficult; avoid planting gayfeathers in areas of known nematode activity. Encourage natural enemies of nematodes by adding organic matter to soil; solarize infested soil as described on page 424; rotate nonsusceptible crops into infested sites. Soil drenches of neem and applications of chitin also help control root knot nematodes.

Ligustrum

Privet. Shrubs.

Privets are opposite-leaved, deciduous or evergreen shrubs. They are easy to grow and readily lend themselves to pruning, which helps explain their popularity for hedges.

Privets will grow in full sun or partial shade and tolerate almost all soils except poorly drained ones. Set out in spring or fall. They benefit from a deep organic mulch.

Problems

Leaves wrinkled and discolored. Cause: Aphids. For control measures, see "Leaves wrinkled and discolored" on page 235.

Leaves tunneled. Cause: Leafminers. See

"Leaves tunneled" on page 237 for controls.

Leaves yellow; stems and leaves covered with small bumps. Cause: Scales. For controls, see "Leaves yellow; stems and leaves covered with small bumps" on page 237.

Leaves with notched margins. Cause: Japanese weevils. The 1/4″, brown adults attack foliage, and the legless white grubs feed on roots. Apply a sticky coating, such as Tanglefoot, to lower stems to prevent adults from climbing up the plants. Drench the soil around the plant base with a solution of parasitic nematodes. Spray leaves several times with pyrethrin for major infestations.

Leaves stippled with yellow; foliage webbed. Cause: Spider mites. See "Leaves stippled with yellow; foliage webbed" on page 236 for suggested controls.

Leaves and shoots blackened; leaves with moist or brown sunken spots. Cause: Anthracnose. See "Leaves and shoots blackened; leaves with moist or brown sunken spots" on page 238 for controls.

Leaves with powdery white coating. Cause: Powdery mildew. See "Leaves with powdery white coating" on page 237 for controls.

Lilium

Lily. Bulbs.

Hybridizers have created a glorious mix of lilies with 3″–12″ flowers in a variety of distinctive shapes, resembling peaked caps, turbans, bowls, trumpets, or broad, curly stars. They bloom in shades and combinations of white, pink, red, yellow, orange, lilac, and green, many dotted in maroon or near-black. Plants bear a few to 2 dozen or more flowers atop 2′–7′, upright stems clothed in narrow leaves. Most lilies are hardy in Zones 3–8, with protection in the North, especially during the first winter.

Lilies thrive in sun or part shade in deep, fertile, moist but well-drained, humus-rich soil out of strong winds. Unlike the hard, dense bulbs of tulips and daffodils, a lily bulb is a fragile package of individual scales joined together rather loosely, making it quite prone to damage and drying out. Also, lilies never go completely dormant; plant them carefully, as soon as possible after you receive them. It is preferable to plant them in fall, although spring planting is quite common. Many specialists prepare the soil to 1½′ deep on a warm fall day and mulch the site heavily to keep the soil unfrozen and ready for planting in very late fall, which is when many dealers ship lilies. Most lilies produce roots along the length of their below-ground stems, which help to feed and support the large plants. Therefore, even tiny lily bulbs should have no less than 6″ of soil above the top of the bulb.

Mark the planting sites to help avoid injuring the newly emerging shoots. As they appear in spring (sometimes surprisingly late), carefully cultivate and scratch in the first of 2 organic fertilizer meals for the season. Fertilize again before bloom; avoid excessive nitrogen applications. Mulch with several inches of compost or finely shredded bark to keep the soil cool. Water during dry spells. Stake tall lilies and deadhead after bloom. After the tops die, cut the stems down to a few inches to mark the spot for next year. Clear away faded foliage and plant debris to remove overwintering pests and diseases.

Move or divide lilies only when overcrowding makes it absolutely necessary; their fragile bulbs and fleshy roots resent any disturbance. Dig bulbs after stalks die back; replant immediately or wrap them in a plastic bag of barely moist perlite to keep the roots from drying out, and store in a cool place. Discard any bulbs that appear diseased or damaged. Expect minimal growth the first year after transplant. Small bulbils or bulblets may form in leaf axils or near the base of stalks; harvest and plant these at the end of the growing season to produce new plants.

Problems

Leaves yellow; plant wilts; bulbs decay. Cause: Bacterial or fungal rots. Control these rots with proper culture: Plant lilies in well-drained soil; select a site with good air circulation; avoid excess water in the soil or on foliage; dig and cultivate with care to prevent injury to bulbs.

Leaves yellow or distorted; bulbs decayed. Cause: Bulb mites. These mites are especially problematic when bulbs are injured by careless digging or cultivation. See "Leaves yellow or distorted; bulbs decayed" on page 50 for controls.

Leaves, stems, and buds distorted, sticky; clusters of small insects. Cause: Aphids. Several species of aphids attack lilies, damaging plants by feeding and also by transmitting viral diseases. Use a strong spray of water to knock aphids from plants; do this early in the day to allow time for leaves to dry before nightfall. Encourage beneficial insects such as lacewings and lady beetles. Sprays of insecticidal soap may give control; treat severe infestations with pyrethrin or rotenone.

Leaves with orange or reddish brown spots; buds rotted. Cause: Botrytis blight. This most common disease of lilies progresses from leaf spots to limp, blackened foliage starting with the lower leaves and moving up. Distorted flowers may have brown flecks. In wet weather gray mold forms on blighted plant parts—the disease is also called gray mold. Fungus spores are present in most soils and are activated by cool, wet conditions.

Some lily species and cultivars are more resistant to Botrytis than others, but this information is not always readily available. Choose resistant lilies, if identified, when buying bulbs; otherwise rely on good culture and healthy plants to limit disease. For tips on buying healthy bulbs, see the illustration on page 49. In fall, clean up debris where spores overwinter. Thin plantings to encourage good air circulation. Remove and destroy infected plants and plant parts. Spray weekly with a copper fungicide until the disease is under control.

Leaves mottled; plant stunted. Cause: Viral diseases. A variety of viruses cause yellowed, mottled, or streaked leaves and stunted stems on lilies. Flowers may be discolored. Aphids and other sucking insects carry viruses from infected lilies or from carriers, such as tiger lilies, that show no disease symptoms. Tulips and cucurbits also carry viruses that affect lilies. Infected plants eventually wilt and die.

There is no cure for viral diseases. Remove and destroy infected plants immediately. Wash tools and hands after working around diseased plants. To limit the spread of viruses, routinely check lilies for aphids, and control the pests if necessary; see "Leaves, stems, and buds distorted, sticky; clusters of small insects" above for control measures. Buy bulbs and plants from reputable sources—some will certify stock as "virus-free." Do not plant lilies in sites where diseases have occurred on bulbous plants. Separate lilies from tulips, cucurbits, and wild lilies.

Leaves pale above with dusty blisters on undersides. Cause: Rust. Powdery pustules on leaves' lower sides correspond to pale areas on the upper surfaces. Provide good air circulation around foliage; avoid wetting leaves when watering. Remove infected leaves; cut plants back in fall and dispose of leaves and stems. Dust or spray leaves with sulfur beginning early in the season to prevent rust or to treat a mild infection.

Shoots disappear or do not emerge.

Cause: Animal pests. Deer and groundhogs relish lily shoots; rodents enjoy the bulbs. Plant lilies where regular human activity will discourage wildlife; pet cats or dogs also deter animal pests. Line planting beds with hardware cloth to exclude burrowing rodents; cover beds with screen wire in winter. Experiment with repellents such as dried blood, human hair, or garlic sprays. Some gardeners use a few musky-smelling fritillary bulbs planted among the lilies to repel animal pests. Keep flower beds free of brush and plant debris, which offer shelter to hungry wildlife.

Liquidambar

Sweet gum. Trees.

Sweet gums are native deciduous trees bearing alternate, star-shaped leaves that turn brilliant scarlet in autumn. While a handsome tree for the large lawn, sweet gum is intolerant of air pollution and is a poor choice for urban sites.

Set out in spring as balled-and-burlapped plants, in full sun. A deep, slightly acid, moist soil of average fertility is best. Allow plenty of room for the tree's extensive root system. The plants are slow to establish; a thick layer of mulch and regular watering for the first few years will promote root development. Once they settle in, sweet gums are fairly problem-free.

Problems

Leaves skeletonized or with large holes; branches may be webbed. Cause: Caterpillars. See "Leaves skeletonized or with large holes; branches may be webbed" on page 236 for controls.

Leaves yellow; stems and leaves covered with small bumps. Cause: Scales. For controls, see "Leaves yellow; stems and leaves covered with small bumps" on page 237.

Trunk or branches with sunken, oozing areas. Cause: Bleeding necrosis. It is natural for the tree to exude some sticky "gum" from the trunk as it grows. However, excessive amounts of sap from dark, sunken areas in the bark may indicate a disease problem; left unchecked, it can quickly kill the tree. Cut out diseased areas as soon as you find them.

Plant defoliated; branches bear cocoonlike bags. Cause: Bagworms. See "Plant defoliated; branches bear cocoonlike bags" on page 236 for control measures.

Liriodendron

Tulip tree, yellow poplar. Trees.

Tulip trees are tall-growing with alternate, 3-lobed leaves. Yellow flowers appear near the top of the tree in late spring or early summer. Tulip tree makes a fine specimen tree for large properties.

Set out in spring as a balled-and-burlapped plant. Full sun and slightly acid, deep, well-drained but moisture-retentive soil suits it best.

Problems

Leaves wrinkled and discolored. Cause: Aphids. The small green tuliptree aphid can be abundant on the undersides of leaves, secreting copious amounts of honeydew. For control information, see "Leaves wrinkled and discolored" on page 235.

Leaves yellow and drop early. Cause: Lack of water. Leaf yellowing is a common problem on newly planted trees as well as established ones that don't get enough water. Avoid by watering during dry periods and using a thick layer of organic mulch.

Leaves yellow; stems and leaves covered with small bumps. Cause: Scales. See "Leaves yellow; stems and leaves covered with small bumps" on page 237 for controls.

Leaves with spots. Cause: Tuliptree spot gall. Circular brown or purple spots surrounded by a circle of yellow on the leaves reveals the presence of tuliptree spot gall, caused by a kind of midge. Damaged leaves may fall early. While unsightly, the problem is not serious and is best countered by removing all affected fallen leaves.

Leaves with powdery white coating. Cause: Powdery mildew. See "Leaves with powdery white coating" on page 237 for controls.

Lobelia

Lobelia. Perennials.

Lobelia's numerous species of annual and perennial flowers offer gardeners a broad array of plant heights and flower types. Cardinal flower (*Lobelia cardinalis*), perhaps the best-known species, features spikes of brilliant red, 3-lobed flowers that hummingbirds find irresistible. Other species bear star-shaped blooms in shades of blue and lavender; lobelias with red or bronze foliage are also available. Plants grow 2′-5′ tall with a spread of about 2′.

Lobelias prefer shady sites where the soil is consistently moist but well-drained. Plants tolerate full sun in cool climates, if ample moisture is provided, but generally need at least partial shade to thrive. Summer mulch helps retain much-needed soil moisture; a light winter mulch helps protect crowns. Renovate plantings of these short-lived perennials by division or reseeding every 2-3 years. Cover seeds lightly to keep them from drying out.

A number of pests may feed on lobelias, but damage is rarely significant enough to require treatment.

Problems

Leaves, stems, and buds distorted, sticky; clusters of small insects. Cause: Aphids. Aphids are sometimes troublesome; see "Leaves, stems, and buds distorted, sticky; clusters of small insects" on page 177 for control information.

Leaves with rotted patches or spots. Cause: Fungal rots; leaf spots. The moist soil conditions favored by lobelias encourage a variety of fungal rots and leaf spots that are best treated culturally. Remove infected plants and plant parts; keep gardens free of plant debris and weeds where diseases can overwinter; space plantings to permit good air circulation.

Lobularia

Sweet alyssum. Annuals.

Alyssums are popular, mound-forming plants that bloom from spring until frost. Plant size is 3″-8″ high and to 10″ wide. Flowers are pink, purple, or white and appear in multiple 3/4″ clusters. In numbers, the sweet honey smell of alyssum reminds some gardeners of fresh-mowed hay. Bees like the aroma, too; use alyssum near fruit trees that need pollinating. Alyssum is ideal for borders or as groundcovers for small areas.

Alyssum likes full sun and well-drained average soil. Direct-sow in loose soil 5 weeks before last frost. Do not cover seeds; they require light to germinate. Seedlings appear within 2 weeks. Thin to 5″ apart; cut, rather than pull out, to avoid damage to intertwined roots of other seedlings. Plants may flower within 6 weeks of seeding.

Water regularly for best growth, even though alyssum can tolerate temporary drought. Use a general-purpose fertilizer once in spring. Excess fertilizing gives ample foliage but few flowers. Flowering mounds tend to become sparse and rangy toward midsummer; shear the plants back and they will resume blossoming.

Problems

Seedlings die. Cause: Damping-off. For control information, see "Seedlings die" on page 20.

Stems and roots rot. Cause: Fungal wilt. Lower leaves and stems of mature alyssum will rot; white fungal strands may be visible around the plant base. Plants growing in well-drained soil are less susceptible. Discard damaged plants. To avoid transferring the fungus to healthy plants, wash your hands and gardening tools after handling infected plants.

Lonicera

Honeysuckle. Shrubs and vines.

Honeysuckles are alternate-leaved shrubs and vines. They bear usually fragrant flowers that give way to colorful fruit.

Plant honeysuckles in spring or fall in sun or partial shade. They can tolerate a range of soil conditions, but moist, well-drained sites suit them best.

Problems

Leaves wrinkled and discolored. Cause: Aphids. See "Leaves wrinkled and discolored" on page 235 for control measures.

Leaves with powdery white coating. Cause: Powdery mildew. For controls, see "Leaves with powdery white coating" on page 237.

Leaves yellow; stems and leaves covered with small bumps. Cause: Scales. See "Leaves yellow; stems and leaves covered with small bumps" on page 237 for controls.

Leaves rolled and chewed. Cause: Leafrollers. These ½″, green-bodied, brown-headed caterpillars form small webs on shoot tips and feed on leaves and buds inside. Break open webs and handpick larvae; spray BTK at the first sign of damage.

Leaves with sunken, discolored spots. Cause: Four-lined plant bugs. These ¼″, yellow bugs have 4 black stripes on their wings. They are active in early summer. Control by handpicking; use rotenone or pyrethrin on leaves for severe infestations.

Trunk or branches with oozing lesions; branch tips die back. Cause: Canker. See "Trunk or branches with oozing lesions; branch tips die back" on page 238 for controls.

Plant defoliated. Cause: Honeysuckle sawflies. The 1″, gray-and-yellow caterpillar-like larvae feed ravenously on foliage. Handpick, or spray shoots with insecticidal soap; use pyrethrin or rotenone as a last resort.

Lupinus

Lupine. Perennials.

Two-foot-long stalks, thick with blue, yellow, rose, cream, or bicolored, ½″–1″, pealike flowers make lupine plantings early summer showpieces. Bright or grayish green, palmlike, compound leaves add a tropical effect.

Several species occur worldwide, but cultivars, especially the Russell hybrids, are favored for the array of colors and color combinations they offer. Plants range from 1½′ to 3′ tall and form 2′ wide clumps.

Lupines are best adapted to areas where summers are cool and not too dry. They suffer in heat, but serve well as annuals where conditions are not optimal. Full sun and moist, well-drained, neutral to acidic soil promote healthy growth. Water and mulch to keep soil moist. Lupines are short-lived, but can be perpetuated through division. Self-sowing also occurs, but seedlings often have different colors than the parent plants.

Lupines attract their share of pests and diseases, although most problems stem from unfavorable growing conditions. Aphids and powdery mildew may appear; see "Leaves, stems, and buds distorted, sticky; clusters of small insects" and "Leaves covered with white powder" on page 177 for controls.

Magnolia

Magnolia. Trees.

Magnolias are alternate-leaved, deciduous or evergreen trees with bold leaves; conspicuous, cup-shaped flowers; and interesting bright red fruit. They make beautiful specimen or accent plants.

Set out in spring as balled-and-burlapped plants. Magnolias grow well in full sun or light shade. They prefer moist soil enriched with organic matter.

Problems

Leaves yellow; stems and leaves covered with small bumps. Cause: Scales. Scale insects are the chief enemy of magnolias. For control measures, see "Leaves yellow; stems and leaves covered with small bumps" on page 237.

Blossoms brown and limp. Cause: Frost damage. The flowers of early blooming magnolias are often subject to frost damage. Avoid planting sites with southern exposures, which can encourage buds to open while there is still danger of frost. Or plant later-blooming species or cultivars.

Leaves discolored, wilted, or dropping; lacking vigor. Cause: Mealybugs. The tiny, powdery white adults generally congregate on the trunks and interior branches, making them difficult to see until the plant starts to weaken. Control by spraying the plant (especially the bark) with superior oil or repeated applications of insecticidal soap.

Leaves stippled with yellow; foliage webbed. Cause: Spider mites. See "Leaves stippled with yellow; foliage webbed" on page 236.

Trunk or branches with oozing lesions; branch tips die back. Cause: Canker. For more details, see "Trunk or branches with oozing lesions; branch tips die back" on page 238.

Leaves with spots. Cause: Leaf spots. See "Leaves with spots" on page 237 for controls.

Mahonia

Mahonia, holly grape. Shrubs.

Mahonias are broad-leaved, evergreen shrubs. They are used for hedges, foundation plantings, and shrub borders.

Set out in spring as balled-and-burlapped or container-grown plants. Choose a partially shaded location to avoid winter leaf burn. Almost any good, well-drained garden soil will suit them; moist, acid soil is ideal.

Problems of mahonias are few and relatively unimportant. Barberry aphid, a small yellowish green insect, can be abundant on the West Coast; see "Leaves wrinkled and discolored" on page 235 for controls. Scale insects can attack twigs and sometimes foliage; for control information, see "Leaves yellow; stem and leaves covered with small bumps" on page 237.

Malus

Apple, crab apple. Trees.

Apples and crab apples are trees with alternate, deciduous leaves. They produce beautiful spring flowers and attractive red or yellow fruit. Apples are common in home orchards. Crab apples are valued as specimen trees; smaller species can be used in shrub borders. Birds are fond of crab apple fruit.

Plant in spring in moist but well-drained soil in full sun. They adapt well to a range of soil conditions. To avoid removing the following year's flower buds, it is best to prune soon after flowering. Keeping the center of the plant open to light and air will help reduce disease problems. Crab apples commonly produce suckers from the roots; cut these shoots down to the ground as soon as you see them.

Pests and disease problems on apple trees are covered in the Apple entry beginning on page 21. Crab apples share many of the same problems, including fire blight, cedar-apple rust, powdery mildew, and apple scab. Fortunately, disease-resistant cultivars are available. Some of the best include 'Adams', with reddish pink flowers; 'Coral Cascade', with white flowers; 'Prarifire', with purplish red flowers; and 'Professor Sprenger', with white flowers. There are also many other resistant cultivars, so check with your local nursery owner or extension agent to learn about the best ones for your area.

Be aware that even resistant cultivars may have problems if the conditions promoting disease are very favorable. Planting trees on sites with good air circulation and keeping trees growing vigorously are easy ways to avoid problems. For more information on crab apple pests and diseases, see the Apple entry beginning on page 21.

Melon

Cucumis melo and other genera (Cucurbitaceae)

Melons can be challenging to grow, especially in cooler climates. But if you have the patience, there is nothing more satisfying than a fragrant, sun-ripened melon.

Culture

Melon seeds need 60°F soil to germinate. In northern areas start plants indoors 2-3 weeks before the last frost date and transplant outside once temperatures are reliably warm. Melons need lots of sunlight and warm temperatures—90°F is ideal. Melons are especially vulnerable to pests and cool temperatures when plants are young. Plants exposed to temperatures below 50°F can be permanently injured and fail to set fruit. Cover plants with floating row cover or clear plastic tunnels as soon as they are set out. If temperatures exceed 90°F inside the tunnels, vent them by making a 6″ cut in the plastic directly over each plant. Remove row cover when melons begin to flower so insects can pollinate the blossoms, or you will not get any fruit. In the fall temperatures below 50°F cause cold stress and rapid wilting. Cover plants on cool nights.

Melons do best in well-drained, loose-textured soils with lots of organic matter. They prefer a pH between 6.0 and 6.8, but can

tolerate a pH as high as 7.6. Melons are shallow-rooted and may wilt on hot, dry days even when they are not diseased. Keep them well-watered, but do not let the soil become saturated. Wet soil can cause stems to rot at soil level. Overwatering, or uneven watering, can cause fruit to split. Potassium deficiency can also cause split fruit.

Prevent disease problems by keeping the leaves dry. Water carefully or use drip irrigation. Mulch melons to help conserve water: Black plastic is a good choice for central and northern areas, but in extremely warm areas it can warm the soil too much. Organic mulches are good, too, but also provide shelter for pests like squash bugs.

To reduce problems, rotate crops so that no member of the cucurbit family (squash, cucumbers, or melons) is grown in the same place more often than every 4 years.

The opposite page illustrates some insects and diseases that attack melons.

Caution: Melon leaves are easily burned by insecticidal soap and copper sprays. Use the most dilute spray recommended and apply sparingly. Do not spray plants in direct sun, or if temperatures are above 80°F, and don't spray drought-stressed plants.

Leaf and Vine Problems

Leaves with chewed holes. Cause: Cucumber beetles. Adults are 1/4" long, greenish yellow beetles with black stripes or spots. They attack young leaves and should be controlled immediately, as they spread bacterial wilt or viruses. Spray plants with sabadilla or a commercial pyrethrin/rotenone mix.

Leaves with pale green patches; afflicted leaves wilt and blacken. Cause: Squash bugs. Adults are brownish black, 1/2" long bugs. Immature bugs are whitish green with dark heads and legs. Both emit a strong, sharp smell when crushed. Eggs are bright orange and are laid on the undersides of leaves. Handpick adults and eggs. Trap bugs by placing a board on the ground near plants. Lift it each morning and destroy the squash bugs hiding underneath.

Leaves with yellow patches; older leaves mottled and distorted. Cause: Mosaic. For an illustration of this virus, see page 214. Afflicted plants are unproductive and fruit is bitter. Remove and destroy diseased plants. Control aphids and cucumber beetles, because they spread the virus.

Leaves yellow, curled, and wilted. Cause: Aphids. Look for small, green, pink, gray, black, or white fluffy-coated, soft-bodied insects feeding on plants. Aphids can also transmit viral diseases. Control them by knocking them off the plant with a blast of water, or use with a weak insecticidal soap spray; see the caution above before spraying. Use foil mulch to keep aphids from finding plants.

Leaves yellow and puckered, becoming bronzed. Cause: Mites. These tiny, red, yellow, or green, spiderlike creatures are worst in dry, hot weather. In severe cases leaves dry out and drop off. There may be a fine webbing on the undersides of leaves. Spray plants with a weak insecticidal soap spray to control; see the caution above before spraying.

Leaves mottled yellow between veins; leaf undersides have purple spots. Cause: Downy mildew. As the disease progresses, older leaves turn brown and die, and younger leaves become infected. Treat plants with a dilute solution of copper spray to reduce the spread of the disease; see the caution above before spraying. Prevent problems by planting resistant cultivars such as 'Morning Dew', 'Saticoy', and 'Tam Honeydew'.

Leaves with powdery white spots, especially on upper surfaces. Cause: Powdery mildew. As the disease progresses, leaves turn brown and dry, and plants may die. Treat plants with a dilute solution of copper spray to reduce the spread of the disease; see the caution above before spraying. Prevent problems by planting resistant cultivars such as 'Hales Best Jumbo', 'Quick Sweet', and 'Starship'.

Leaves with spots, blotches, or brown areas. Causes: Alternaria leaf blight; anthracnose; angular leaf spot; scab; gummy stem blight; other fungal and bacterial diseases. Various diseases attack melons. Reduce problems by keeping the foliage dry when watering and by not touching plants when wet. Spray infected plants with a dilute solution of copper spray to control outbreaks; see the caution on page 149 before spraying.

Dark brown spots with concentric rings, usually appearing on older leaves first, are caused by Alternaria leaf blight. As the disease progresses, leaves curl down and eventually drop off. Fruit infected with this disease has brown, concentrically ringed, sunken spots. Prevent problems by planting resistant cultivars such as 'Earligold', 'Pulsar', and 'Saticoy'.

Yellow spots that turn brown are caused by anthracnose. Infected leaves eventually die. This disease causes fruit with circular black cankers. Prevent problems by planting anthracnose-resistant muskmelon cultivars, such as 'Passport', and resistant watermelon cultivars, such as 'All Sweet', 'Crimson Sweet', 'Dixie Lee', and 'Sweet Favorite'.

Water-soaked spots that turn gray, die, and drop out leaving shotholes are caused by angular leaf spot or scab. Fruit infected with angular leaf spot has small, cracked, white spots. Scab causes fruit with sunken, brown spots with a gummy ooze, and damage is worst in cool, moist weather. Pale, round leaf spots with dark margins are caused by Cercospora leaf spot.

Brown to gray spots on leaves and stems, and dark, gummy stems are caused by gummy stem blight. Infected leaves turn yellow and die. Disease begins as spots on stems, which become streaks; stems then turn dark and gummy. When it attacks fruit, the disease is called black rot: Infected fruit has round, black spots and the fruit flesh collapses. For an illustration of this disease, see page 148.

Vines wilt suddenly. Cause: Squash vine borers. Check for fat, white, 1″ long larvae burrowing into stems, and masses of yellow-green, sawdustlike excrement. Slit stems lengthwise above injury with a sharp knife and kill larvae. Cover cut stems with moist soil so they will form new roots. Injecting stems with BTK or parasitic nematodes may also control borers. Spraying stem bases with BTK once a week in late spring and early summer may prevent damage.

Vines wilt at midday; leaf margins brown. Cause: Fusarium wilt. As the disease progresses, vines fail to recover, and die. Destroy infected plants. Prevent problems by planting resistant muskmelon cultivars, such as 'Pulsar', 'Saticoy', 'Savor', and 'Starship', and resistant watermelon cultivars, such as 'All Sweet', 'Crimson Sweet', 'Dixie Lee', and 'Sweet Favorite'.

Vines wilt at midday, starting with younger leaves; leaves remain green. Cause: Bacterial wilt. As the disease progresses, leaves fail to recover, and die. Cut wilted stem and touch the tip of your knife to the sap. If it is milky, sticky, and astringent, your plant is infected. For an illustration of this disease and technique, see page 148. Destroy infected plants immediately. Prevent problems by controlling cucumber beetles, since they spread the disease.

Fruit Problems

Fruit with spots; flesh may rot. Causes: Angular leaf spot; Alternaria blight; scab; black rot; anthracnose. Several fungal and bacterial diseases cause these symptoms on melon fruit. For complete symptoms, controls, and resistant cultivars, if available, see "Leaves with spots, blotches, or brown areas" above.

Fruit flesh not sweet. Causes: Premature harvest; overwatering; nitrogen deficiency. Harvest muskmelons and honeydews when they smell ripe and the stem separates easily from the vine when the fruit is gently lifted. Harvest watermelons when the bottom of the fruit turns from pale yellow to golden yellow. Excess rain or irrigation while fruit is swelling can dilute sugars. Keep soil evenly moist throughout the season. If leaves are small and yellow, and growth is stunted, plants are nitrogen starved. Spray foliage and drench roots with fish emulsion.

Fruit rots on underside. Cause: Damp soil or mulch. To prevent rot, support fruit on scraps of wood, or tuna cans with ends removed. See page 148 for an illustration of this technique.

Mertensia

Bluebells. Perennials.

Native Virginia bluebells (*Mertensia virginica*) seem to materialize from the floors of eastern forests in early spring. Clusters of pink buds that open to tubular blue flowers are carried on graceful stems over rounded, medium green foliage. Plants grow 1′-2′ tall and spread to about 1′ wide.

Rich, moist, well-drained soil and partial shade are bluebells' requirements. Plantings increase slowly but steadily in sites that resemble their woodland habitat. Leaves yellow and die back in summer; plants seem to have disappeared by July. Do not cut fading foliage; allow it to die back naturally. Root rot may damage bluebells in poorly drained soil; they are otherwise trouble-free.

Mint

Mentha spp. (Labiatae)

Mints are hardy (Zone 5) perennial herbs grown for their fragrant leaves. They are usually grown from cuttings. Plant in rich, moist, well-drained soil with a pH between 5.0 and 7.5, in full sun or part shade.

Mints are vigorous, trouble-free plants. They can be quite invasive once established. Let them take over a wild area or plant them in containers or in bottomless buckets sunk into the garden to prevent them from spreading.

Fungi can cause dark, sunken spots on leaves. Spray foliage with fish emulsion, or with sulfur if disease is severe. Tan to red blisters on leaves are caused by rust. Avoid wetting leaves to prevent its spread. Pale stippled or bronzed leaves are caused by mites—tiny, spiderlike creatures that thrive in hot, dry weather. Spray with insecticidal soap. See the Herbs entry beginning on page 116 for other possible problems.

Monarda

Bee balm. Perennials.

Brightly colored, spidery flower heads and dense, dark green, aromatic foliage attract gardeners to these native plants. Hummingbirds, bees, and butterflies are also drawn to bee balms' blossoms, especially the red ones. Plants form dense clumps, 2′-5′ tall and 3′ wide; the mintlike foliage remains attractive after flowers fade.

Found in the wild along shady stream banks, bee balms favor similar sites in the garden—consistently moist, well-drained soil and light shade. Giving such favorable surroundings, however, encourages bee balms' invasive nature; plantings can spread out rapidly. Planted in full sun, bee balms still grow nicely but are more easily controlled. Water and mulch in summer to prevent water stress, which quickly disfigures plantings. Clumps tend to die out in the center; divide every 3 years to keep growth compact. Remove spent flower heads to prolong bloom.

Problems

Leaves covered with white powder. Cause: Powdery mildew. This fungus is almost guaranteed to appear on bee balm foliage, even under the best of conditions. Plants suffering drought or overcrowding succumb readily and are soon covered with white or grayish mildew. Don't let bee balms dry out, but avoid wetting leaves, and water early in the day so foliage can dry before nightfall. Space plantings for optimum air circulation, and keep gardens free of plant debris that may host fungal spores. Sprays of sulfur and antitranspirants help prevent powdery mildew infections; use sulfur, fungicidal soap, or garlic sprays to prevent the spread of mild infections.

Narcissus

Daffodil, narcissus. Bulbs.

Today's much-hybridized daffodils are welcome harbingers of spring. Hundreds of cultivars, classified by flower type and bloom time, blossom in shades of yellow, orange, white, and pink; many are bicolored. Each flower has a trumpet or cup-shaped center, surrounded by 6 petal-like structures; many are fragrant. Bluish green leaves are straplike and upright, 1'-1½' long and ¾" wide. Plants grow from 6" to 1½' tall, depending on cultivar.

Buy daffodils from a reputable source to ensure that you get healthy, pest-free bulbs. Look for double- or triple-"nosed" bulbs, so-named because of their multiple growing points that produce more flowers than single-nosed bulbs. Plant bulbs in fall, at least 1 month before the ground freezes. Select a site in full sun or light shade with humus-rich, well-drained soil. Set bulbs 4"-8" deep—roughly 1½ times bulb height. A location behind annuals or perennials helps hide the yellowing foliage that remains after flowering ends. Remove spent flowers, but allow leaves to die back naturally; this lets bulbs store food for next year's floral display. Tying, cutting, or covering the leaves reduces the amount of light they receive and jeopardizes flowering in subsequent years.

Daffodil bulbs multiply gradually to form clumps. While plants will grow undisturbed for years, overcrowding eventually reduces flower size. Dig bulbs after foliage fades and shake away loose soil. Place them out of the sun to dry for a few days, then gently separate offsets from parent bulbs and replant. Small offsets may take a couple of years to bloom.

Don't worry about the animal pests that eat most other bulbs; daffodils' bitter taste deters feeding wildlife. Choose bulbs carefully to avoid most other pests and diseases; see the illustration "Buying Healthy Bulbs" on page 49.

Problems

Flowers and leaves with silvery flecks. Cause: Thrips. Flower thrips rasp daffodil leaves and flowers, causing silvery or whitish flecks; leaves eventually turn brown and die. They reproduce and spread rapidly, especially in dry areas. Other symptoms include leaf browning. These pests burrow into plant tissue and are difficult to control. Inspect bulbs carefully. Remove and destroy severely infested plant parts. Use blue sticky traps to monitor and trap pests. Applications of insecticidal soap may provide some control once pests are spotted in traps.

Leaves with white or brown streaks; foliage wilted. Cause: Viral decline. Plants suffering from viral decline develop white or brown streaks on the leaves late in their growing season. Foliage may wilt and topple over. There is no cure for decline. Remove diseased plants. Control aphids, which spread decline and other viral diseases; for control techniques, see "Leaves, stems, and buds distorted, sticky; clusters of small insects" on page 51.

Plant fails to appear; bulbs soft. Cause: Bulb flies. Narcissus bulb flies resemble small bumblebees; lesser bulb flies are 1/3″ long, blackish green flies. Both lay their eggs at the base of bulb leaves; larvae hatch and bore into bulbs to feed—usually 1 maggot per bulb in the case of narcissus bulb flies and several per bulb in the case of lesser bulb flies. Their holes admit rot organisms into bulbs. Discard infested bulbs. If bulb flies are a problem in your area, try covering plants with floating row cover during the egg-laying period (late spring). Dust plants weekly with a commercial pyrethrin/rotenone mix if pests have been a problem in past years.

Plant stunted; flowers deformed. Cause: Basal rot. Infection by *Fusarium* basal rot causes the base of the bulb to turn soft and brown. The rot spreads up through the bulb. Infected plants are stunted and have few flowers. Dig and destroy bulbs that show signs of disease. Avoid injuring bulbs during planting or cultivation; fungi most often infect bulbs through wounds.

Plant fails to grow and/or flower; leaves with swollen spots. Cause: Bulb and stem nematodes. Feeding by microscopic roundworms causes deformed leaves with yellow-green spots and small, swollen areas. Bulbs develop dark internal circles or blotches and may fail to grow or bloom in spring. Dig and destroy severely infected bulbs and foliage. Soak mildly infested bulbs in hot (110°F) water for 3 hours, then plunge them immediately into cold water. Let them dry and store in a cool, dark place until fall when they can be replanted. Solarize infected soil or treat with a chitin source.

Nicotiana

Flowering tobacco. Annuals.

For both fragrance and color, nicotiana belongs in every summer garden. Clusters of sweetly fragrant, starlike, tubular flowers appear in red, pink, lime, lavender, or white. Group these 1′-3′ tall plants together for best showing. Older kinds bloom in evening only, but new offerings bloom in daylight. Night-blooming whites are most fragrant; day-blooming cultivars are much less fragrant. New blossoms open each day to replace spent ones.

Sow seed indoors in April; bottom heat will speed germination. Or, direct-seed outdoors in late spring. Do not cover very small seeds, which need light to germinate. Seedlings appear within 3 weeks and grow rapidly. Thin or transplant to 9″ apart. Nicotiana grows best in average soil in full sun or partial shade. Water well during hot, dry weather. Give less fertilizer and water in late summer to keep plants blooming longer. Nicotiana self-sows but may not come true to color.

Do not grow nicotiana in dusty areas; the sticky, fuzzy leaves attract and hold dirt. Also, do not plant near tomatoes. Nicotianas may attract insect pests and diseases that seldom seriously bother this hardy plant, but will quickly move on to any nearby tomatoes.

Problems

Leaves, stems, and buds distorted. Cause: Aphids. For controls, see "Leaves, stems, and buds distorted" on page 20.

Leaves yellow; plant weakened. Cause: Whiteflies. See "Leaves yellow; plant weakened" on page 18 for controls.

Seedlings or young plants cut off at soil level. Cause: Cutworms. For control measures, see "Seedlings or young plants cut off at soil level" on page 20.

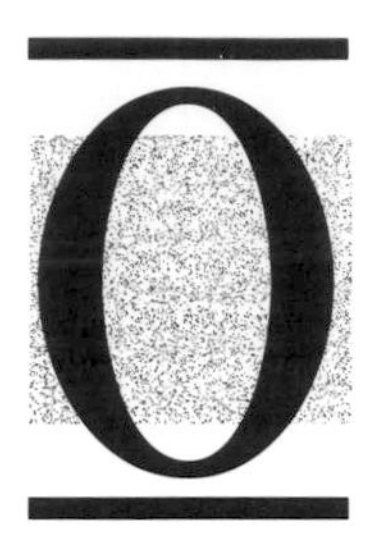

Okra

Abelmoschus esculentus (Malvaceae)

Okra is an annual vegetable grown for its fleshy seed pods. Okra does best in loose, well-worked soil with a pH between 6.0 and 7.0, and full sun. It needs lots of phosphorus, so work in plenty of bonemeal and compost before planting.

Okra grows best at temperatures between 70° and 85°F and does not tolerate temperatures below 60°F.

Warm soil at planting time is very important. Wait until the soil is at least 60°F before planting. Okra does not germinate well and is more susceptible to pests in cooler soil. If you have cool soils with high clay content, cover the planting area with clear plastic to warm up the soil at least 2 weeks before you want to plant. Cover planted seeds with clear plastic or row cover for a few weeks to give them a good start. Remove the clear plastic as soon as seeds germinate to avoid damaging the seedlings. Okra does well with black plastic mulch because it warms the soil.

Okra is susceptible to several common vegetable crop problems, including damping-off, aphids, fungal wilts, leafminers, and mites. See the Vegetables entry beginning on page 245 for symptoms and controls.

Problems

Leaves turn yellow, wilt, and fall off; stem rotted at soil line. Cause: Southern blight. White to pink fungal growth may spread over the soil around the base of infected plants. Severely infected plants die. Destroy infected plants and dig out and dispose of the top few inches of soil within 6″ of the stem. Adding lots of compost to the soil helps prevent this disease.

Plant yellow and stunted, wilts during bright, hot days; roots have swollen galls. Cause: Root knot nematodes. Galls may be up to 1″ in diameter. Destroy infected plants. Control pest nematodes by adding chitin and parasitic nematodes to the soil.

Buds, flowers, and seed pods malformed and drop prematurely. Cause: Stink bugs. Seed pods may have hard, callused bumps. Adults are flat, shield-shaped, green, blue, or red insects that emit a sharp odor when crushed. Handpick to control, or spray plants with a commercial pyrethrin/rotenone mix if injury is serious.

Flowers turn brown; seed pods rot. Causes: Choanephora blight; gray mold. Infected parts may be covered with white or gray growth. Spray plants with compost tea as soon as plants start to bloom and repeat every few weeks to help prevent these fungal diseases, or with sulfur to keep mild infections from spreading.

Seed pods with chewed holes. Cause: Caterpillars. Leaves may also have holes. Handpick or spray plants with BTK if you see caterpillars.

Seed pods pitted, discolored, or with water-soaked areas. Cause: Cold injury. Prevent damage by protecting plants with row cover when cold nights are expected.

Onion

Allium cepa and other species
(Liliaceae)

Onions are biennial vegetables grown for their sweet to pungent bulbs and greens. Dried or fresh, raw or cooked, onions are an indispensable ingredient in a variety of soups, salads, breads, and casseroles.

Culture

Onions grow best in full sun and deep, fertile, well-drained soil with lots of organic matter. Work in a generous amount of compost before planting. Onions need high levels of nitrogen and potassium and moderate to high levels of phosphorus, so do a soil test and amend soil as needed before planting. Onions grow well in raised beds or ridges, especially if soil is clayey.

Onions grow best between 55° and 75°F, and will tolerate temperatures as low as 45° and as high as 85°F. They prefer cool temperatures early in their growth and warm temperatures near maturity.

Keep the soil moist since onions have shallow roots, but don't allow soil to become saturated because onions are susceptible to several root rot diseases. Mulching onions with composted leaves or straw is highly recommended to maintain soil organic content, help prevent disease, and keep down weeds. Wait until soil warms to apply mulch.

To help prevent populations of disease organisms from building up in the soil, avoid planting onions where onion family members have been grown during the previous 3 years. In general, white onions are more prone to problems than yellow or red ones.

You can grow onions from seeds, transplants, or sets. Discard any diseased sets or transplants. Soak sets, roots, or seeds in compost tea for 15 minutes before planting to help prevent disease. Dust the roots of sets and plants with bonemeal after soaking and before planting to give them a good start.

Onion bulb formation is controlled by day length, so selecting suitable cultivars for your area is crucial. In the North choose "long-day" cultivars, and in the South choose "short-day" cultivars.

Allow tops to fall over naturally, then pull bulbs and let them air-cure for 2 weeks. After curing, sort out damaged bulbs and those with thick necks and put aside for immediate use. Store others at temperatures just above 32°F.

Plant and Leaf Problems

Seedlings fall over. Cause: Damping-off. Prevent problems by planting in raised beds and presoaking seed in compost tea.

Plant wilts rapidly. Cause: Cutworms. Check for a hole in the stem at or just below the soil line and fat, 1″-2″ long, dull brown or gray caterpillars in the soil near the base of plants. Sprinkle moist bran mixed with BTK on the soil surface in the evening, or add parasitic nematodes to the soil at least a week before planting to control them.

Plant stunted; leaves may be yellow. Causes: Nitrogen deficiency; waterlogged soil; aphids; pink root. Onions need ample nitrogen, deficient plants are pale and grow slowly. Waterlogged soil damages roots and produces the same symptoms. Spray plants and drench roots with fish emulsion or fish-meal tea to alleviate symptoms. Plant in raised beds to improve drainage and add compost before planting to prevent problems.

Aphids are soft-bodied, pale green, black, gray, pink, or white fluffy-coated, sucking insects. Check for them on young leaves. For mild infestations, knock the pests off the plants with a blast of water. Spray plants with insecticidal soap to control them, or with a commercial pyrethrin/rotenone mix if infestation is severe.

If roots and bulb are pinkish, the plant has pink root. Roots infected with this fungal disease shrivel and die. Destroy infected plants. Prevent problems by ensuring good drainage and adding ample organic matter to the soil. Plant resistant cultivars such as 'Crystal Wax Pickling', 'Early White Supreme', 'Hybrid Big Mac', 'J K Special', 'Tokyo Long White', 'Uno Grande PPR', and 'Voyager'.

Plant yellow and wilted. Causes: Fusarium bulb rot; onion maggots; lesser bulb flies; wireworms. Onions infected with Fusarium bulb rot have soft necks, and entire bulbs may be soft and brown. Destroy infected plants. Prevent problems by planting cultivars, such as 'Long White Summer Bunching', 'Northern Oak', 'Sentinel', and 'Valiant', that are somewhat resistant.

Onion maggots feed on roots, killing seedlings and older plants. They also burrow into bulbs, making them unfit for use. Onion maggots are 1/4" long, white, and taper to a point at the head. Adults are small gray flies that lay eggs early in the spring. Destroy infested plants. Do not compost them; onion maggots thrive in compost piles. To prevent problems, apply parasitic nematodes to the soil before planting and cover plants with row cover as soon as they come up to prevent flies from laying eggs. Trap maggots by planting a few onions at scattered points around the garden a few weeks before the main planting date. These larger plants attract egg-laying adults, remove and destroy them when infested. 'Egyptian Tree' onions are tolerant of onion maggots. In general, white cultivars are more susceptible to maggot attack than are yellow or red cultivars.

Lesser bulb fly larvae are 1/2" long, wrinkled, and yellow-gray. They cause injury similar to onion maggots and are controlled the same way.

Yellow to reddish brown, slender, tough-bodied, segmented worms up to 1 1/2" long feeding on roots and bulbs are wireworms. Adults are dark-colored, elongated click beetles. Apply parasitic nematodes to the soil before planting to control them.

Leaves with white streaks or blotches. Causes: Precipitation damage; thrips. Excessive rain or hail can spot onion leaves. Spots are various sizes, and damage doesn't spread. Maintain good air circulation and make sure soil is well-drained to minimize problems.

If leaf tips are distorted or brown, and leaves are stippled with white, look for onion thrips. Heavy infestations cause plants to wither and turn brown. Adults are tiny, slender, yellow to brown, rapidly moving, winged insects. The larvae can barely be seen with the naked eye. Thrips thrive in hot, dry weather. Trap them with sticky traps hung just above plant level. Try blue, yellow, and white traps to see which work best. Spray plants with insecticidal soap or a commercial pyrethrin/rotenone mix to control severe infestations.

Leaves with water-soaked or papery, white spots with vertical splits. Cause: Onion leaf blight. Leaf tips turn yellow, then brown. Spray plants with sulfur as soon as you see symptoms if weather is cool and humid. Prevent problems by planting tolerant cultivars such as 'Tokyo Long White'.

Leaves with pale green to brown spots. Causes: Downy mildew; ozone injury. Leaf tips turn yellow, then brown, and may be covered with a fuzzy mold. As the disease progresses, spots turn black with a purple, fuzzy mold, and leaves yellow and die. Downy mildew thrives in cool, humid weather and tends to stop spreading when the weather is warm and dry. Spray plants with sulfur to prevent the disease from spreading in cool, wet weather. Don't handle plants when wet to avoid spreading the disease.

Irregular areas with tiny brown flecks are caused by high levels of ozone in the air. Spray plants with seaweed extract and fish emulsion to encourage new growth.

Leaves with sunken, light-colored spots with concentrically ringed, purple centers. Cause: Purple blotch. Spots enlarge and girdle leaves, which wither and fall over. Spray plants with sulfur if disease is present to keep it from spreading. Cool soil makes plants more prone to this fungal disease. Warm soil by covering it with clear plastic for a few weeks before you plant.

Leaves with small reddish orange blisters. Cause: Rust. Leaves infected with this fungal disease may turn yellow and die. Bulbs are small. Clean up and dispose of tops to eliminate overwintering spores. One type of rust infects both onions and asparagus, so keep the 2 crops apart to prevent problems.

Leaves with black streaks filled with dark brown powder. Cause: Smut. Young plants are usually affected. Cool soil makes plants more prone to this fungal disease. Warm soil by covering it with clear plastic for a few weeks before you plant. Prevent problems by planting tolerant cultivars such as 'Evergreen Hardy White' and 'Tokyo Long White'.

Bulb Problems

Bulbs small and soft; roots pinkish or shriveled. Cause: Pink root. See "Plant stunted; leaves may be yellow" on page 155 for controls.

Bulbs rot in the ground. Cause: White rot. Bulbs may be covered with a white fluffy growth. Destroy infected plants; don't compost them. Prevent problems by providing good drainage and presoaking seeds in compost tea.

Bulbs with gray, water-soaked outer layers. Cause: Heat or cold injury. If temperatures are above 85°F, protect plants with a thick straw mulch and keep soil moist. Protect overwintering onions with mulch, or dig and store just above freezing.

Bulbs with bleached, soft patches. Cause: Sunscald. Protect onions from direct sun while curing, especially in hot weather. White cultivars are especially sensitive to sunscald.

Bulbs with thick necks that do not cure well. Causes: Potassium deficiency; seed stalk formation. Confirm deficiency with soil test and amend soil as needed before planting. Use bulbs with thick necks first.

Onions form seed stalks after any dormant period. Improperly stored sets or cultivars not suited to the day length in your area may go to seed before forming bulbs. Fluctuating temperatures or drying and wetting of soil may cause seed stalk formation. Harvest and use bulbs as soon as possible.

Bulbs with dark green or black, concentrically ringed spots. Cause: Onion smudge. This fungal disease usually appears near harvest or in storage. Destroy infected bulbs. Prevent problems by planting tolerant cultivars such as 'Downings Yellow Globe', 'Early Yellow Globe', and 'Southport Red Globe'.

Bulbs with sunken, dry, brown to black areas around neck. Cause: Neck rot. Necks of bulbs rot and gray mold develops inside bulbs. This fungal disease usually appears near harvest or in storage. Destroy infected bulbs. To prevent problems, cut back on watering as onions begin to mature, especially near harvest. Be careful not to injure bulbs while weeding or harvesting. Cure bulbs properly before storing them in a cool place.

Oregano

Origanum heracleoticum (Labiatae)

Oregano is a hardy (Zone 5) perennial herb grown for its aromatic leaves. Start oregano from cuttings or plants rather than from seed. Seedlings are quite variable and may have very little flavor. Oregano does best in average, well-drained soil with a pH between 5.0 and 8.0, and at least 4 hours of sun per day. Oregano needs little added fertilizer or water, but does like a layer of mulch to protect its shallow roots.

Oregano is usually quite trouble-free. Pale, stippled, or bronzed leaves are caused by mites—tiny, spiderlike creatures that thrive in hot, dry weather. Control them with insecticidal soap spray. See the Herbs entry beginning on page 116 for other possible problems.

Pachysandra

Pachysandra, spurge. Perennials.

Low-growing pachysandras top the list of durable, shade-tolerant groundcovers. These hardy, 8″-12″ plants with glossy evergreen foliage spread quickly, providing excellent cover below trees and on slopes. Clusters of tiny, tubular, white flowers appear in late spring.

Plant in slightly acid, moist but well-drained soil rich in organic matter. Mature plants endure drought, but do best with about 1″ of water per week. Full sun may scorch variegated cultivars; give partial to deep shade. Pachysandras compete well with taller plants and prosper in shady sites where few other plants will grow. Cut back by about one-quarter in spring to promote compact growth.

Most problems arise when plantings are stressed by drought, poorly drained soil, or overcrowding. Fungal diseases favor damp, crowded conditions; thin to improve air circulation and remove infected plants. Scale insects also prefer tightly spaced plantings, where their populations swell rapidly if undeterred. Spider mites infest drought-stressed plants when weather is hot and dry; water adequately and use a strong stream of water to knock pests off.

Paeonia

Peony. Perennials.

Attractive, long-lived, bushy plants that bear numerous 3″-6″ blossoms, peonies are the perennial gardener's dream come true. Early summer flowers arrive in shades of pink, red, white, and yellow amid glossy, lobed, green foliage; plants form neatly rounded clumps roughly 3′-4′ tall.

Peonies prefer full sun and moist, well-drained, richly organic soil, although light shade is tolerated and may prolong bloom in the South. A protected site limits wind damage to blossoms. Most are hardy to Zone 5 and do best in cold-winter climates.

Easy to care for in most respects, peonies are finicky about planting. Set rootstocks so that the reddish buds or "eyes" are no more than 1″-2″ below the soil surface. Mulch after the ground freezes the first winter after planting to prevent heaving. Divide roots in fall, if necessary, leaving at least 3 buds on each section. Cut stems back to below ground level in fall.

Problems

Flower buds absent. Causes: Improper planting; excess shade; immature plant; large, old crown; excess nitrogen; disturbed roots. Choose planting sites carefully; set roots at the proper depth; be patient with new plants. Do not apply high-nitrogen fertilizers. If mature peonies stop blooming, rule out other possible problems and divide if needed—division and other root disturbances also reduce bud formation.

Flower buds don't open. Causes: Late spring frost; drought; high temperatures; low soil fertility. Weather extremes notwithstanding, water adequately and feed peonies with compost or a slow-acting, general-purpose fertilizer in spring. If summer heat is the problem, plant early flowering cultivars.

Flower buds die or petals distorted. Cause: Flower thrips. These 1/25″ insects feed on buds, stem tips, and flowers, causing distortion or white, brown, or red flecks. Thrips are hard to control because they burrow into plant tissue. Remove and destroy infested plant parts. Encourage natural predators such as pirate bugs, lacewings, and lady beetles. Use blue sticky traps to monitor and trap pests. Insecticidal soap sprays may provide some control once pests appear on traps.

Stems with sunken lesions. Cause: Anthracnose. Sunken lesions with pink blisters appear on stems. Plants may die. Cultural controls such as regular fall cleanup and thinning stems to improve air circulation are effective. Treat severe problems with copper fungicide sprays.

Shoots wilt, collapse; crowns with gray mold. Cause: Fungal diseases. Several fungi cause blights or stem and crown rots in peonies. Botrytis blight causes shoots to wilt suddenly and fall over. Stem bases blacken and rot; gray mold may appear near soil; buds may wither and blacken. Flowers and leaves may turn brown and develop mold. Remove and destroy infected plant parts. Don't put manure near plant crowns; clear mulch from crowns in spring to let soil dry. Avoid overwatering and wet, poorly drained soil. If problems persist, scrape away the top 2″ of soil around plants and replace with clean sand. In spring, spray shoots with bordeaux mix.

Plant stunted; leaves yellow, spotted; roots with tiny galls. Cause: Root knot nematodes. Feeding by these microscopic pests reduces roots' ability to take up water and nutrients. Control root knot nematodes by applying chitin or parasitic nematodes to soil. Also promote natural nematode controls by increasing soil organic matter. Drenching soil with neem may provide some control. In severe cases, remove plants, solarize the soil, and replant with nematode-free stock.

Papaver

Poppy. Perennials.

History has heaped symbolism upon the brightly colored blooms of poppies, yet these broad, crinkled flowers burst forth undeterred in spring and early summer. The diminutive alpine poppy (*Papaver alpinum*) offers 1″ blooms and 8″-10″ height, while Iceland poppy (*P. nudicaule*) and Oriental poppy (*P. orientale*) can exceed 1′ in height and spread, with 3″-7″ flowers. Deeply divided, hairy, gray-green foliage surrounds leafless flower stems, each with a single bloom.

Most poppies are very hardy and perform best in cool summers and cold winters; grow plants as annuals in warmer climates. Full sun to light shade and well-drained soil satisfy poppies' needs; established plants tolerate some drought, but soggy soil guarantees rotting of fleshy roots. Foliage dies back after flowering ends, and plants disappear by late summer. Mark the spot to avoid digging injury to roots. New leaves appear in fall. Divide every 5 years, in late summer, to maintain vigor. Poppies self-sow if allowed to set seed; seedlings may not come true to parents.

Problems

Leaves, stems, and buds distorted, sticky; clusters of small insects. Cause: Aphids. See "Leaves, stems, and buds distorted, sticky; clusters of small insects" on page 177 for controls.

Leaves with water-soaked spots; foliage and flowers blacken. Cause: Bacterial blight. Infected plants turn brown and lose leaves; girdled stems die. Dig and destroy infected plants and surrounding soil. Solarize soil before replanting with disease-free seed; water early in the day so leaves dry quickly.

Parsley

Petroselinum crispum (Umbelliferae)

Parsley is a biennial herb grown as an annual for its leaves. Sow seed outdoors once the soil has reached 50°F; it takes 3 weeks to germinate. Plant in cool, moist, well-drained soil with a pH between 5.0 and 8.0. It needs at least 3 hours of full sun daily.

Parsley has few problems. Handpick leaf-eating caterpillars or spray plants with BTK. Dwarfed or abnormally crinkled leaves can be caused by viruses. Destroy infected plants, and control aphids and other sucking insects that spread viral diseases. Dark, yellow-bordered spots are leaf blight caused by fungi. Spray with fish emulsion or with sulfur if disease is severe. Carrot weevils eat roots and make leaves yellow; for controls, see "Roots with dark tunnels" on page 61. See the Herbs entry beginning on page 116 for other possible problems.

Parsnip

Pastinaca sativa (Umbelliferae)

Parsnips are biennial vegetables grown for their long, white, sweet-flavored roots. Culture is much like that of carrots, except that parsnips require less fertilizer and prefer a pH between 6.0 and 6.8. They grow best at temperatures between 60° and 64°F, poorly above 75°F, and tolerate temperatures as low as 40°F. Parsnip seed can take 3 weeks to germinate. Soak seed overnight before planting, keep seedbed moist, and always use fresh seed to prevent germination problems.

Parsnips are in the same family as carrots and are troubled by many of the same pests and diseases. See the Carrot entry beginning on page 60 for culture and information on problems.

Problems

Roots with dark cankers; leaves with water-soaked or dark spots. Cause: Brown fungal canker. Interior of root may be discolored, and other diseases can enter root through cankers and can rot flesh. Destroy severely infected plants and spray remaining plants with copper or bordeaux mix. Prevent problems by not planting parsnips where they have been grown for at least 2 years, providing good drainage, keeping soil over shoulders of roots, and planting resistant cultivars such as 'Andover' and 'Cobham Improved Marrow'.

Parthenocissus

Boston ivy, Virginia creeper, woodbine. Vines.

These are deciduous, alternate-leaved vines that climb by means of rootlike holdfasts.

Plant container-grown creepers in spring. A moist, loamy soil is best, in sun or shade. These plants are very adaptable to difficult conditions.

Japanese beetles can attack foliage; see "Leaves skeletonized" on page 236 for controls. For details on controlling scales, see "Leaves yellow; stems and leaves covered with small bumps" on page 237. Powdery mildew may disfigure leaves; see "Leaves with powdery white coating" on page 237 for controls. If leaf spots are a problem, see "Leaves with spots" on page 237.

Pea

Pisum sativum (Leguminosae)

Peas are annual vegetables grown for their tasty seeds and, in some cases, seedpods. Fresh peas are a special treat you have to grow yourself to enjoy.

Culture

Peas grow well in almost any soil, but do best in soil with lots of organic matter and a pH between 5.5 and 6.8. A 1" layer of compost worked well into the soil before planting will provide sufficient nutrients for a good crop.

Peas are a cool-season, moisture-loving crop. They grow best between 60° and 75°F, poorly at temperatures above 75°F, but will tolerate temperatures as low as 45°F. Pea foliage can withstand a light frost, but pods and flowers will be damaged unless they are covered.

Most disease problems in peas can be avoided with proper culture. Do not plant in wet soils. Plant in raised beds and add plenty of compost to loosen the soil. Rapid germination is essential to avoid root rot problems. Choose lighter soils for earliest plantings if you have a choice, and keep soil moist, but not wet. Avoid touching plants when they are wet. Dispose of vines after harvest and till soil to reduce future problems. Plant peas where no peas or beans have grown for at least 3 years.

Soak seed in compost tea for 15 minutes or as long as overnight to help prevent disease and speed germination. Treat seed with an inoculant labeled for garden peas before planting to promote nitrogen fixation. Be sure to buy fresh inoculant each year, or check the date on the package for viability.

Peas are susceptible to certain micronutrient deficiencies. Spray young plants with seaweed extract every 2 weeks to help prevent deficiencies and boost production.

Leaf and Whole Plant Problems

Seeds do not germinate; seedlings stunted or dying. Causes: Seedcorn maggots; damping-off; root rot. Seedcorn maggots are 1/4″ long, yellow-white, spindle-shaped seed-eaters. Adults are small flies. Seedlings that do come up are deformed and spindly. Remove damaged seedlings and plant fresh seed about a week after applying parasitic nematodes to the soil to control maggots. Seedcorn maggots thrive in cool, wet soil, so wait until soil is warm to plant to help avoid problems with them.

Damping-off is caused by soil-dwelling fungi that thrive in cool, wet conditions. Keep soil moist, but not soggy, thin seedlings to improve air circulation, and spray them with compost tea as soon as the first true leaves open to prevent problems.

Pea root rot can kill seedlings. Older infected plants are stunted and have shrunken, discolored roots and stems near the soil line. Prevent problems by planting in well-drained soil with lots of organic matter. Beans are susceptible to the same fungus, so keep the 2 crops apart. Prevent problems by planting cultivars, such as 'Bolero' and 'Sprite', that are somewhat tolerant to pea root rot.

Seedlings clipped off at soil line. Cause: Cutworms. Check for fat, 1″–2″ long, dull brown or gray caterpillars in the soil near the base of plants. Sprinkle moist bran mixed with BTK on the soil surface in the evening, or apply parasitic nematodes to the soil at least a week before planting to control them.

Leaves yellow; growth slow. Causes: Nitrogen deficiency; waterlogged soil. Drench soil and spray foliage with compost tea, fish emulsion, or fish-meal tea, or side-dress plants with compost to alleviate deficiency symptoms. Waterlogged soil damages roots and prevents them from using nutrients available in the soil. Prevent problems by choosing well-drained sites, adding organic matter to the soil to improve drainage, and planting in raised beds.

Leaves yellow and distorted. Causes: Tarnished plant bugs; pea aphids; potato leafhoppers. Tarnished plant bugs are oval, light green or brown, 1/4″ long bugs that inject a plant-deforming toxin as they feed on young leaves. Trap them with white sticky traps or spray plants with a commercial pyrethrin/rotenone mix in the evening to control severe infestations.

Pea aphids are soft-bodied, small, light to dark green, sucking insects usually found on new growth. Infested leaves are thickened and curled and may be covered with a sticky material. For light infestations, knock pests off plants with a blast of water. Spray plants with insecticidal soap in the evening to control, or with a commercial pyrethrin/rotenone mix if infestation is severe. Repel aphids with reflective mulch or by planting cultivars with silvery leaves.

Potato leafhoppers are green or brown, spindle-shaped, 1/10″–1/2″ long, winged insects. Nymphs are smaller and wingless. Infested leaves have curled margins, and flowers or pods may fall off. Trap leafhoppers with yellow sticky traps or spray as for pea aphids above. Cover seedlings with row cover if leafhoppers have been a problem in the past.

Leaves yellow; plant wilting and stunted. Cause: Fusarium wilt. Stem near soil line is yellow-orange to black when cut open. If pods form, they contain few seeds. Destroy plants infected with this fungal disease. To prevent problems, plant resistant cultivars such as 'Bounty', 'Daybreak', 'Green Arrow', 'Knight',

'Maestro', 'Oregon Sugar Pod 2', 'Snowflake', and 'Sparkle'.

Leaves mottled and distorted. Cause: Mosaic viruses. Destroy infected plants. Control aphids and cucumber beetles that spread viruses, and leguminous weeds, such as vetch, that can harbor viruses. Prevent problems by planting tolerant cultivars such as 'Knight' and 'Maestro'.

Leaves stippled with white. Cause: Mites. Leaves become bronzed when severely infested. These tiny, spiderlike insects thrive in hot, dry weather. Look for tiny moving specks on the undersides of leaves. Spray plants with insecticidal soap in the evening to control mites.

Leaves with water-soaked or white spots. Causes: Downy mildew; powdery mildew. Downy mildew is common in damp weather. Leaves and pods are covered with a thick, white growth that turns violet-black. Powdery mildew is more common in dry weather. The whole plant may be covered with white powdery growth. Spray plants with sulfur in the evening, as soon as you notice either disease. Prevent problems by planting cultivars resistant to downy mildew, such as 'Green Arrow' and 'Knight', or to powdery mildew, such as 'Bounty', 'Knight', 'Maestro', 'Oregon Sugar Pod 2', and 'Snowflake'.

Leaves with light brown to purple spots. Cause: Blight. Stems and pods are also spotted. Leaves may turn yellow and plants may die. Various fungi and bacteria can cause these disease symptoms. Spray plants with copper if weather is wet. Remove severely infected plants. Presoak seed in compost tea and don't touch plants when they are wet to help prevent problems.

Leaves with wandering, white or translucent tunnels. Cause: Leafminers. Larvae are white maggots that tunnel through leaves. Adults are tiny black-and-yellow insects. Once tunnels appear, the larvae are inside leaves and spraying will not kill them. Destroy infested leaves. Trap future generations of adults with yellow sticky traps, and spray plants with a commercial pyrethrin/rotenone mix if large numbers of adults are trapped. Prevent problems by protecting plants with row cover as soon as they come up to exclude egg-laying adults. Certain parasitic nematodes can attack leafminer larvae inside leaf tunnels.

Leaves with small holes. Cause: Cucumber beetles. Damage usually occurs on young plants. Beetles are yellow or greenish, 1/4" long, with spots or stripes. Spray plants with a commercial pyrethrin/rotenone mix if infestation is severe. Cover emerging seedlings with row cover to prevent problems.

Leaves with large holes. Cause: Caterpillars. Many caterpillars feed on leaves and pods. Handpick, or spray plants with BTK if worms are feeding.

Pod Problems

Blossoms drop; no pods form. Causes: Weather extremes; nutrient imbalances. Excessive heat or rain can cause blossoms to drop. Wait for new blossoms to form. Copper and/or molybdenum deficiency cause the same symptoms. Spray plants with seaweed extract to help prevent deficiencies. If plants are very dark green and no blossoms form, suspect too much nitrogen. Wait for blossoms to form. Avoid high-nitrogen fertilizers.

Young pods distorted and withered; older pods with water-soaked or purplish spots. Cause: Blight. See "Leaves with light brown to purple spots " above for controls.

Pods mottled, deformed, and rough. Cause: Mosaic viruses. See "Leaves mottled and distorted" above for controls.

Pods with white spots. Causes: Downy mildew; powdery mildew. See "Leaves with water-soaked or white spots" above for controls.

Pods with chewed holes. Cause: Caterpillars. Various caterpillars eat pea pods. Handpick, or spray plants with BTK if caterpillars are feeding. Cover young plants with row cover to prevent moths from laying eggs.

Seeds with brown spots or cavities. Cause: Manganese deficiency. Spray plants with seaweed extract every 2 weeks to prevent deficiencies.

Seeds with small, round holes. Cause: Pea weevils. Seeds may be hollow. Fat, white, 1/3″ long larvae feed on seeds. Adults are 1/5″ long, dark beetles with light markings that may feed on pea flowers. Discard infested seeds; do not compost them. Cover seeded areas with row cover to prevent adults from laying eggs. Spray plants with a commercial pyrethrin/rotenone mix if adults are present.

Peach

Prunus persica (Rosaceae)

Peaches are deciduous trees growing from 4′ to 20′, depending on soil, rootstock, and cultivar. In very early spring pink blossoms appear on last year's shoots. Nectarines belong to the same species as peaches and have the same cultural requirements as well as the same diseases and pests. Both are hardy in Zones 5-9.

Culture

Plant in full sunlight in an area with well-drained soil and no late spring frosts. Prune each winter to stimulate growth, to thin next year's fruit, and to allow sunlight to penetrate the tree. Train trees to a framework of well-spaced, wide-angled branches. Prune bearing trees each winter to admit light into the tree and encourage good air circulation. As you prune, remove diseased and spindly wood and crossed branches. Where growth is too dense, remove extra shoots at their bases. To develop growth on spindly shoots, prune off the end of the shoot just above an outward-facing bud. For more pruning information, see "Pruning and Training" on page 101. Thin fruit each spring so it has room to develop.

Like other fruit trees, peach trees need a period of cold weather rest or dormancy. The number of hours of cold between 32° and 45°F each cultivar needs before it breaks dormancy is referred to as chill hours. (Cold below 32°F doesn't count toward meeting the dormancy requirement.) Once the number is reached, the tree assumes winter is over and starts growing the next warm day. Peaches bloom rapidly once their requirement has been met, which makes them more prone to frost damage than other fruit trees that are slower to burst into bloom. Call your local extension service to find out how many chill hours your area receives and what cultivars match that requirement. If you choose a cultivar that needs fewer chill hours than you normally receive, an unseasonable winter thaw in your area may bring the tree well into flower weeks before spring actually arrives. But if you choose one that needs more chill hours than your climate supplies, the tree won't get enough chilling to stimulate normal bloom.

Most peaches are self-pollinating, but a few cultivars, such as 'Indian Free' and 'J. H. Hale', require cross-pollination. For more information on setting fruit, see "Setting Fruit" on page 101.

Fruit Problems

Young fruit with crescent-shaped scars. Cause: Plum curculios. These beetles, common east of the Rockies, leave characteristic scars as they lay eggs in fruit. Damaged fruit often drops. For control, spread a dropcloth under the tree and jar the trunk and branches with a padded mallet. Collect and destroy beetles that fall onto the sheet. For best results,

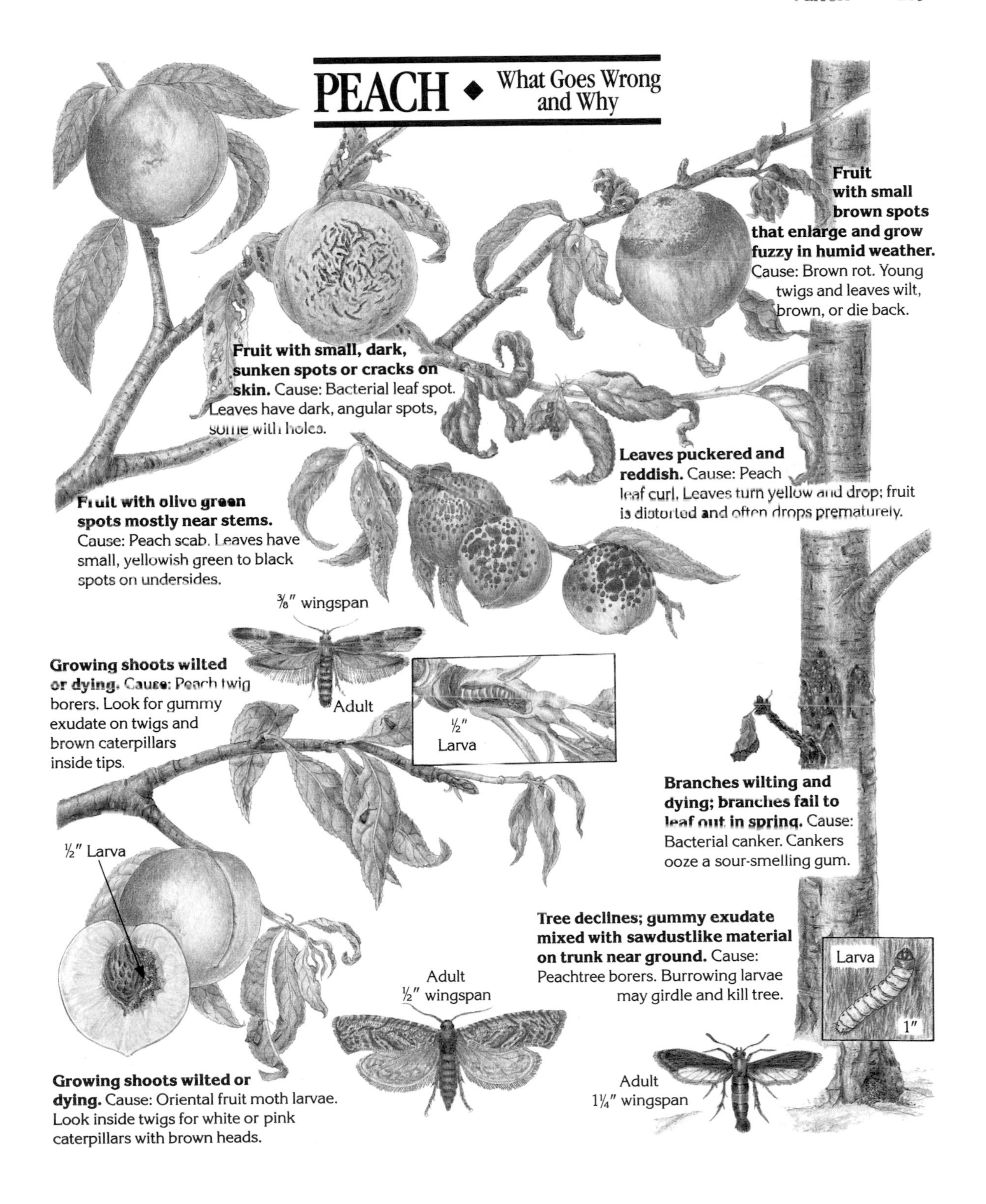
PEACH ◆ What Goes Wrong and Why
Fruit with small brown spots that enlarge and grow fuzzy in humid weather. Cause: Brown rot. Young twigs and leaves wilt, brown, or die back.
Fruit with small, dark, sunken spots or cracks on skin. Cause: Bacterial leaf spot. Leaves have dark, angular spots, some with holes.
Leaves puckered and reddish. Cause: Peach leaf curl. Leaves turn yellow and drop; fruit is distorted and often drops prematurely.
Fruit with olive green spots mostly near stems. Cause: Peach scab. Leaves have small, yellowish green to black spots on undersides.
⅜″ wingspan
Adult
½″ Larva
Growing shoots wilted or dying. Cause: Peach twig borers. Look for gummy exudate on twigs and brown caterpillars inside tips.
Branches wilting and dying; branches fail to leaf out in spring. Cause: Bacterial canker. Cankers ooze a sour-smelling gum.
½″ Larva
Tree declines; gummy exudate mixed with sawdustlike material on trunk near ground. Cause: Peachtree borers. Burrowing larvae may girdle and kill tree.
Larva
1″
Adult ½″ wingspan
Adult 1¼″ wingspan
Growing shoots wilted or dying. Cause: Oriental fruit moth larvae. Look inside twigs for white or pink caterpillars with brown heads.

do this twice a day, beginning when you see the first scarred fruit. Also, to prevent curculio eggs from hatching, collect and discard dropped fruit. For more information, see "Young fruit with crescent-shaped scars" on page 186.

Fruit with small brown spots that enlarge and grow fuzzy in humid weather. Cause: Brown rot. This fungus may also cause blossoms to wither and die. Fruit is most infection-prone 3 weeks before ripening. Infected fruit may drop early or turn soft and brown, then wither into hard, black, shriveled fruit (known as mummies) that remains on the tree. For control, inspect trees before growth begins in the spring. Remove and destroy both mummies and twigs or branches with gummy lesions. For more control, spray sulfur early to protect blooms, then again later to protect fruit. Copper sprays also help. Since damaged fruit is more prone to infection, control insects such as plum curculios, which puncture fruit and allow infection to enter. Resistant cultivars include 'Elberta', 'Orange Cling', 'Red Bird', and 'Sunbeam'. See page 165 for an illustration of this disease.

Fruit with olive green spots mostly near stems. Cause: Peach scab. Spots first appear on immature fruit and then turn brown and velvety. Fruit skin cracks; fruit is distorted or dwarfed. Infection arises from twig lesions, but damp weather spreads the fungus throughout leaves, fruit, and twigs. Infection is worst in warm climates and in late-fruiting cultivars. See page 165 for an illustration of this disease. To control scab, remove infected fruit and clean up fallen leaves and fruit. For persistent infection, spray lime-sulfur or sulfur only every 10-21 days throughout the growing season. If spring weather is unusually warm and wet, spray sulfur weekly from the time flower buds first show green until blossoms begin to open.

Fruit with small, dark, sunken spots or cracks on skin. Cause: Bacterial leaf spot. This disease, common east of the Rockies, spreads in the spring from oozing cankers. Twig cankers appear water-soaked. Leaves turn yellow and drop. Infection weakens trees, makes them prone to winter injury, and reduces fruit quality and yields. See page 165 for an illustration of this disease. Control with copper spray. Resistant cultivars include 'Belle of Georgia', 'Candor', 'Com-Pact Red Haven', 'Dixieland', 'Earlired', 'Early-Red-Free', 'Harbrite', 'Harken', 'Loring', 'Madison', and 'Red Haven'.

Fruit with sunken, corky lesions. Cause: Tarnished plant bugs. Bugs hibernate in nearby weeds and move into trees in the spring. For control, remove weeds and plant debris. For persistent infestation, hang white sticky traps in lower tree branches.

Fruit with pinkish worms. Cause: Oriental fruit moth larvae. For more information, see "Growing shoots wilted or dying" below.

Leaf and Branch Problems

Growing shoots wilted or dying. Causes: Oriental fruit moth larvae; peach twig borers. Both pests tunnel into growing shoots and cause wilting. With oriental fruit moths, the tree may look unusually bushy from growth of new lateral shoots below wilted parts. To find oriental fruit moth larvae, slit stems below the wilted sections and look for a pinkish white caterpillar, up to ½" long. Later, a second larval generation bores into and ruins fruit. Controls that help keep this pest in check include repeated, timely sprays of BTK and release of the parasitic wasp *Macrocentrus ancylivorus*. Use pheromone traps to monitor and control the pests. Mating disruption pheromones, such as Isomate-M, are also effective. See page 165 for an illustration of an oriental fruit moth larva.

Wilting plus a gummy exudate from twigs may indicate peach twig borers. This pest also damages fruit. The second generation of these brown, ½" caterpillars tunnels into fruit, usually near the stem. For control, find borer entry holes and cut off wilted branches just below the hole. Destroy infested prunings. For large branches, slide a wire into the hole to kill the borer. Peach twig borers prefer weak trees; keep trees strong through proper fertilization, pruning, and irrigation. See page 165 for an illustration of this pest.

Growing shoots covered with a white powdery coating. Cause: Powdery mildew. This disease also causes new growth to be stunted and distorted. Rainy weather does not cause the disease fungus to spread. In fact, it is most common in weather patterns featuring cool nights changing to warm days. For control, apply sulfur spray or lime-sulfur spray.

Young twigs and leaves wilted, brown, or dying back. Cause: Brown rot. Shoots and leaves may turn brown and decay. Gummy branch or twig lesions may also form. For more information, see "Fruit with small brown spots that enlarge and grow fuzzy in humid weather" above.

Leaves puckered and reddish. Cause: Peach leaf curl. Later in the season, infected leaves may yellow, shrivel, and drop. New growth is stunted and swollen and often dies. Fruit often drops prematurely and may have a reddish, irregular, rough surface. See page 165 for an illustration of this disease. You can't cure this fungal disease during the current season, but copper sprays or lime-sulfur sprays help control it. Resistant cultivars include 'Candor', 'Clayton', 'Com-Pact Red Haven', 'Correll', 'Dixieland', 'Elberta', 'Red Haven', and 'Stark EarliGlo'.

Leaves with dark, angular spots; some spots with holes. Cause: Bacterial leaf spot. Eventually, infected leaves may yellow and drop. For more information, see "Fruit with small, dark, sunken spots or cracks on skin" above.

Leaves with small purple spots, some spots with centers missing. Cause: Shothole disease. Spots also appear on fruit and then turn scabby. Centers of leaf spots often enlarge to about 1/4", then fall out. You can't cure this fungus during the current season, but copper spray aids control. Keeping irrigation water away from foliage also helps. Dormant season pruning of infected buds and twigs, which have a varnished appearance, helps prevent the disease next year.

Leaves stippled yellow. Cause: European red mites. These extremely tiny spider mites suck juices from leaves, causing yellow speckles on foliage. For light infestations, knock mites off leaves with a strong spray of water; for heavy infestations, spray with insecticidal soap. Lime-sulfur spray applied early in the growing season also aids control. Or, to kill overwintering mite eggs, apply superior oil in the spring, just when leaf buds are about 1/2" long. Mites are controlled by natural enemies, including predatory mites. If simple controls don't work, purchase predatory mites (*Metaseiulus occidentalis*) from insectaries or organic farm and garden supply catalogs.

Branches wilting and dying, fail to leaf out in spring. Causes: Bacterial canker; Valsa canker. Look on the branches for sunken, elliptical lesions, often oozing a reddish gum. With bacterial canker, this gum smells sour, and leaves may have small, angular spots. See page 165 for an illustration of this disease. For both cankers, prune wilted or dying branches off below the infected area. Sterilize pruning tools in isopropyl alcohol or a 10 percent bleach solution (1 part bleach to 9 parts water) between cuts. On large limbs, cut out the canker into healthy wood, which has a lighter color. Copper sprays help control bacterial canker.

Valsa canker is a fungal disease that enters through injured bark. To prevent bark damage, make sure trees harden off in the fall by avoiding late-season fertilization or pruning. Also plant in well-drained soil and paint trunks with white latex paint diluted with an equal amount of water to reduce bark-damaging temperature fluctuations. Train trees to a good form with strong, wide-angled scaffold limbs. Also prune during bloom time, when wounds heal fastest. Cultivars adapted to cold climates or that drop leaves early are most resistant to Valsa canker. Choose 'Brighton', 'Elberta', 'Harbrite', 'Harken', 'Madison', or 'Reliance'.

Whole Plant Problems

Leaves yellow; dieback of whole limbs. Cause: Scales. Colonies of these sucking insects cling to the bark and appear as small bumps that can be easily scraped off with a fingernail. Control scale by spraying trees with dormant oil in the winter. Also control ants, which encourage scale insects, by wrapping a strip of paper or plastic around the trunk and covering it with a commercial sticky coating such as Tanglefoot.

Tree declines; gummy exudate mixed with sawdustlike material on trunk near ground. Cause: Peachtree borers. Inspect the trunk near or just below the ground; you may find holes and a gummy exudate made by peachtree borers, which are the larvae of a clear-winged moth that bore the inner bark. See page 165 for an illustration of this pest. Dig borers out, being careful to minimize tree damage, or kill them by inserting a wire into holes. Spreading a ring of tobacco dust around the base of the tree discourages these pests. Peachtree borers are attracted to weak trees. Keep trees vigorous and avoid mechanical damage. Resistant cultivars include 'Dixie Red', 'Elberta', and 'Jubilee'.

Leaves unusually small; leaves yellow or oddly shaped. Cause: Viral infection. A few viral diseases cause leaf anomalies in peach trees. One particular virus, known as peach rosette, causes trees to produce shoots that have abnormally short distances between the leaf nodes. Avoid viruses by starting with clean stock. Avoid planting near possible virus carriers, such as old peach trees or wild chokecherries (*Prunus virginiana*). Remove and destroy infected trees.

Pear

Pyrus communis and hybrids (Rosaceae)

Pears are deciduous, upright-growing trees ranging in size from about 8′ to more than 30′, depending on the cultivar and rootstock. The white blossoms appear in spring mostly on spurs—short growths that elongate less than an inch each year. Fruit ripens from late summer into autumn. Most gardeners are familiar with European pears, including the familiar 'Bartlett' and 'Bosc' pears. However, Asian pears, which have crisp, juicy, almost-round fruit, will also grow well in most parts of the United States and Canada. Pears are hardy in Zones 4-9.

Culture

Pears require full sun. They enjoy well-drained, moderately fertile soil, but will tolerate heavy or poorly drained soils more easily than most other fruits. Since pears flower later than most other common fruits, they do not require a site completely free of spring frost. Train European pears to a central leader system and Asian pears to an open-center system. See the illustration on the opposite page. Prune bearing trees a moderate amount, enough to allow light into the tree and to stimulate enough new growth to replace old spurs or branches that have been removed. For more pruning information, see "Pruning and Training" on page 101. To set fruit, most pear cultivars require cross-pollination from a second compatible cultivar planted nearby. Before buying any cultivar, be sure you can meet the pollination requirements. For more information, see "Setting Fruit" on page 101.

Fruit Problems

Fruit with olive-brown, corky spots that turn dark brown. Cause: Pear scab. This fun-

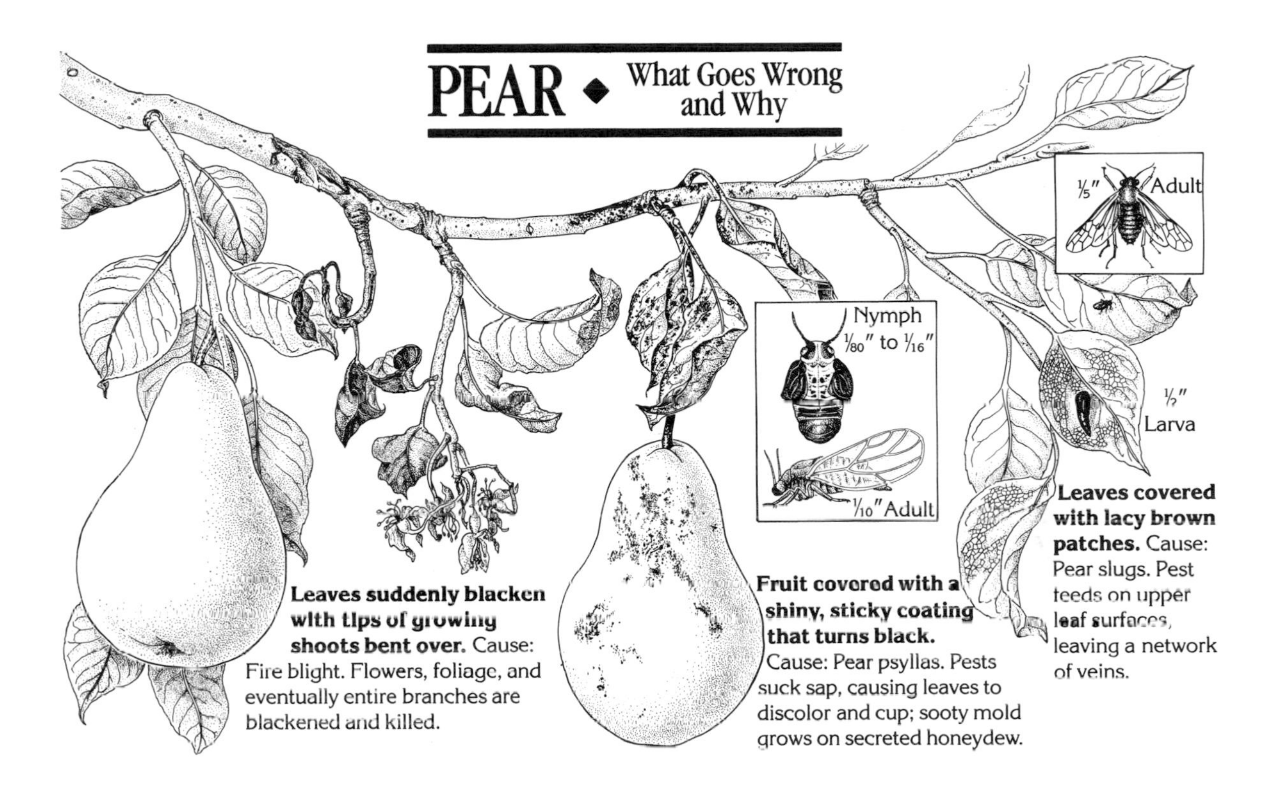

gal disease may also cause malformed fruit. It overwinters on fallen leaf litter and infected twigs and spreads to the tree in the spring when warm air currents and splashing rain move fungal spores from the ground into the tree. To control pear scab, remove old leaves from beneath trees and compost or bury them. Prune out infected twigs, which bear small, blisterlike pustules. Do both of these chores in late winter or early spring before growth begins. Lime-sulfur applied early in the growing season also helps control the disease. Scabby fruit may be unattractive, but it is still edible once diseased portions are cut away. If scab is prevalent in your area, plant scab-resistant cultivars, such as 'Bartlett'.

Fruit with holes surrounded by brown, crumbly excrement. Cause: Codling moth larvae. These fat, white or pinkish, 7/8″ caterpillars tunnel through fruit and may have departed by the time you discover the holes, which may be filled with what looks like moist sawdust. Infested fruit may drop prematurely. For light infestations, kill eggs by spraying superior oil on leaves and twigs within 2–6 weeks of blossoming. For heavy infestations, kill larvae before they tunnel into fruit by spraying the tree canopy with ryania at petal fall and again 10–14 days later. Pyrethrin spray mixed with a synergist (see "The Other Ingredients" on page 469) also kills larvae. For more information, see "Fruit with holes surrounded by brown, crumbly excrement" on page 22.

Fruit covered with a shiny, sticky coating that turns black. Cause: Pear psyllas. These 1/10″, red to green, winged insects resemble cicadas. They overwinter in tree crevices and ground litter and emerge in early spring to lay eggs in trees. Newly hatched psylla nymphs pierce and suck juices from both foliage and fruit. Most regions support at least 4 generations of this pest per year, allowing plenty of opportunity for infestation throughout the growing season.

The sticky coating on fruit is honeydew excreted by the psyllas, and the black blotches are mold that grows on the honeydew. Frequently, yellow jackets congregate around honeydew-coated leaves. You can clean the honeydew or mold off the fruit and eat it. Excessive psylla feeding will weaken the tree. To control psyllas, smother overwintering adults with dormant oil sprays applied in late winter before spring growth begins. During the growing season, kill eggs and mature young with superior oil spray or insecticidal soap.

Leaf Problems

Leaves suddenly blacken, with tips of growing shoots bent over. Cause: Fire blight. Don't confuse this bacterial disease with sooty mold, a black fungus that rubs off easily. Fire blight bacteria enter the tree at the growing tips and may travel down toward the roots and kill the whole tree.

Fire blight most readily attacks very succulent growth; to make trees a little more blight-resistant, avoid heavy pruning or nitrogen fertilizer, both of which induce vigorous growth. When pruning, thin out whole branches rather than heading them back. This pruning strategy reduces the total number of cuts and won't stimulate the growth of soft, blight-susceptible, sideshoots. To prevent rampant growth on mature trees, grow grass right up to the trunk; if needed, allow the grass to grow longer than normal.

Control fire blight during the growing season by pruning off branches a foot below infected sections. Between each cut, dip pruning shears into isopropyl alcohol or a 10 percent bleach solution (1 part bleach to 9 parts water) to prevent spreading the disease as you prune. As the growing season progresses, fire blight bacteria grow within blighted shoots down branches and toward the roots. During the fall, the bacteria form a sunken, dark canker in which to overwinter. Over the winter, inspect branches for these cankers and prune off damaged branches at least 6″ below the cankers.

One way to prevent fire blight from killing an entire tree is to purchase trees in which a blight-susceptible cultivar has been grafted onto blight-resistant framework. Good rootstocks include the 'Old Home' × 'Farmingdale' series, *Pyrus betulifolia,* and *P. calleryana.* Copper sprays applied in spring may also reduce the incidence of fire blight, but spraying alone will not control the disease. Cultivars fairly resistant to fire blight include 'Duchess', 'Garber', 'Moonglow', 'Orient', 'Seckel', 'Seuri', 'Shinko', 'Starking Delicious', and 'Yali'.

Leaves suddenly blacken in autumn. Cause: Pseudomonas blight. You might mistake this disease for fire blight, but fire blight symptoms appear in warm spring weather, and its cankers ooze in the spring. Pseudomonas blight first appears during cool fall weather and its cankers do not ooze in the spring. Control it by pruning, as for fire blight (see "Leaves suddenly blacken, with tips of growing shoots bent over" above).

Leaves with small, round, dark spots having purple margins. Cause: Fabraea leaf spot. This disease, most common east of the Mississippi River, may cause leaves to turn yellow and drop, weakening the tree. Since the disease overwinters in twig cankers as well as fallen leaves, garden sanitation does little to control the disease. Copper spray, applied when leaves are half out and then once or twice again at 2-week intervals, offers some control.

Leaves with dark, velvety patches. Cause: Pear scab. The spots are mostly on leaf undersides, and infected leaves often pucker and drop. For more information, see "Fruit with olive-brown, corky spots that turn dark brown" on page 168.

Leaves covered with lacy brown patches. Cause: Pear slugs. These green-black, slimy creatures are not true slugs; they're the larvae of the pear sawfly, a black-and-yellow insect slightly larger than a housefly. In the spring, sawflies emerge from soilborne cocoons and lay eggs on leaves. Eggs hatch into sluglike larvae that skeletonize leaves. Remove them by handpicking or with blasts of water. For heavy infestations, use insecticidal soap spray; on a dwarf tree, apply wood ashes or diatomaceous earth. (Caution: Wood ashes may damage leaf tissue.) Spider mites multiply quickly in dry, dusty conditions, so wash ashes off after a few days to avoid encouraging a rise in spider mite population.

Leaves with small brown blisters on undersides. Cause: Pearleaf blister mites. Inside these blisters are small, elongated, white mites that you need a magnifying glass to see. Damage usually is cosmetic and can be ignored, but if control is necessary, spray lime-sulfur before buds open or superior oil as buds swell. Be sure to spray oil thoroughly to kill adults hiding in bud scales.

Leaves turn yellow; limbs die back. Cause: San Jose scale. If limb dieback is preceded by extensive leaf yellowing, look for the small gray bumps of San Jose scale clinging to the bark. Colonies of these sucking, nearly immobile insects cling to bark and weaken trees by sucking sap. Control scale with dormant oil spray applied in late winter.

Pecan

Carya illinoinensis (Juglandaceae)

Pecans are large, deciduous trees with separate male and female flowers borne on the same plant. These plants are hardy in Zones 6-9. Trees grown at the northern hardiness limit may not have a long enough season to ripen nuts.

Plant pecans in deep, well-drained soil in full sun. Trees need almost no pruning. Since male and female flowers on a given tree often mature at different times, plant 2 different cultivars to ensure pollination.

Problems

Nuts contain insect larvae. Causes: Pecan weevil grubs; hickory shuckworms; pecan nut casebearer larvae. Pecan weevils, snouted beetles similar to plum curculios, lay eggs within newly formed nut kernels; larvae that hatch from these eggs feed on the nut and emerge several weeks later through tiny holes in the shell and husk. Infested nuts are worthless and generally cling to the tree instead of splitting normally from the hulls. To control this pest, place dropcloths beneath the tree and jar the limbs with padded poles. Collect and destroy the fallen adults. Do this every 2 weeks beginning in midsummer until weevils no longer drop.

You will find cream-colored, 3/8" hickory shuckworms eating the kernels of immature nuts that drop early as a result of the infestation. These pests are the larvae of gray moths, and the final larval generation overwinters on the ground in dropped shucks. To control shuckworms, keep dropped hulls picked up or bury them at the end of the season.

Pecan nut casebearers, 1/2", green larvae of a gray moth, spin webs around nuts and then enter to feed. Infested nuts may drop prematurely. Overwintering larvae leave cocoons in early spring and bore into growing shoots, causing wilting. Damage is worst early in the season. To help control casebearers, pick up and destroy all dropped infested nuts. The parasitic wasp *Trichogramma minutum* attacks and parasitizes casebearer eggs. For heavy infestations of casebearers, release 5,000 of these wasps per infested tree.

Leaves with olive-brown spots on undersides; shucks with small, velvety, olive-brown spots. Cause: Pecan scab. Leaf spots may enlarge to large black areas. The spots on the shucks may spread into large, sunken, black lesions. Nuts may drop prematurely. This fungal disease is worst in humid areas and overwinters on infected twigs, shucks, and leaves. To reduce infection the following season, clean up plant debris in autumn and knock off old leaf stems and shucks before trees leaf out in the spring. Scab-resistant cultivars include 'Cape Fear', 'Chocktaw', 'Curtis', 'Desirable', 'Gloria Grande', and 'Stuart'.

Tree declines; leaves yellow, die, but remain on tree. Cause: Cotton root rot. Roots of infected trees are brownish, rotted, and soft. Cotton root rot fungus lives in the soil for 5 years or more, thriving especially in areas with heavy, moist, alkaline soil. An acid-type fertilizer might help mildly affected trees, but dig up dying trees and do not replant with pecans.

Pelargonium

Pelargonium, geranium.
Tender perennials grown as annuals.

Pelargoniums, commonly known as geraniums, are easy-to-grow plants that are popular with both beginner and advanced gardeners. Most gardeners grow flowering geraniums for the clusters of red, pink, lilac, salmon, or white flowers that appear all season over 1′-3′ plants. Other geraniums are grown primarily for their fragrant leaves, which release powerful scents when brushed. Geraniums can also have attractively shaped and variegated foliage, with up to 4 colors in just 1 leaf. In mild climates geraniums persist throughout the year. They are delightful in beds, borders, pots, and hanging baskets.

Buy new nursery-grown plants each year, or start your own from seed. You can direct-seed outdoors in late spring, although indoor seeding in late winter or early spring gets better results. Good window or artificial light is needed indoors. Cover seeds lightly. Germination is erratic within 3-8 weeks. When seedlings have 2-3 true leaves, transplant them to individual 3″ pots. After the last frost, plant outdoors 10″ apart.

Geraniums usually need full sun. In the Deep South, give them light shade, or plants may burn. Soil should be well-drained, medium rich, and slightly acid; if it is alkaline, add peat moss to planting area. Water regularly, letting the soil dry out slightly between waterings. Fertilize regularly but lightly; overfeeding results in large leaves and fewer flowers. Remove spent flower heads to prolong blooming. Pinching back growing tips in early stages promotes fullness. Cut the plant back if it becomes leggy.

Problems

Leaves yellow; plant weakened. Cause: Whiteflies. For controls, see "Leaves yellow; plant weakened" on page 18.

Leaves stippled with yellow; foliage webbed. Cause: Spider mites. For control measures, see "Leaves stippled with yellow; foliage webbed" on page 18.

Leaves, stems, and buds distorted. Cause: Aphids. See "Leaves, stems, and buds distorted" on page 20 for controls.

Stems with rotted sections; leaves wilt. Cause: Stem rot. This fungal disease starts at the base of plants and works upward. It commonly affects cuttings, but can also injure full-grown plants. Stems often turn black. Only take cuttings from healthy plants, and stick them in a sterile medium. Remove and destroy affected parts or plants.

Leaves with spots. Cause: Leaf spots. See "Leaves with spots" on page 19 for control information.

Leaves with large, ragged holes. Cause: Slugs and snails. See "Leaves with large, ragged holes" on page 18 for controls.

Pepper

Capsicum annuum var. *annuum* (Solanaceae)

Peppers are tender perennials that are grown as annuals in temperate climates for their sweet to fiery hot fruit. Pick peppers when they are unripe and green (or sometimes yellow or purple-black) or after they ripen to red, orange, yellow, or brown, depending on the cultivar. In general, ripe peppers are sweeter or less hot than unripe ones of the same cultivar.

Culture

Peppers require deeply worked, well-drained soil with lots of organic matter. They do best at a pH between 6.0 and 6.8, but tolerate pH as low as 5.5. Peppers require a moderate to high level of nitrogen and moderate levels of phosphorus, potassium, and calcium. Have the soil tested and amend as needed before planting. Peppers grow best between 65° and 80°F. Temperatures above 85° or below 60°F can cause blossoms to drop without setting fruit.

Peppers tolerate drought, but do best in soil that is evenly moist, but not soggy. Plant in raised beds to improve drainage, if needed. Stake peppers to keep fruit from touching the ground and use mulch to control weeds and prevent soilborne diseases from splashing up on the fruit.

Do not plant peppers where tomatoes, potatoes, eggplants, or peppers have been planted within the past 3-5 years. Also, try to plan your planting scheme to separate these crops in the garden. Compost or till under all plant residues at the end of the season, and till the soil to reduce overwintering pests.

Purchase sturdy, insect- and disease-free plants, or start your own from seed indoors. Soak seed in a 10 percent bleach solution (1 part bleach to 9 parts water) for 10 minutes, and rinse in clean water before planting to reduce seed-borne diseases. Pepper seeds germinate best above 80°F. Once seedlings are up, they grow best at 70°F during the day and 60°F during the night. Wait until soil temperatures reach 65°F before setting out transplants. Spray transplants with an antitranspirant to help reduce disease problems, and water them with seaweed extract or compost tea to give them a good start. To improve fruit set, spray plants with seaweed extract with 1 teaspoon of Epsom salts added to 1 gallon when the first flowers open.

Leaf and Whole Plant Problems

Seedlings fall over; stems girdled or rotted at soil line. Cause: Damping-off. Disinfect reused pots and flats by dipping them in a 10 percent bleach solution and letting them air-dry before filling them with fresh seed-starting mix. Sow seeds thinly to allow for air movement around seedlings. Cover seed with a thin layer of soilless mix or vermiculite. Water only enough to keep soil moist, not soggy. Thin seedlings and spray with compost tea as soon as first true leaves open to help prevent the problem.

Seedlings clipped off at soil line. Cause: Cutworms. Check for fat, 1"-2" long, dull brown or gray caterpillars in the soil near the base of plants. Once they chew off a seedling, there is nothing you can do except protect the remaining seedlings from nocturnal cutworm attacks. Place cutworm collars around transplants, sprinkle moist bran mixed with BTK on the soil surface in the evening, or add parasitic nematodes to the soil at least a week before planting to control them.

Leaves pale green and small. Cause: Nitrogen deficiency. Spray plants and drench roots with fish emulsion to alleviate symptoms, and side-dress with compost.

Leaves yellow, distorted, and sticky. Cause: Aphids. These small green, pink, black, gray, or white fluffy-coated insects suck plant sap. For mild infestations, knock pests off plants with a blast of water. Spray plants with insecticidal soap in the evening to control, or with a commercial pyrethrin/rotenone mix if infestation is severe.

Leaves mottled with yellow; young growth malformed. Causes: Tobacco mosaic virus; other viral diseases. Destroy diseased plants. Presoak seed in a 10 percent bleach solution before planting, or choose resistant cultivars to prevent problems. Wash hands after handling tobacco and before touching peppers to prevent tobacco mosaic virus (TMV). Control aphids, because they spread viral diseases as they feed. Prevent problems by planting resistant cultivars such as 'Ace', 'Bell Captain', 'Bell Tower', 'Early Wonder', 'Elisa', 'Galaxy', 'Goldcrest', 'Gypsy', 'Lady Bell', 'Lasto', 'Northstar', 'Orobelle', and 'Yolo Wonder'.

Leaves yellow; plant stunted and wilts in hot weather. Cause: Nematodes. Plants eventually die. Roots may have swollen galls. Destroy infested plants, do not compost them. To control these pests, apply chitin or parasitic nematodes to the soil.

Older leaves yellow; shoots or whole plant wilts. Cause: Fusarium or Verticillium wilt. Fusarium wilt and Verticillium wilt are both fungal diseases and are difficult to tell apart. Both Fusarium and Verticillium wilt begin as a yellowing and wilting of the lower leaves. Plants are stunted and do not recover when watered. Cut open a stem near the soil line and look for internal discoloration. Verticillium fungi are active between 68° and 75°F, while Fusarium is active between 80° and 90°F. Destroy infected plants. Pepper Fusarium infects only peppers, while Verticillium infects a wide range of plant species, making effective rotation control difficult. Prevent problems by presoaking seed in a 10 percent bleach solution. Control pest nematodes to help reduce wilt problems. Few wilt-resistant cultivars are available; 'Giant Szegedi' is tolerant of Verticillium.

Leaves stippled yellow, or bronzed. Cause: Mites. Leaves turn dry and papery. These tiny, spiderlike insects thrive in hot, dry weather. Spray plants with insecticidal soap if populations are high (more than 1-2 pests per leaf).

Leaves with small, sunken, yellow-green spots. Cause: Bacterial spot. Spots eventually turn brown with lighter centers. Spray plants with copper as soon as symptoms appear to prevent further symptom development. Presoak seed in a 10 percent bleach solution to disinfect, and avoid touching wet plants.

Leaves with gray-brown spots. Cause: Cercospora leaf spot. This fungal disease only occurs in very warm climates. Spots develop a "frog-eye" appearance with light centers and dark edges. Spray plants with copper as soon as symptoms appear to prevent further symptom development. Presoak seed in a 10 percent bleach solution and plant resistant cultivars, such as 'California Wonder', to prevent problems.

Leaves with wandering, white or translucent tunnels. Cause: Leafminers. White, maggotlike larvae feed inside leaves, leaving empty tunnels behind them. Once larvae have entered leaves, spraying will not control them. Destroy mined leaves. Cover plants with row cover until flowers open to prevent adults from laying eggs on plants. Certain nematodes can attack leafminer larvae inside leaf tunnels.

Leaves with small holes. Cause: Flea beetles. Young transplants are the most susceptible. These tiny, black, brown, or bronze insects hop when disturbed. Spray plants with a commercial pyrethrin/rotenone mix if infestation is severe. Protect transplants with row cover until they start to flower.

Leaves with large holes. Causes: Hornworms; other caterpillars. Hornworms are 3″-4½″ caterpillars with white diagonal stripes. The tobacco hornworm has a red horn projecting from the rear, while the tomato hornworm has a black horn. Handpick or spray plants with BTK to control them. Do not spray caterpillars that are covered with small white cocoons; these cases contain the larvae of parasitic wasps that are natural hornworm predators.

Other caterpillars such as European corn borers and corn earworms sometimes feed on pepper leaves and fruit. Handpick, or spray plants with BTK if many caterpillars are feeding.

Flower and Fruit Problems

Few flowers form; flowers may drop without setting fruit. Causes: Excess nitrogen; extreme temperatures; pepper weevils. Plants with excess nitrogen are dark green and vigorous, but produce few flowers. Wait for flowers to form. Prevent problems by avoiding high-nitrogen soil amendments.

Temperatures over 85° or below 60°F can damage flowers and cause them to fall without setting fruit. Wait for new flowers to form. Protect plants with row cover until night temperatures remain above 60°F.

Check flowers for pepper weevils, which are ⅛″, reddish brown to black insects. These pests may also cause misshapen and discolored fruit. Destroy infested fruit; dust with rotenone as a last resort.

Fruit misshapen. Causes: Extreme temperatures; pepper maggots; viral diseases. Temperatures over 85° or below 60°F can damage flowers and prevent complete pollination. Resulting fruit is lopsided and deformed, but ripens evenly.

Pepper maggots are white with pale heads and grow up to ¼″ long. Adults are small yellow-and-brown flies. Ripening fruit is blotchy. Destroy infested fruit. Spray plants with a commercial pyrethrin/rotenone mix when they begin to set fruit, and repeat if damage occurs if you have had severe problems in the past.

If ripening fruit is blotchy but no maggot feeding is found, the plants may have a virus. See "Leaves mottled with yellow; young growth malformed" above for controls.

Fruit with green to dark brown, raised, wartlike spots. Cause: Bacterial spot. Spots can provide entry for more destructive rot fungi in wet weather. Spotted fruit is edible if not rotted. See "Leaves with small, sunken, yellow-green spots" above for controls.

Fruit with small, rotten spots or shallow depressions. Cause: Pepper maggots. See "Fruit misshapen" above for description and controls.

Fruit with faded or gray-white, sunken patches or pits. Causes: Sunscald; cold injury. Green or ripe fruit can be sunscalded. Damage shows up as a large, sunken patch on the exposed side. Patches turn dry and may develop black mold. Control leaf diseases to prevent defoliation, so fruit will be shaded and protected from direct sun. Stake plants.

Pepper fruits are damaged by temperatures below 37°F and develop small, sunken pits or large, discolored areas. Cover plants with row cover if temperatures near freezing are predicted.

Fruit with water-soaked, sunken areas at the blossom end. Cause: Blossom end rot. Seen on green or ripe fruit. Affected area becomes dark and shriveled. This condition is caused by calcium deficiency in the fruit. It is aggravated by drought or uneven soil moisture, root damage, high salt levels in the soil, and excess nitrogen. If soil test indicates deficiency, add high-calcium lime to the soil. Prevent problems by keeping soil evenly moist and by spraying plants with seaweed extract when the first flowers open and again when green fruit is visible.

Fruit with water-soaked areas near stem; entire fruit collapses into a slimy mess. Cause: Bacterial soft rot. Pick fruit as soon as water-soaked areas appear, and discard soft portions. Destroy rotted fruit. Prevent problems by controlling insect damage, staking and spacing plants so they dry out rapidly, and mulching to prevent soilborne bacteria from splashing up on plants during watering. Spray fruiting plants with copper if the weather is wet and you have had severe problems in the past.

Fruit with concentrically ringed, sunken spots. Cause: Anthracnose. Spots appear on green or ripe fruit. Destroy infected fruit. Spray fruit with copper when they are almost full-size and begin to feel solid if you have had problems in the past. Presoak seed in a 10 percent bleach solution to prevent seed-borne disease.

Fruit colors prematurely; small holes and sawdustlike material visible near stems. Causes: European corn borers; corn earworms. Once caterpillars are feeding in fruit, it is too late to control them by spraying. Pick and use or destroy infested fruit. Both of these pests feed on leaves for a few days before they enter fruit, so reduce damage by spraying plants with BTK if caterpillars are feeding on leaves. Prevent adults from laying eggs on plants by covering them with row cover.

Perennials

Pests and diseases that trouble ornamental perennial plants may warrant different treatment than those same problems appearing on annuals. Perennials are plants you hope to have around for a long time, so eliminating hungry insects and disfiguring diseases becomes more important. With few exceptions, the pests and diseases that plague perennials are best controlled through good site selection and cultural practices, as described in the individual plant entries.

The symptoms listed below represent damage caused by pests and diseases that attack a wide variety of herbaceous perennials. Controls for these problems do not vary significantly among plant species.

Problems

Leaves and/or flowers with holes. Causes: Japanese beetles; other leaf-eating beetles. Perennial beds attract a number of hungry beetles, including 3/4" black blister beetles, 1/10" flea beetles, 1/3", reddish brown rose chafers, and nocturnal, 1/3", gray Fuller rose beetles. Iridescent blue-green with bronze wing covers, 1/2" Japanese beetles top this list of troublesome pests; the adults devour the leaves, stalks, and flowers of nearly 300 plant species, while the 3/4", C-shaped grub larvae feed on roots and are a major pest of lawn grasses.

Injury caused by adult beetles ranges from small holes in leaves to skeletonized foliage to complete defoliation. Feeding by weevils, such as long-nosed Fuller rose beetles, resembles "ticket punches" around leaf margins. Japanese beetles are fond of flowers and often do most of their damage there. Most beetles are active during the day and are large enough to be readily visible while they feed. Weevils more often prefer night-time feeding; a flashlight reveals their activity.

Handpick adult beetles into a can of soapy water or a 5 percent solution of isopropyl alcohol. Wear gloves if your pests are blister beetles; contact with crushed beetles can cause burns and blisters on skin. Shake infested plants in early morning to knock beetles onto a dropcloth, then scoop them up and destroy.

Coordinate a community effort to set up Japanese beetle traps over a large area; otherwise traps will just attract these pests to your yard. Treat lawns with milky disease spores to control beetle larvae. Apply parasitic nematodes to the soil to limit all beetle grub populations. As a last resort, spray with neem, pyrethrin, or rotenone.

Leaves with large, ragged holes. Cause: Slugs and snails. Differing only in the presence or absence of a shell, slugs and snails range in size from 1/8" to 8" and may be gray, tan, green, black, yellow, or spotted. The cool, moist soil conditions favored by many perennials make plantings appealing to these busy mollusks. Mulch and plants with low-growing leaves provide shady hiding places from which they emerge to feed nocturnally. All species rasp large holes in leaves, stems, and bulbs; slimy trails of mucus also signal their presence.

Since slugs and snails travel on a flat, muscular foot, you can impede their progress with barriers of materials that irritate their soft bodies. Push 4"-8" copper strips into the soil around beds as edging. Sprinkle bands of coarse, dry, scratchy materials such as cinders, wood ashes, cedar sawdust, and diatomaceous earth around plants or beds; renew frequently. Lay boards, cabbage leaves, or overturned clay pots—anything that offers a cool, damp daytime haven—around the garden; destroy pests that congregate underneath. Handpick slugs and snails from plants at night; kill the pests by dropping them into soapy water or sprinkling them with table salt. Set shallow pans into the soil, placing the lip flush with the soil surface, and fill with stale beer or any fermenting liquid; remove drowned pests daily. Encourage predatory ground beetles by maintaining clover, sod, or stone walkways.

Leaves, stems, and buds distorted, sticky; clusters of small insects. Cause: Aphids. Several species of pear-shaped, 1/32"-1/8" aphids plague perennials. Aphids can be green, pink, black, gray, or with a white fluffy coating and have long antennae and 2 short tubes projecting from the rear of their abdomens. These pests cluster under leaves and on growing tips. Aphids suck plant sap, causing leaf and bud distortion and blossom and leaf drop. Their feeding may spread diseases such as aster yellows. As they feed, they excrete sticky honeydew on which sooty mold grows; see "Leaves with black coating" below for more information.

Wash aphids from plants with a strong spray of water; repeat as needed to control infestations. Encourage natural predators and parasites such as aphid midges, assassin bugs, lacewings, lady beetles, and spiders. Apply homemade garlic or tomato-leaf sprays. If water sprays fail, try sprays of alcohol, citrus oil, insecticidal soap, or neem. Use boric acid bait to control ants that herd aphids onto plants. As a last resort, spray pyrethrin, nicotine, or rotenone.

Leaves with black coating. Cause: Sooty mold. This fungus grows on the sugary, sticky honeydew produced by aphids and other sucking pests such as scales, whiteflies, and mealybugs. The black fungal coating doesn't harm leaves directly, but it does shade the leaves and reduce growth. The best control is to deal with the pests that are producing the honeydew. Identify the pest and apply the appropriate control. (If the plant itself doesn't show signs of pest damage, the honeydew may be dripping down from an overhanging plant.) On small plants, wipe the leaves with a damp cloth to remove the honeydew and the sooty mold.

Leaves covered with white powder. Cause: Powdery mildew. Powdery white or grayish fungal patches grow on upper leaf surfaces of plants subject to poor air circulation or hot, humid weather. As the disease spreads, leaves become distorted and may drop off. Powdery mildew is more severe when growing conditions are unsatisfactory.

To prevent powdery mildew, spray with sulfur fungicide approximately every 10 days during warm, wet weather. If infection occurs, remove and destroy infected leaves. Water early in the day so foliage dries before nightfall. Avoid wetting leaves when watering. Thin plants to improve air movement. Avoid ongoing problems by cleaning up debris in the fall and destroying severely infected plants. To protect plants, spray with antitranspirants, baking soda solution, a commercial fungicidal soap, garlic, copper, or sulfur.

Buds and leaves deformed or dwarfed. Cause: True bugs. Several of the generally shield-shaped true bugs (order Hemiptera), including tarnished plant bugs, four-lined plant bugs, lace bugs, and lygus bugs, feed on perennials. While many species of true bugs are beneficial insects, the pests in this group pierce plant tissue to drink sap, injecting toxins that deform plant tissues. In addition to dwarfing and deforming plants, feeding by true bugs causes sunken, rounded, tan to dark brown spots on foliage. As leaves grow, the dead areas tear into small holes.

Many of these pests overwinter in garden refuse; remove weeds and debris from area in spring and fall. Plant groundcovers and pollen plants to encourage populations of predatory insects. Handpick pests and drop them into a jar of soapy water. Spray every 3-5 days with insecticidal soap, making sure to treat the undersides of leaves, for 2 weeks. Treat severe infestations with rotenone.

Seedlings or young plants cut off at soil level. Cause: Cutworms. The fat, gray or dull brown, 1″ caterpillars of 1½″ brown or gray moths, cutworms feed on stems at night, cutting off transplants and sometimes eating entire seedlings. Severed plants are found lying on the ground the next morning. Some species climb plant stems, cutting off leaves and sideshoots as they feed. Most active in May and June, cutworms can destroy several plants each night. These caterpillars remain below the soil surface during the day.

Protect plants by placing a collar, such as a toilet paper roll or an open-ended can, around each young plant. Make each collar 2″-3″ wide and 1½″-2″ in diameter; push them into the soil so that about half the collar is below the soil surface. To kill some larvae, scatter moist bran mixed with molasses and BTK over soil in the evening a week before planting. Also, apply parasitic nematodes to the soil to reduce future populations.

Petunia

Petunia. Annuals.

Showy petunias are popular in bedding schemes all over the country. Single or double, plain or ruffled, trumpet-shaped flowers bloom all summer on 8″-18″ plants. Flower colors include lavender, yellow, red, pink, purple, and white, alone or in combinations. Plant in masses for eye-catching effect in beds, borders, containers, or hanging baskets.

Sow seed indoors 9 weeks before the last frost. Do not cover the extremely tiny seeds. Transplant carefully outdoors when frost danger passes. Many gardeners prefer to buy bedding plants. Get sturdy young plants, not leggy ones that won't grow well.

Single-flowered cultivars do well in average soil and are hardier in general; fancy ruffled types grow best in richer soil. Petunias need sunshine at least half the day, and prefer more. Water frequently during hot weather. Monthly light feedings encourage bigger and better flowers; overfertilization gives stem and leaf growth rather than blooms. In midsummer plants may begin looking straggly. Cut back to 6″ high, fertilize, and give a good soaking to revitalize.

Problems

Flowers with gray or brown spots; leaves with brown spots. Cause: Botrytis blight. This fungus is a real problem in humid climates. As the fungus spreads, the spots may become fuzzy with mold. The initial attack is on weak plants but spreads to healthy ones on contact, or by wind, rain, or handling. Remove infected flowers and leaves. Thin plantings to improve air circulation. Use drip irrigation rather than overhead watering. Look for Botrytis-resistant cultivars.

Leaves mottled and crinkled. Cause: Viral diseases. Several viruses can attack petunias. They are spread by contact or by insects, such as aphids and leafhoppers. Remove and destroy infected plants. Wash your gardening tools and hands after touching infected plants. Also, don't smoke in the garden, to avoid introducing viruses from the tobacco.

Leaves and/or flowers with round or irregular holes. Causes: Caterpillars; beetles; slugs and snails. Both moth and butterfly caterpillars feed on petunia leaves and buds. Handpick, or spray with BTK on appearance. Many types of beetles, including flea beetles, Japanese beetles, Colorado potato beetles, and cucumber beetles, feed on petunias. Handpick larger beetles. For serious infestations, apply rotenone in early morning. Slugs and snails like petunias, too. For control information, see "Leaves with large, ragged holes" on page 18.

Philadelphus

Mock orange. Shrubs.

Mock oranges are deciduous shrubs, grown primarily for their white, late-spring flowers. They are useful in shrub borders.

Set out in fall or spring in full sun or light shade. While they can adapt to varying soil conditions, mock oranges prefer a moist but well-drained site rich in organic matter.

Mock oranges are remarkably free of serious problems. Aphids sometimes infest the leaves; see "Leaves wrinkled and discolored" on page 235 for controls. Leafminer larvae make curved mines in the leaves; see "Leaves tunneled" on page 237. Plants crowded together are susceptible to powdery mildew; for controls, see "Leaves with powdery white coating" on page 237.

Phlox

Phlox. Perennials.

Showy, long-flowering plants that produce clusters of trumpet-shaped blooms in many colors, phlox range from tall, upright plants to low, trailing ones. Bountiful flowers of white, purple, pink, and blue—often with contrasting centers—cover the various species: Garden phlox (*Phlox paniculata*) grows 3′-4′ tall and blooms mid- to late summer; creeping phlox (*P. stolonifera*) is a 9″-12″ spring bloomer; 5″-6″ tall moss pink (*P. subulata*) flowers in early spring. Lance-shaped, 3″ long, light green leaves are typical of most phlox, although moss pink's fine, semi-evergreen foliage is needlelike.

In general, plant upright phlox in full sun and give trailing species partial shade. Garden phlox tolerates light shade, but weaker plants result. Moist, well-drained, and richly organic soils are favored by all phlox; add compost to soil at planting. Choose sites with good air movement to limit diseases; taller phlox need shelter from wind and may require staking. Phlox do not endure drought well;

give ½″-1″ of water weekly, applying it at soil level to avoid wetting leaves. Mulch to retain moisture in summer.

Remove spent flowers to prolong blooming and reduce seed set. Seedlings gradually take over garden phlox plantings if plants are allowed to self-sow; flower color and plant size may change dramatically over time, as seedlings are not true to parents. Shear low-growing moss pink back by half to encourage bushy growth. Divide as necessary to improve air circulation and reduce clumps.

The severity of pest and disease attacks varies among phlox species; most are durable plants that easily withstand problems. Garden phlox, however, is injured by a number of pest and disease organisms, as described below.

Problems

Buds and leaves deformed or dwarfed. Cause: Phlox plant bugs. See "Buds and leaves deformed or dwarfed" on page 178.

Leaves covered with white powder. Cause: Powdery mildew. See "Leaves covered with white powder" on page 177 for controls.

Leaves deformed; stems swollen or deformed; plant stunted. Cause: Bulb and stem nematodes. Plants infested with these tiny roundworms have crinkled, thin, or thread-like leaves, distorted shoot tips, and swollen stems. Stunted plants fail to flower and may die. Nematodes travel over wet plants on a film of water or on garden tools and gardeners moving among plants. There is no cure for infested plants; remove and destroy them. Do not grow phlox in sites where nematodes are present. Remove debris from gardens in fall to control pest populations.

Leaves stippled, reddish to yellow, with fine webbing. Cause: Spider mites. These 1/50″ pests feed on the undersides of foliage, causing it to yellow, dry up, and die; excessive feeding turns foliage almost white, and fine webs appear on leaves. Mites seek out water-stressed plants in hot, dry weather; adequate watering guards against these pests. While spraying plants daily with a strong stream of water often gives control, this practice promotes fungal leaf disorders in phlox. Introduce predatory mites (1,000 per 500 square feet); spray severe cases with avermectins, citrus oils, insecticidal soap, neem, pyrethrin, or rotenone.

Leaves with brown spots. Cause: Fungal diseases. Several fungi cause leaf spots that enlarge, run together, and form blotches. Don't overcrowd plants; avoid working with phlox when leaves are wet; remove and destroy infected plant parts and severely infected plants. If leaf spots are a regular problem, spray with sulfur or bordeaux mix.

Picea

Spruce. Trees.

Spruces are evergreen, needle-leaved, cone-bearing trees. In general, they grow as rigidly upright, pyramidal trees, reaching more than 100′ in height. They make impressive specimen trees on large properties. Their large size, formal habit, and tendency to lose their lower branches make spruces less desirable for the average home landscape.

Before you plant a spruce, be sure you have enough room to let the tree reach its normal size; it may be small when you bring it home, but it will grow surprisingly fast! Set out in spring as balled-and-burlapped plants on a sunny or lightly shaded site. Spruces need a moist but well-drained, slightly acid soil with lots of organic matter. They grow best in cool, northern climates.

Caution: When controlling pests, be aware

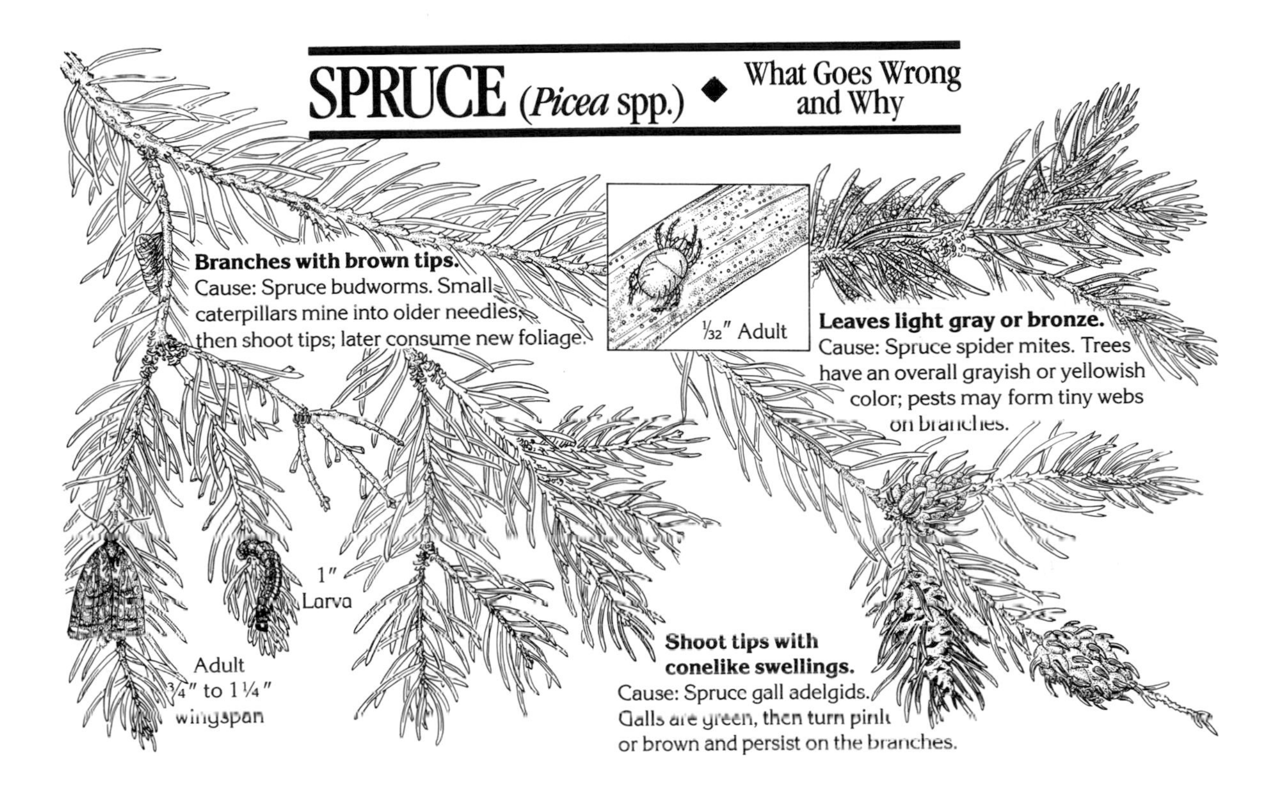

that oil sprays will remove the blue color from the foliage of blue spruces until new normal foliage is produced. Other spruce species may also be sensitive to oil sprays; read the product label, apply at a reduced rate if recommended, and test spray on a branch before treating the whole plant.

Problems

Branches with brown tips. Cause: Spruce budworms. This pest is widely distributed over the northern United States. The brown, 1″ caterpillars have 2 rows of white dots along their backs; they tunnel into opening buds, destroying the terminals and gradually migrating to the younger foliage, leaving the tree looking as if it had been swept by fire. See above for an illustration of this pest.

Handpicking works if only a few caterpillars are present; usually, however, they are so numerous as to require an application of BTK. Remember to reapply after rain.

Leaves light gray or bronze. Cause: Spruce spider mites. These tiny, spiderlike pests damage needles as they feed; the injury weakens the plant and can kill young trees. Mites start feeding on lower, older needles and progress upward and outward; tiny webs may be visible on needles. See above for an illustration of this pest.

Control in spring with dormant oil before growth starts; see caution above before spraying. During active growth, spray plants repeatedly with strong jets of water. For severe infestations, make 2 applications of insecticidal soap 7-10 days apart, or apply pyrethrin as a last resort.

Shoot tips with conelike swellings. Cause: Spruce gall adelgids. The pests themselves are seldom seen; in the spring the young insects feed on the new needles, causing the spruce to produce galls that enclose the pests. The adelgids mature in the galls, and the adults emerge in mid- to late summer. They overwinter as nymphs or adults. See page 181 for an illustration of this pest.

If there are only a few galls, prune them off before they open in July. For serious infestations, spray the plant with insecticidal soap as new growth starts the following spring. Or apply a dormant oil spray over the whole plant in late winter to kill overwintering pests; see caution above before spraying.

Leaves discolored; branch tips webbed. Cause: Spruce needleminers. The ½", greenish caterpillars feed at the base of needles, mining as they go, and sometimes web together the needles at the tips of twigs. Control by spraying the plant with a strong blast of water in fall or late winter to knock the webs and loose needles from the tree; then clean up all fallen debris.

Plant defoliated; branches bear cocoonlike bags. Cause: Bagworms. For controls, see "Plant defoliated; branches bear cocoonlike bags" on page 236.

Plant defoliated. Causes: Gypsy moths; sawflies. See "Plant defoliated" or "Leaves skeletonized" on page 236 for controls.

Terminal shoot curled and brown. Cause: White pine weevils. For more information, see "Terminal shoot curled and brown" on page 183.

Trunk or branches with small holes; limbs die or break off. Cause: Borers. For controls, see "Trunk or branches with small holes; limbs die or break off" on page 238.

Leaves yellow; stems and leaves covered with small bumps. Cause: Scales. See "Leaves yellow; stems and leaves covered with small bumps" on page 237 for controls.

Trunk or branches with oozing lesions; branch tips die back. Cause: Canker. For more information, see "Trunk or branches with oozing lesions; branch tips die back" on page 238.

Branch tips die back. Cause: Blight. This fungal disease attacks mostly during cool, wet weather. Good air circulation among trees discourages its onset. Leaves of infected plants shrivel and twigs die back. Cut off and destroy dead or diseased wood. In a very cool, wet spring, you can spray with bordeaux mix, starting when new growth appears on the trees and repeating at 2-week intervals until the weather gets warmer and drier.

Leaves yellow and drop. Cause: Rust. Whitish blisters on the undersides of the leaves followed by yellowing and leaf drop indicates rust. Prune out and destroy infected branch tips. To reduce the spread of mild infections, spray with sulfur, repeating 2 or 3 times at weekly intervals.

Pieris

Pieris. Shrubs.

Pieris are broad-leaved, evergreen shrubs. The waxy, white flowers are borne in terminal clusters and open in early spring. Pieris are used in foundation plantings and are ideal for woodland gardens.

Set out in spring as balled-and-burlapped or container-grown plants. Moist but well-drained, acid soils with lots of organic matter suit them best. Apply an organic mulch to keep down weeds and maintain a cool, moist root zone. Pieris thrive in sun or partial shade; plants will survive in dense shade, but flower less.

Pieris are subject to some of the insects and diseases that attack rhododendron; see the Rhododendron entry beginning on page 200 for symptoms and controls.

Pinus

Pine. Shrubs and trees.

Pines are needle-leaved, cone-bearing evergreens important both for their wood and ornamental value. In the landscape, some are useful as informal screens, others as hedges and foundation plants, and the more picturesque ones as specimens in the lawn or rock garden.

Set out in spring as balled-and-burlapped plants. A well-drained, somewhat acid soil enriched with organic matter suits them best. Pines prefer full sun, but will grow, although with a somewhat more open habit, in partial shade.

Problems

Branches dead; bark with yellow-orange blisters. Cause: White pine blister rust. This fungal disease needs 2 kinds of plants to complete its life cycle: a pine and a currant or gooseberry plant. While it does little damage to currants or gooseberries, this disease can quickly kill infected pines. It enters through the needles and grows into the bark, where it causes cankers that can girdle stems and branches. These cankers develop orange-yellow blisters that release spores. See page 184 for an illustration of these symptoms.

Avoid planting white or other 5-needled pines within 200 feet of currants or gooseberries. For other controls, see "Trunk or branches with oozing lesions; branch tips die back" on page 238.

Leaves brown at base; shoots ooze resin. Cause: Tip blight. This fungal disease causes needles to brown in early and midsummer, when they are about half their normal size. Entire shoot tips may be killed, most commonly on lower branches. Resin oozing from infected twigs is another common symptom. See page 184 for an illustration of these symptoms.

Tip blight most commonly attacks old or stressed trees. To avoid problems, keep plants growing vigorously with proper pruning, mulching, and watering (during drought). Remove the source of infection by pruning out dead branches during dry autumn weather. (Sterilize pruners between cuts in a solution of 1 part bleach to 9 parts water.) Also clean up fallen cones and other debris around the base of the tree. The following spring, spray with bordeaux mix when the new shoots start to grow; repeat when the shoots are half grown and again 2 weeks later.

Branches with brown tips. Causes: Pine tip and pine shoot moths; spruce budworms. For an illustration of pine shoot moths, see page 184; for spruce budworms, see page 181.

The larvae of these pests all cause similar damage, boring into the bases of needles and tunneling into shoot tips all over the plant. Their damage differs from that of white pine weevils in that the weevils only attack the top shoot (leader) of a plant. Handpicking works if only a few caterpillars are present. Pruning off and destroying infested tips in winter is a very effective control. Spraying with BTK before the caterpillars enter the shoots may be effective.

Terminal shoot curled and brown. Cause: White pine weevils. The 1/3", pale yellow larvae bore into the terminal shoot, distorting and eventually killing it. The 1/4", mottled brown adult weevils emerge in late summer to feed

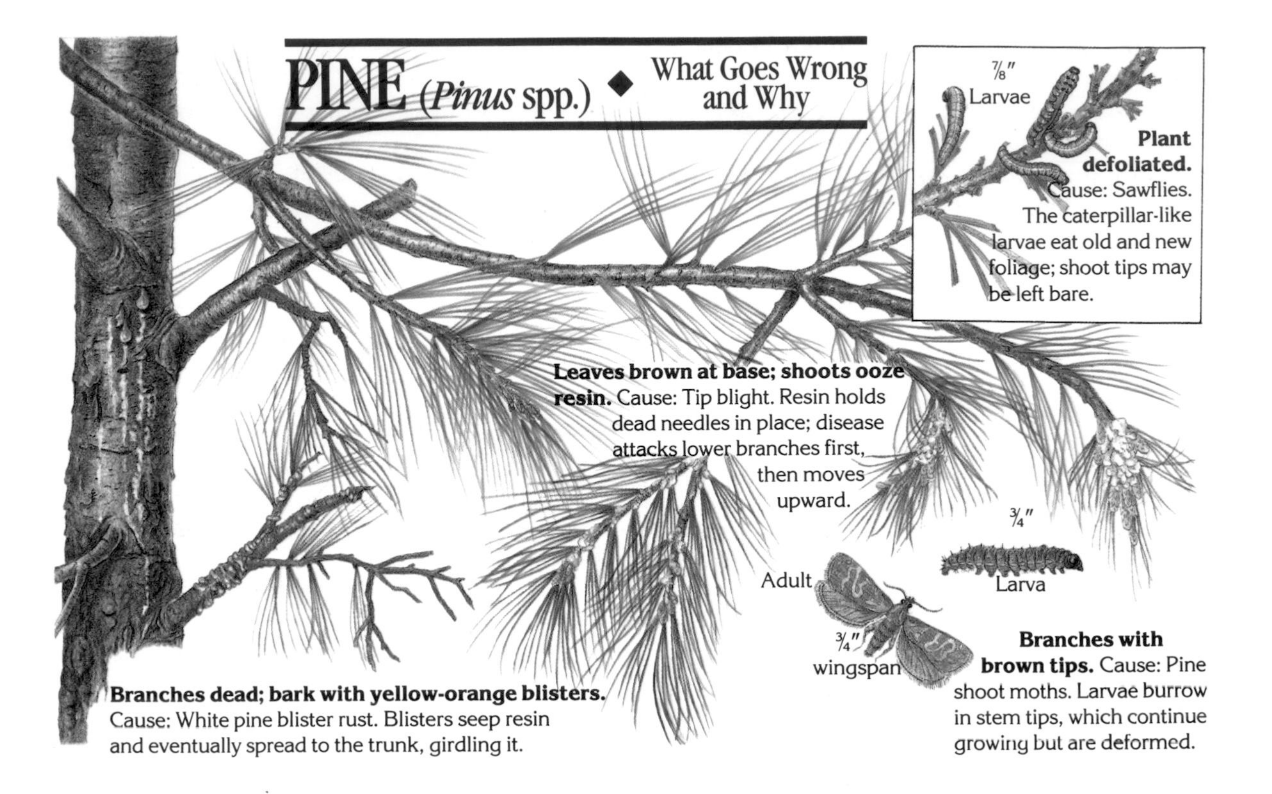

on buds and bark. Reduce populations of adults by spraying weekly with pyrethrin as long as adults are visible. Early in the season, cut off damaged shoots several inches below the affected area. To replace the damaged leader, select a horizontal branch right below the cut and fasten it to a stake in an upright position.

Bark damaged; leaves yellowish. Causes: Beetles; weevils; pine bark adelgids. A wide variety of pests attack the bark of pines. Beetles and weevils make small holes in the trunk where they bore through the bark and tunnel underneath. Pine bark adelgids are small, cottony, white insects that congregate on the bark of the trunk and limbs.

Beetles and weevils most commonly attack already stressed trees, so the best control for these pests is prevention. Keep trees growing vigorously by watering during dry spells; fertilize if plants are injured or stressed.

Control adelgids by spraying the plant several times, a day or two apart, with a strong blast of water. For more serious infestations, spray the trunk and branches with insecticidal soap every 3 days until the pests are gone.

Leaves light gray or bronze. Cause: Spruce spider mites. For an illustration of this pest, see page 181. For control measures, see "Leaves light gray or bronze" on page 181.

Leaves discolored and drop. Cause: Needlecasts. Several kinds of fungi attack pine foliage, causing spots or bands or total browning of the needles. These symptoms usually appear in early spring the year after the needles are infected. Planting trees where they will have good air circulation is the best way to avoid problems. Cleaning up fallen needles and pruning out damaged tips will help reduce the source of infection. To avoid symptoms the following year, spray with bordeaux mix when the new shoots are half-grown; repeat 2 weeks later.

Trunk or branches with oozing lesions; branch tips die back. Cause: Canker. Several kinds of fungi cause cankers on pines. For more information, see "Trunk or branches with oozing lesions; branch tips die back" on page 238.

Leaves skeletonized or with large holes; branches may be webbed. Cause: Caterpillars. Several kinds of caterpillars feed on pines. For control information, see "Leaves skeletonized or with large holes; branches may be webbed" and "Plant defoliated" on page 236.

Plant defoliated. Cause: Sawflies. Several species of sawflies cause similar damage in midsummer or early fall. For an illustration of the pests and their damage, see the opposite page. Natural parasites usually keep them in check. If necessary, handpick the white, green, or brown, caterpillar-like larvae. For severe infestations, spray the plant with insecticidal soap, or pyrethrin as a last resort.

Leaves yellow; stems and leaves covered with small bumps. Cause: Scales. See "Leaves yellow; stems and leaves covered with small bumps" on page 237.

Plant defoliated; branches bear cocoonlike bags. Cause: Bagworms. For control measures, see "Plant defoliated; branches bear cocoonlike bags" on page 236.

Trunk or branches with small holes; limbs die or break off. Cause: Borers. For more information, see "Trunk or branches with small holes; limbs die or break off" on page 238.

Leaves tunneled. Cause: Leafminers. The brown larvae of the pine leafminers enter the leaves and excavate them. Injured tips turn yellow and dry up. See "Leaves tunneled" on page 237 for control measures.

Branches with knotlike swellings. Cause: Gall rusts. Large galls, up to several inches in diameter, form on branches and eventually release yellowish spores. Pine-pine gall rusts produce spores that only infect pines; pine-oak gall rusts also attack oak trees, causing galls on their leaves. Control both kinds by pruning the galls from pine trees as soon as you spot them.

Leaves yellow and wilt. Cause: Root rot. For more information, see "Leaves yellow and wilt" on page 237.

Platanus

Sycamore, planetree. Trees.

Sycamores are deciduous, alternate-leaved trees with colorful, exfoliating bark and ball-shaped fruit. They are widely used as street trees since they withstand urban conditions.

Set out sycamores in spring or fall, planting in full sun or very light shade. They are remarkably tolerant of a range of soil conditions.

Problems

Leaves and shoots blackened; leaves with moist or brown sunken spots. Cause: Anthracnose. This fungal disease is very common on American planetree (*Platanus occidentalis*); London planetree (*P.* × *acerifolia*) is generally more resistant. For control information, see "Leaves and shoots blackened; leaves with moist or brown sunken spots" on page 238.

Trunk and branches with elongated, sunken, cracked areas; leaves yellow or sparse. Cause: Canker stain. This is a serious disease that can spread quickly, girdling and killing branches or the whole tree. Remove and destroy seriously affected trees. Prevention is the best control; avoid creating any wounds that allow the disease to enter the tree. Summer pruning, for example, often encourages canker stain on the branches; winter pruning is best. Also, be careful when using lawn mowers or other equipment around the tree to avoid damage to the trunk.

Leaves stippled with red or yellow spots. Cause: Sycamore lace bugs. This widespread and serious pest feeds on the undersides of planetree foliage. As the 1/10″, silvery white lace bugs feed, they puncture the leaves and then exude a toxin that turns the area around these punctures red or yellow. Spraying leaves with insecticidal soap just after the leaves unfold should control this pest.

Leaves with powdery white coating. Cause: Powdery mildew. For control information, see "Leaves with powdery white coating" on page 237.

Trunk or branches with oozing lesions; branch tips die back. Cause: Canker. See "Trunk or branches with oozing lesions; branch tips die back" on page 238.

Leaves with spots. Cause: Leaf spots. See "Leaves with spots" on page 237.

Leaves yellow; stems and leaves covered with small bumps. Cause: Scales. Heavy infestations can seriously weaken the tree; see "Leaves yellow; stems and leaves covered with small bumps" on page 237 for controls.

Trunk with shelflike growths. Cause: Wood rots. A number of fungus-caused wood rots can occur; see "Trunk with shelflike growths" on page 238.

Plum

Prunus spp. (Rosaceae)

Plums are deciduous trees that grow from 5′ to 20′ depending on soil, rootstock, and cultivar. Japanese plums (*Prunus salicina*) bear fruit on spurs—short branches that elongate only a fraction of an inch per year—1 year old or older. European (*P. domestica*) and hybrid plums bear fruit on spurs 2 years old or older. Plums are hardy in Zones 4–10, depending on species and cultivar.

Plant plums in a well-drained, sunny site free of late spring frosts. Prune European plums to a central leader and Japanese plums to a modified central leader (see the illustration on page 102). For more pruning information, see "Pruning and Training" on page 101. Some plums need cross-pollination; for more information, see "Setting Fruit" on page 101.

Fruit Problems

Young fruit with crescent-shaped scars. Cause: Plum curculios. These beetles appear around bloom time and leave a characteristic scar as they lay eggs in fruit. Infested fruit usually drops. Egg-laying ceases by early summer, but insects return later in the season to feed.

Botanical sprays don't control curculios adequately. To control this pest, spread a dropcloth beneath the tree and jar it with a padded mallet. Collect and destroy curculios that fall onto the sheet. For best results, jar the tree twice a day, beginning as soon as you see the first scarred fruit. In addition, collect and discard dropped fruit to prevent newly laid eggs from hatching. A traditional control is to keep chickens beneath the trees to consume adult curculios and grubs from fallen fruit.

Fruit with small brown spots that enlarge and grow fuzzy in humid weather. Cause: Brown rot. Leaves and flowers may also turn brown. The cultivars 'AU-Rosa' and 'Crimson' resist this fungus. For more information, see "Fruit with small brown spots that enlarge and grow fuzzy in humid weather" on page 166.

Fruit with brown, sunken spots on the surface. Cause: Bacterial leaf spot. Brown or black angular leaf spots may also appear. This disease isn't curable, although copper spray offers partial control. Resistant cultivars include 'AU-Amber', 'Crimson', and 'Simon'. For more information and control measures, see "Fruit with small, dark, sunken spots or cracks on skin" on page 166.

Leaf and Branch Problems

New leaves twisted or curled and covered with a sticky coating. Cause: Aphids. For heavy infestations, apply a strong spray of water or insecticidal soap solution to trees. For more control information, see "New leaves twisted or curled and covered with a sticky coating" on page 67.

Twigs and branches bearing tarry, black galls. Cause: Black knot. To control this fungal gall, prune out branches a few inches below galls. Resistant cultivars include 'AU-Producer,' 'Crimson', 'Milton', 'President', and 'Shiro'.

Branches wilting and dying, fail to leaf out in spring. Causes: Bacterial canker; Valsa canker. Look on the branches for sunken, elliptical lesions, often oozing a reddish gum. For more information and controls for these diseases, see "Branches wilting and dying, fail to leaf out in spring" on page 167. The cultivars 'AU-Amber', 'California', 'Crimson', 'Duarte', 'Homeside', and 'President' are resistant. 'Myrobalan' rootstock confers some resistance.

Tree stunted or dying. Cause: Plum leaf scale. This disease is common in the Southeast. There is no cure. Dig up infected plants and replant with a resistant cultivar such as 'AU-Rosa' or 'AU-Amber'.

Populus

Poplar. Trees.

Poplars are deciduous trees, generally fast-growing and weak-wooded. Male and female flowers appear on different plants. Females produce cottony seeds, so choose male cultivars to minimize the mess.

Easy to grow, poplars need full sun but can adapt to a range of soil conditions. They have spreading roots that can invade drains and water pipes; make sure you site plants away from these features. The extensive root system makes poplars tolerant of drought. These trees are naturally short-lived and are prone to many pest problems.

Problems

Trunk or branches with oozing lesions; branch tips die back. Cause: Canker. This is a widespread and serious problem on poplars. For more information, see "Trunk or branches with oozing lesions; branch tips die back" on page 238.

Trunk or branches with small holes; limbs die or break off. Cause: Borers. A number of different borers attack poplars; see "Trunk or branches with small holes; limbs die or break off" on page 238 for controls.

Leaves skeletonized or with large holes; foliage may be webbed. Cause: Caterpillars. For control information, see "Leaves skeletonized or with large holes; foliage may be webbed" on page 236.

Leaves with powdery white coating. Cause: Powdery mildew. See "Leaves with powdery white coating" on page 237.

Leaves yellow; stems and leaves covered with small bumps. Cause: Scales. For controls, see "Leaves yellow; stems and leaves covered with small bumps" on page 237.

Leaves wrinkled and discolored. Cause: Aphids. See "Leaves wrinkled and discolored" on page 235 for controls.

Portulaca

Moss rose, garden portulaca. Annuals.

These bright-flowered, mat-forming annuals come in several colors, including red, purple, pink, white, and yellow. The ruffled flowers are 1″ wide. Plants grow 4″-8″ high and can spread to 1½′. Flowers open in sunlight and close with dusk or cloudy skies.

Sow seed indoors 7 weeks before the last frost date. Set plants out in late spring when the soil is warm. Direct-seed outdoors in warm soil and full sun after last frost. Mix the tiny seed with sand to get an even distribution. Press seeds down firmly, but don't bury them. Water gently, keeping the soil moist until seedlings appear (about 10 days).

Drought-tolerant moss rose thrives in rocky soil or sand. It enjoys full sun, high temperatures, and good drainage. Pests and diseases are seldom a problem.

Potato

Solanum tuberosum (Solanaceae)

Potatoes are annual vegetables grown for their nutritious, starchy tubers. They are commonly known as white potatoes, but different cultivars have white, yellow, pink, or even bluish flesh, and yellow, brown, red, or purple skin.

Culture

Potatoes require deeply worked, well-drained soil with lots of organic matter, and a pH between 5.0 and 6.8. Potatoes require moderate to high levels of nitrogen, phosphorus, potassium, calcium, and sulfur. Have the soil tested and amend as needed before planting. Gypsum is a good source of calcium and sulfur for potatoes.

Keep soil moist, but not soggy, and do not allow it to dry out. Alternating dry and wet soil can cause cracked or knobby tubers. Once tops begin to yellow near harvest, you can let the soil dry out without damaging tubers.

Do not plant potatoes where tomatoes, potatoes, eggplants, peppers, strawberries, or brambles have been planted within the past 4-5 years. Also, try to plan your planting scheme to separate these crops in the garden. Don't plant potatoes where sod or small grains were grown the previous year: Wireworms, a common sod pest, also feed on potato tubers. Compost or till under all plant residues at the end of the season. Tilling the soil helps prevent pests from overwintering.

Potatoes are usually grown from seed potatoes (tubers) or "buds" (tiny tissue-cultured tubers), but a few cultivars, such as 'Explorer' and 'Homestead Hybrid', are grown from true seeds. Prevent problems by planting only certified disease-free tubers. Planting true seeds or buds also helps avoid many tuber-borne diseases.

Pre-condition tubers by storing them between 65° and 70°F for 2 weeks before planting to encourage rapid growth. Soak pieces in compost tea for several hours before planting to help prevent disease problems. Plant them out when soil is at least 40°F.

Leaf and Whole Plant Problems

Sprouts fail to emerge; seed pieces rot. Cause: Various fungal and bacterial rots. Soak seed pieces in compost tea before planting to help suppress diseases. Plant in well-drained, warmed soil and cover lightly to encourage rapid sprouting. Cold, wet soil encourages rot.

Leaves discolored and puckered or curled. Causes: Leafhoppers; aphids; viral diseases; bacterial ring rot. If leaves have yellow patches and brown edges, look for leafhoppers—tiny green or brown insects that suck plant sap. Leafhoppers hop, scuttle sideways, or fly when disturbed. Spray plants with insecticidal soap or sabadilla to control. Prevent infestations by covering plants with row cover when they come up.

If leaves are stippled yellow and stunted, look for aphids. Potato aphids are tiny pink

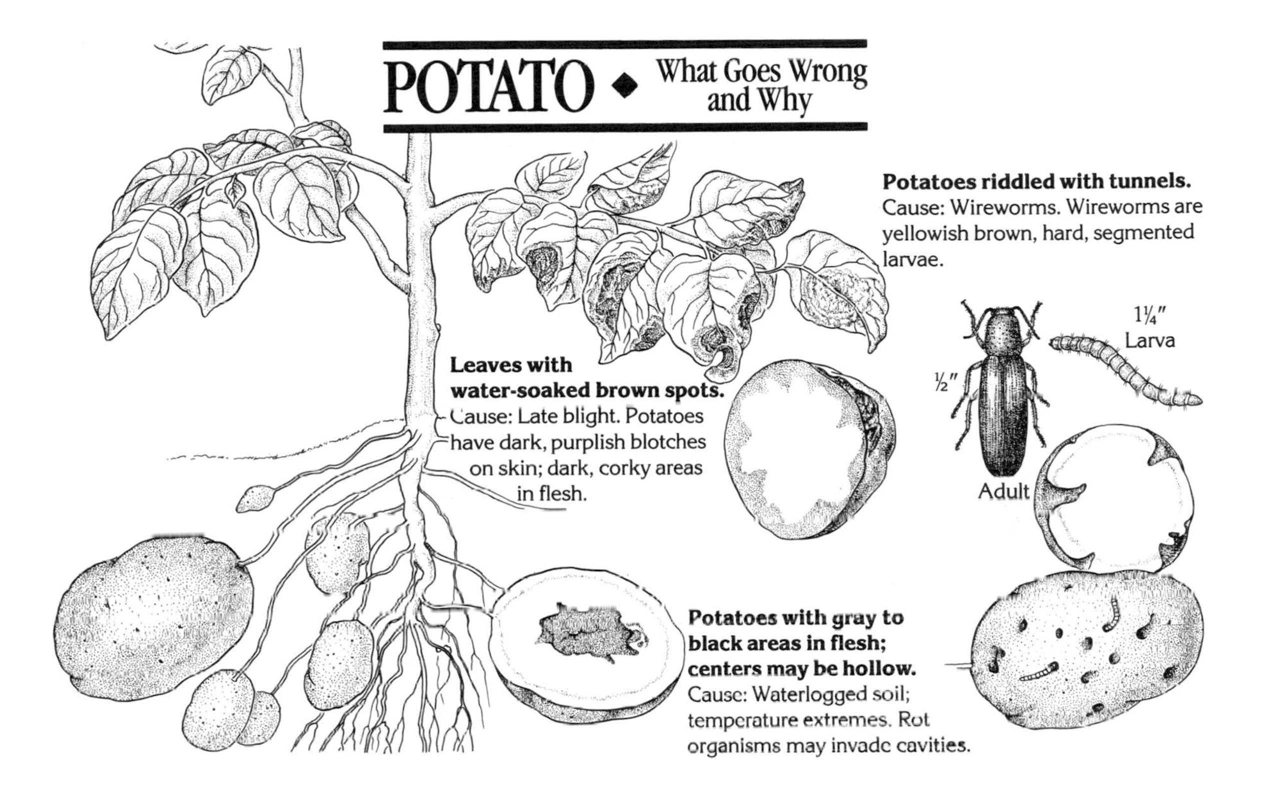

insects and are often found on young leaves. For mild infestations, knock pests off plants with a blast of water. Spray plants with insecticidal soap in the evening to control. Prevent infestations by covering plants with row cover when they come up.

If leaf edges roll upward and are yellow-green, the plants may have leafroll virus. Destroy infected plants. Prevent problems by controlling aphids, which spread the disease as they feed, or by planting resistant cultivars such as 'Katahdin' and 'Yukon Gold'. If leaves are dark green or mottled with yellow, the plants have one of many viral diseases. Destroy infected plants. Prevent problems by controlling aphids that spread viruses.

If leaves turn yellow between the veins and curl upward, the plant is probably suffering from bacterial ring rot. Shoots are stunted at the tip and may wilt. Wilted stems cut near the soil exude a whitish ooze. Destroy infected plants. Prevent problems by planting certified disease-free seed and by washing knife after each tuber while cutting seed pieces.

Leaves yellow; plant stunted and may be wilted. Causes: White grubs; tuberworms; wireworms; black leg; root knot nematodes; Verticillium wilt. If plants wilt suddenly, look for fat white grubs with brown heads, pinkish white larvae, or yellow to reddish brown, hard, segmented larvae chewing on roots or tubers. See "Potatoes riddled with tunnels" on page 191 for descriptions and controls.

If the stem is black and shrunken for a few inches above the soil line, the plant has black leg, a bacterial disease. Destroy diseased plants. To avoid black leg, plant potatoes in well-drained soil, and don't overwater. Cultivars such as 'Atlantic', 'Katahdin', 'Kennebec', and 'Red Pontiac' are resistant.

Stunted roots with swollen galls and tubers with warty skin are symptoms of root knot nematode infestation. Tubers are edible if peeled. Destroy infested plants; do not compost them. To control these pests, apply chitin or parasitic nematodes to the soil.

If roots and stem are undamaged, the plants may have Verticillium wilt. Stems cut near the soil line are discolored inside. This fungal disease usually appears when the plants flower. Destroy infected plants. Prevent problems by planting tolerant cultivars such as 'Beltsville' and 'Rhinered'.

Leaves with gray-brown, concentrically ringed spots. Cause: Early blight. Spots may merge and cover entire leaf, which then yellows and falls off. This fungal disease is active during warm, rainy weather. Spray plants with copper or bordeaux mix if symptoms are present and weather is warm and wet. Prevent problems by spraying plants with an antitranspirant before symptoms appear if you have had problems in the past, or by planting resistant cultivars such as 'Butte', 'Kennebec', and 'Krantz'.

Leaves with water-soaked brown spots. Cause: Late blight. Spots expand rapidly and develop a light halo. A white, velvety growth appears on the undersides of spots in wet weather. Stems are also infected and become dark and water-soaked. This fungal disease is active during moderately warm, rainy weather. Spray plants with copper if symptoms are present and weather remains wet to stop further symptom development. Prevent disease problems by spraying plants with compost extract when they come up and every 2 weeks throughout the season or by planting tolerant cultivars such as 'Butte', 'Cherokee', 'Katahdin', 'Kennebec', 'Krantz', 'Onoway', 'Rosa', and 'Sebago'.

Leaves with small holes. Cause: Flea beetles. Spray plants with a commercial pyrethrin/rotenone mix to control them if damage is severe. These tiny black, brown, or bronze insects hop when disturbed. Prevent problems by covering newly planted potatoes with row cover until they get large and able to tolerate insect damage. Apply parasitic nematodes to the soil to help control overwintering larvae in the soil.

Leaves with large, ragged holes or leaves missing. Causes: Colorado potato beetles; blister beetles. Colorado potato beetles are oval, yellowish orange, hard-shelled, ⅓" long beetles with black stripes. Larvae are soft-bodied, humpbacked, dark orange grubs with 2 rows of black spots down each side of their bodies. Eggs are orange and laid in rows on undersides of leaves. Overwintering adults appear on young plants in spring; handpick beetles zealously to reduce subsequent generations. Squash any eggs you see as well. Spray plants with BTSD as soon as you see any larvae. Prevent problems by covering plants with row cover before beetles appear. A thick mulch of loose straw around the plants may help prevent Colorado potato beetle damage.

Blister beetles are ¾" long, elongated and thin, metallic, black, blue, purple, or brown insects. Wear gloves to handpick, since these beetles secrete a substance that may cause blisters. Spray plants with a commercial pyrethrin/rotenone mix if damage is severe. Blister beetle larvae help control grasshoppers, so think twice before spraying to control them if grasshoppers are a problem in your area.

Tuber Problems

Potatoes small. Causes: Poor growing conditions; calcium deficiency. See "Culture" on page 188 for growing guidelines. If soil test indicates calcium deficiency, amend soil as needed.

Potatoes with green patches on skin. Cause: Exposure to light. Tubers exposed to light often turn green and develop a toxic substance known as solanine. Peel off green tissue before eating. Prevent greening by hilling plants with soil or mulch and renewing it as necessary. Store potatoes in the dark.

Potatoes with brown or black spots or patches on skin. Cause: Scurf. Peel off spots before using tubers. Soak seed pieces in compost tea before planting to help suppress fungal diseases. Do not plant spotted tubers: The resulting plants will produce only small tubers, many of which will rot.

Potatoes with rough, corky spots on skin. Cause: Scab. Trim out spots before using tubers. Prevent this fungal disease by keeping soil pH below 5.5 and planting resistant cultivars such as 'Beltsville', 'Norland', 'Onoway', 'Pungo', 'Rhinered', 'Russet Burbank', 'Russian Banana', and 'Superior'.

Potatoes with wartlike bumps. Cause: Root knot nematodes. See "Leaves yellow; plant stunted and may be wilted" on page 189 for controls.

Potatoes knobby. Cause: Uneven soil moisture. If tuber growth is interrupted because of lack of soil moisture, tubers are often deformed. Keep soil moist, but not soggy, and never let it dry out.

Potatoes with a black, rotted ring at stem ends. Cause: Bacterial ring rot. Tubers have a soft, light brown ring in the flesh near the skin. See "Leaves discolored and puckered or curled" on page 188 for controls.

Potatoes with dark blotches on skin; flesh with dark, corky areas. Causes: Early blight; late blight. Trim out corky areas before using; discard severely affected tubers. See "Leaves with gray-brown, concentrically ringed spots" and "Leaves with water-soaked brown spots" above for controls.

Potatoes with gray to black areas in flesh; centers may be hollow. Causes: Waterlogged soil; extreme temperatures; potassium or phosphorus deficiency; mechanical injury; viral disease. Potatoes grown in wet, poorly drained soil often grow too fast and may develop a discolored and/or hollow area in the center. Prevent problems by keeping soil moist, but not soggy.

Extremely hot weather or cold snaps before potatoes are harvested can cause similar symptoms. Prevent problems by keeping tubers well-covered with hilled-up soil or mulch.

Phosphorus or potassium deficiency can cause spots or patches of dark flesh. Oversize tubers with hollow centers may also indicate potassium deficiency. If deficiencies are suspected, have soil tested and amend as needed. If plants show symptoms of phosphorus deficiency, raise soil pH to 6.0 so the mineral will be more available to the plants.

Rough handling can cause mechanical injury, such as bruising, that appears as discolored areas on the flesh. Handle tubers gently when harvesting and storing to prevent problems. Plants suffering from viral diseases may produce deformed or discolored tubers. See "Leaves discolored and puckered or curled" on page 188 for controls.

Potatoes riddled with tunnels. Causes: Tuberworms; wireworms. If tubers have browned, silk-lined tunnels, look for tuberworms. These pinkish white larvae are ½" long and feed in tubers, stems, and leaves. Destroy infested tubers and plants. Prevent problems by keeping tubers hilled with soil as they grow and by removing the dead vines before digging tubers. Cover plants with row cover to prevent adult moths from laying eggs.

Wireworms are yellow to reddish brown, hard, segmented larvae up to 1½" long that tunnel into tubers and chew on roots. Adults are dark-colored, elongated click beetles. Apply parasitic nematodes to the soil before planting to control them. Avoid planting potatoes where sod or grain grew the previous season, because wireworms are often numerous there.

Potatoes spoil in storage. Cause: Bacterial or fungal rots. Poor growing conditions or improper curing or storage may encourage various rot diseases. See "Culture" on page 188 for growing guidelines.

Let vines yellow and dry before carefully digging tubers. Sort out any bruised, cut, cold injured, or diseased potatoes and keep them cool (around 40°F) until they can be used. Cure healthy tubers by storing them between 50° and 60°F for 2-3 weeks, then store them between 35° and 45°F in a humid place.

Primula

Primrose. Perennials.

This large genus includes more than 400 species of mostly low-growing plants noted for numerous brightly colored and often fragrant blossoms. The many species, hybrids, and cultivars range from 5″ to over 2′ tall. Spring flowers bloom in a wide range of colors. Primrose foliage is coarse and crinkled; large, bright green leaves form a rosette below the flowers.

Primroses are hardy (some to Zone 3), but prefer humid regions without extreme heat or cold. Moisture is critical for primroses; few tolerate any drought. Plant in partial shade in well-drained, rich, organic soil. Healthy primroses have few pest problems. Low leaves and moist soil attract slugs; see "Leaves with large, ragged holes" on page 177 for controls.

Pyracantha

Firethorn. Shrubs.

Firethorns are thorny, evergreen plants grown for their handsome glossy foliage and the persistent red, orange, or yellow fruit. Firethorns adapt well to foundation plantings; upright cultivars are excellent for hedges.

Set out in spring as balled-and-burlapped or container-grown plants in full sun or light shade. Avoid alkaline soils and locations subject to strong winter winds. When possible, choose disease-resistant cultivars like 'Apache', 'Mohave', or 'Teton'.

Problems

Leaves, flowers, and branches blackened. Cause: Fire blight. This bacterial disease causes new shoots to wilt suddenly, turn dark, and die back. The disease eventually spreads, killing the whole plant. Lush new growth is particularly susceptible, so avoid overfertilizing. Prune out diseased tissue, cutting back at least 6″ beyond the discolored area; destroy or dispose of prunings. Disinfect pruners after each cut in a 10 percent bleach solution (1 part bleach to 9 parts water). Spray bordeaux mix the following spring during flowering.

Leaves and berries with black, scabby areas. Cause: Scab. Infected leaves eventually turn yellow and brown before falling off the plant. Damaged fruit is unsightly. Clean up fallen leaves and berries. The following year, spray plants twice, 2 weeks apart, with bordeaux mix.

Leaves skeletonized; branches webbed. Cause: Webworms. These 1″, pale green to nearly black caterpillars have a dark or light stripe down the back and 3 dark spots on the side of each segment. They feed on leaves and spin silken webs around leaves and stems. Young plants may die; older ones can be seriously weakened. Break up and remove the webs; spray leaves with BTK.

Leaves wrinkled and discolored. Cause: Aphids. See "Leaves wrinkled and discolored" on page 235 for controls.

Leaves pale and mottled. Cause: Lace bugs. These 1/10″, dark-colored bugs have lacy-patterned wings. They feed on the undersides of leaves and deposit small black spots of excrement. Severe infestations can cause early leaf drop. Control by spraying leaves (especially the undersides) with superior oil or insecticidal soap.

Leaves yellow; stems and leaves covered with small bumps. Cause: Scales. See "Leaves yellow; stems and leaves covered with small bumps" on page 237 for more information.

Leaves stippled with yellow; foliage webbed. Cause: Spider mites. See "Leaves stippled with yellow; foliage webbed" on page 236 for controls.

Quercus

Oak. Trees.

Oaks typically have large, lobed leaves and rough bark; they are mostly slow-growing and long-lived, valued both for their ornamental effect and for their wood. Those suitable for northern planting are deciduous; there are a number of evergreen oaks grown in the South. The larger species are primarily used as ornamentals on large properties; some of the shorter-growing ones are valued as street trees and for smaller gardens.

Set out in spring as balled-and-burlapped plants. Full sun and deep, fertile, well-drained, somewhat acid soils suit them best. Oaks vary in their moisture requirements; ask your local nursery owner or extension agent about the needs of the oak trees you have.

Problems

Plant defoliated. Cause: Gypsy moths. The larvae of the gypsy moth are particularly fond of oaks and can easily defoliate a tree. See page 194 for an illustration of the pest and the damage it causes. For control measures, see "Plant defoliated" on page 236.

Leaves skeletonized or with large holes; branches may be webbed. Cause: Caterpillars. Many caterpillars feed on oak foliage. See "Leaves skeletonized or with large holes; branches may be webbed" on page 236.

Leaves or stems with brown, white, or green swellings. Cause: Galls. A number of tiny mites and insects feed on oaks, causing swellings on leaves and twigs. For an illustration of common oak galls, see page 194. Damage is rarely serious. Prune off and destroy affected leaves and twigs.

Leaves with roundish, puckered areas. Cause: Leaf blister. This fungal disease causes yellowish white blisters, up to ½″ in diameter,

OAK (*Quercus* spp.) ◆ What Goes Wrong and Why

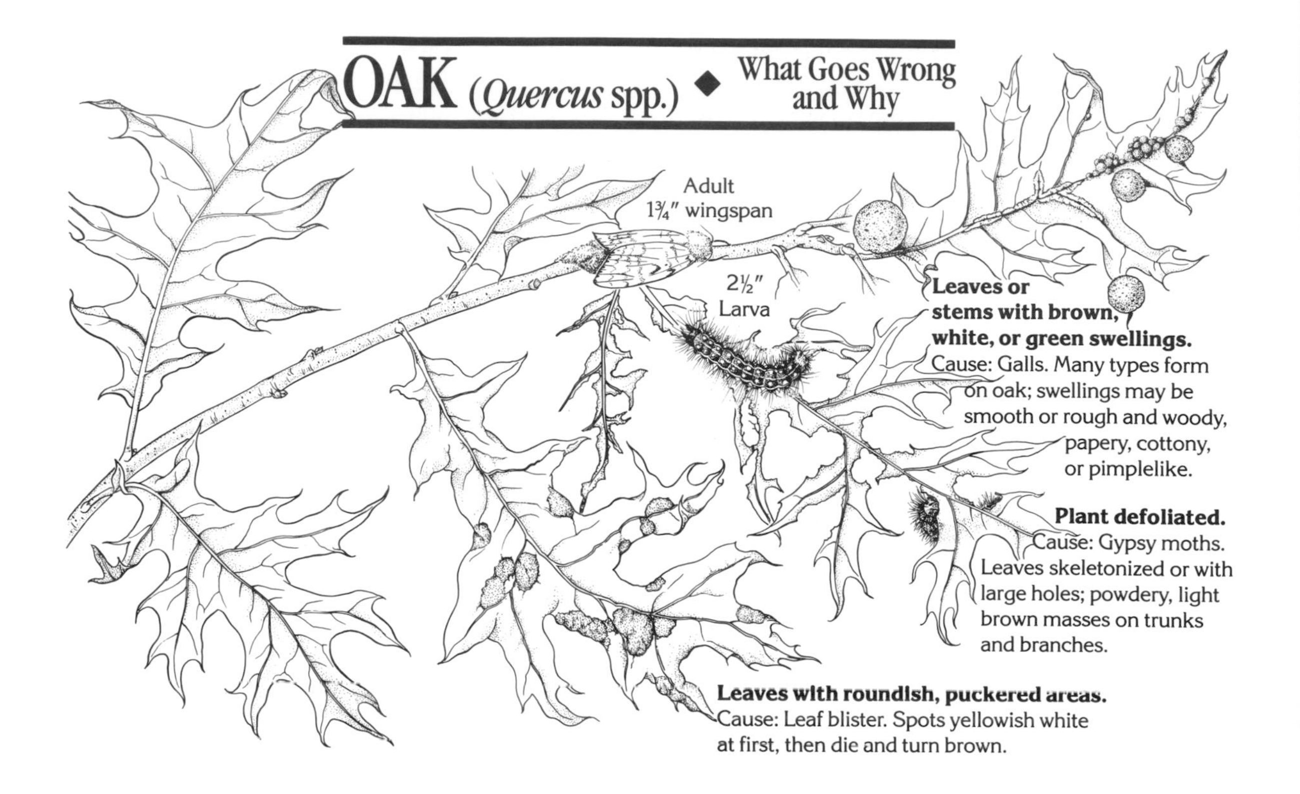

on oak foliage. For an illustration of leaf blister, see above. Although symptoms may appear serious, little actual harm is done to the tree, so control isn't necessary. To prevent damage the following year, spray branches with lime-sulfur or bordeaux mix before the buds open in spring.

Leaves with powdery white coating. Cause: Powdery mildew. See "Leaves with powdery white coating" on page 237 for controls.

Leaves and shoots blackened; leaves with moist or brown sunken spots. Cause: Anthracnose. For controls, see "Leaves and shoots blackened; leaves with moist or brown sunken spots" on page 238.

Leaves yellow; stems and leaves covered with small bumps. Cause: Scales. See "Leaves yellow; stems and leaves covered with small bumps" on page 237.

Trunk or branches with oozing lesions; branch tips die back. Cause: Canker. For more information, see "Trunk or branches with oozing lesions; branch tips die back" on page 238.

Leaf edges browned; leaves wilt and drop early. Cause: Oak wilt. Symptoms of this fungal disease usually show up first on the top of the tree, eventually spreading to the lower branches. Infected trees die quickly, often within 1 year of infection. There is no cure; remove and destroy infected trees.

Leaves stippled with white. Cause: Oak lace bugs. Adult lace bugs are 1/10″, flattened, dark insects with lacy-patterned, silvery white wings. Both the adults and the tiny nymphs feed on the undersides of leaves, sucking the sap and producing a gray, splotched or stippled appearance to the upper sides of the foliage. Leaves may curl, turn brown, and drop early. At the first sign of damage, spray with insecticidal soap, paying particular attention to the undersides of the leaves. Repeat the application if pests reappear in mid- to late summer. For severe infestations, spray with superior oil, or pyrethrin as a last resort.

Leaves rolled and chewed. Cause: Leafrollers. These ½″, green caterpillars form small webs on shoot tips and feed on leaves and buds inside. Break open webs and handpick larvae; spray BTK at first sign of damage.

Leaves with spots. Cause: Leaf spots. For controls, see "Leaves with spots" on page 237.

Leaves tunneled. Cause: Leafminers. See "Leaves tunneled" on page 237 for controls.

Leaves stippled with yellow; foliage webbed. Cause: Spider mites. For controls, see "Leaves stippled with yellow; foliage webbed" on page 236.

Radish

Raphanus sativus (Cruciferae)

Radishes are annual and biennial vegetables grown for their crisp, peppery roots. Certain cultivars do not have fleshy roots, but are grown for their crunchy seed pods. Some Daikon radishes grow 2′ roots.

Most radishes do best in cool, moist conditions. They need a pH between 5.5 and 6.8 and light, relatively rich soil. Plant radishes as soon as soil can be worked in spring. Make small plantings weekly until early summer for a continuous supply of radishes. Temperatures between 50° and 65°F produce the best radishes; growth above 75°F is poor. Some cultivars of Daikon radishes are designed for summer planting and will flower without forming large roots if planted too early.

The secret to mild, tender radishes is rapid growth. Water heavily the first 2 weeks after they come up if soil is dry. A light application of compost is usually enough for a good radish crop. Radishes will not tolerate soils high in salt.

Radishes are related to cabbage and suffer from many of the same problems. Since leaves are not harvested, more insect damage can be tolerated than in cabbage plants. See the Cabbage entry beginning on page 52 for descriptions and controls. Prevent problems with Fusarium yellows by planting resistant cultivars such as 'Fancy Red', 'Fuego', 'Red Devil B', 'Red King', and 'Red Pak'.

Root Problems

Roots enlarged and clublike. Cause: Club root. Destroy plants suffering from this fungal disease. Prevent problem by rotating crops and providing good drainage. 'Red King' and 'Saxafire' are tolerant cultivars.

Roots riddled with slimy, winding tunnels. Cause: Cabbage maggots. Maggots are white and ¼″ long. Adults look somewhat like houseflies and lay eggs on the soil near the base of the plants. Frequent, light cultivation when plants are young helps to decrease maggot populations. Remove and destroy infested plants.

Roots small and imperfect. Causes: Nitrogen deficiency; phosphorus deficiency; lack of water; excessive heat; wrong season for cultivar. Yellow or pale leaves suggest nitrogen deficiency. Purple leaves suggest phosphorus deficiency. Spray leaves with compost tea or fish emulsion, or side-dress with compost to correct either deficiency. If soil was dry, water subsequent plantings well. Do not try to grow most radishes in the hottest summer months, wait till cooler fall weather. Planting radishes in the shade of other plants may extend the spring season somewhat. Check proper planting season before planting Daikon cultivars.

Roots tough and dry; flesh pithy with white spots. Cause: Excessive heat. Few radishes tolerate high summer temperatures. Select heat-tolerant cultivars or plant earlier or later.

Roots with rough, dark spots on skin. Cause: Scab. This fungal disease is a problem in dry soil when pH is high and magnesium is low. Keep soil moist and maintain a pH below 6.5. Spray foliage with Epsom salts (1 tablespoon per gallon of water) or side-dress with compost to add magnesium.

Roots cracked; skin and flesh normal. Causes: Overmaturity; uneven soil moisture. Radishes are at their best for only a few days. If left unharvested, they quickly develop a harsh flavor and often crack open. Make small plantings every week until early summer for a continuous supply, and pull them as soon as ready. Keep soil evenly moist.

Roots cracked; skin rough; flesh dark. Causes: Downy mildew; black root. Prevent these fungal diseases by providing well-drained soil and by using a 4-year rotation.

Roots soft and shriveled. Cause: Cold injury. Protect roots with mulch below 32°F.

Raspberry

Rubus spp. (Rosaceae)

Raspberries are perennials that usually bear fruit on second-year canes. Canes of black (*Rubus occidentalis*), purple (*R.* × *neglectus*), and summer-bearing red (*R. idaeus*) raspberries bear fruit in their second season of growth. Everbearing red raspberries bear in late summer on new canes and again the next summer farther down the same canes. Red raspberries spread by underground runners. Black and purple raspberries spread by taking root where cane tips touch the ground. Raspberries are hardy in Zones 3-9, depending on species and cultivar.

Culture

Plant in a sunny site with good air circulation and well-drained soil. Start with disease-free stock. Plant in hills or rows well away from wild or abandoned raspberries, which may carry diseases. Provide posts or a wire fence to support the canes.

For black, purple, and summer-bearing red raspberries, cut off all fruit-bearing canes at ground level as soon as harvest is over, or when growth begins in the spring. In late winter or early spring thin out new canes that emerged the previous season; save the sturdiest ones and leave 6 canes per hill or 6″ spacing between row-planted canes. Shorten lanky canes to 4′-5′. Since black and purple raspberries fruit most heavily on side branches, induce side branching during the summer by pinching the growing tips of canes when they reach 2½′. The following late winter or early spring, shorten the side branches to about 1′.

For everbearing red raspberries, remove fruiting canes each summer as soon as the second fruiting is complete. Or sacrifice the second berry crop (which may be light anyway) and cut the entire planting to the ground as soon as leaves drop in the fall. Although this approach yields only 1 crop instead of 2, it has several advantages. Pruning all canes to the ground eliminates winter injury to canes, results in vigorous new canes for next fall's crop, and cuts down on overwintering pests that will appear next spring.

Raspberries are self-pollinating. A well-maintained planting may fruit heavily for many years, but disease often appears as plants age.

Plan on establishing a new raspberry bed every 10 years (less if plants begin to decline).

Fruit Problems

Fruit covered with light gray fuzz. Cause: Fruit rot. Especially common during wet weather, this fungal disease appears less frequently where plants have good air circulation and proper pruning. To control fruit rot, harvest berries often. Also pick and discard infected fruit far away from plants.

Fruit covered with a powdery white coating. Cause: Powdery mildew. This fungal disease, most common on red raspberries, makes fruit inedible and may weaken or kill whole canes. Pruning out old canes to provide good air circulation aids control. To prevent the disease, apply sulfur dusts. To control existing infection, apply lime-sulfur spray. Resistant cultivars including 'Meeker', 'Sumner', and 'Willamette'.

Fruit of red raspberries small and crumbly. Cause: Crumbly berry virus. This disease is incurable. Dig out and destroy infected plants and replant at a new site with virus-free (also called virus-indexed) plants.

Fruit small, tasteless, and dry. Cause: Verticillium wilt. For more information, see "New canes die in midsummer" on page 199. Drought can also cause these symptoms.

Fruit infested with slender white grubs. Cause: Raspberry fruitworms. These 1/4″ larvae feed in ripening fruit and may cause berries to drop prematurely. Adults are 5/32″, brown beetles that feed on flowers and leaves before laying eggs on fruit. Collect and destroy worm-infested fruit to prevent larvae from dropping to the ground and overwintering. Spray plants with rotenone as blossom buds appear, and again when flowers open, if you have had problems in the past.

Leaf and Branch Problems

Canes with dark blotches; sideshoots may wilt and canes may die. Causes: Spur blight; cane blight; anthracnose. Spur blight causes reddish brown blotches around leaf bases. This fungal disease appears mostly in midsummer on new canes of red raspberries. The following spring, blotches will be gray and leaf buds of affected cane areas will be dead. Sideshoots may wilt and entire canes may die. For prevention, plant resistant cultivars, including 'Boyne', 'Festival', 'Haida', and 'Newburgh'.

Cane blight causes canes with large, brownish purple areas extending over several buds. This fungal disease, most common on black raspberries, usually enters through a wound. Sideshoots may wilt and entire canes may die.

Anthracnose causes canes and leaves with gray spots surrounded by red or purple margins. Infected leaves have small yellowish white spots, the centers of which may dry and fall out. Sideshoots may wilt and entire canes may die. To help prevent this fungal disease, plant anthracnose-resistant cultivars, including 'Black Hawk', 'Jewel', 'Lowden', 'Mac Black,' and 'Munger'. See page 197 for an illustration of this disease.

To control all 3 of these diseases, also use cultivation methods that help prevent fungal infections. Maintain good air circulation by selecting a sunny, well-drained site. Each fall, remove diseased or crowded canes at ground level. Prune when plants are dry and rain is not expected for the next 3 days. For persistent fungal infection, apply lime-sulfur spray just as leaf buds break in the spring.

Leaves lightly stippled, curled, and dry. Cause: Mites. Tap a leaf over a sheet of white paper. Mites will appear on the paper as moving red, green, or yellow specks. Mites are most prevalent under dusty conditions and on water-stressed plants. For prevention, keep plants mulched and well-watered. For light infestations, spray plants with water; for heavy infestations, spray with insecticidal soap. Lime-sulfur spray applied early in the growing season also aids control. Mites are controlled by natural enemies, including predatory mites. If simple controls don't work, you can purchase predatory mites from insectaries or organic farm and garden supply catalogs.

Leaves and growing tips of canes covered with powdery white coating. Cause: Powdery mildew. For more information, see "Fruit covered with a powdery white coating" on page 197.

Leaves with greenish black spots that later turn gray. Cause: Raspberry leaf spot. This fungal disease may cause leaf spots to develop holes and leaves to drop early. Good air circulation aids prevention. Choose a sunny, well-drained planting site and thin plantings each fall or winter to prevent overcrowding of canes. For persistent infection, apply lime-sulfur spray just as leaf buds begin to green in the spring.

Leaf undersides with bright orange pustules. Cause: Orange rust. This incurable fungal disease appears only on black or purple raspberries, and blackberries. Diseased plants never recover; infection spreads quickly to neighboring plants. Dig up and destroy plants as soon as you notice disease symptoms. Install new plantings well away from wild brambles, which are a source of infection.

Leaves skeletonized. Causes: Raspberry sawfly larvae; Japanese beetles; raspberry fruitworm beetles. The prickly, pale green, ½" larvae of raspberry sawflies usually cluster on leaf undersides. They feed on leaves for up to 2 weeks before descending to the ground, where they pupate and emerge the following spring as adult sawflies. For heavy infestations, handpick larvae or spray with BTK. Or apply rotenone or pyrethrin sprays.

Japanese beetles are ½", metallic blue-green insects with bronze wing covers. For light infestations, check plants in the morning while beetles are sluggish and knock them off leaves into jars filled with soapy water. Traps baited with floral or fruit scents may reduce damage if placed at some distance from plants. Some gardeners find that these traps attract additional beetles into the garden, so be sure to place traps well away from plantings you wish to protect. Japanese beetle populations naturally decline by midsummer, but for heavy infestations, spray with rotenone as a last resort.

Milky disease spores, available commercially and applied to lawns to control Japanese beetle grubs, is a traditional control that works in some gardens and not in others. The disease does not always infect the entire grub population, and beetles that emerge from neighboring properties may fly to your plants.

If Japanese beetles are a consistent problem in your area, avoid planting raspberry cultivars particularly attractive to these pests. Cultivars to avoid include 'Fall Gold', 'Festival', 'Heritage', 'Latham', 'Newburgh', 'Reveille', 'Ruby', and 'Skeena'. In the future, gardeners may be able to buy parasitic nematodes that can be applied to the soil to control the grub stage of this pest.

Raspberry fruitworm beetles also chew leaves into a lacy pattern. For more information, see "Fruit infested with slender white grubs" on page 197.

Whole Plant Problems

New canes die in midsummer. Cause: Verticillium wilt. Typically leaves yellow, wilt, and fall before the entire cane dies. There is no cure for this soilborne fungal disease, which may be carried by other host plants such as tomatoes, potatoes, peppers, or eggplants. Avoid planting raspberries where other Verticillium host plants formerly grew. For extra precaution, solarize the soil to kill the fungus before planting (see "Soil Solarization" on page 424). See page 197 for an illustration of this disease.

Canes with wartlike or corky swellings; canes dying. Cause: Crown gall. Look for irregular corky swellings on the roots and crown area as well as the canes. Crown gall bacteria live in the soil and enter plants through wounds, so avoid injury to roots or crowns of plants. When working with infected plants, use isopropyl alcohol or a 10 percent bleach solution (1 part bleach to 9 parts water) to disinfect shovels and other tools before you move from one plant to the next.

Avoid introducing the infection into your soil by planting only healthy stock from a reputable grower. Many different types of plants are susceptible to crown gall, including roses, melons, and chrysanthemums. If your soil is infected, don't plant susceptible plants on that site for at least 3 years. If you must plant raspberries in soil with a history of crown gall, pretreat plants with the biological control agent Galltrol-A. The cultivar 'Willamette' has some crown gall resistance.

Canes dying, break off easily. Cause: Raspberry crown borers. The adult form of this insect, a clear-winged moth, lays eggs at cane bases. Eggs hatch into white, ½" grubs that feed on crowns, roots, and cane bases. Heavily infested canes may be hollowed out entirely. Dig out and destroy infested crowns. Vigorous canes are less likely to be attacked.

Cane tips wilted and dying. Causes: Rednecked cane borers; raspberry cane borers; raspberry horntails. All 3 of these insects lay eggs near cane tops, causing tips to wilt and die. Gall-like, cigar-shaped swellings on canes indicate rednecked cane borers. Canes girdled with spiral-shaped galls have been attacked by horntails. In all 3 cases, the adult insect lays eggs in the canes. Eggs hatch into grubs that feed and overwinter inside canes. For control, prune out and destroy infected canes as soon as you notice the injury. For heavy infestations, apply rotenone just before plants come into bloom. See page 197 for an illustration of the rednecked cane borer.

Leaves mottled with irregular yellow spots. Cause: Mosaic virus. This disease is transmitted by aphids and appears only on black or purple raspberries. Red raspberries may carry the virus but will show no symptoms. Infected purple or black raspberries eventually become stunted and develop small, misshapen leaves with green blisters. The only control is to dig up infected plants and replant with virus-free (also called virus-indexed) stock. Also avoid planting raspberries in sites near wild or old, neglected raspberries. Since red raspberries may carry the disease, keep plantings of black and purple raspberries separate from red raspberries. 'Royalty' purple raspberry is less susceptible to mosaic than other purple raspberries because it is seldom infested with the species of aphid that transmits the disease.

Rhododendron

Rhododendron, azalea. Shrubs.

Rhododendrons and azaleas, both members of the genus *Rhododendron,* are deciduous, semi-evergreen, or evergreen, broad-leaved shrubs. Rhododendron leaves are usually large and evergreen, while azalea leaves are generally small and deciduous. Gardeners prize both for the colorful flowers. Best adapted to cool, moist climates, they are commonly used in foundation and woodland plantings.

Set out in spring as balled-and-burlapped or container-grown plants. Sun or partial shade suits them well; the further south, the more shade they need. Rhododendrons and azaleas require cool, moist, highly organic, acid soil. Choose a site where they won't be exposed to hot summer or drying winter winds. Their feeding roots are close to the surface, so be sure to set out plants at the same level at which they were growing in the nursery.

Cultivation can damage the shallow roots, so apply a thick layer of organic mulch to help keep down weeds and maintain a cool, moist root zone. Rhododendrons and azaleas don't tolerate drought, so take care to water them thoroughly during dry spells. Some azaleas may be damaged by insecticidal soap sprays, so test the product on a few leaves before spraying the whole plant.

Problems

Leaves mottled, with shiny, black flecks underneath. Cause: Lace bugs. These pests are most common on plants growing in full sun. Adults are 1/8", flattened, dark insects with lacy-patterned, silvery white wings. Both adults and the tiny nymphs feed on the undersides of leaves, sucking the sap and producing a gray, splotched or stippled appearance to the upper sides of the foliage. The undersides of the leaves are spotted with tiny dots of brown excrement. For an illustration of this pest and the damage it causes, see the opposite page.

At the first sign of damage, spray with insecticidal soap, paying particular attention to the undersides of the leaves; test soap on azaleas before spraying. Repeat the application if pests reappear in mid- to late summer. For severe infestations, spray with superior oil, or pyrethrin as a last resort.

Leaves with notched edges. Cause: Black vine weevils. The wingless, 1/3", brownish black adults feed on the leaves and bark and can kill branches. See the opposite page for an illustration of this pest. To control them, try spreading a dropcloth around your plants, then shaking the plants; the adults will drop and can be removed and destroyed. Spray leaves several times with pyrethrin for major infestations. The small, curved, white grubs of these weevils feed on the roots; drench the soil around

RHODODENDRON ◆ What Goes Wrong and Why

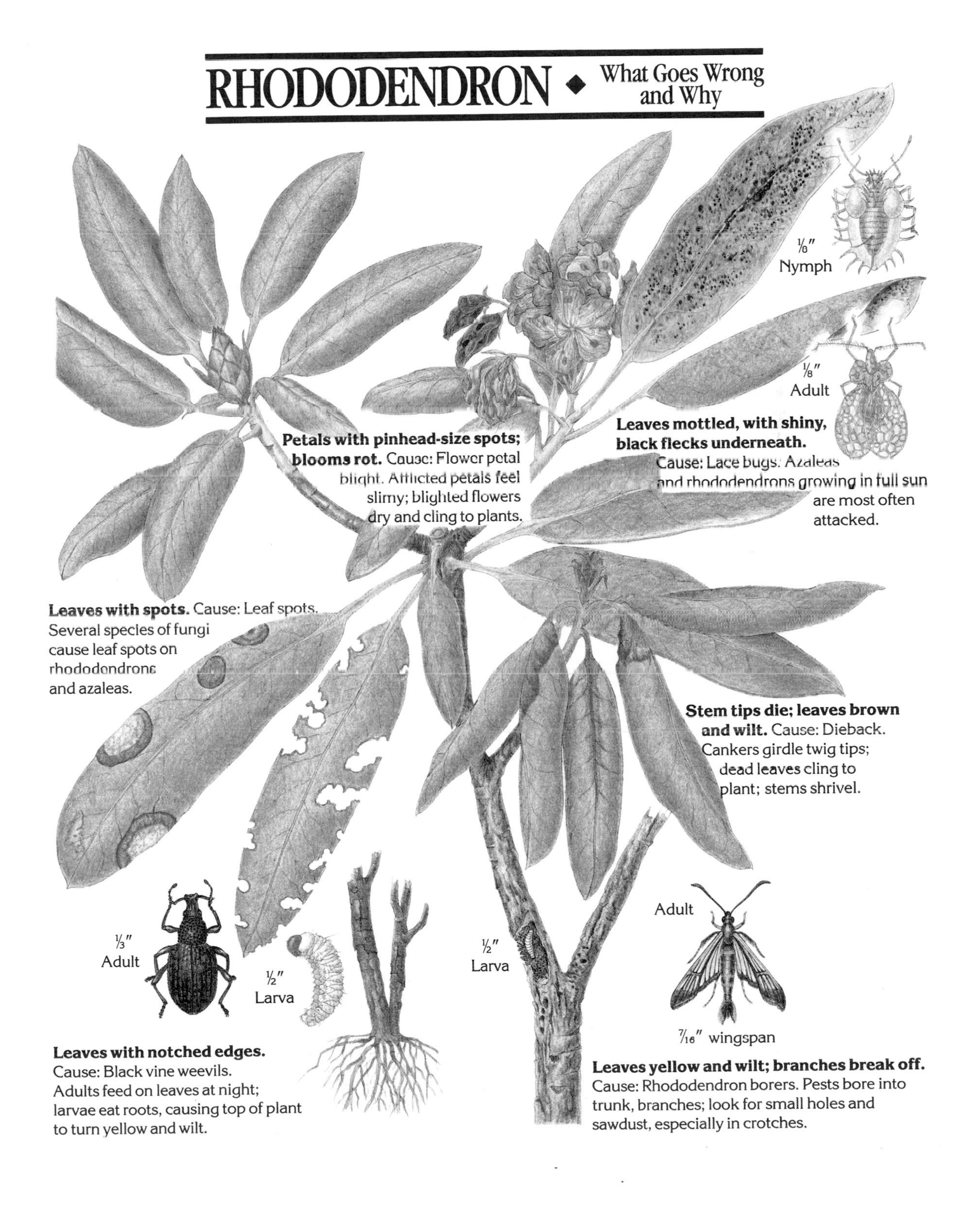

Petals with pinhead-size spots; blooms rot. Cause: Flower petal blight. Afflicted petals feel slimy; blighted flowers dry and cling to plants.

Leaves mottled, with shiny, black flecks underneath. Cause: Lace bugs. Azaleas and rhododendrons growing in full sun are most often attacked.

Leaves with spots. Cause: Leaf spots. Several species of fungi cause leaf spots on rhododendrons and azaleas.

Stem tips die; leaves brown and wilt. Cause: Dieback. Cankers girdle twig tips; dead leaves cling to plant; stems shrivel.

Leaves with notched edges. Cause: Black vine weevils. Adults feed on leaves at night; larvae eat roots, causing top of plant to turn yellow and wilt.

Leaves yellow and wilt; branches break off. Cause: Rhododendron borers. Pests bore into trunk, branches; look for small holes and sawdust, especially in crotches.

the base of the plant with a solution of parasitic nematodes for control.

Leaves yellow and wilt; branches break off. Cause: Rhododendron borers. These small, clear-winged, wasplike moths lay eggs on leaves, twigs, and bark. The eggs hatch into ½", whitish larvae that burrow into and weaken stems and branches. For an illustration of the pest and its damage, see page 201. Control borers by pruning and destroying affected branches. Seal the cut branches with putty or grafting wax. Water and fertilize the plant to help it recover quickly.

Leaves with spots. Cause: Leaf spots. Several kinds of fungi cause spots on azalea and rhododendron leaves. For an illustration of this problem, see page 201. For control measures, see "Leaves with spots" on page 237.

Petals with pinhead-size spots; blooms rot. Cause: Flower petal blight. For an illustration of this problem, see page 201. Remove and destroy infected flowers and branch tips. Rake debris from around the base of the plant and replace with fresh mulch.

Stem tips die; leaves brown and wilt. Cause: Dieback. For an illustration of this problem, see page 201. Besides causing leaves to brown, roll, and wilt, dieback may kill stem tips and cause cankers to form on stems. Avoid dieback by keeping plants healthy, with regular watering and good winter protection. Prune out diseased tips several inches below the damaged areas. Spray the plant with bordeaux mix after the plant blooms, and again 10 days later.

Leaves yellow; stems and leaves covered with small bumps. Cause: Scales. For controls, see "Leaves yellow; stems and leaves covered with small bumps." on page 237.

Leaves yellowed. Cause: High soil pH. Chlorosis (yellowed leaves) is most common on azaleas and rhododendrons used in foundation plantings. Lime leaches from the foundation of the house, making the soil too alkaline. Restore soil acidity by applying 2-3 pounds of sulfur per 100 square feet of growing area; scratch it lightly into the soil. Mulching with evergreen needles or chopped leaves will help maintain the proper pH.

Leaves mottled yellow. Cause: Whiteflies. These tiny, white, mothlike insects and their even smaller larvae feed on the undersides of leaves, weakening the plant. Spray leaves (make sure you get the undersides, too!) with insecticidal soap, or pyrethrin as a last resort; repeat weekly until pests are under control. Test soap on azaleas before spraying.

Leaves tunneled. Cause: Leafminers. See "Leaves tunneled" on page 237 for controls.

Leaves with green, white, or brown galls. Cause: Azalea leaf gall. This fungal leaf gall attacks both azaleas and rhododendrons. Pick off and destroy infected leaves as soon as you spot the galls.

Rhubarb

Rheum rhabarbarum (Polygonaceae)

Rhubarb is a hardy perennial grown for its red or green leaf stalks that are sweetened and used in pies, or stewed. Rhubarb leaves contain toxic quantities of oxalic acid. Don't eat them: Trim them off harvested stalks and compost them.

Culture

Rhubarb grows best in areas with cool, moist summers and requires a winter cold enough to freeze the top few inches of soil and induce a dormancy period. Choose a weed-free location with deep, sandy (if you have it), well-drained soil, with lots of organic matter. Rhubarb grows best with a pH between 6.0 and 6.8. It is a heavy feeder and will not produce large stalks if underfertilized. Work in plenty of compost before planting and mulch with several inches of a high-nitrogen compost each spring. Mix compost with a high-nitrogen supplement such as blood meal or soybean meal to boost its nitrogen content.

Most disease problems can be avoided with proper culture. Do not plant in wet or clayey soils. Use disease-free crowns and soak them in compost tea for 5 minutes before planting. Dispose of dead leaves in the fall to prevent pests and diseases from overwintering.

Do not harvest new plantings for at least 2 years. When you do start to harvest, choose only thick stalks and let the thin stalks feed the roots for the following year. Harvest for only 5-7 weeks, then stop so that plants can store energy for the next season. Cut out seed stalks as soon as they form to prevent them from using the plant's energy.

Problems

Stalks thin and small; leaves yellow. Causes: Young plants; overharvested plants; nutrient deficiency; overcrowding; crown or foot rot; waterlogged soil. Spray young plants with fish-meal tea to promote strong early growth. Follow the information under "Culture" above to prevent overharvesting or nutrient deficiency.

Rhubarb roots spread vigorously and need to be divided every 5 years or so. As plants become crowded, they produce thinner and thinner stalks.

If the roots or crown are also soft and rotted, the plant is suffering from crown or foot rot. Destroy plants infected with these fungal diseases. Prevent both diseases by planting in raised beds or hills.

Waterlogged soil will produce the same symptoms. Do not let soil get saturated; plant in raised beds.

Leaves yellow; stalks collapse. Causes: Verticillium wilt; crown or foot rot. Destroy dying leaves and plants suffering from these fungal diseases. Do not plant rhubarb where any Verticillium-susceptible crops have grown within the last 3-5 years. Prevent fungal disease problems by planting in raised beds.

Leaves yellow with curled margins; plant wilted. Cause: Leafhoppers. These green or brown, spindle-shaped, 1/10"-1/2" long insects suck plant sap. Spray plants with insecticidal soap in the evening to control, or with a commercial pyrethrin/rotenone spray if infestation is severe.

Leaves with small, round, brown spots. Cause: Leaf spots. Fungal leaf spots rarely reduce yields. Destroy diseased leaves after harvest. Spray plants with sulfur if disease is severe.

Stalks with soft, watery areas. Cause: Anthracnose. Leaves wilt and die. Destroy diseased stalks. Follow recommendations under "Culture" above to prevent this fungal disease.

Stalks with small, dark spots or bored holes. Cause: Rhubarb curculios. These yellowish gray, powder-covered, 1/2"-3/4" snout beetles damage stalks by boring holes in which to lay eggs. Handpick adults. Eliminate dock plants from weedy areas; curculios feed on it.

Stalks with brown, sunken spots at base. Cause: Crown or foot rot. Destroy plants infected with these fungal diseases. Prevent both by planting in raised beds or hills.

Stalks and/or leaves with chewed holes. Causes: Caterpillars; Japanese beetles. Ignore damage unless it is severe. Handpick insects. Spray plants with BTK if caterpillars are feeding. Protect plants with row cover early in the spring if caterpillars or beetles have been a problem in the past.

Rosa

Rose. Shrubs.

Roses are flowering shrubs with compound leaves, usually thorny stems, and flowers in a wide range of forms, colors, and fragrances.

ROSE (*Rosa* spp.) ◆ What Goes Wrong and Why

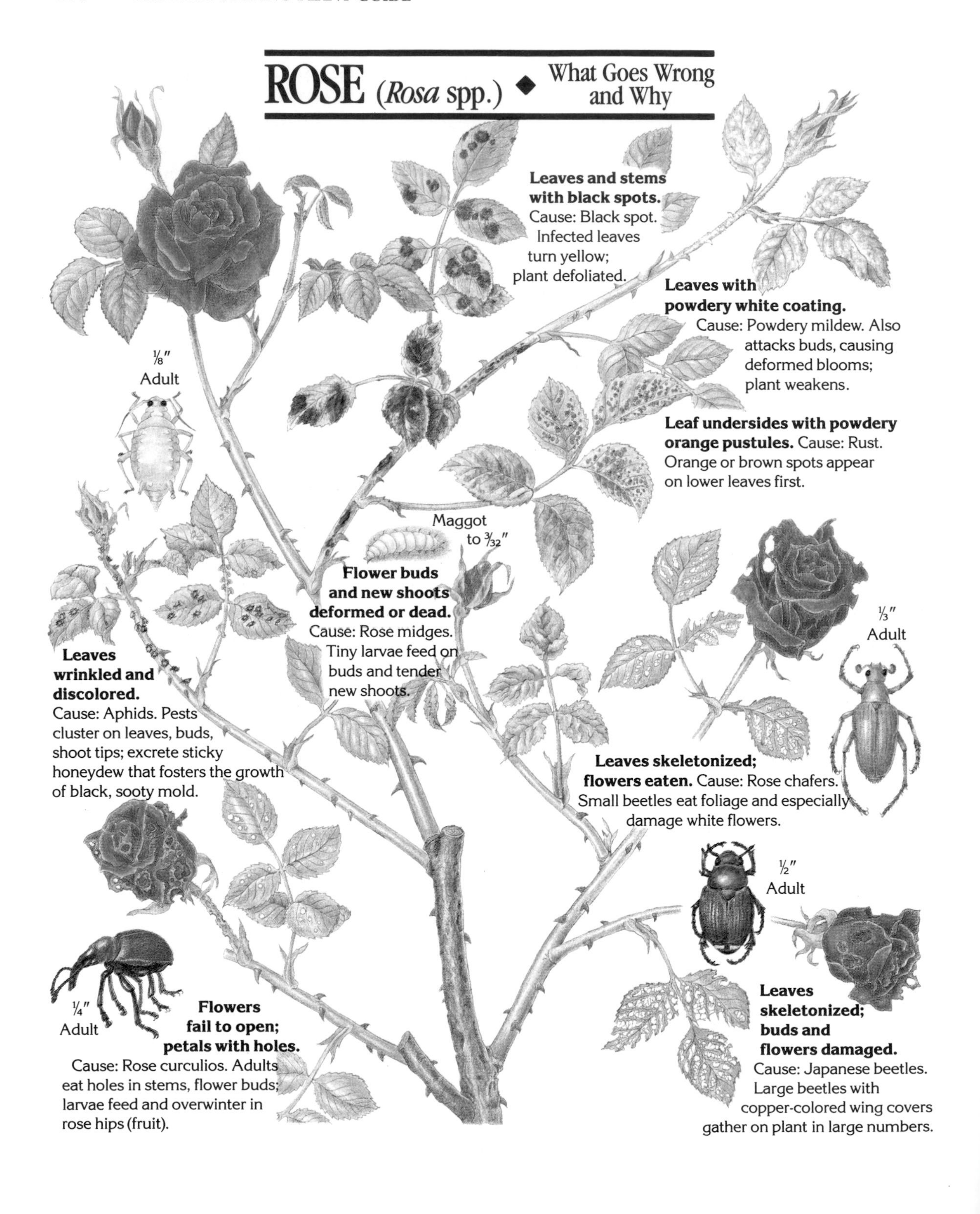

Most species are deciduous; there are a few evergreen or semi-evergreen species. In many instances their attractive fruit offers fall and winter interest.

Use roses in the shrub border, in small groups for landscape emphasis, or as single specimens. Climbers can be grown against walls, on trellises and arbors, on fences, or even trained up small trees. Hybrid teas, floribundas, and grandifloras, because they are so demanding in the care they require, are generally grown in beds devoted to roses alone; ironically, this massing contributes to some of the diseases to which they are prone. They also are effectively grown in large containers.

Culture

Set out roses in late winter or spring as bare-root or container-grown plants. They grow best in full sun (at least 6 hours per day) and a deep, rich, well-drained soil high in organic matter. Roses tend to do better in loamy or clayey, rather than sandy, soils. Allow plenty of room between plants for good air circulation.

In planting the modern hybrids, particular attention should be paid to the bud union (the knob where the graft is made). In severe climates, set the bud union 2″ below the soil surface; in moderate climates, even with the soil surface; and in frost-free or nearly frost-free areas, set the bud union 2″ above the soil surface. Roses are heavy feeders, so fertilize liberally. Don't feed after midsummer, though, or the plant may produce soft growth that will be subject to winter damage.

Problems

Leaves wrinkled and discolored. Cause: Aphids. Several kinds of aphids attack roses, often congregating on the stem tips just below the buds. Their feeding damages leaves and buds. See the opposite page for an illustration of the pests and their damage. For controls, see "Leaves wrinkled and discolored" on page 235.

Leaves skeletonized; buds and flowers damaged. Causes: Japanese beetles; sawflies; bristly rose slugs. Japanese beetle is an all-too-familiar pest of roses in the Northeast, attacking leaves, buds, and flowers. For an illustration of the pest and its damage, see the opposite page. For controls, see "Leaves skeletonized" on page 236.

Rose sawfly, curled rose sawfly, and, most important, bristly rose slug, have sluglike larvae that skeletonize foliage. They are especially destructive early in the growing season. Spray leaves with insecticidal soap, or dust with rotenone to control serious infestations. Handpicking is effective, but be sure to wear gloves; handling these pests can severely irritate your skin.

Leaves and stems with black spots. Cause: Black spot. While most of the species and shrub roses are little troubled by diseases, such is not the case with hybrid teas, floribundas, and grandifloras. The most serious disease is black spot, caused by a fungus. Black spots appear on the leaves, surrounded by yellow patches. The disease, prevalent throughout the country, is most likely to occur under warm, moist conditions. Left unchecked, the plant may become entirely defoliated. For an illustration of the damage, see the opposite page.

To some extent, a mulch will prevent the spores from splashing up from the ground onto the leaves during rains; pick off and destroy all diseased leaves and clean up all fallen ones. If possible, avoid wetting the leaves when watering; otherwise, water early in the day so the leaves can dry by evening. A 0.5 percent solution of baking soda may help control blackspot. Dissolve 1 teaspoon of baking soda in 1 quart of water; spray infected plants thoroughly. For severe infections, the recommended control is weekly applications of sulfur.

Leaves with powdery white coating. Cause: Powdery mildew. This is a common and serious disease of roses. A grayish white powdery deposit forms, first on young leaves, then spreading to the older ones and buds, and even sometimes to the canes. See page 204 for an illustration of the damage. At the first sign of the disease, pick off affected parts; commence a weekly spraying program with sulfur.

Flower buds fail to open. Cause: Thrips. When flower buds have brown edges and fail to open, thrips may be at work. They may also cause spots or streaks on open blooms. These pests are tiny, slender, light brown, fast-moving insects. A number of natural predators normally help keep them in check. If damage is severe, spray weekly with insecticidal soap, or rotenone if necessary.

Leaves stippled with yellow; foliage webbed. Cause: Spider mites. For control measures, see "Leaves stippled with yellow; foliage webbed" on page 236.

Canes with discolored or dead areas. Cause: Cankers. A number of fungal cankers attack roses. Pruning off and destroying diseased canes is the best approach to this problem.

Leaf undersides with powdery orange pustules. Cause: Rust. In the western United States, rust causes reddish orange bubbles to appear on the undersides of the leaves. Later, they spread to the upper surfaces. See page 204 for an illustration of the damage. Pick off and destroy infected leaves; in areas where the disease is common, start a spraying program early in the spring with weekly applications of sulfur.

Leaves with yellow-green mottling. Cause: Viral diseases. Besides discoloring leaves, viruses may also stunt the plant's growth. A number of insects spread viruses as they feed, so keeping insect pests under control will reduce the chances of viral problems. Destroy infected plants immediately.

Leaves skeletonized; flowers eaten. Cause: Rose chafers. Rose chafers, also known as rose bugs, skeletonize foliage and damage flowers. For an illustration of the pest and the damage it causes, see page 204. This 1/3", reddish brown beetle appears in late spring. It is especially active in areas with sandy soils. Handpicking is the best control. Reduce populations of larvae by treating lawn areas with milky disease spores. Spray plants with pyrethrin or rotenone as a last resort.

Flowers fail to open; petals with holes. Cause: Beetles. Rose curculios are 1/4", bright red, black-beaked insects; for an illustration, see page 204. Rose leaf beetles are 1/8", shiny, blue or green pests. Both of these insects bore into flower buds, preventing them from opening. If there are only a few pests, handpicking is the best control. Remove and destroy infested buds. For severe infestations, spray leaves and buds with pyrethrin.

Flower buds and new shoots deformed or dead. Cause: Rose midges. The white larvae of this tiny yellow-brown insect feed on flower buds and tender shoot growth, causing the injured parts to turn brown and die. See page 204 for an illustration of this pest and its damage. Cut off and destroy all infested buds to reduce future damage.

Shoot tips wilted; leaves with large holes. Cause: Leafcutter bees. Cleanly cut holes in the leaves, either round or oval, suggest the activity of leafcutter bees. After damaging the leaves, these pests bore into canes to lay their eggs, causing the shoots to wilt. Control by pruning out the injured tips several inches below the damaged area. Seal the cut end of the cane with grafting wax or putty.

Shoot tips die back; canes swollen or with small holes. Cause: Stem girdlers and borers. Rose stem girdler causes spiral swellings in the bark; raspberry cane borer causes the tips to die back; and rednecked cane borer and flatheaded appletree borer burrow in the canes. Cut off and destroy all dead and dying wood.

Leaves stippled with white. Cause: Leafhoppers. Foliage whitened or stippled indicates leafhopper activity. These lively, 1/10″-1/2″, green or brown, wedge-shaped insects hop around, sucking the juices from the leaves and serving as the vectors for several viral diseases. Spraying with insecticidal soap usually will control them; repeat as necessary. To control serious infestations, spray with pyrethrin.

Leaves yellow; stems and leaves covered with small bumps. Cause: Scales. For controls, see "Leaves yellow; stems and leaves covered with small bumps" on page 237.

Buds fail to open, turn brown. Cause: Botrytis blight. Should the buds on your roses turn brown and decay instead of opening normally, it may indicate a fungal blight. Pick off and destroy diseased blooms; spray plants weekly with sulfur.

Leaves skeletonized or with large holes; branches may be webbed. Cause: Caterpillars. See "Leaves skeletonized or with large holes; branches may be webbed" on page 236 for controls.

Trunk or roots with swollen, wartlike growths. Cause: Crown gall. For more information, see "Trunk or roots with swollen, wartlike growths" on page 238.

Rosemary

Rosmarinus officinalis (Labiatae)

Rosemary is a half-hardy (Zone 8) perennial commonly grown for its aromatic, needlelike leaves. This herb also bears attractive white, pink, or pale or dark blue flowers. In the South, rosemary is an evergreen shrub that can grow 3″-6′ tall. In the North, grow it in pots and move it indoors during the winter. Start rosemary from cuttings or purchase plants. Seedlings grow very slowly. Rosemary does best in full sun with very well drained soil and a pH between 6.0 and 8.2. Keep the soil evenly moist.

Rosemary has very few problems except root rot and powdery mildew. Avoid root rot by being careful not to overwater. When planting in containers, choose a very porous potting mix. See "Leaves yellow; plant wilts, stems, crowns, or roots water-soaked and rotted" on page 117 for additional controls. Powdery mildew can cause fuzzy white growth on leaves and stems. Spray affected plants repeatedly with compost tea for control.

Rudbeckia

Coneflower, black-eyed Susan. Perennials.

Cheery, daisylike coneflowers provide masses of warm yellow, gold, or orange blossoms from summer up until frost. A dozen or so slightly droopy petals—actually ray flowers—surround the rounded, raised, dark brown or greenish center that gives black-eyed Susan (*Rudbeckia hirta*) its name. Plants range from 1½′ to 6′ tall with rather coarse, toothed, dark green foliage.

Plant coneflowers in full sun to partial shade in average, well-drained soil. Plants are hardy and moderately tolerant of drought; excess moisture promotes fungal diseases. Coneflowers self-sow but are not invasive. Divide clumps every 4-5 years to maintain vigorous, trouble-free plantings.

Sage

Salvia officinalis (Labiatae)

Sage is a hardy (Zone 5) perennial herb grown for its pleasantly bitter-tasting leaves. Start sage from cuttings or purchased plants, as seedlings are quite variable. It does best in moderately rich, well-drained soil with a pH between 5.0 and 8.0, and at least 4 hours of sun per day.

Sage is normally quite trouble-free. Aster yellows can cause dwarfed, abnormally colored leaves and bushy growth. Destroy infected plants, and control leafhoppers and other sucking insects that spread diseases. Tan or red blisters on leaves are caused by rust. Destroy infected leaves and avoid wetting leaves to prevent its spread. See the Herbs entry beginning on page 116 for other possible problems.

Salix

Willow. Trees and shrubs.

Willows are deciduous trees and shrubs with simple, alternate leaves. The tree-forming species are principally valued for their graceful, pendulous form; among the shrubs, the immature catkins are valued for cutting and forcing.

Willows love moisture and have questing roots that can invade drains and water pipes. Take care not to locate willows near such structures. Set willows out in spring or fall, in moist (even swampy) soil with sun or light shade.

Problems

Leaves skeletonized or with large holes; branches may be webbed. Cause: Caterpillars. For controls, see "Leaves skeletonized or with large holes; branches may be webbed" and "Plant defoliated" on page 236.

Leaves skeletonized. Causes: Japanese beetles; imported willow leaf beetles. Japanese beetles are ½″ long, with metallic blue-green bodies and bronze wing covers. Adult imported willow leaf beetles are ¼″ long, with shiny, blue-black wings. Their ¼″ larvae, which also feed on leaves, are black and sluglike. For Japanese beetle controls, see "Leaves skeletonized" on page 236. Attract beneficial insects to your garden to help control leaf beetles; as a last resort, dust small plants with rotenone.

Trunk or branches with small holes; limbs die or break off. Cause: Borers. See "Trunk or branches with small holes; limbs die or break off" on page 238 for controls.

Leaves with brown blotches or tiny holes; shoot tips damaged. Cause: Willow flea weevils. The tiny black adult weevils overwinter in debris on the ground, feed on shoot tips in late spring, and lay eggs on leaves. The larvae emerge in early summer and feed by tunneling within the leaves. The pests come out of the leaves as adults and chew small holes in leaves until winter. Control willow flea weevils in the adult stage (spring or fall) by spraying leaves with pyrethrin, or rotenone as a last resort.

Leaves wrinkled and discolored. Causes: Aphids; willow lace bugs. See "Leaves wrinkled and discolored" on page 235 for more information on aphids.

Willow lace bug adults and larvae also feed on willow leaves, causing severe discoloration. These ¼″–⅓″ pests have distinctive lacy wings and commonly feed on the undersides of leaves. Control by spraying upper and lower leaf surfaces with insecticidal soap.

Leaves wilted and discolored; branches die back. Cause: Blight. Both a bacterial and a fungal blight attack willows, causing similar symptoms. Prune out and destroy infected branches during the dormant season. Disinfect pruners after each cut in a 10 percent bleach solution (1 part bleach to 9 parts water). In spring, spray twice with bordeaux mix at 10-day intervals, starting when the leaves first unfold.

Trunk or roots with swollen, wartlike growths. Cause: Crown gall. For more information, see "Trunk or roots with swollen, wartlike growths" on page 238.

Trunk or branches with oozing lesions; branch tips die back. Cause: Canker. See "Trunk or branches with oozing lesions; branch tips die back" on page 238 for details.

Leaves yellow; stems and leaves covered with small bumps. Cause: Scales. See "Leaves yellow; stems and leaves covered with small bumps" on page 237.

Salvia

Sage, salvia. Perennials.

This genus of the mint family contains hundreds of perennial and annual species grown for flowers, foliage, and herbal uses. Tubular flowers are carried on upright spikes over bushy, smooth to densely hairy, and often fragrant foliage. Red-flowered plants, such as scarlet sage (*Salvia splendens*), are often annuals or tender perennials, while most perennial species, including garden sage (*S. officinalis*), flower in shades of blue or purple. Silver sage (*S. argentea*) bears unimpressive white flowers that pale next to the large, crinkly, woolly leaves. Plants grow from 1′ to 4′ tall with similar spread.

Grow sages in full sun or light shade in average, well-drained soil. Some species tolerate heat and drought, but most flower more freely if kept evenly watered. Very hot weather may interrupt blooming, even if plants are well-tended. A summer mulch helps retain moisture in the soil and keep roots cool. In areas where temperatures fall below 0°F, protect perennial species by applying a winter mulch after the soil freezes. Remove faded flower spikes to encourage branching. Few pests or diseases cause significant damage to sages.

Sedum

Sedum, stonecrop. Perennials.

Hundreds of species of durable, succulent plants make up this genus. Some are low-growing creepers that spread vigorously, even in the poorest of soils, while others form 2′ tall, upright clumps. Light green, fleshy leaves tinged with white, red, or bronze in some species, may upstage the blooms; however, showy yellow, pink, or red flowers are the hallmark of sedums such as showy stonecrop (*Sedum spectabile*) and hybrid 'Autumn Joy', which bear profuse, flattened clusters of blooms. Flowering occurs from spring through summer, depending on the species.

Favored as rock garden plants, sedums endure poor, dry soil, but perform well in most well-drained soils. Full sun is best; sedums tolerate some shade, but resulting stems will be weaker. Although drought-tolerant, sedums bloom better if watered regularly. Excess moisture encourages rots, especially in winter. Sedum stems and leaves form roots readily.

Sedums are relatively untroubled by pests and diseases; the ease with which new plants are propagated reduces the likelihood that problems will decimate a planting. Aphids are occasionally troublesome; see "Leaves, stems, and buds distorted, sticky; clusters of small insects" on page 177 for control information.

Senecio

Dusty miller.
Tender perennials grown as annuals.

Although they sometimes bear 1″, yellow flowers, dusty millers are generally grown for their foliage. The lacy leaves of the plant are covered with white hairs, giving the foliage a silvery appearance. Plants grow 1½′-2′ tall. Dusty millers are ideal for edging beds and borders, or in window boxes.

Sow seeds indoors in late winter. Don't cover seeds; they need light to germinate. Seedlings appear in 10 days. Because seedlings develop slowly, some gardeners prefer to start with nursery-grown plants. Set out after the last frost, 8″-10″ apart.

Dusty millers grow best in full sun with average to fertile, well-drained soil. Pinch off flower heads as they appear. Few pests or diseases bother dusty millers.

Sorbus

Mountain ash. Trees.

Mountain ashes are fast-growing, alternate-leaved, deciduous trees grown principally for the profuse clusters of white flowers that are followed by brightly colored fruit in autumn. Although rather weak-wooded and very subject to pests and diseases, they are popular as ornamental specimen trees.

Set out in spring as balled-and-burlapped plants. Mountain ashes need sun and a slightly acidic, well-drained location. They are rather intolerant of the air pollution common to cities. Mountain ashes are best adapted to northern gardens; high summer temperatures stress the plant and make it more susceptible to problems.

Problems

Leaves, flowers, and branches blackened. Cause: Fire blight. For more information and controls, see "Leaves, flowers, and branches blackened" on page 192.

Leaves skeletonized. Causes: Japanese beetles; sawflies. For controls, see "Leaves skeletonized" on page 236.

Trunk or branches with oozing lesions; branch tips die back. Cause: Canker. See "Trunk or branches with oozing lesions; branch tips die back" on page 238 for details.

Leaves distorted, orange-yellow spots on leaf undersides. Cause: Rust. See "Leaves distorted, orange-yellow spots on leaf undersides" on page 99.

Leaves and berries with black, scabby areas. Cause: Scab. Infected leaves eventually turn yellow and brown before falling off the plant. Damaged fruit is unsightly. Clean up fallen leaves and berries. The following year, spray plants twice, 2 weeks apart, with bordeaux mix.

Leaves wrinkled and discolored. Cause: Aphids. See "Leaves wrinkled and discolored" on page 235 for controls.

Trunk or roots with swollen, wartlike growths. Cause: Crown gall. See "Trunk or roots with swollen, wartlike growths" on page 238 for more information.

Leaves with brownish blisters underneath. Cause: Pear leaf blister mites. These microscopic mites live and feed on tissue inside leaves. They overwinter on buds and infest new leaves in spring. Spray plants in late winter with dormant oil or lime-sulfur.

Trunk or branches with small holes; limbs die or break off. Cause: Borers. See "Trunk or branches with small holes; limbs die or break off" on page 238 for controls.

Leaves yellow; stems and leaves covered with small bumps. Cause: Scales. See "Leaves yellow; stems and leaves covered with small bumps" on page 237 for more information.

Spinach

Spinacia oleracea (Chenopodiaceae)

Spinach is a cool-season annual vegetable grown for its tender, green leaves. It can be grown in the spring or fall, and may even survive over winter. Cultivars come in 2 types: savoy-leaved (crinkled) and flat-leaved.

Culture

Grow spinach in well-drained soil with lots of organic matter and a pH between 6.0 and 7.0. Spinach seed germinates best at soil temperatures between 45° and 75°F, but will germinate as low as 35°F. Mature spinach can survive temperatures of 20°F if gradually hardened. However, prolonged exposure of young plants to temperatures below 45°F will cause bolting—production of a flower stalk—and plants will produce few, low-quality leaves. Temperatures above 75°F and long days also cause bolting. In warmer climates, plant spinach in filtered shade to extend its season into the warmer months.

Keep soil moist, but not soggy. Do not allow it to dry out, or plants may bolt. Spread a thin layer of mulch around plants to conserve moisture, suppress weeds, and keep soil cool.

Soak seed in compost tea for 30 minutes before planting to speed germination, and help suppress soilborne diseases.

Spinach requires moderate levels of potassium and phosphorus and high levels of nitrogen. It is also sensitive to low levels of calcium and boron. Have the soil tested and amend as necessary. Fast-acting sources of nitrogen, such as bloodmeal and soybean meal, are good fertilizers for spinach.

Problems

Plant sends up a flower stalk. Causes: Extreme temperatures; long days. Prolonged exposure of young plants to temperatures below 45°F causes plants to bolt, or send up a flower stalk at the expense of succulent leaves. Temperatures above 75°F and long days also cause bolting. Discard bolted plants. Prevent problems by covering plants with row cover until temperatures are stable and by planting bolt-resistant cultivars. Try planting New Zealand spinach *(Tetragonia tetragonioides)* in the summer. Its flavor is similar to that of regular spinach, and it thrives in hot weather.

Leaves yellow; plant stunted and may be wilted. Causes: Nitrogen deficiency; waterlogged soil; Fusarium wilt. If leaves are pale or yellowish, and plants are stunted but not wilted, they may be suffering from nitrogen deficiency or waterlogged soil. Spray plants and drench roots with compost tea or fishmeal tea to encourage the production of dark green leaves. Prevent problems by choosing well-drained sites, adding organic matter to the soil to increase fertility and improve drainage, and planting in raised beds.

Wilted plants have Fusarium wilt. Destroy infected plants. This fungal disease thrives in warm (70°-80°F) soil, so avoid problems by raising spinach while the soil is cool in spring or fall.

Leaves yellow and deformed. Causes: Aphids; curly top virus; mosaic virus. Yellow curled leaves and stunted growth can be caused by aphids. Check for these green, pink, black, gray, or white fluffy-coated, soft-bodied, small insects on the undersides of lower leaves, often near the leaf midrib. Spray plants, especially the undersides of the leaves, with water to discourage aphids, or with insecticidal soap if infestation is severe.

If young leaves are yellow, deformed, and stunted, the plant has curly top virus. Deformed leaves may die. If older leaves are also mottled, the plant is suffering from mosaic virus, also called blight or yellows. Destroy infected plants. Control aphids, which spread mosaic as they feed, and beet leafhoppers, which transmit curly top. Prevent some problems by planting mosaic-tolerant cultivars such as 'Indian Summer', 'Melody', and 'Winter Bloomsdale'.

Leaves with pale yellow patches on upper surfaces. Causes: Downy mildew; white rust. If spots develop a grayish mold on the undersides of the leaves, they are suffering from downy mildew, a fungal disease. Destroy infected leaves or plants. Prevent problems by thinning plants to increase air circulation and by planting resistant or tolerant cultivars such as 'Crystal Savoy', 'Fall Green', 'Gladiator', 'Indian Summer', 'Kent', 'Melody', 'Olympia', 'Seven R', 'Tyee', and 'Winter Bloomsdale'.

If spots develop white blisters on the undersides of the leaves, they have white rust, another fungal disease. Destroy infected leaves or plants. Prevent problems by thinning plants to increase air circulation and by planting tolerant cultivars such as 'Fall Green'.

Leaves with water-soaked or brown spots. Causes: Anthracnose; other fungal leaf spots. Spots may enlarge rapidly, especially in wet weather. Destroy infected leaves or plants. Spray plants with sulfur to prevent disease from spreading, especially if weather is wet. Prevent problems by thinning plants to provide good air movement.

Leaves with light-colored tunnels or blotches. Cause: Leafminers. Larvae are creamy white, 1/8" long maggots that feed on leaf tissue. Destroy mined leaves. Prevent problems by tilling soil after harvest.

Leaves with small holes. Cause: Flea beetles. These tiny, black, brown, or bronze beetles hop when disturbed. Larvae are small and white, feed mostly on the undersides of leaves, and drop to the ground and play dead when disturbed. Control severe infestations by spraying plants with a commercial pyrethrin/rotenone mix. Prevent problems by covering young plants with row cover, which can be left on until harvest as long as temperatures are moderate. Flea beetles can also be discouraged by planting in partial shade.

Leaves with large, ragged holes. Cause: Caterpillars. Various caterpillars will feed on spinach. Handpick, or spray plants with BTK if caterpillars are feeding, or protect plants with row cover.

Spiraea

Spirea. Shrubs.

Spireas are alternate-leaved, deciduous shrubs that flower in late spring. They are best in shrub borders or massed on banks.

Set out in spring or fall in full sun or light shade. A pH below 6.5 suits spireas best, and ample organic matter will ensure the moisture they require. Make sure they are well-mulched and well-watered going into winter.

Problems

Leaves wrinkled and discolored. Cause: Aphids. For control measures, see "Leaves wrinkled and discolored" on page 235.

Leaves rolled and chewed. Cause: Leafrollers. These ½", green caterpillars with brown heads form small webs on shoot tips and feed on leaves and buds inside. Break open webs and handpick larvae; spray BTK at first sign of damage.

Leaves skeletonized or with large holes; branches may be webbed. Cause: Caterpillars. For controls, see "Leaves skeletonized or with large holes; branches may be webbed" on page 236.

Leaves yellow; stems and leaves covered with small bumps. Cause: Scales. See "Leaves yellow; stems and leaves covered with small bumps" on page 237 for controls.

Leaves, flowers, and branches blackened. Cause: Fire blight. See "Leaves, flowers, and branches blackened" on page 192 for controls.

Whole plant stunted and lacking vigor. Cause: Nematodes. See "Whole plant stunted and lacking vigor" on page 239 for controls.

Leaves with powdery white coating. Cause: Powdery mildew. For controls, see "Leaves with powdery white coating" on page 237.

Squash

Cucurbita spp. (Cucurbitaceae)

Squash are frost-tender annuals grown for their fleshy fruit. Summer squash, such as zucchini, are eaten before the seeds and rinds harden. Winter squash are harvested after the fruit is mature. The name pumpkin is used for some winter squash. Species include: *Cucurbita maxima, C. mixta, C. moschata,* and *C. pepo.*

Culture

Squash seeds need 60°F soil to germinate, so wait until warm weather to plant. Cover plants with floating row cover to protect them from insects and late cold snaps. Remove row cover when plants begin to flower so insects can pollinate the blossoms, or you will not get any fruit.

Squash do best in well-drained, loose-textured soils with lots of organic matter. They will grow in soils with a pH between 5.5 and 6.8, but prefer a pH above 6.0. Squash need lots of water, but don't let soil become saturated. Prevent disease problems by keeping the leaves dry. Mulch squash to help conserve water. Black plastic is a good choice for northern areas, but in the extremely warm areas it can warm the soil too much. Organic mulches are good, too, but may provide shelter for pests like squash bugs. Foil mulches help prevent aphid problems. To prevent rot, support fruit on scraps of wood.

Rotate crops so that no member of the cucurbit family (cucumbers, melons, and squash) is grown in the same place more often than every 4 years.

Caution: Squash leaves are easily burned by insecticidal soap and copper sprays. Use the most dilute spray recommended and use sparingly. Do not spray plants in direct sun or

SQUASH ◆ What Goes Wrong and Why

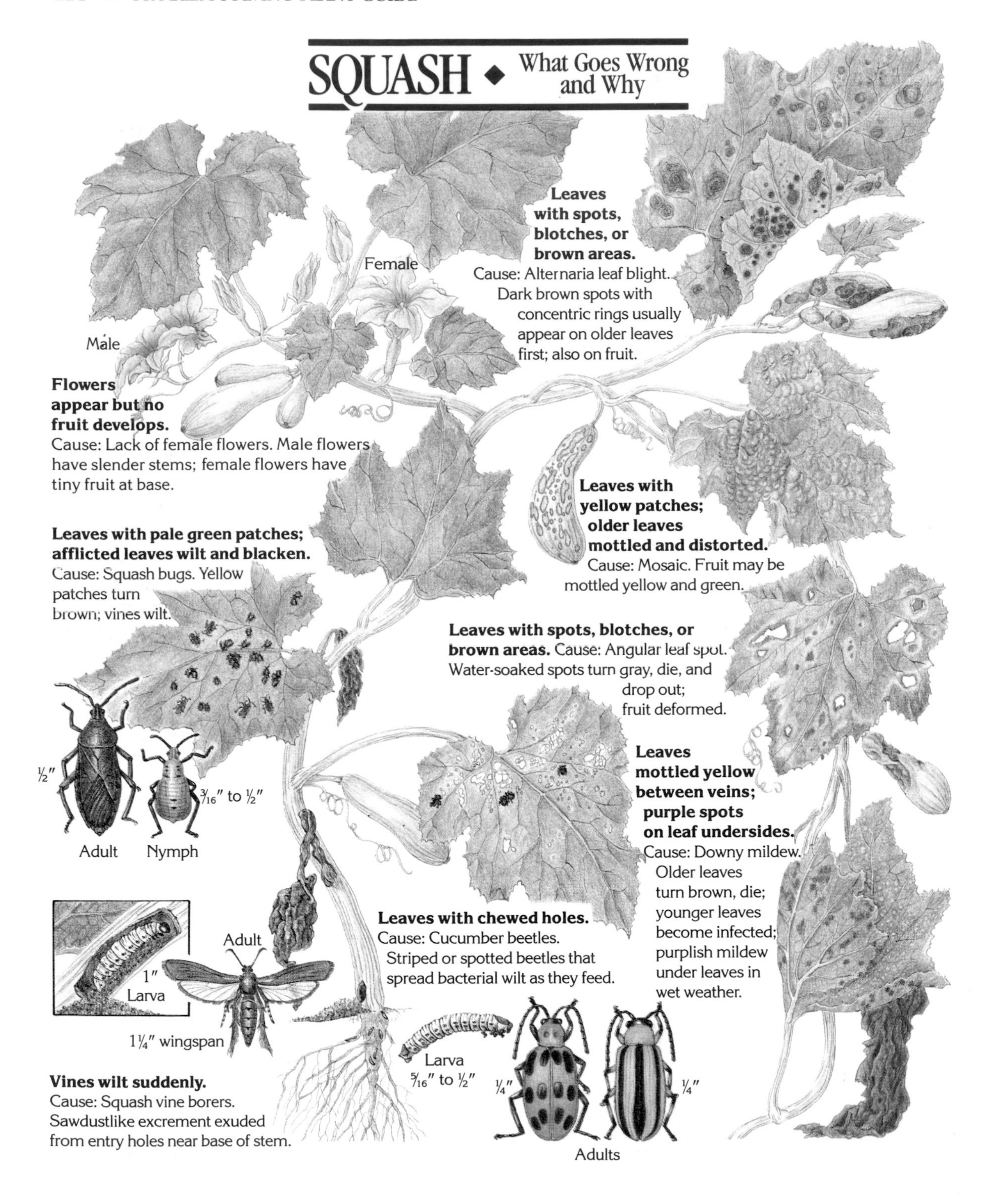

if temperatures are above 80°F, and don't spray drought-stressed plants.

Leaf and Vine Problems

Leaves with chewed holes. Cause: Cucumber beetles. Adults are 1/4" long, greenish yellow beetles with black stripes or spots. Larvae chew on roots. For an illustration of this pest, see the opposite page. They attack young leaves and should be controlled immediately as they can spread bacterial wilt or viral diseases. Spray plants with sabadilla or a commercial pyrethrin/rotenone mix. Reduce problems by planting cultivars, such as 'Bennings Green Tint', 'Blue Hubbard', 'Early Butternut Hybrid', 'Seneca', and 'Table King', that tolerate beetles.

Leaves with pale green patches; afflicted leaves wilt and blacken. Cause: Squash bugs. Adults are brownish black, 1/2" long bugs. Immature bugs are whitish green with dark heads and legs. Eggs are bright orange and laid on undersides of leaves. For an illustration of this pest, see the opposite page. Handpick adults and eggs. Trap bugs by laying a board near plants. Squash bugs will hide underneath it and can be destroyed each morning. To reduce problems, plant cultivars that tolerate squash bugs such as 'Early Prolific', 'Early Summer', 'Royal Acorn', and 'Table Queen'.

Leaves with yellow patches; older leaves mottled and distorted. Cause: Mosaic. Several types of mosaic viruses are found on squash. Besides affecting leaves, mosaics may also cause deformed fruit that is mottled with yellow and green. For an illustration of this disease, see the opposite page. Remove and destroy diseased plants. Control aphids and cucumber beetles that spread it. Reduce problems by planting cultivars, such as 'Multipik', 'Napolini', 'Superpik', and 'Superset', that tolerate mosaic.

Leaves yellow, curled, and wilted. Cause: Aphids. Look for small, green, pink, gray, black, or white fluffy-coated, soft-bodied insects feeding on plants. Aphids can also transmit viral diseases. Control aphids by knocking them off the plants with a strong blast of water. Or spray with weak insecticidal soap; read the caution above before spraying. Prevent problems by using a foil mulch or by planting silver-leaved cultivars, such as 'Cocozelle', that confuse or don't attract aphids.

Leaves yellow and puckered, becoming bronzed. Cause: Mites. These tiny, red, yellow, or green, spiderlike creatures are worst in dry, hot weather. In severe cases leaves dry out and drop off. There may be a fine webbing on the undersides of leaves. Spray plants with a weak insecticidal soap spray to control; see the caution above before spraying.

Leaves mottled yellow between veins; purple spots on leaf undersides. Cause: Downy mildew. For an illustration of this disease, see the opposite page. As the disease progresses, spots enlarge, older leaves turn brown and die, and younger leaves become infected. Treat plants with a dilute solution of copper spray to reduce the spread of the disease. Prevent problems by planting tolerant cultivars such as 'Super Select' and 'Zucchini Select'.

Leaves with spots, blotches, or brown areas. Causes: Powdery mildew; angular leaf spot; scab; Alternaria leaf blight; other fungal and bacterial diseases. Various diseases attack squash. Reduce problems by keeping foliage dry when watering and by not touching plants when wet. Spray infected plants with a dilute solution of copper spray to reduce the spread of the disease.

Powdery white spots, especially on upper leaf surfaces, are caused by powdery mildew. As the disease progresses, leaves turn brown and dry, and plants may die. Prevent problems by planting resistant cultivars such as 'Multipik' and 'Zucchini Select'.

Water-soaked spots that turn gray, die, and drop out leaving shotholes are caused by angular leaf spot. Fruit infected with angular leaf spot has small, cracked, white spots. For an illustration of this disease, see page 214. Water-soaked spots can also be caused by scab. Scab causes sunken, brown spots with gummy ooze on fruit; damage is worst in cool, moist weather.

Dark brown spots with concentric rings, usually appearing on older leaves first, are caused by Alternaria leaf blight. As the disease progresses, spots enlarge and merge, and leaves curl down and eventually drop off. Infected fruit has dark, concentrically ringed, sunken spots. For an illustration of this disease, see page 214.

Vines wilt suddenly. Cause: Squash vine borers. These fat, white, 1″ long larvae burrow into the stems and exude masses of yellow-green, sawdustlike excrement. For an illustration of this pest, see page 214. Slit stems lengthwise above injury with a sharp knife and kill larvae. Cover cut stems with moist soil so they will form new roots. Injecting stems with BTK or parasitic nematodes may also control borers. To reduce problems, plant the cultivar 'Sweet Mama Hybrid', which is resistant to vine borers. Or spray base of stems with BTK once a week in late spring and early summer.

Vines wilt at midday, starting with younger leaves; leaves remain green. Cause: Bacterial wilt. For an illustration of this disease, see page 150. As the disease progresses, leaves fail to recover, and die. Cut wilted stems and press out drops of sap. If it is milky, sticky, and astringent, your plant is infected. Destroy infected plants immediately. Control cucumber beetles since they spread the disease.

Fruit Problems

Flowers appear but no fruit develops. Causes: Male flower; lack of pollination. Male flowers open a week or more before female flowers and don't form fruit. See page 214 for an illustration of male and female squash flowers. If female flowers fail to set fruit, or if small fruit turns black and rots starting at blossom end, they haven't been pollinated. Pollinate open female flowers by hand.

Fruit misshapen. Causes: Diseases; poor pollination. Many diseases cause misshapen fruit; use leaf symptoms (if any) to diagnose the problem (see "Leaf and Vine Problems" on page 215). If leaves are healthy, high temperatures may have damaged pollen, or bees may not have been active.

Fruit with spots; flesh may rot. Causes: Angular leaf spot; Alternaria blight; scab. Several fungal and bacterial diseases cause these symptoms on squash fruit. For complete controls, see "Leaves with spots, blotches, or brown areas" on page 215.

Fruit tunneled. Cause: Pickleworms. Larvae are pale green with black, and up to ¾″ long. Keep fruit off ground or mulch, since worms feed at soil level.

Strawberry

Fragaria × *ananassa* (Rosaceae)

Strawberries are herbaceous perennials growing from crowns that send forth whorls of leaves, flowers, and surface runners with new daughter plants at their nodes. June-bearing strawberries bloom in spring for only 1 crop. Everbearing strawberries bloom and fruit in the spring and again in the fall. Day-neutral strawberries are unaffected by day length and bear heavily from June through frost in northern areas; January through August in milder climates. Day-neutrals are somewhat more difficult to grow than the other types; they are fragile and sensitive to heat, drought, and weed competition.

Culture

Plant strawberries in well-drained soil rich in organic matter. The ideal location is in full sun on high or sloping ground. Avoid frost-prone, low-lying areas.

To prevent diseases associated with overcrowding, allow a square foot of space for each plant. Choose 1 of 3 different planting systems: hill, matted row, or spaced runner. For a hill system, space plants 1′ apart each way in double rows, with 2′-3′ between each double row. Remove every runner so plants channel their energy into producing large berries. Since plants are well-spaced, the hill system minimizes diseases associated with crowding. For a matted-row system, space plants 1½′-2′ apart in rows 4′ apart. Allow the runners and daughter plants to grow in all directions to form a wide, solid row. For the spaced-runner system, set plants closer than the matted row; remove all but a few runners. Pin down runner tips so daughter plants are about 8″ apart in every direction.

Even well-managed strawberries will decline after a few seasons. Start a fresh bed in a new site with new plants every few years. Renovate June-bearers each year right after harvest. Cut off and rake away leaves; dig out old, woody plants; and thin out remaining plants. Then fertilize and water.

Flower and Fruit Problems

Fruit deeply furrowed or gnarled (cat-faced). Causes: Tarnished plant bugs; frost damage. Tarnished plant bugs can damage strawberries by injecting a toxin into fruit while sucking fluids from stem tips, buds, and fruit. Since they overwinter in dead garden refuse, the simplest control is to clean out dead plant tops at the end of the season. A floating row cover applied during the growing season also helps minimize damage. As a last resort, apply sabadilla to flowering plants. See page 218 for an illustration of this pest.

Flowers damaged by frost will produce deformed, cat-faced fruit. If spring frost threatens, cover beds overnight with a blanket.

Fruit with holes. Causes: Slugs and snails; earwigs; birds. Silvery trails near holes indicate slugs or snails. Keep these pests out with barriers of copper flashing, dry ashes, or diatomaceous earth. After rain renew ashes or diatomaceous earth. Or trap slugs under boards, in overturned clay pots, or in saucers of beer. See page 218 for an illustration of this pest.

Earwigs, leathery brown insects with pincers at their abdomen tips, nibble holes in fruit. Since they hide in dark places, trap them in short lengths of hose or rolled-up newspaper. Check traps and destroy captured earwigs daily.

The best defense against birds is a net well-secured at the edges. See "Stopping Animal Pests" on page 408 for controls.

Fruit rotted. Causes: Gray mold; leather mold. Both diseases strike during rainy weather. Fruit that rots rapidly and then turns into fuzzy balls is infected with gray mold. Blossoms infected with gray mold turn brown and die. See page 218 for an illustration of this disease.

Leather mold causes fruit to turn dark and leathery. Infected fruit is bitter-tasting. To minimize these fungal diseases, thin plants to avoid overcrowding and mulch beds to keep fruit off the soil. Pick and dispose of infected fruit as soon as you notice them. Annual bed renovation helps control gray mold.

Leaf Problems

Leaves with spots. Causes: Leaf spot; leaf blight; leaf scorch. Leaf spot causes small purple spots that develop tan centers on foliage. See page 218 for an illustration of leaf spot. Leaf blight is characterized by oval or V-shaped spots with purple centers and tan borders. Irregular, purplish blotches are the symptoms of leaf scorch. When severe, these diseases kill leaves, which weakens plants. Berries are

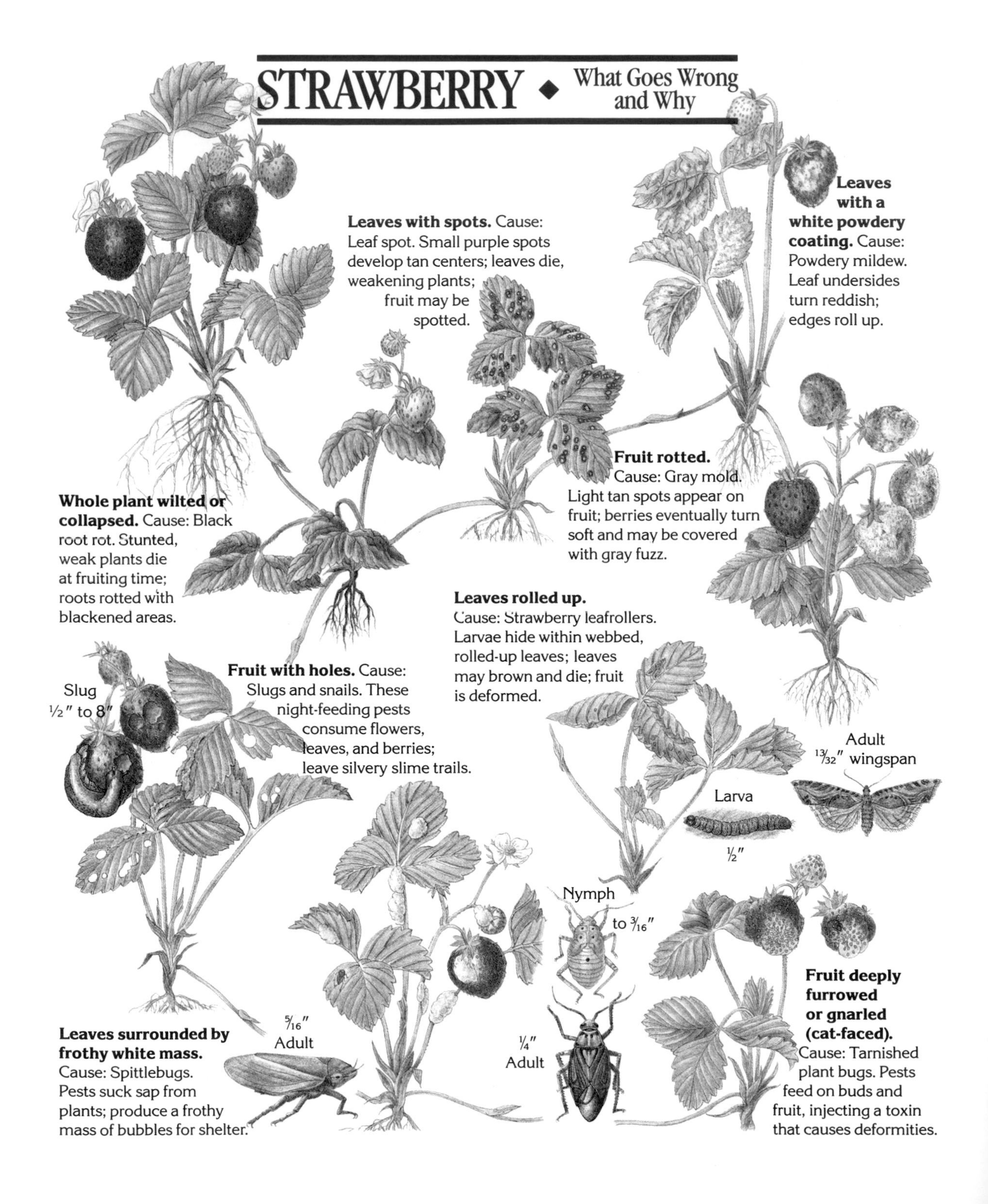
STRAWBERRY ◆ What Goes Wrong and Why
Leaves with spots. Cause: Leaf spot. Small purple spots develop tan centers; leaves die, weakening plants; fruit may be spotted.
Leaves with a white powdery coating. Cause: Powdery mildew. Leaf undersides turn reddish; edges roll up.
Fruit rotted. Cause: Gray mold. Light tan spots appear on fruit; berries eventually turn soft and may be covered with gray fuzz.
Whole plant wilted or collapsed. Cause: Black root rot. Stunted, weak plants die at fruiting time; roots rotted with blackened areas.
Leaves rolled up. Cause: Strawberry leafrollers. Larvae hide within webbed, rolled-up leaves; leaves may brown and die; fruit is deformed.
Slug
½″ to 8″
Fruit with holes. Cause: Slugs and snails. These night-feeding pests consume flowers, leaves, and berries; leave silvery slime trails.
Adult
13/32″ wingspan
Larva
½″
Nymph
to 3/16″
5/16″
Adult
¼″
Adult
Leaves surrounded by frothy white mass. Cause: Spittlebugs. Pests suck sap from plants; produce a frothy mass of bubbles for shelter.
Fruit deeply furrowed or gnarled (cat-faced). Cause: Tarnished plant bugs. Pests feed on buds and fruit, injecting a toxin that causes deformities.

spotted as well. All 3 are fungal diseases that you can control by annual bed renovation. Cultivars resistant to all 3 of these diseases include 'Albritton', 'Blakemore', 'Earlibelle', 'Fairfax', and 'Midland'. 'Cardinal', 'Delite', 'Redchief', and 'Surecrop' are resistant to leaf blight and leaf scorch.

Leaves with a white powdery coating. Cause: Powdery mildew. The powdery coating may be less apparent on strawberries than with other mildew-afflicted plants. The undersides of infected leaves turn reddish and the edges roll up. Infected fruit is stunted, rotted, or fails to ripen. See the opposite page for an illustration of this disease. Keeping plants adequately spaced and cleaning up dead plant debris help minimize this fungal disease. For persistent mildew, apply sulfur spray or dust. Or plant mildew-resistant cultivars, such as 'Albritton', 'Catskill', 'Puget Beauty', 'Sparkle', 'Sunrise', or 'Surecrop'.

Leaves rolled up. Cause: Strawberry leafrollers. These green or brown, ½" caterpillars mine leaves in the early spring and later form webs and roll leaves as they feed. Leaves may brown and die; fruit is deformed. For light infestations, destroy the rolled leaves along with the caterpillars inside. For a very heavy infestation, apply BTK or rotenone. See the opposite page for an illustration of this pest.

Leaves surrounded by frothy white mass. Cause: Spittlebugs. These tiny tan, brown, or black insects, which hide inside the frothy masses of bubbles they produce, suck sap from leaves, stems, and flowers. See the opposite page for an illustration of this pest. Spittlebugs rarely cause significant damage. To control them, wash them off with a strong spray of water.

Leaves with brown, dry undersides and fine webbing. Cause: Spider mites. Heavy pesticide use often kills naturally occurring mite predators, resulting in an abundance of spider mites. Spider mites also multiply in dry, dusty conditions. Repeated sprays with plain or soapy water usually control mites. For heavy mite infestation, purchase and release predatory mites (*Metaseiulus occidentalis*).

Whole Plant Problems

Whole plant wilted or collapsed. Causes: Black root rot; red stele disease; Verticillium wilt; strawberry crown moth larvae; strawberry crown borers. To discover which cause applies, remove a plant from the soil and examine its roots. Rotting black roots indicate black root rot; plants are stunted, produce few fruits, and may die at fruiting time. See the opposite page for an illustration of black root rot.

If roots look like "rattails" (few or no side roots) and are red inside when slit lengthwise, red stele is the problem. Both fungal diseases survive in soils for up to 10 years without a host plant; the only control is to plant new plants in well-drained soil at a new site. Cultivars resistant to some strains of red stele include 'Columbia', 'Earliglow', 'Guardian', 'Hood', 'Rainier', 'Redchief', 'Sparkle', and 'Surecrop'.

Wilted plants with no root damage may have Verticillium wilt. Verticillium-infected plants look stunted and may collapse during their first summer in the ground; inner leaves may remain green until the plant dies. There is no cure for Verticillium wilt. You must plant a new bed where you haven't grown strawberries or other Verticillium-susceptible plants, such as tomatoes, peppers, potatoes, or eggplants, for the past few years. Verticillium-resistant cultivars include 'Blakemore', 'Catskill', 'Guardian', 'Robinson', 'Sunrise', and 'Surecrop'.

Wilted plants with healthy roots may be infested with strawberry crown moth larvae or strawberry crown borer larvae. Crown borers are ¼" snout beetles and crown moths are large, clear-winged moths. Both pests lay eggs on strawberries. The resulting grublike larvae burrow into plant crowns and cause wilting and death. Cut crowns of wilted plants in half. If you find a large tunnel, your plants are infected with one of these pests. The only cure is to dig up and destroy these plants immediately. Crown borers don't fly and can't crawl more than 300 feet, so starting with clean plants planted far from old infested beds keeps this pest in check.

Whole plant stunted; roots and crown chewed. Cause: Strawberry root weevils. The brown-headed, white larvae feed on plant crowns and roots; the black, 1/4" adult weevils may feed on leaves. Apply parasitic nematodes to the soil to control larvae. For severe infestations of adults, dust with rotenone.

Sweet Potato

Ipomoea batatas (Convolvulaceae)

Sweet potatoes are perennial vegetables grown as annuals in temperate regions for their tuberous roots. While they are tropical natives, you can grow them if you get at least 100 frost-free days. Cultivars have light yellow to purplish red skin, and white to deep orange flesh.

Culture

Sweet potatoes prefer loose, well-drained soil with a pH between 6.0 and 6.5. They require moderate amounts of nitrogen and boron, moderate to high levels of phosphorus, and high levels of potassium. Have the soil tested and amend as necessary before planting. Sweet potatoes do well in raised beds. Work in lots of organic matter before planting. Avoid top-dressing after early summer, or root formation may be interrupted.

Keep soil moist, but not soggy, until the vines begin to spread. After that, water only if vines wilt. When the roots begin to enlarge in late summer, keep the soil moist again until harvest. Mulch plants to suppress weeds and conserve moisture. Black plastic mulch will also warm the soil.

Plant sweet potatoes where they have not been grown for at least 2 years. After harvest, cut vines and let dry, then compost or till under to reduce disease buildup.

Purchase disease-free plants, or start your own from healthy, overwintered roots. Plant out when nights stay above 60°F. Soak plant roots in compost tea for 5 minutes before planting to help reduce disease problems. Water in with a fish emulsion or fish-meal tea after planting to give the plants a good start.

Dig potatoes gently before the first frost, after foliage starts to yellow. Dry them for 2-3 hours in the garden. Sort out any damaged or diseased potatoes to use as soon as possible. Cure healthy potatoes for 10 days in a 80°-85°F humid place. Gradually reduce temperature and store them in a 55°-60°F humid room.

Leaf and Whole Plant Problems

Leaves yellow between veins. Causes: Stem rot; Fusarium wilt. Symptoms of stem rot, also known as bacterial soft rot, usually appear when temperatures are above 90°F. Leaves turn yellow, vines wilt, and the base of stems are black and shiny. Destroy diseased plants. Roots may already be rotted or will rot in storage, so use any healthy potatoes as soon as possible. Prevent problems by planting disease-free plants or tolerant cultivars such as 'Goldmar'.

If young leaves are yellow and wilted, and older leaves drop, the plant is suffering from Fusarium wilt. The stems may have a faint purple coloration just below the soil line. Destroy plants infected with this fungal disease. Prevent problems by planting disease-free plants or resistant cultivars such as 'Allgold'.

Leaves with purple-bordered yellow spots. Cause: Internal cork. Destroy infected plants. See "Potatoes with hard, dark, corky spots in the center of flesh" below for controls.

Leaves yellow with dead, brown spots; plant stunted. Cause: Nematodes. See "Potatoes with rough, rotted pits" below for controls.

Leaves thin and pale; plant stunted. Cause: Pox. See "Potatoes with rough, rotted pits" below for controls.

Leaves riddled with small holes. Cause: Flea beetles. These tiny, black, brown, or bronze insects hop like fleas when disturbed. Adults can transmit disease when feeding, and larvae damage roots. Control flea beetle adults by covering plants with row cover as soon as they are planted out. Apply parasitic nematodes to the soil to help control larvae. Reduce problems by planting tolerant cultivars such as 'Centennial' and 'Jewell'.

Leaves with large, round holes. Causes: Tortoise beetles; caterpillars. Tortoise beetles are 1/4" long, oval, flattened beetles with varying colors and patterns on their wing covers. Larvae are 1/2" long and have flattened bodies with spiny margins and a forked horn at the tail end. Both adults and larvae damage young plants. Handpick, or spray plants with a commercial pyrethrin/rotenone mix if damage is severe.

Caterpillars sometimes feed on sweet potato leaves. Spray plants with BTK if caterpillars are feeding.

Root Problems

Potatoes long and spindly. Causes: Potassium deficiency; growing season too short. Check for deficiency with a soil test and amend soil as necessary. Sweet potatoes need a long growing season. Prevent problems by planting cultivars suited to your region.

Potatoes cracked. Cause: Uneven soil moisture. If soil is alternately dry then wet, roots may split their skins. Prevent problems by keeping soil evenly moist and storage humidity constant. Some cultivars, such as 'Jewell' and 'Puerto Rico 198', are somewhat resistant to cracking.

Potatoes with dark discolored patches on skin. Cause: Scurf. Symptoms often appear in storage. Initial damage is only skin deep, but subsequently, skin may split open, causing the flesh to shrivel up. Prevent this fungal disease by planting disease-free plants and by keeping storage temperatures above 50°F.

Potatoes with black, circular, corky depressions. Cause: Black rot. Symptoms of this fungal disease may develop in the garden or in storage. Flesh under spots is brown to green and has a bitter taste, so trim well before using. Prevent problems by starting with disease-free plants, controlling root-feeding insects, and by planting resistant cultivars such as 'Allgold'.

Potatoes with hard, dark, corky spots in the center of flesh. Cause: Internal cork. This condition develops if storage conditions are too warm and may be caused by a viral disease. Store potatoes between 55° and 60°F to reduce problems. Prevent problems by planting tolerant cultivars such as 'Allgold', 'Centennial', and 'Nemagold'.

Potatoes with rough, rotted pits. Causes: Pox; nematodes. Plants with pox, a bacterial disease, may be pale and stunted, and roots often resemble a string of irregular beads. Destroy infected plants and use healthy roots as soon as possible. Prevent problems by planting disease-free plants. Adjust the soil pH to below 5.2 by adding sulfur if pox has been a major problem in the past. While this is below the optimal pH range for sweet potatoes, they will tolerate it and the bacteria will be inactive.

Root knot and other pest nematodes cause poorly colored, deformed potatoes with rotted areas under the skin, surface blemishes, and surface cracks. Control pest nematodes by applying a chitin or parasitic nematodes to the soil. Prevent problems by planting nematode-resistant cultivars such as 'Heart', 'Jasper', 'Jewell', 'Kandee', 'Nemagold', and 'Nugget'.

Potatoes with small holes, tunnels, or shallow splits. Causes: Flea beetle larvae; sweet potato weevils; wireworms. Sweet potato flea beetle larvae tunnel just under the skin. As the roots grow, the skin over the tunnels splits open, leaving shallow scars. Larvae are 3/16", slender, and white. Striped flea beetle larvae are white and up to 3/4″ long; they tunnel into the center of roots leaving wandering, branched tunnels. See "Leaves riddled with small holes" on page 221 for controls.

Sweet potato weevil adults are small, reddish, antlike insects. Larvae are white with pale brown heads and grow up to 3/8″ long. Larvae tunnel into potatoes in the field or in storage. Weevils do not hibernate and must have food to survive. Destroy all plant residue and weeds after harvest to control them. Apply parasitic nematodes to the soil if weevils have been a problem in the past.

Wireworm larvae are yellow to reddish brown, slender, tough-bodied, segmented worms with brown heads and grow up to 1 1/2″ long. Adults are dark-colored, elongated click beetles. To prevent problems, don't plant sweet potatoes in soil where grass or grain grew the previous season. Apply parasitic nematodes to the soil before planting, and plant tolerant cultivars such as 'Centennial'.

Potatoes with large, shallow feeding scars or hollow cavities. Causes: White grubs; cutworms; cucumber beetles. White grubs are fat, whitish larvae that tend to feed over the whole surface of the potato. Adults are Japanese or June beetles. Apply parasitic nematodes to the soil before planting to control.

Cutworms are grayish or dull brown caterpillars that curl up when disturbed and are active at night. Cutworms tend to feed near the ends of potatoes. Sprinkle moist bran mixed with BTK on the soil surface in the evening, or add parasitic nematodes to the soil before planting to control cutworms.

Spotted cucumber beetle larvae (also called southern corn rootworms) are white with brown heads and grow up to 1/2″ long. Adults are 1/4″, greenish yellow beetles with black spots. Spray plants with a commercial pyrethrin/rotenone mix if damage is severe. Prevent problems by covering plants with row cover.

Potatoes turn spongy in storage. Cause: Cold injury. Cool temperatures can damage sweet potatoes. Harvest when air is above 50°F and store above 50°F to prevent injury.

Potatoes rot in storage. Cause: Various bacterial and fungal diseases. Prevent problems by planting disease-free plants and curing and storing potatoes at the recommended temperatures and humidity levels; see "Culture" on page 220 for instructions. Soak potatoes in a 10 percent bleach solution (1 part bleach to 9 parts water) for a few minutes before curing if storage rots have been a past problem.

Syringa

Lilac. Shrubs and small trees.

Lilacs are among the most popular of deciduous flowering shrubs. They bloom in late spring in a number of colors; borne in large clusters, the flowers are exquisitely fragrant. Lilacs are plants of cool weather; with few exceptions, they do not flourish south of Zone 7. Lilacs are best used in the mixed shrub border, grouped for emphasis in the landscape, or planted for screening.

Plant in spring or fall in full sun with neutral or slightly alkaline soil enriched with ample organic matter. Prune and deadhead them immediately after blooming. Lilacs are available either on their own roots or grafted onto privet roots; the former are much preferred. Cut back to the ground any suckers that form.

Problems

Leaves with powdery white coating. Cause: Powdery mildew. This is a very common problem on lilacs. See "Leaves with powdery white coating" on page 237 for controls.

Leaves tunneled. Cause: Leafminers. For controls, see "Leaves tunneled" on page 237.

Leaves yellow; stems and leaves covered with small bumps. Cause: Scales. See "Leaves yellow; stems and leaves covered with small bumps" on page 237 for control measures.

Trunk or branches with small holes; limbs die or break off. Cause: Borers. For details, see "Trunk or branches with small holes; limbs die or break off" on page 238.

Leaves wilted and discolored; branches die back. Cause: Blight. Both a bacterial and a fungal blight attack lilacs, causing similar symptoms. Prune out and destroy infected branches during the dormant season. Disinfect pruners after each cut in a 10 percent bleach solution (1 part bleach to 9 parts water). In spring, spray twice with bordeaux mix at 10-day intervals, starting when the leaves first unfold.

Leaves with spots. Cause: Leaf spots. See "Leaves with spots" on page 237 for controls.

Leaves pale and drop early; branches wilt and die. Cause: Wilt. If plants are affected by this fungus, prune off dead and diseased branches. Feeding with a high-nitrogen fertilizer may help plants recover. Remove and destroy badly infected shrubs, and don't plant lilacs in the same soil.

Leaves skeletonized or with large holes; branches may be webbed. Cause: Caterpillars. For control measures, see "Leaves skeletonized or with large holes; branches may be webbed" on page 236.

Shoots clustered tightly, with small leaves. Cause: Witches' broom. This problem is usually not serious; merely prune out and destroy the dense, bushy growth.

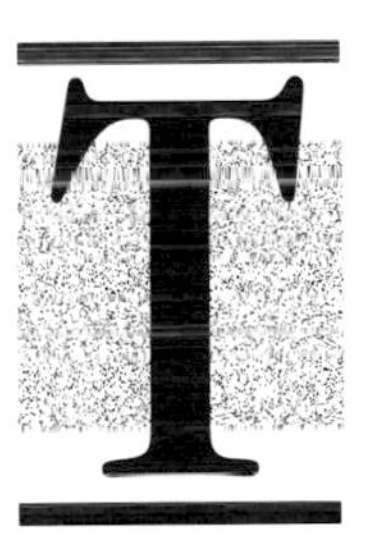

Tagetes

Marigold. Annuals.

Hearty marigolds are a mainstay of the early summer to late-fall garden. Flower colors range from palest yellow and near-white to red and mahogany red; some are even bicolored. Single or double blooms may be 1″–6″ wide. The 6″–36″ plants are covered with finely cut leaves. Marigolds may have a strong odor; if you dislike the scent, look for an odorless cultivar. Marigolds are good bedding plants, pot plants, and cut flowers.

It's easy to raise marigolds by sowing the seed directly into the garden 2–3 weeks before the last frost. For best results, hold the white tuft of each seed and place the long dark section into the ground at an angle. Indoors, seeds sown in flats germinate within 2 weeks at about 70°F. Move outdoors only after all frost danger has passed. Thin or plant 8″-18″ apart.

Marigolds thrive in full sun, except in southern and southwestern areas where afternoon shade prolongs bloom. They demand regular watering and well-drained, average soil. Pick off old flower heads. Taller cultivars need staking. Avoid overhead watering, as even the small blossoms hold water like cups, and the weight may cause the brittle stems to snap. If you work compost into the soil at planting time, no extra fertilizer is necessary.

Problems

Leaves skeletonized; flowers eaten. Cause: Japanese beetles. This ½″ long pest has a metallic blue-green body and bronze wing covers. Besides feeding on leaves, the adult beetles often chew on buds and flowers. Handpick in early morning (when the beetles move more slowly) if there are only a few.

Leaves stippled with yellow; foliage webbed. Cause: Spider mites. For control measures, see "Leaves stippled with yellow; foliage webbed" on page 18.

Flowers covered with gray mold. Cause: Botrytis blight. Brown spots may appear on leaves, and petals may turn black. Botrytis is promoted by overhead watering, continuing cool, wet weather, or crowded plantings that impede air circulation. Once this fungal disease appears, it may spread to healthy plants on contact, or by wind, rain, or handling. Remove and discard infected blooms. Avoid wetting flowers when watering.

Stems turn black at base; plant wilts. Cause: Stem rot. Caused by either bacteria or fungi, stem rot starts at the soil level and works upward. Remove and destroy infected plants. Don't replant marigolds in that area unless you solarize the soil.

Leaves greenish yellow; growth poor. Cause: Aster yellows. For more information, see "Leaves greenish yellow; growth poor" on page 19.

Taxus

Yew. Shrubs and trees.

Yews are evergreen shrubs and trees with needle-shaped leaves arranged densely and spirally on the branches. The male and female flowers are separate on the plants, with the females producing fleshy red berries, each surrounding a toxic seed. Yews can be utilized as foundation plants, hedges, and even specimens.

Set out in spring as balled-and-burlapped or container-grown plants. Yews prefer a well-drained but moisture-retentive soil and grow well in full sun or partial shade.

Problems

Leaves with notched edges. Cause: Black vine weevils. The wingless, ⅓″, brownish black adults feed on the leaves and bark and can kill branches. Try spreading a dropcloth around your plants, then shaking the plants; the adults will drop and can be removed and destroyed. Spray leaves several times with pyrethrin for major infestations. The small, curved, white grubs of these weevils feed on the roots; drench the soil around the base of the plant with a solution of parasitic nematodes for control.

Leaves stippled with yellow; foliage webbed. Cause: Spider mites. See "Leaves stippled with yellow; foliage webbed" on page 236 for controls.

Leaves discolored, wilted, or dropping; plant lacking vigor. Causes: Mealybugs; blights. The tiny, powdery-looking, white adults generally congregate on the trunks and interior branches, making them difficult to see until the plant starts to weaken. Control by thoroughly spraying the plant with superior oil or repeated applications of insecticidal soap.

Various blights can cause similar symptoms on yews. If you don't see the fluffy white masses that are a sign of mealybugs, suspect blight. Cutting off the affected portion of the plant is the best way to deal with this disease.

Leaves yellow; stems and leaves covered with small bumps. Cause: Scales. See "Leaves yellow; stems and leaves covered with small bumps" on page 237 for controls.

Leaves yellow. Causes: Low pH; waterlogged soil. Yews prefer soils with a neutral or slightly acid pH (around 6.5). They do not grow well in highly acid soils, so avoid interplanting them with rhododendrons or other acid-loving plants. Raise the pH around your yews by adding 3 pounds of ground limestone per 100 square feet of soil area around the plants.

Too much water around plant roots causes similar symptoms. Avoid overwatering and planting in poorly drained areas. Improving the drainage around plants will help them recover.

Thalictrum

Meadow rue. Perennials.

Tall plants with blue-green, fernlike foliage and loosely fluffy clusters of lavender or yellow flowers, meadow rues range from 1′ to over 5′ tall. Columbine meadow rue (*Thalictrum aquilegifolium*) has leaves resembling columbine foliage; other species' leaves look like those of maidenhair ferns.

Grow meadow rues in moist, well-drained, richly organic soil and partial shade. Tall flower stems need shelter from strong winds and may require staking. Water regularly to maintain soil moisture. Divide every 4-5 years to reduce crowding.

Meadow rues have few insect pests. Fungal diseases such as powdery mildew may damage foliage and are best controlled with cultural practices. See "Leaves covered with white powder" on page 177 for controls.

Thuja

Arborvitae. Shrubs and trees.

Arborvitaes are evergreen trees and shrubs with flat sprays of scalelike leaves. They are excellent in foundation plantings and used as hedges; upright cultivars make strong specimen plants.

Set out in spring or fall as container-grown or balled-and-burlapped plants. Arborvitaes thrive in deep, moist, well-drained soil in full sun. If possible, choose a site protected from strong winter winds to avoid damage.

Problems

Plant defoliated; branches bear cocoonlike bags. Cause: Bagworms. These pests are one of the most common problems on arborvitae. For control measures, see "Plant defoliated; branches bear cocoonlike bags" on page 236.

Leaves yellow; stems and leaves covered with small bumps. Cause: Scales. For controls, see "Leaves yellow; stems and leaves covered with small bumps" on page 237.

Leaves stippled with yellow; foliage webbed. Cause: Spider mites. See "Leaves stippled with yellow; foliage webbed" on page 236.

Plant defoliated. Cause: Hemlock loopers. The 1″ long, greenish yellow, black-spotted caterpillars usually appear in June. They feed on the needles, starting from the branch tips and working toward the center of the plant. Handpick small populations; control large infestations with BTK.

Leaves tunneled. Cause: Leafminers. See "Leaves tunneled" on page 237 for controls.

Trunk or branches with small holes; limbs die or break off. Cause: Borers. For more information, see "Trunk or branches with small holes; limbs die or break off" on page 238.

Leaves with spots. Cause: Leaf spots. See "Leaves with spots" on page 237 for controls.

Branch tips die back. Causes: Blight; winter injury; sun scorch; drought stress. Blight, a fungal disease, attacks mostly during cool, wet weather; good air circulation among plants discourages its onset. Leaves of infected plants shrivel and twigs die back. Cut off and destroy dead or diseased wood. To prevent the spread of mild infections, spray with bordeaux mix at 2-week intervals until the weather gets warmer and drier.

If you see the same symptoms in very early spring or during hot, dry weather, your arborvitae may be suffering from winter injury, sun scorch, or drought stress. Minimize damage by watering deeply and regularly during dry periods. A thick mulch may also help by retaining moisture and keeping roots cool.

Thunbergia

Thunbergia. Annuals.

Thunbergias are showy, twining climbers ranging from 10′ to 20′ high. The 1½″-3″ flowers may be orange with dark brown centers, white with red centers, pure orange, or purple-blue. All need some type of support unless used in hanging baskets.

Sow seeds indoors 6 weeks before the last frost. Germination takes 2-3 weeks. Seedlings grow slowly. Place outdoors when night temperatures are above 50°F. Thunbergias have little cold tolerance, even as mature plants.

Grow plants in full sun to light shade with fertile, well-drained soil. Avoid sites with reflected light, which can cause leaf sunburn. If spider mites attack leaves, see "Leaves stippled with yellow; foliage webbed" on page 18. Powdery mildew can also be a problem; see "Leaves with powdery white patches" on page 19 for controls.

Thyme

Thymus vulgaris
and other species (Labiatae)

Thyme is a hardy (Zone 5) perennial herb grown for its tiny, aromatic leaves. It does best in light, dry, well-drained soil with a pH between 5.0 and 8.0. Thyme needs at least 4 hours of full sun per day. Start pinches of seed indoors in 70°F planting mix, purchase plants, or grow them from cuttings.

Thyme is normally quite problem-free. Dark spots on leaves are caused by fungal leaf spot. Spray foliage with fish emulsion or sulfur to prevent the spread of mild infections. Tan or red blisters on leaves are caused by rust, a fungal disease. Destroy infected leaves and avoid wetting leaves to prevent its spread. See the Herbs entry beginning on page 116 for other possible problems.

Tilia

Linden. Trees.

Lindens are deciduous shade trees, principally valued for their foliage. Their summer flowers are fragrant and attractive to bees. Lindens are widely used as street trees and, because they adapt well to pruning, for tall hedges.

Set out in spring or fall in full sun. Lindens tolerate drought and a wide range of soil conditions, although they perform best in deep, moist soils.

Problems

Leaves wrinkled and discolored. Cause: Aphids. These pests, and the black, sooty mold that grows on the sticky honeydew they produce, are a common problem on lindens. For control measures, see "Leaves wrinkled and discolored" on page 235.

Leaves skeletonized. Causes: Japanese beetles; sawflies. See "Leaves skeletonized" on page 236 for controls.

Plant defoliated; branches bear cocoonlike bags. Cause: Bagworms. For controls, see "Plant defoliated; branches bear cocoonlike bags" on page 236.

Leaves skeletonized or with large holes; branches may be webbed. Cause: Caterpillars. Numerous caterpillars feed on foliage to varying degrees. See "Leaves skeletonized or with large holes; branches may be webbed" on page 236 for suggested controls.

Leaves yellow; stems and leaves covered with small bumps. Cause: Scales. See "Leaves yellow; stems and leaves covered with small bumps" on page 237 for control measures.

Trunk or branches with small holes; limbs die or break off. Cause: Borers. See "Trunk or branches with small holes; limbs die or break off" on page 238 for controls.

Leaves and shoots blackened; leaves with moist or brown sunken spots. Cause: Anthracnose. See "Leaves and shoots blackened; leaves with moist or brown sunken spots" on page 238 for control measures.

Leaves with powdery white coating. Cause: Powdery mildew. See "Leaves with powdery white coating" on page 237 for controls.

Leaves tunneled. Cause: Leafminers. See "Leaves tunneled" on page 237 for controls.

Trunk or branches with oozing lesions; branch tips die back. Cause: Canker. For more information, see "Trunk or branches with oozing lesions; branch tips die back" on page 238.

Leaves pale and drop early; branches wilt and die. Cause: Wilt. Prune off dead and diseased branches. Feeding with a high-nitrogen fertilizer may help plants recover. Remove and destroy badly infected trees, and don't plant new trees in the same area; they may also be attacked.

Tomato

Lycopersicon esculentum (Solanaceae)

Tomatoes are tender perennials that are grown as annuals in temperate climates. Their fruits can be pale greenish white, yellow, orange, red, or pink, and anywhere from currant size to well over a pound apiece. There are 2 main types: Determinant cultivars grow to a certain height and stop, putting all their energy into producing fruit heavily over a 4–6 week period; indeterminant cultivars grow and produce fruiting clusters until frost.

Culture

Tomatoes require full sun and deep soil with a pH between 6.0 and 6.8. Work in plenty of compost before planting to add organic matter. Tomatoes require moderate levels of nitrogen and phosphorus, and moderate to high levels of potassium and calcium. Tomatoes grow best between 75° and 90°F. Temperatures over 100°F can kill blossoms, while temperatures below 50°F can cause chilling injury.

Keep soil moist, but not soggy, and do not allow it to dry out. Avoid wetting leaves when watering to help prevent diseases. Tomatoes do well in raised beds with drip irrigation and mulch. Black plastic is a good mulch in cool areas because it helps warm the soil as well as suppresses weeds and conserves water. Organic mulch helps keep the soil cooler in very warm areas while adding organic matter. Mulch also helps prevent disease by preventing the fruit from touching the ground or being splashed with soil containing disease-causing organisms.

Choose cultivars that are adapted to local growing conditions. Many are resistant to one or more problems. Resistant cultivars are usually denoted in seed catalogs as follows: F = Fusarium-resistant, V = Verticillium-resistant, T = tobacco mosaic virus-resistant, and N = nematode-resistant.

Do not plant tomatoes where tomatoes, potatoes, eggplants, or peppers have been planted within the past 3-5 years. Also, try to plan your planting scheme to separate these crops in the garden. Compost or till under all plant residues at the end of the season to reduce overwintering pests. After tilling, spread 2-4 pounds of bloodmeal or soybean meal per each 100 square feet to encourage rapid breakdown of plant material.

Purchase stocky, insect- and disease-free plants, or start your own from seed indoors. Soak seed in a 10 percent bleach solution (1 part bleach and 9 parts water) for 10 minutes and rinse in clean water before planting to reduce seed-borne diseases. Tomato seeds germinate best between 75° and 90°F. Once seedlings are up, they grow best between 60° and 70°F. When plants are set out, add 1 cup each of bonemeal and kelp to each hole. Water the transplants thoroughly with fish emulsion or compost tea to give them a good start. Spray young plants with seaweed extract to help prevent transplant shock and nutrient deficiencies.

Leaf and Whole Plant Problems

Seedlings fall over; stems girdled or rotted at soil line. Cause: Damping-off. Disinfect reused pots and flats by dipping them in a 10 percent bleach solution and letting them air-dry before filling them with fresh seed-starting mix. Sow seed thinly to allow for air movement around seedlings. Cover seed with a thin layer of soilless mix or vermiculite. Water only enough to keep soil moist, not soggy. Thin seedlings and spray with compost tea as soon as first true leaves open to help prevent the problem.

Seedlings clipped off at soil line. Cause: Cutworms. Check for fat, 1"-2" long, dull brown or gray caterpillars in the soil near the base of plants. Once they chew off a seedling, there is nothing you can do except protect the remaining seedlings from nocturnal cutworm attacks. To prevent cutworm damage, place cutworm collars around transplants, sprinkle moist bran mixed with BTK on the soil surface in the evening, or add parasitic nematodes to the soil at least a week before planting.

Leaves yellow or pale. Cause: Nutrient deficiency. Spray young plants with seaweed extract to help prevent deficiencies, and add compost to the soil. Have the soil tested and amend as necessary.

If young leaves are yellow with green veins, suspect iron deficiency. Reduce soil pH to help make iron more available. If dark spots develop in the yellow areas, and leaves are

TOMATO ◆ What Goes Wrong and Why

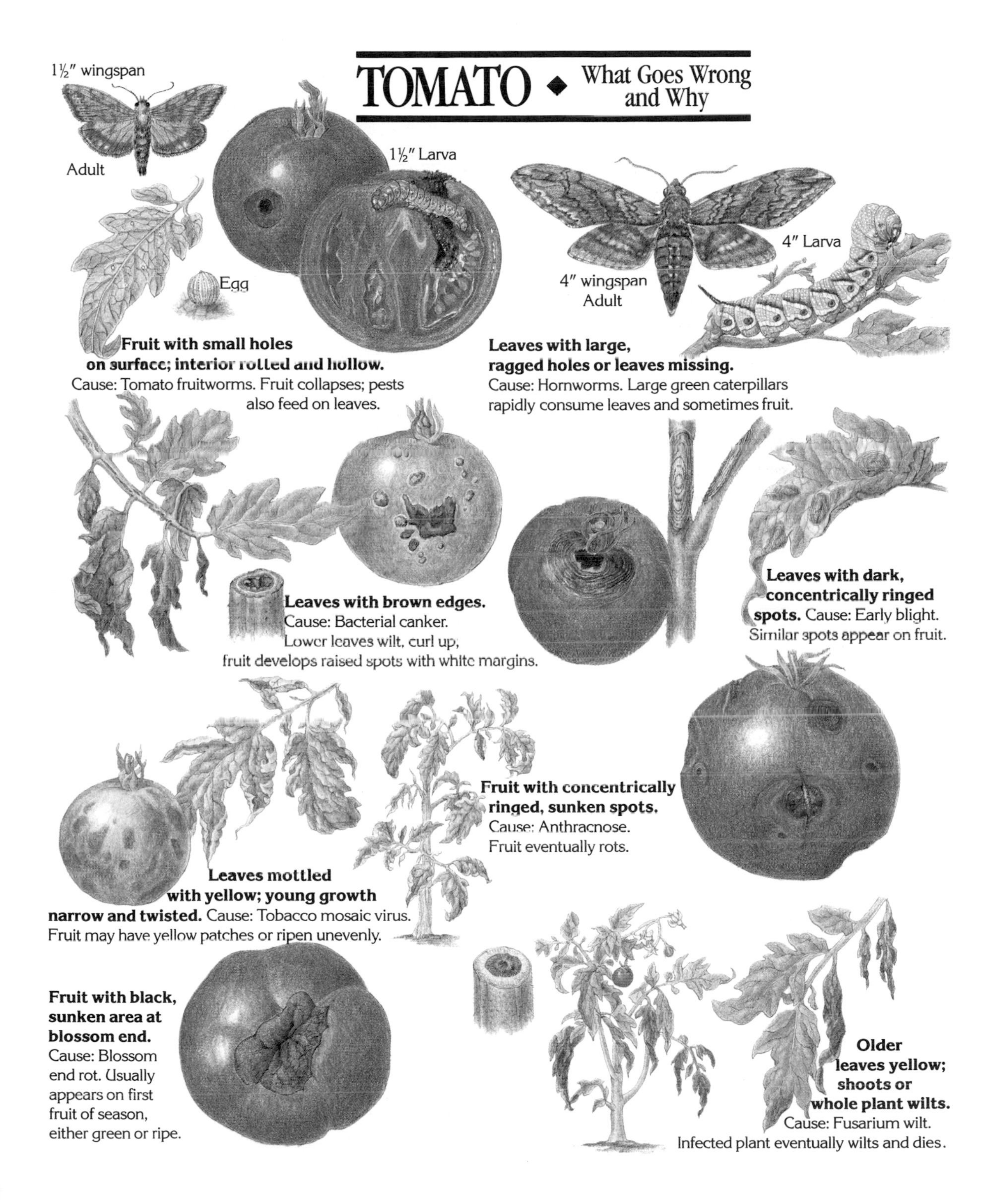

small and narrow, the problem may be zinc deficiency. If older leaves are yellow with green veins, and then become bronzed, suspect potassium deficiency.

If young leaves are pale and growing tips die, the problem may be calcium deficiency. Add high-calcium lime, dolomitic lime if magnesium is also low, wood ashes, or gypsum to the soil.

Plants that are stunted and have yellow or pale older leaves may be nitrogen deficient.

Leaves yellow, distorted, and sticky. Cause: Aphids. Leaves may develop brown spots. These small green, black, gray, pink, or white fluffy-coated insects suck plant sap. For mild infestations, knock pests off plants with a blast of water. Spray plants with insecticidal soap in the evening to control, or with a commercial pyrethrin/rotenone mix if infestation is severe.

Leaves of young plant purple. Causes: Phosphorus deficiency; lack of dark period. Phosphorus is unavailable in cool soil; symptoms usually fade as soil warms. Spray plants with seaweed extract to alleviate symptoms.

Seedlings grown under lights will be purple if the lights are left on continuously. Give seedlings 8 hours of darkness each night to reverse or prevent purpling.

Leaves stippled or bronzed. Cause: Mites. A fine webbing may be present below or between leaves. Infested leaves dry out and fall off. These tiny, spiderlike insects feed on plant sap and thrive in hot, dry weather. Spray plants with insecticidal soap or sulfur to control mites.

Leaves mottled with yellow; young growth narrow and twisted. Cause: Tobacco mosaic virus. See page 229 for an illustration of tobacco mosaic virus (TMV). Destroy diseased plants. Presoak seed in 10 percent bleach solution before planting, or choose resistant cultivars to prevent problems. Wash hands after handling tobacco and before touching tomatoes to prevent TMV, and control aphids, which spread viruses as they feed.

Leaves yellow; plant stunted and wilts in hot weather. Cause: Root knot nematodes. Plants eventually die. Check roots for swollen sections or galls up to 1″ in diameter. Destroy infested plants; do not compost them. Apply chitin or parasitic nematodes to the soil to control the pests.

Older leaves yellow; shoots or whole plant wilts. Cause: Walnut, Fusarium, or Verticillium wilt. Walnut wilt occurs in soil containing black walnut roots, which secrete a substance that is toxic to many kinds of plants. Tomatoes within 50′ of black walnut trees or stumps wilt and die suddenly. The toxic compound remains in the soil for some years after trees are cut down. Plant your tomatoes at least 50′ from walnut trees. If this is impossible, grow tomatoes in containers in a good organic potting mix.

Fusarium wilt and Verticillium wilt are both fungal diseases and are difficult to tell apart. See page 229 for an illustration of Fusarium wilt. Both Fusarium and Verticillium wilts begin as a yellowing and wilting of the lower leaves. Plants are stunted and do not recover when watered. Cut open a stem near the soil line and look for internal discoloration. Verticillium wilt usually affects the whole plant while Fusarium may infect individual shoots before the whole plant is affected. Fusarium thrives in warmer temperatures (80°-90°F) than Verticillium does (68°-75°F). Destroy infected plants. Tomato Fusarium infects only tomatoes; Verticillium infects a wide range of plant species, making effective rotation control difficult. Prevent problems by presoaking seed in a 10 percent bleach solution or by planting resistant cultivars. Control pest nematodes to help reduce wilt problems.

Whole plant wilts; leaves remain green. Cause: Southern bacterial wilt. Plants do not recover when watered. This disease is most damaging in the Deep South. Destroy infected plants. Prevent problems by planting tolerant cultivars such as 'Saturn' and 'Venus'.

Leaves with small dark spots. Causes: Bacterial spot; bacterial speck. Centers of spots may dry and fall out. Leaves may turn yellow, then brown, and fall off. Spray infected plants with copper to prevent development of further symptoms.

Leaves with dark, water-soaked patches. Causes: Late blight; Septoria leaf spot. Water-soaked patches that turn brown, dry, and papery are symptoms of late blight. In wet weather the patches may develop a ring of white mold. Stems may have blackened areas. Infected fruit have large, irregular, firm, greasy-looking brown spots. This fungal disease often occurs during periods of humid weather with cool nights (below 60°F) and warm days (70°-85°F). Spray plants with compost tea to help prevent further symptom development, or with copper if the disease is severe. Prevent problems by planting tolerant cultivars such as 'Pieraline'.

If the patches develop into circular, dark spots with light centers peppered with dark specks, suspect Septoria leaf spot, another fungal disease. Older leaves are affected first. Spray plants with copper to prevent further symptom development. Spraying transplants with an antitranspirant may help prevent Septoria. However, don't spray with antitranspirant when tomatoes are blooming heavily.

Leaves with dark, concentrically ringed spots. Cause: Early blight. Lower leaves and stems affected first. See page 229 for an illustration of this fungal disease, also known as Alternaria blight. Disease occurs when plants are loaded with fruit and during humid, warm (75°-85°F) weather. Spray infected plants with copper and/or sulfur to prevent further disease development. Prevent problems by planting resistant cultivars such as 'Kotlas'. Spraying transplants with an antitranspirant may help to prevent this disease.

Leaves with brown edges. Cause: Bacterial canker. Lower leaves wilt and curl up; stems develop light-colored streaks and are brown and mealy inside. See page 229 for an illustration of this disease. Destroy infected plants. To prevent problems, presoak seed in 10 percent bleach solution, and avoid wounding plants. Don't touch plants when they are wet to avoid spreading the disease.

Leaves with small holes. Cause: Flea beetles. Young transplants are the most susceptible. These tiny black, brown, or bronze insects hop when disturbed. Spray plants with a commercial pyrethrin/rotenone mix if infestation is severe. Protect transplants with row cover until they start to flower.

Leaves with large, ragged holes or leaves missing. Causes: Colorado potato beetles; hornworms; other caterpillars. Colorado potato beetles are yellowish orange, oval, hard-shelled, 1/3" long beetles with black stripes. Larvae are soft-bodied, humpbacked, dark orange grubs with a row of black spots down each side of their bodies. Eggs are orange and laid in rows on undersides of leaves. Severe infestations can defoliate plants. Handpick insects or spray plants with BTSD if young larvae are feeding.

Hornworms are 3"-4½" caterpillars with white diagonal stripes. The tobacco hornworm has a red horn projecting from the rear, while the tomato hornworm has a black horn. See page 229 for an illustration of these pests. Handpick or spray plants with BTK to control them. Do not spray if caterpillars are covered with small, white, cigar-shaped projections, which are actually parasitic wasp cocoons.

Other caterpillars, such as cabbage loopers and beet armyworms, sometimes feed on tomato leaves. Handpick or spray plants with BTK if many caterpillars are feeding.

Flower and Fruit Problems

Few flowers form; flowers may drop without setting fruit. Causes: Excess nitrogen;

shading; extreme temperatures; drought stress. Plants with excess nitrogen are dark green and vigorous, but produce few flowers. Wait for flowers to form. Prevent problems by avoiding high-nitrogen soil amendments.

Tomatoes need at least 6 hours of sun a day to produce flowers; prevent problems by not planting them in shaded areas.

Temperatures over 100° or below 55°F can damage flowers and cause them to fall without setting fruit. Wait for new flowers to form. Protect plants with row cover until night temperatures remain above 55°F.

Dry soil may also cause blossom drop. Keep soil evenly moist, but not soggy, and do not allow it to dry out.

Fruit with large, faded or gray-white, sunken patches. Cause: Sunscald. Green or ripe fruit can be affected. Control leaf diseases to prevent defoliation so fruit will be shaded and protected from direct sun.

Fruit gnarled and malformed with dry scars near the blossom end. Cause: Cold injury. Cat-facing, as this symptom is also called, is caused by prolonged cool weather during blossoming. Poor pollination may be partially responsible. Protect plants with row cover until nights remain above 55°F.

Fruit ripens unevenly and has grayish yellow blotches. Cause: Graywall. Green fruit has grayish skin blotches and interior is discolored. Ripe fruit has green or brown areas in interior. This condition can be caused by dense shade from crowded plants, cool temperatures (below 60°F) during ripening, wet or compacted soil, excess nitrogen, potassium deficiency, or various diseases, including tobacco mosaic virus. Prevent problems by providing good growing conditions.

Fruit with green, water-soaked spots. Cause: Late blight. Spots expand into large, greasy-looking, brown areas but remain firm. See "Leaves with dark, water-soaked patches" on page 231 for controls.

Fruit with pale yellow spots just under the skin; spots may have a central puncture. Cause: Stink bugs. Flesh under spots is white and spongy. These brown, tan, gray, or green, ½" long, shield-shaped bugs inject a toxin when they feed on green fruit. Many species of weeds are also host plants. Keep garden well-weeded to prevent problems.

Fruit with small, raised spots. Causes: Bacterial canker; bacterial spot; bacterial speck. If spots are tan with white margins and look like "bird's eyes", the plants are suffering from bacterial canker. See "Leaves with brown edges" on page 231 for controls.

If spots are brown, scabby, and rough with sunken centers, the plants have bacterial spot. If spots are tiny, dark brown, and surrounded with white borders, the plants have bacterial speck. See "Leaves with small dark spots" on page 231 for controls.

Fruit with concentrically ringed, sunken spots. Causes: Anthracnose; early blight. If these spots appear on ripe fruit, and fruit eventually rots, the problem is probably anthracnose. See page 229 for an illustration of this disease. Keep plants dry when watering to prevent spread. Pick fruit promptly, as overripe fruit is more susceptible. Spray plants with copper when first fruit develops if you have had problems in the past.

Spots that appear near the stem while fruit is green are symptoms of early blight. See "Leaves with dark, concentrically ringed spots" on page 231 for controls.

Fruit with black, sunken area at blossom end. Cause: Blossom end rot. Seen on green or ripe fruit. First fruit to ripen is more likely to be affected than fruit that ripens later. See page 229 for an illustration of this problem. This condition is due to calcium deficiency in the fruit. It is aggravated by drought or uneven soil moisture, root damage, high salt levels in the soil, and excess nitrogen. If soil test indicates deficiency, add high-calcium lime to the soil. Prevent problems by keeping soil evenly moist and by spraying plants with seaweed extract when the first flowers open and again when green fruit is visible.

Fruit with cracks around the stems; shoulders may be green or yellow. Cause: Uneven irrigation. Cracks start at the stem and extend out, or are semi-circular splits on the shoulders of the fruit. Rots may invade fruit through cracks. Prevent problems by keeping soil evenly moist throughout the season and by planting crack-resistant cultivars such as 'Early Girl', 'Jet Star', and 'Roma'.

Ripe fruit rots. Cause: Various fungal and bacterial diseases. Remove infected fruit from plants. Stake plants to keep them off the ground and mulch to prevent soil from splashing up on fruit. Keep plants dry when watering and avoid touching them when wet. Grow crack-resistant cultivars and harvest ripe fruit promptly. Spray plants with copper after fruit forms if rots have been a problem in the past and weather is warm and wet.

Fruit with narrow, black tunnels through flesh and small holes near stem. Cause: Tomato pinworms. Larvae are small and gray and may have reddish markings. Destroy infested fruit. Till soil after harvest to prevent pests from overwintering.

Fruit with small holes on surface; interior rotted and hollow. Cause: Tomato fruitworms. Fruit appears to collapse like a deflated balloon. Larvae of tomato fruitworm, also called corn earworm, are light yellow, green, pink, or brown and grow up to 2" long with spines and lengthwise stripes. See page 229 for an illustration of this pest. Once larvae are inside fruit, there is nothing to do but destroy the infested fruit. If you see larvae feeding on leaves before attacking fruit, spray plants with BTK to control. Prevent eggs from being laid on plants by covering them with row cover until they flower.

Green fruit with large, chewed holes. Cause: Hornworms; Colorado potato beetles; other caterpillars. Ripe fruit may have brown, calloused pits. See "Leaves with large, ragged holes or leaves missing" on page 231 for descriptions and controls.

Ripe fruit with large, chewed holes. Causes: Slugs and snails; various animal pests. Slugs and snails can eat into a tomato quite rapidly. They often leave a shiny slime trail on the plant or ground marking their passage. Slugs hide under objects during the day. Place inverted flowerpots around the garden, check them daily, and destroy slugs hiding under them.

Various furry pests like tomatoes, too. See "Stopping Animal Pests" on page 408 for controls.

Trees, Shrubs, and Vines

Controlling pests and diseases on landscape plants is largely a matter of common sense and simple preventive measures. Choosing the right plants and supplying their simple needs will help make pest control a minimal part of your gardening chores.

Searching Out Stress

When your plants are showing signs of insects or disease, the first step to control is figuring out what attracted the problem in the first place. Plants stressed by any of several causes, like drought, extreme cold, or soil compaction, are most prone to insects and disease.

The solution to these problems often depends on your identifying the stress that is weakening the plant. Once you know the source of the stress, you can take steps to control it. Sometimes it's easy to determine what's causing the problem. During dry spells, for instance, lack of water is an obvious possibility. Other noticeable stresses include wounds caused by equipment or root damage from construction activity.

In other cases, the initial problem may be more subtle. Plants growing near streets and walkways are often damaged in winter by the de-icing salts that are washed into the soil, although symptoms may not appear until spring.

Topping means trouble. *Topping, or cutting tree limbs back drastically to the main branches, can expose trees to a wide range of problems. The large wounds heal slowly and provide a perfect entrance point for diseases. Many insect pests are attracted to the wealth of succulent shoots that sprout from the stumps.*

Keep your eyes open for potential problems, and you may be able to minimize the damage to your plants before pests or diseases attack.

Getting a Good Start

When you are buying new plants, you can avoid a lot of future problems by choosing locally adapted plants or resistant cultivars. Plants that are native to an area are often less prone to problems because they are growing in the environment to which they are best adapted. Read about the plants you intend to buy, and avoid very pest-prone species. If you have your heart set on a plant that is especially vulnerable to some problem, consider buying a resistant cultivar if one is available. Look for this information in catalogs, or ask your local nursery owner or extension agent for more information on the plants best adapted to your area.

Once you get your plant home, some basic care will help it get established quickly. Good soil preparation will provide the ideal conditions for strong root development. Providing ample water for the first few years after planting also encourages vigorous growth. A 2″-3″ thick layer of organic mulch helps keep the soil moist and weeds down; just be sure to keep the mulch a few inches away from the trunk or main stem to discourage animal and insect pests from attacking the base of the plant. Do any necessary pruning or staking carefully, and avoid making wounds in the stems with lawn mowers or string trimmers.

Choosing a Control

If a problem does require control, try the least drastic solution, like handpicking insects or pruning off diseased parts. Observing your plants often will help you catch problems before they require more severe controls.

When dealing with large plants like trees, realize that controlling insect and disease problems may be impractical. If the tree is other-

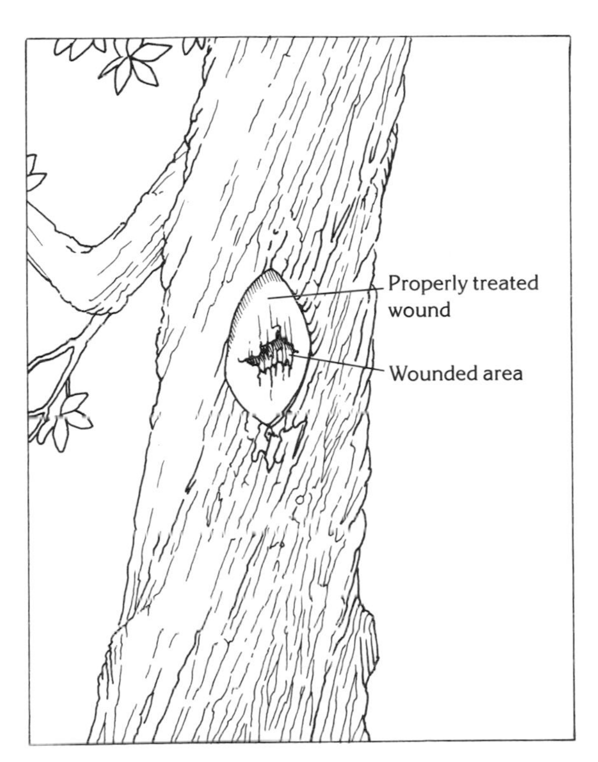

Treating tree wounds. *Wounds in tree bark, caused by equipment or animals, are an easy target for pests and diseases. Treat the wound by using a sharp knife to smooth the wound edges and shape the area into an elliptical form. This promotes the development of callus tissue and helps the wound close quickly.*

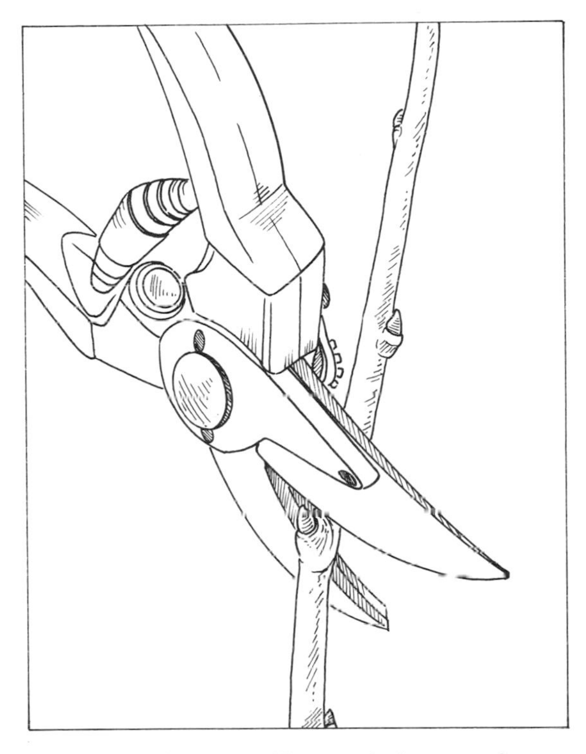

Proper pruning cuts. *You can help your plants stay healthy and vigorous by using correct pruning techniques. Always make your pruning cuts on a 45-degree angle just above a bud. Cutting too close may damage or kill the bud. If you cut too far above the bud, the cut may not heal over, and the wounded tissue is an easy target for disease spores.*

wise healthy, most disorders will not cause permanent damage. To deal with serious infestations or infections on large plants, consider getting the advice of your local extension agent or a qualified tree-care professional; this is especially true before attempting a drastic control, like removing the plant.

Below is a discussion of some of the most common insect and disease problems on trees, shrubs, and vines.

Leaf Problems

Leaves wrinkled and discolored. Cause: Aphids. These 1/32″–1/8″, pear-shaped, green, pink, black, dusty gray, or white fluffy-coated insects cluster on leaves, buds, and young stems. As they feed, they drop sticky honeydew on lower leaves. Sooty mold fungi often grow on the honeydew on aphid-infested foliage—as well as on the plants growing beneath them; see "Leaves with black coating" below for more information. Heavy infestations may result in summer leaf drop. Avoid overfeeding, which promotes succulent growth attractive to aphids. Spray plants vigorously with water, several times a day for 2-3 days to knock off the pests. If that doesn't control them, spray them every 2-3 days with insecticidal soap until the pests are under control. Use pyrethrin as a last resort.

Leaves stippled with yellow; foliage webbed. Cause: Spider mites. These tiny, spiderlike pests generally feed on the undersides of plant leaves. They suck sap from plant leaves, initially causing a yellow flecking on the upper leaf surfaces. Severe infestations can cause leaves to turn yellow or white; damaged leaves will eventually turn brown and drop. Tiny webs may be evident on leaves and stem tips. The leaf damage these pests cause may stunt the growth of the plant. Control spider mites by spraying plants thoroughly with water (especially the undersides of the leaves) 2-3 times a day for several days. For severe infestations, spray plants with insecticidal soap, or pyrethrin as a last resort.

Leaves skeletonized or with large holes; branches may be webbed. Cause: Caterpillars. Several kinds feed on foliage, including tent caterpillars and webworms. Control all of these pests by manually destroying their nests or egg cases, spraying leaves with BTK at the first sign of damage, or, as a last resort, applying pyrethrin spray.

Plant defoliated. Cause: Gypsy moths. Gypsy moth larvae are up to 2½" long, gray-brown, hairy caterpillars with red and blue spots on their backs. Unlike tent caterpillars and webworms, gypsy moth larvae do not make webs.

These caterpillars feed mainly at night and crawl down the trunk every morning. To trap them on their way back up the tree, wrap a wide piece of burlap around the trunk, tie it in the center, and fold over the top half. Check the band every afternoon, and collect and destroy the larvae that are trapped there. From late April through mid-June, spray leaves with BTK every 10-14 days and after each rain to provide more widespread control. The whitish adult female moth lays eggs in tan masses on trunks and branches. In winter, crush the egg masses or scrape them into a bucket of soapy water.

Plant defoliated; branches bear cocoonlike bags. Cause: Bagworms. These pests are actually ¾"-1", brown caterpillars, although you'll seldom see them. They feed from inside bags they create out of silk and the leaves of whatever plant they're feeding on. That's why the bags look different on maples, for example, than they do on pine or spruce trees. On some needle-leaf evergreens, you might even mistake the bags for cones at first glance.

Handpick and destroy the bags; as soon as you notice the caterpillars, start a program of spraying with BTK applied every 10 days (or after a heavy rain) through midsummer.

Leaves skeletonized. Causes: Japanese beetles; sawflies. Japanese beetles have metallic blue-green bodies with bronze wing covers. Besides feeding on leaves, the adult beetles often chew on buds and flowers.

Pheromone traps are available, but they are most helpful if you can convince your neighbors to use them, too; otherwise, your traps may just attract beetles from their yards into yours. Apply milky disease spores to your lawn for long-term larval control. Handpick in early morning (when the beetles move more slowly) if there are only a few.

Sawflies may also cause similar damage. They are closely related to bees and wasps. The larvae are the most destructive stage, feeding on and skeletonizing plant leaves or completely defoliating the plant. They range in size from ⅕" to 1½" and often closely resemble caterpillars, although sawfly larvae have more than 5 pairs of "legs." Damage may occur throughout the summer because different species feed at different times. Control larvae as soon as you spot them by handpicking or spraying leaves with insecticidal soap. Use rotenone for severe infestations.

Leaves tunneled. Cause: Leafminers. The larvae of some flies, moths, sawflies, and beetles feed in between upper and lower leaf surfaces; these pests are collectively known as leafminers. They may cause narrow, curved tunnels in the leaves or large, silvery brown blotches. Severe infestations can cause leaves to brown and wilt or drop. The most effective control is handpicking and destroying infested leaves. Chemical controls are usually ineffective because the larvae are protected within the leaves for most of their lives.

Leaves with black coating. Cause: Sooty mold. This fungus grows on the sugary, sticky honeydew produced by aphids, scales, whiteflies, and mealybugs. The black fungal coating doesn't harm leaves directly, but it does shade the leaves and reduce growth. The best control is to deal with the pests that are producing the honeydew. Determine what pests your plant has and apply the appropriate control. (If the plant itself doesn't show signs of pest damage, the honeydew may be dripping down from an overhanging plant.) On small plants, you can wipe the leaves with a damp cloth to remove the honeydew and the mold.

Leaves with spots. Cause: Leaf spots. A large number of fungi and bacteria cause spots on plant leaves, in a variety of colors, shapes, and sizes. In some cases the spots may spread to cover entire leaves, stunting plant growth. Other leaf spots have centers that die and fall out of the leaf, giving a "shothole" effect.

Fortunately, the same controls are effective against many leaf spot diseases. Pick off infected leaves. Rake up and destroy fallen leaves and branches in autumn to eliminate overwintering spores. To prevent the spread of mild infections, spray leaves with bordeaux mix; repeat twice at 10-day intervals.

Leaves with powdery white coating. Cause: Powdery mildew. Commonly attacking the foliage of many kinds of plants, powdery mildew may also appear on buds and shoot tips. Although it is unsightly, this fungal disease seldom causes serious damage, especially if it occurs late in the season. Reduce the chances of disease by leaving plenty of room around plants for good air circulation. Clean up and destroy or dispose of infected leaves.

Leaves yellow and wilt. Cause: Root rot. Root rots are caused by various species of fungi. Besides the common leaf symptoms, root rots can also cause reduced growth, branch dieback, and the ultimate death of the plant. At the soil level, the stem wood may be discolored, or stringlike fungal structures may be present. Yellow-orange mushrooms sometimes appear at the base of dying trees. Remove infected trees as soon as possible to reduce the spread of the fungus to other trees.

Leaves yellow; stems and leaves covered with small bumps. Cause: Scales. As they feed, these tiny pests cover themselves with 1/10" long shells in a range of shapes and colors. Some scales have hard, shiny shells, others form cottony, white coatings. These insects often feed on the undersides of the leaves, causing a generally unhealthy plant appearance and yellowish blotches on the upper leaf surfaces. Other types of scales feed on twigs and branches. Sooty mold fungi often grow on the honeydew on scale-infested foliage—as well as on the plants growing beneath them; see "Leaves with black coating" above for more information.

If possible, scrape off minor infestations of scale with your fingernail. Prune out badly infested growth, or use a soft brush and soapy water to gently scrub the scales off the stems (if the plant isn't too bushy or spiny.) Apply a dormant oil spray to the trunk and branches before growth starts in spring, or a superior oil during the growing season.

Trunk and Branch Problems

Trunk or branches with small holes; limbs die or break off. Cause: Borers. Numerous borers attack woody plants, mining the inner bark and wood of branches and trunk. To make plants less susceptible to attack, keep them healthy with proper pruning, mulching, and watering (during drought). Avoid wounding bark unnecessarily. Be especially careful when using a lawn mower or string trimmer around trunks of woody plants. Prune off borer-infested branches. If you see borer holes in your trees, probe into them with a flexible wire or inject a solution of parasitic nematodes; after treatment, seal holes with putty.

Trunk or branches with oozing lesions; branch tips die back. Cause: Canker. Several kinds of fungi cause cankers on twigs, trunk, and branches. As they spread, these sunken areas can girdle stems, killing the branch tips and stunting growth.

Mildly affected plants may recover from an attack of this fungus. Remove and destroy affected branches. If possible, cut away and destroy the cankered area, along with 2″ of healthy bark around the edge of the damaged area. Heavily diseased plants cannot be cured and should be removed and destroyed. The best prevention is to provide good growing conditions; healthy plants resist attacks. Avoid damaging plants with lawn mowers, string trimmers, or pruning tools; wounds are a common place for cankers to start.

Trunk or roots with swollen, wartlike growths. Cause: Crown gall. This bacterial disease causes wartlike swellings on plant roots, stems, or branches. It can enter the plant through wounds caused by lawn-maintenance equipment or chewing insects. If the plant is only slightly infected, prune out diseased growth. Afterward, disinfect your pruners with a 10 percent bleach solution (1 part bleach to 9 parts water). Destroy severely infected plants. This disease remains viable in the soil for several years without a host, so avoid replanting susceptible plants, such as euonymus and forsythia, in that area.

Trunk with shelflike growths. Cause: Wood rots. Wood rots near the base of the tree, caused by various species of fungus, are indicated by the appearance of shelflike growths. Prevent the disease by maintaining general health and treating injuries to the tree. Once established over a large area of the trunk, there is little that can be done; remove the tree.

Whole Plant Symptoms

Leaves and shoots blackened; leaves with moist or brown sunken spots. Cause: Anthracnose. This fungal disease is particularly a problem in cool, wet springs. In severe cases, twigs die back and defoliation can occur. Gather and dispose of fallen diseased leaves, and cut off affected branches several inches below the damaged area. A single spraying with bordeaux mix should provide control on young trees. No control is needed on larger trees; they'll produce new healthy leaves when the weather becomes warmer and drier.

Leaves yellow, sparse, distorted, or with brown edges; branches die; growth stunted. Cause: Decline. Decline does not refer to a particular pest or disease organism; rather, it relates to a general loss of plant vigor that is not due to a specific cause. It is usually a result of a number of stresses acting on a tree over a period of years. Plants that are mismatched to their sites, soil compaction, root damage, trunk injuries, repeated attack by insects or disease, and improper pruning are among the factors that can cause stressed plants and lead to decline.

If you can identify and eliminate the sources of the problem, you may be able to restore the plant's health. A seriously weakened plant may be too far gone for recovery, and you'll need to remove it. If you have identified the source of the original problem and have taken steps to resolve it, consider replanting with a suitable species.

Whole plant stunted and lacking vigor. Cause: Nematodes. Plant-parasitic species of these microscopic, wormlike creatures attack the roots and make plants look sickly and stunted. If you dig up the plant, you may find knotlike galls on the roots. The best control measures are preventive: Mulch regularly with compost or other organic mulch to ensure that soil organic matter levels remain high. When preparing soil for an entire bed or shrub border, incorporate plenty of compost into the soil. Do not amend the soil in individual planting holes: See the illustration "Proper Planting" at right for recommendations. Remove seriously infected plants and avoid replacing them with nematode-susceptible species (contact your local extension office for recommendations).

Whole tree falls over. Cause: Windthrow. If an otherwise healthy tree suddenly falls over, roots and all, the problem may be due to improper planting. The roots of trees planted in small holes and backfilled with heavily amended soils may not extend out of the planting area in search of nutrients. The small root system that develops may not be enough to anchor the tree if a sudden strong wind gust comes along.

To avoid damage, follow the guidelines given in the illustration "Proper Planting" at right. If windthrow does happen to a young tree, you may be able to save the plant. While the roots are exposed, use your fingers or a tool to gently loosen the congested root mass. Loosen the soil to a depth of about 8″, in a circle a few feet out from the trunk. Then carefully pull the tree upright, and stake it on several sides to hold it upright. Water the tree thoroughly, and mulch well with compost to promote new root growth.

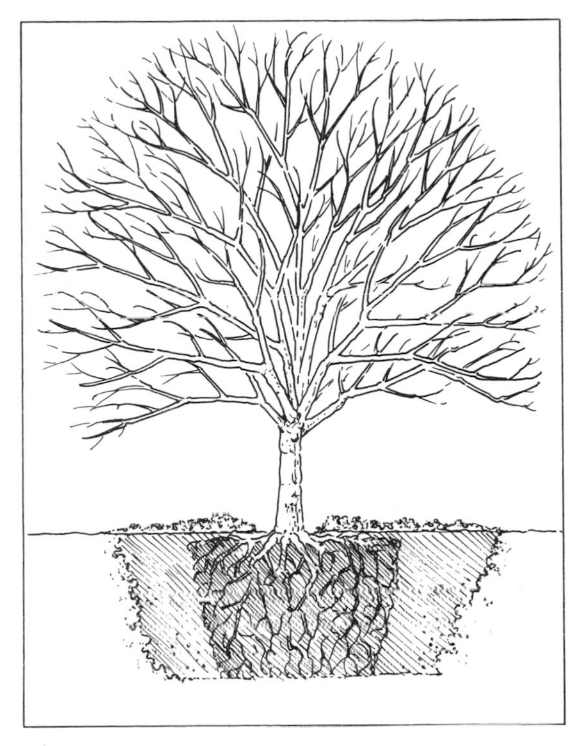

Proper planting. *Following a few simple planting tricks will get your plants off to a good start. First, dig a wide but shallow planting area. Ideally, the tree should rest on undisturbed soil, with the top of the root ball the same level at which it was growing in the nursery. Scratch the sides of the hole with a spading fork to loosen the soil and encourage root penetration. Don't add peat moss or other amendments to the hole; simply replace the original soil, water thoroughly, and apply a 2″-3″ layer of mulch.*

Leaves yellow; growth stunted; top of plant breaks off from roots. Cause: Graft incompatibility. This problem occurs when the root system and the top of a grafted plant do not join properly. This lack of connection interferes with the flow of water and nutrients, and the top of the plant may die and break off at the graft union. This problem most commonly occurs within a few months of grafting, but it can also happen after several years of apparently healthy growth. Fortunately, this problem does not happen often. If your plant is damaged, remove it; there is no cure.

Tropaeolum

Nasturtium. Annuals.

Available in all shades of red, yellow, orange, and white, these easy-to-grow plants thrive in warm, sunny weather. Flowers are 2″ wide, highly visible, and born singly. Some cultivars have a bushy habit; others climb quickly if given any type of support. Otherwise, they sprawl along the ground with a 1′-6′ wide spread.

Direct-sow nasturtium seeds in late spring; they germinate in 1-2 weeks. Plants prefer sun, but some afternoon shade is beneficial in very hot areas. They are not fussy about the soil conditions, but too much fertilizer encourages ample foliage growth and fewer flowers. Space plants 8″ apart. Water only as needed, to prevent wilting.

Problems

Leaves, stems, and buds distorted. Cause: Aphids. For controls, see "Leaves, stems, and buds distorted" on page 20.

Leaves with tan or brown blotches or serpentine tunnels. Cause: Leafminers. See "Leaves with tan or brown blotches or serpentine tunnels" on page 27 for controls.

Plant yellows, wilts, and dies. Cause: Bacterial wilt. The stems have black streaks, and roots will rot. There is no cure. Avoid planting nasturtiums near other susceptible crops, such as eggplants, tomatoes, or peppers. Solarize the soil before replanting with susceptible plants.

Leaves with many tiny holes. Cause: Flea beetles. Infested leaves quickly dry out and may wither. These tiny black, brown, or bronze insects hop readily when disturbed. Keep nasturtiums sufficiently watered. Eliminate garden debris where flea beetles reside. Spray serious infestations with pyrethrin or rotenone.

Tsuga

Hemlock. Trees.

Hemlocks are narrow-leaved, evergreen, cone-bearing trees. Often used as attractive hedges, they can be allowed to grow naturally or be sheared to any height and width.

Hemlocks are best adapted to moist, cool climates. They are shallow-rooted and easily transplanted; set out in spring as balled-and-burlapped plants. Hemlocks prefer moist, acid soil rich in organic matter. They will flourish in sun or partial shade; if planted in sun, make sure to water during drought.

Problems

Leaves drop; plant weak or dead. Cause: Hemlock woolly adelgids. The most noticeable sign of this pest is the cottony, white egg masses that appear in late winter and early spring. The white-fringed nymphs emerge in early spring and feed on twigs and leaf bases throughout the summer; the mature adelgids then overwinter and repeat the cycle. Control by spraying the trunk and branches in early spring (before growth starts) with dormant oil to reduce the population of overwintering adults. Repeat the spray in June and October with insecticidal soap or superior oil.

Leaves yellow; stems and leaves covered with small bumps. Cause: Scales. Several kinds of scale attack hemlocks. For information on controlling these pests, see "Leaves yellow; stems and leaves covered with small bumps" on page 237.

Branches with brown tips. Cause: Spruce budworms. For more information, see "Branches with brown tips" on page 181.

Plant defoliated. Causes: Hemlock loopers; gypsy moths. The 1″ long, greenish yellow, black-spotted hemlock loopers usually appear in June. They feed on the needles, starting from the branch tips and working toward the center of the plant. Handpick small populations; control large infestations with BTK.

Gypsy moths cause similar damage, for a description and control measures, see "Plant defoliated" on page 236.

Plant defoliated; branches bear cocoonlike bags. Cause: Bagworms. See "Plant defoliated; branches bear cocoonlike bags" on page 236.

Leaves light gray or bronze. Cause: Spruce spider mites. See "Leaves light gray or bronze" on page 181 for controls.

Trunk or branches with small holes; limbs die or break off. Cause: Borers. See "Trunk or branches with small holes; limbs die or break off" on page 238 for details.

Branch tips die back. Causes: Blight; sun scorch; drought stress. Blight, a fungal disease, attacks mostly during cool, wet weather. Good air circulation among trees discourages its onset. Leaves of infected plants shrivel and twigs die back. Cut off and destroy dead or diseased wood. In a very cool, wet spring, you can spray with bordeaux mix, starting when new growth appears on the trees and repeating at 2-week intervals until the weather gets warmer and drier.

If you see the same symptoms during hot, dry weather, your hemlock may be suffering from sun scorch or drought stress. Minimize damage by watering deeply and regularly during dry periods. A thick mulch may also help by retaining moisture and keeping roots cool.

Leaves yellow and drop. Cause: Rust. Whitish blisters on the undersides of the leaves, followed by yellowing and leaf drop, indicates rust. Prune out and destroy infected branch tips. To prevent the spread of the disease, spray with sulfur, repeating 2 or 3 times at weekly intervals.

Trunk or branches with oozing lesions; branch tips die back. Cause: Canker. See "Trunk or branches with oozing lesions; branch tips die back" on page 238 for details.

Tulipa

Tulip. Bulbs.

Perhaps the best-loved of the spring-flowering bulbs, tulips have been admired and prized for centuries. Plant breeders have developed thousands of hybrids and cultivars from the more than 100 species in this genus. Divided into 15 divisions, based on bloom time, flower form, and parentage, tulips bear cup-shaped blossoms on 6″-24″ tall flower stalks. Flowers come in all colors except true blue; bicolors are common. Thick, 6″-8″ long, straplike leaves surround flower stems.

Hardy bulbs, tulips require winter chilling to bloom. Where temperatures don't fall low enough, several weeks in a refrigerator at 40°F provides the necessary cold period. In cold-winter climates, plant tulips in fall, at least 1 month before the ground freezes. Keep bulbs cool until planting; exposure to temperatures over 70°F reduces flower size. Plant

6″-10″ deep in full sun and well-drained soil. Poor drainage promotes bulb rots. Mulch lightly in winter. Top-dress with compost and bonemeal in spring, about 1 month before bloom. Remove spent flowers to promote bulb growth. As clumps enlarge, flower size may decline. Dig crowded clumps after foliage fades, shake off loose soil, and air-dry bulbs in shade for a few days. Divide offsets and parent bulbs and replant.

Problems

Leaves with large, ragged holes. Cause: Slugs and snails. Slugs and snails may feed on foliage; see "Leaves with large, ragged holes" on page 50 for controls.

Leaves yellow or distorted; bulbs decayed. Cause: Bulb mites. These mites may arrive on new bulbs. See "Leaves yellow or distorted; bulbs decayed" on page 50 for controls.

Plant fails to appear in spring; bulbs missing. Cause: Animal pests. Rodents like to feed on tulip bulbs; droppings or disturbed soil may appear in flower beds. Plant bulbs where human activity will discourage wildlife; pet cats or dogs also deter animal pests. Line planting beds with hardware cloth to exclude burrowing rodents; cover beds with screen wire in winter. Try repellents such as dried blood, human hair, or garlic sprays. Keep flower beds free of debris where pests may hide.

Leaves yellow; plant stunted. Cause: Tulip bulb aphids. These aphids infest both the bulbs and aboveground portions of the plant. They suck sap from leaves, stems, and flowers, causing foliage to curl, pucker, and yellow. Buds may be stunted. Inspect bulbs carefully when you buy; look for clusters of gray, waxy aphids under the bulb coat. Spray infested plants with insecticidal soap. Destroy seriously infested bulbs or dust with pyrethrin or a commercial pyrethrin/rotenone mix.

Leaves streaked or spotted; flowers rotted. Cause: Botrytis blight. Also known as tulip fire, this fungus causes red-brown leaf spots that later turn gray. Plants may be stunted or pale yellow-green with deformed flowers and rotting stems. Dark spots form on bulbs; gray mold may be present. Dig and destroy infected plants. Limit disease spread by watering early in the day so leaves have time to dry before evening. Trim off leaves as soon as they turn yellow; remove spent flowers. To protect tulips, apply bordeaux mix when shoots appear in spring; repeat 1 week later.

Leaves and flowers streaked or mottled; foliage spindly or deformed. Cause: Viral diseases. Several viruses infect tulips. Some tulip cultivars get their flower color from viral infections; the resulting blooms are said to be broken. Aphids and leafhoppers may spread virus from broken tulips to solid-colored ones and also to lilies. Viruses weaken tulips without killing them; remove infected plants to halt disease spread to healthy tulips. Wash tools used around infected plants; control sucking insects; see "Leaves, stems, and buds distorted, sticky; clusters of small insects" on page 51 for information on controlling aphids.

Turnip

Brassica rapa
Rapifera group (Cruciferae)

Turnips are a cool-season vegetable grown for their crisp roots and tasty greens. They require cool, moist, rich soil with a pH between 5.5 and 6.8. Plant seed directly in the garden. Turnips grow best between 60° and 65°F. They grow poorly above 75°F, but will tolerate temperatures as low as 40°F. Harvest greens and roots when they are small and tender.

Turnips are in the same family as cabbage

and are troubled by many of the same pests and diseases. See the Cabbage entry beginning on page 52 for descriptions and controls.

Root Problems

Roots are riddled with slimy, winding tunnels. Cause: Cabbage maggots. Maggots are white and 1/4″ long. See "Leaves yellow; plant stunted; plant wilts on bright, hot days" on page 54 for description and controls.

Root flesh black but firm; skin rough and cracked. Cause: Downy mildew. Control this fungal disease by spraying plants with a baking-soda-and-soap spray (1 teaspoon baking soda, 1 teaspoon liquid dish soap, 1 quart water) or copper if you have had problems in the past. Prevent problems by planting tolerant cultivars such as 'Crawford' and 'Scarlet Queen Hybrid'.

Roots black and rotted. Cause: Black rot. Leaves have yellow, V-shaped spots on margins. Destroy infected plants. Prevent this bacterial disease by treating seed with 122°F water for 25 minutes before planting. (Be aware that this treatment can injure seed viability; for complete instructions, see page 422.)

Roots with dry, sunken spots. Cause: Anthracnose. Avoid this fungal disease by planting in cool soil (early spring or fall).

Roots with small, water-soaked spots or pits on surface. Cause: Cold injury. Protect plants with mulch if temperatures are below 30°F.

Ulmus

Elm. Trees.

Elms are alternate-leaved, deciduous trees valued for their ornamental use as specimen or street trees. There was a time when virtually every town and city in the northeastern United States was dominated by the majestic, vase-shaped American elm (*Ulmus americana*); unfortunately, with the appearance and spread of Dutch elm disease (DED), many of them have died.

Because most elms are susceptible to so many insects and diseases, you might want to avoid using them in the landscape. If you do decide to grow elms, consider planting lacebark elm (*U. parvifolia*), which is more resistant to Dutch elm disease and elm leaf beetles. Plant elms in spring or fall as bare-root or balled-and-burlapped trees. Full sun and well-drained soils are best.

Problems

Leaves with rectangular holes or skeletonized. Cause: Elm leaf beetles. This is a 1/4", yellow-green beetle with a dark line on the outer edge of each wing cover. It lays its eggs in spring on the undersides of leaves; these eggs hatch in June. One to several generations may occur each year.

Both the adults and the 1/2", black-spotted, yellow larvae feed on the leaves, eating everything but the veins. For an illustration of the damage, see below. Trees are often defoliated and so weakened that they are susceptible to other insect and disease problems. Control beetles by spraying leaves with BTSD, particularly in June.

Leaves wilted and yellow or brown, drop early; branches show symptoms one at a time. Cause: Dutch elm disease. Caused by a fungus, Dutch elm disease is spread by the feeding of elm bark beetles, and by natural root grafts between trees growing in the same area. Keep trees healthy with proper pruning, mulching, and watering (during drought). Quickly repair all wounds to help prevent insect attacks and subsequent infection. Once the disease is established, there is no effective remedy. Remove and immediately destroy all diseased or dying elms. Remove the stump if possible, or peel the bark off to below the soil line to deter elm bark beetles from feeding there.

Bark tunneled. Cause: Elm bark beetles. These 1/10", dark reddish brown beetles attack weakened elm trees and serve as vectors of Dutch elm disease. The adult beetles bore small holes through the bark and lay eggs in

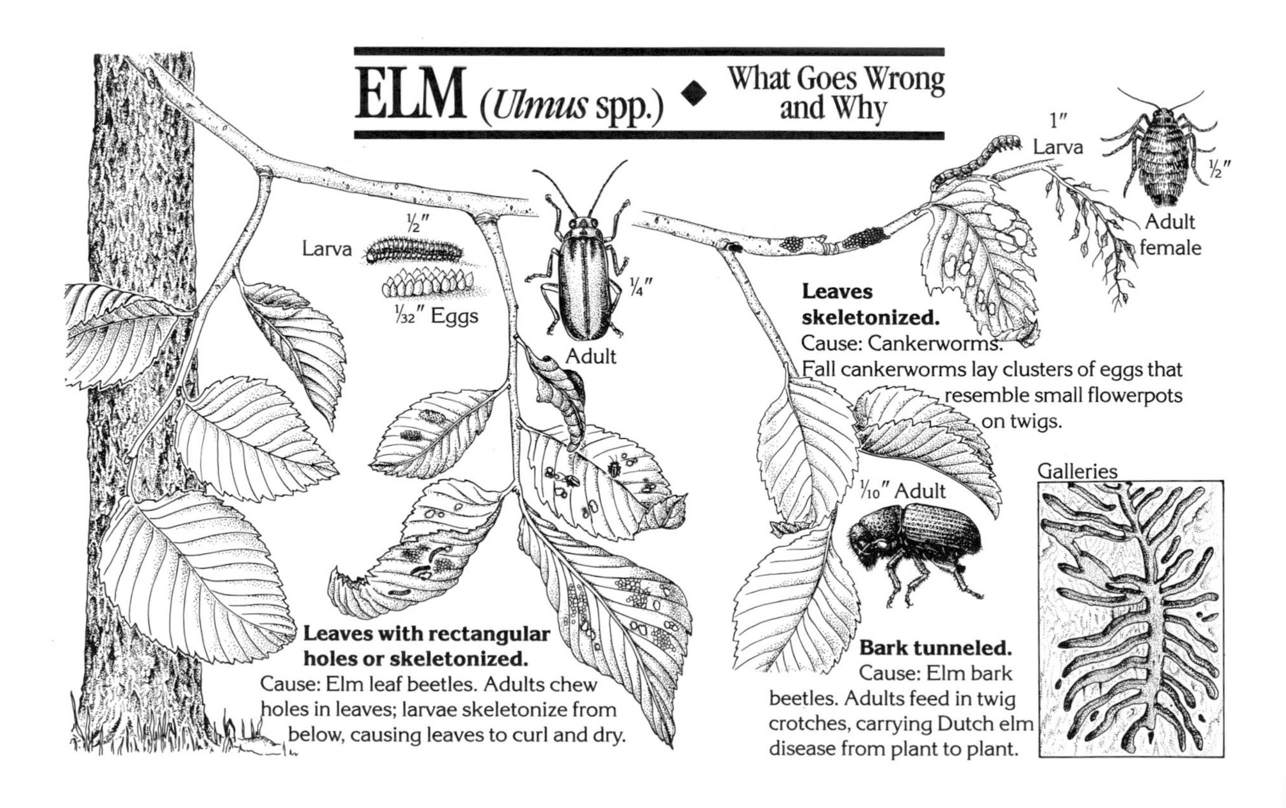

the wood beneath. The eggs hatch into 1/4″, curved, white grubs that continue to mine underneath the bark, creating winding tunnels, known as galleries. See the opposite page for an illustration of their damage. The grubs transform into adult beetles, which emerge through small holes in the bark. If the tree they emerge from was infected with DED, the beetles will carry the disease spores as they fly away to feed on other elm trees.

Make trees less attractive to beetles by promoting vigorous, healthy growth with proper pruning, mulching, and watering (during drought). Eliminate breeding areas by removing and destroying weak and dying trees.

Leaves skeletonized. Cause: Cankerworms. Both spring and fall cankerworms are 1/2″-1″, yellow or greenish caterpillars that feed on the foliage of birches, often defoliating the tree. See the opposite page for an illustration of fall cankerworm. For control measures, see "Leaves skeletonized" on page 42.

Leaves wilted and yellow or brown, drop early; entire crown of tree affected. Cause: Phloem necrosis. In later stages, the inner bark (phloem) is discolored and smells faintly of wintergreen. This disease, also known as elm yellows, kills plants quickly, often in a single growing season. It is spread from tree to tree by leafhoppers.

Controlling leafhoppers by spraying infested plants with insecticidal soap or pyrethrin may reduce the spread of phloem necrosis. If trees are close together, however, the disease can also spread underground by means of natural root grafts. Once plants are infected, there is no control; remove and destroy infected trees.

Leaves skeletonized. Cause: Japanese beetles. This 1/2″ long pest has a metallic blue-green body and bronze wing covers. Handpick in early morning if there are only a few. Apply milky disease spores to your lawn for long-term control.

Leaves wrinkled and discolored. Cause: Aphids. See "Leaves wrinkled and discolored" on page 235 for controls.

Leaves tunneled. Cause: Leafminers. For more information and controls, see "Leaves tunneled" on page 237.

Leaves yellow; stems and leaves covered with small bumps. Cause: Scales. See "Leaves yellow; stems and leaves covered with small bumps" on page 237 for controls.

Trunk or branches with small holes; limbs die or break off. Cause: Borers. For more information, see "Trunk or branches with small holes; limbs die or break off" on page 238.

Trunk or branches with oozing lesions; branch tips die back. Cause: Canker. For more information, see "Trunk or branches with oozing lesions; branch tips die back" on page 238.

Leaves with spots. Cause: Leaf spots. See "Leaves with spots" on page 237 for controls.

Leaves with powdery white coating. Cause: Powdery mildew. For controls, see "Leaves with powdery white coating" on page 237.

Vegetables

Vegetables are plants grown for their edible roots, stems, leaves, seeds, fruit, and other plant parts. Most vegetables are annuals; a few are biennials or perennials grown as annuals; and some are hardy perennials. Vegetables grow best in deeply worked, well-drained soil with lots of organic matter. Most require full sun to produce well. See individual vegetable entries for specific cultural requirements and problems. The following problems affect many vegetables.

Problems

Seedlings fall over; stems girdled or rotted at soil line. Cause: Damping-off. Disinfect reused pots and flats by dipping them in a 10 percent bleach solution (1 part bleach to 9 parts water) and letting them air-dry before filling them with fresh seed-starting mix. Sow seed thinly to allow for air movement around seedlings. Cover seed with a thin layer of soilless mix or vermiculite. Water only enough to keep soil moist, not soggy. Thin seedlings and spray with compost tea as soon as first true leaves open to help prevent the problem.

Seedlings clipped off at soil line. Cause: Cutworms. Check for fat, 1″ long, brown or gray caterpillars in the soil near the base of plants. Once they chew off a seedling, there is nothing you can do except protect the remaining seedlings from nocturnal cutworm attacks. To prevent damage, place cutworm collars around transplants, sprinkle moist bran mixed with BTK on the soil surface in the evening, or add parasitic nematodes to the soil at least a week before planting.

Leaves turn yellow beginning at base of plant; plant stunted. Causes: Nitrogen deficiency; waterlogged soil. Spray foliage with compost tea or fish-meal tea, or side-dress plants with compost. Waterlogged soil damages roots and prevents them from using nutrients available in the soil. Prevent problems by choosing well-drained sites, adding organic matter to the soil to improve drainage, and by planting in raised beds.

Leaves yellow and curled; new growth distorted. Cause: Aphids. These small, soft-bodied, green, black, or pinkish insects suck plant juices. They may spread various diseases or leave a sticky honeydew on leaves and fruit that in turn supports the growth of black, sooty mold. For mild infestations, knock pests off plants with a blast of water. Spray plants with insecticidal soap to control aphids, or with a commercial pyrethrin/rotenone mix if infestation is severe.

Leaves yellow; plant wilts; stems, crowns, or roots water-soaked and rotted. Cause: Fungal or bacterial rots. Various fungi and bacteria cause stem, root, or crown rot. Poor drainage or overwatering encourages the development of these diseases. Destroy infected plants or plant parts. Thin plants to increase air movement and reduce moisture around plants. Spray plants with copper to prevent spread if weather is warm and wet and disease is severe. Plant in well-drained soil or in raised beds to prevent problems.

Plant yellow and stunted, wilts during bright, hot days; roots may have swollen galls. Cause: Root knot and other pest nematodes. These microscopic, wormlike creatures invade and feed on plant roots. Pull and destroy infected plants. Control pest nematodes by adding chitin or parasitic nematodes to the soil. Solarize infested areas to reduce future problems.

Leaves yellow; plant wilts gradually. Causes: Verticillium wilt; Fusarium wilt. Leaves may roll up as the disease progresses. Cut-open stems are discolored. Severely infected plants eventually die. There is no cure for plants with these fungal diseases. Destroy infected plants. There are many species of Fusarium, each of which infects only 1 plant or plant family. One species of Verticillium wilt, however, can infect over 300 species of cultivated plants including eggplants, tomatoes, peppers, potatoes, brambles, fruits, and ornamentals; so preventive rotation is difficult. Avoid planting wilt-susceptible plants where any wilt symptoms have developed within the last 3 years.

Leaves or other plant parts covered with tan to gray, fuzzy growth. Cause: Gray mold. This fungal disease attacks a wide range of edible and ornamental plants. Pick off and destroy moldy parts. Thin plants to increase air movement and reduce moisture around leaves, since the mold thrives in damp conditions. Remove faded flowers promptly.

Spray foliage with compost tea to control mold. Spray plants with sulfur to prevent further symptom development if the weather is wet and cool and the disease is severe.

Leaves with powdery white growth on upper surfaces. Cause: Powdery mildew. This symptom is caused by a number of fungi, each of which attacks only specific plants. Thin plants to increase air movement and reduce moisture around leaves. Spray foliage with compost tea to control. Spray plants with sulfur or bordeaux mix if disease is serious to prevent further symptom development.

Leaves with wandering, white or translucent tunnels. Cause: Leafminers. Larvae are tiny white maggots that feed on leaf tissue. Adults are tiny black-and-yellow insects. Once maggots enter leaves, no spray will control them. Destroy mined leaves. Apply row cover as soon as plants emerge or are set out to prevent problems. Control adults with yellow sticky traps or a commercial pyrethrin/rotenone mix.

Leaves pale and stippled. Cause: Mites. Leaves may become bronzed. These tiny, spiderlike insects thrive in hot, dry weather and feed on the undersides of leaves. Spray plants with insecticidal soap in the evening to control them.

Leaves riddled with small holes. Cause: Flea beetles. These small, shiny, black beetles hop when disturbed. They can transmit viral and bacterial diseases. Control severe infestations by spraying plants with commercial pyrethrin/rotenone mix. Prevent problems by protecting young plants with row cover.

Leaves with large, ragged holes. Causes: Caterpillars; snails and slugs; various animal pests. Many different caterpillars feed on vegetables. Look for dark green excrement at the base of leaves or plants. Handpick, or spray plants with BTK as soon as active caterpillars or feeding are observed. Let nature work for you and avoid handpicking or spraying if caterpillars are sluggish and yellowish (infected with a fatal virus) or covered with the small, white, cigar-shaped cocoons of parasitic wasps.

Slugs and snails eat leaves and fruit, often leaving shiny slime trails on the plants or ground marking their passage. Slugs hide under objects during the day. Place inverted flowerpots around the garden, check them daily, and destroy slugs hiding under them. If slug problems are severe, use a copper strip edging around beds to exclude them.

Various furry pests eat vegetables; see "Stopping Animal Pests" on page 408.

Verbena

Verbena, vervain. Annuals.

Verbenas' hardiness and long growing season make them favorites of beginners as well as long-time gardeners. Compact flower clusters range from white to deep purples and reds; they bloom from summer through autumn. Plants grow 6″-10″ tall with a spread of 1′-2′. Use verbenas to edge beds and borders, or in containers and window boxes.

Where summers are short, start seeds indoors 3 months before last frost. Ample light and 70°F soil for 4 weeks are needed for germination. Move outdoors when night temperatures are above 50°F. Space 1′ apart. Or direct-seed outdoors as soon as weather warms up, although germination may be poor. Many gardeners prefer to buy nursery-grown plants.

Verbenas need full sun, average water, and fertile, well-drained soil. In very hot climates give light shade to prevent drought stress, which reduces flowering.

Problems

Leaves with powdery white patches. Cause: Powdery mildew. See "Leaves with powdery white patches" on page 19 for controls.

Leaves stippled with yellow; foliage webbed. Cause: Spider mites. For control information, see "Leaves stippled with yellow; foliage webbed" on page 18.

Branch tips wilt. Cause: Budworms. These ½", greenish yellow caterpillars feed on new shoots. Prune and destroy infested tips.

Leaves, stems, and buds distorted. Cause: Aphids. See "Leaves, stems, and buds distorted" on page 20 for controls.

Leaves with tan or brown blotches or serpentine tunnels. Cause: Leafminers. See "Leaves with tan or brown blotches or serpentine tunnels" on page 27 for controls.

Veronica

Speedwell, brooklime. Perennials.

Ranging from prostrate to strongly upright in habit and from 3″-48″ in height, speedwells form a varied genus of mostly blue-flowered, summer-blooming plants. Leaves are lance-shaped and green to gray-green; numerous small blossoms cover tall spikes that arise from the plant tops or from the leaf axils. Woolly speedwell (*Veronica incana*) is grown for its 1″-3″, silvery white, fuzzy leaves as well as its blue blossoms.

Planting requirements vary somewhat among species but, in general, speedwells require average, very well drained soil and full sun. Plants tolerate some shade, but most do not appreciate drought, extreme heat, or humidity. Speedwell plantings enlarge via new shoots appearing at edges, but are not invasive. Divide every 4 years to reduce crowding. Remove spent flowers to prolong bloom.

Few insect pests attack speedwells, but fungal diseases are likely, especially when the plants' rather strict moisture requirements are not met. Prevent problems with good drainage, garden sanitation, and air circulation.

Problems

Leaves with spots or blotches. Cause: Various fungi. Brown, black, or yellow leaf spots may enlarge and kill entire leaves. Remove severely infected plants and plant parts. Avoid wetting leaves when watering. Apply sulfur sprays when symptoms appear to avoid further damage.

Leaves with white or gray powdery spots on upper and lower surfaces. Cause: Downy mildew. As the fungus spreads, stems become distorted and die; flowers fail to open. Destroy severely infected plants. Sprays of sulfur or copper fungicide give some protection.

Viburnum

Viburnum. Shrubs and small trees.

Viburnums are opposite-leaved shrubs or small trees valued for their flowers, fruit, foliage, and growth habit. Most are deciduous; some are semi-evergreen or evergreen. They are excellent in shrub borders or woodland plantings.

Set out in fall or spring in well-drained soil amply enriched with organic matter. Viburnums grow well in sun or partial shade. Be aware that sulfur-containing fungicides may harm the foliage of some viburnums. Test the spray on a few leaves before treating the whole plant.

Problems

Leaves wrinkled and discolored. Cause: Aphids. For control measures, see "Leaves wrinkled and discolored" on page 235.

Leaves with spots. Cause: Leaf spots. See "Leaves with spots" on page 237 for controls.

Leaves with powdery white coating. Cause: Powdery mildew. For controls, see "Leaves with powdery white coating" on page 237.

Leaves and shoots blackened; leaves with moist or brown sunken spots. Cause: Anthracnose. See "Leaves and shoots blackened; leaves with moist or brown sunken spots" on page 238 for controls.

Leaves skeletonized. Cause: Japanese beetles. For suggested controls, see "Leaves skeletonized" on page 236.

Trunk or roots with swollen, wartlike growths. Cause: Crown gall. See "Trunk or roots with swollen, wartlike growths" on page 238 for controls.

Leaves yellow; stems and leaves covered with small bumps. Cause: Scales. See "Leaves yellow; stems and leaves covered with small bumps" on page 237 for controls.

Vinca

Periwinkle, vinca, myrtle. Perennials.

Two species of ground-covering vines represent *Vinca* in the United States: Greater periwinkle (*V. major*) and common periwinkle (*V. minor*). Both feature glossy, dark green, opposite leaves and blue, funnel-shaped flowers in spring. Greater periwinkle has larger leaves and is hardy only to Zone 7; common periwinkle's foliage is smaller and plants are hardy to Zone 5, making it the more widely used landscape plant. Stems bearing flowers stick up 6″–18″ above ground; the vines are otherwise prostrate. White-flowered and variegated cultivars are available.

Periwinkles grow rapidly in moist, sunny to lightly shaded spots; roots form along stems touching the ground. These vines spread, but rarely grow out of control. Average, well-drained soil is fine, and periwinkles make good groundcovers for erosion control on slopes—plants tolerate light foot traffic.

In the landscape, periwinkles have few severe insect problems, although several pests trouble greater periwinkles in greenhouse or subtropical conditions. Wet soil causes most periwinkle problems by encouraging fungal diseases; prevent problems by selecting a site with good soil drainage, keeping the garden clean, and thinning plantings to promote air circulation.

Problems

Shoots blacken, wilt, and die back. Cause: Canker. Also known as dieback, this disease is caused by 2 fungal organisms and most often occurs during rainy weather. Prune and destroy infected shoots; thin plantings to improve air circulation. Remove and destroy severely infected plants. If canker has been a problem, use preventive sprays of bordeaux mix.

Viola

Pansy. Biennials grown as annuals.

Colorful pansies help to brighten up the spring garden. The 6″ tall plants bear cheerful blossoms in a wide color range. They bloom from May to July, providing a perfect complement to spring bulbs.

For very early bloom, sow pansy seeds in August and overwinter plants in a cold frame. Otherwise, sow seeds indoors in winter. Make sure seeds are covered; they need darkness to germinate. Seeds sprout in 14 days. After hardening the plants off, set them out into the garden about a month before the last frost. Pansies like cool weather; rich, loose soil; and filtered sun. Mulching keeps roots cool. Water plants regularly, and apply diluted liquid fertilizer every 4 weeks. Remove spent flowers to encourage more blooms. Cut plants back hard in midsummer and they may rebloom in the fall.

Problems

Leaves with large, ragged holes. Cause: Slugs and snails. For controls, see "Leaves with large, ragged holes" on page 18.

Leaves, stems, and buds distorted. Cause: Aphids. See "Leaves, stems, and buds distorted" on page 20 for controls.

Leaves stippled with yellow; foliage webbed. Cause: Spider mites. See "Leaves stippled with yellow; foliage webbed" on page 18 for control information.

Seedlings or young plants cut off at soil level. Cause: Cutworms. For control measures, see "Seedlings or young plants cut off at soil level" on page 20.

Leaves with spots. Cause: Leaf spots. See "Leaves with spots" on page 19 for controls.

Walnut

Juglans spp. (Juglandaceae)

Walnuts are large, deciduous trees bearing separate male and female flowers on the same plant. To get the heaviest nut production, plant 2 different cultivars for cross-pollination. The most commonly grown types are black walnut (*Juglans nigra*) and English walnut (*J. regia*). Walnuts are hardy in Zones 4-8.

Walnuts need a site free of late spring frost; they prefer deep, well-drained soil. Prune lightly in winter to allow sunlight into the tree and to remove dead, diseased, or crossing branches.

Problems

Immature nuts dry up and drop early. Cause: Codling moth larvae. These fat, white or pinkish, ⅞" caterpillars tunnel into nuts and may have departed by the time you discover the damage. Nuts may have a hole filled with what looks like moist sawdust. Control coddling moths with superior oil or ryania. Late-blooming cultivars, such as 'Hartley' and 'Vina', are least susceptible to codling moths. For more information, see "Fruit with holes surrounded by brown, crumbly excrement" on page 22.

Husks have soft, black, smooth spots and maggots inside. Cause: Walnut husk fly maggots. Adult walnut husk flies lay eggs in nut husks and these hatch into small, cream-colored maggots. Although the larvae never eat the nut shells, larval feeding in the husk causes shells to blacken or shrivel. Since nuts themselves are unaffected, you can usually just ignore this pest. To control the flies, collect and dispose of infested husks. For severe infestations, capture adult flies with apple maggot fly traps (4 traps per tree). For information on apple maggot fly traps, see "Fruit dimpled; brown tunnels through flesh" on page 22. Resistant cultivars include 'Ashley', 'Erhardt', 'Payne', and 'Placentia'.

Husks blacken; nuts blacken, shrivel, and drop prematurely. Cause: Walnut blight. Leaves may also bear angular brown spots, and dead, sunken lesions may appear on shoots. This bacterial disease overwinters in attached nuts, diseased buds, and twig lesions. To control blight, keep the canopy dry by pruning to allow good air circulation at the centers of the trees. Also avoid overhead irrigation. For severe infection, use copper spray. Resistant cultivars include 'Hartley' and 'Vina' .

Leaves twisted or curled and covered with a sticky coating. Cause: Aphids. The shiny coating is honeydew, a substance excreted by feeding aphids. Leaves may show a sooty fungus that feeds on honeydew. Sprays of water or insecticidal soap solution help control aphids. Rotenone kills aphids, but reserve this for emergency use since it also kills beneficial insects. For more information, see "New leaves twisted or curled and covered with a sticky coating" on page 67.

Leaves with circular brown spots. Cause: Anthracnose. This fungal disease, common in wet, humid summers, may weaken trees and cause nuts to shrivel and drop early. To control anthracnose, keep trees well-nourished with nitrogen and clean up fallen leaves.

Tree stunted and bears yellow leaves. Causes: Crown rot; blackline. If the trunk near the soil line is discolored or oozing sap, suspect crown rot, a disease caused by too much water and poor soil drainage. Improved drainage may help. If you find small holes or cracks at the graft union, remove some bark around the area and look for a black line. Blackline virus infects English walnuts grafted onto *J. hindsii* rootstocks. There is no cure.

Watermelon

Citrullus lanatus (Cucurbitaceae)

Watermelons belong to the same family as muskmelons and honeydews. See the Melon entry beginning on page 148 for culture and pest information. Harvest watermelons when the bottom of the fruit turns from pale yellow to golden yellow.

Weigela

Weigela. Shrubs.

Weigelas are opposite-leaved, deciduous shrubs grown for their bright flowers, which appear in late spring and early summer. Because they lack interesting fruit and autumn color, weigelas are best used in the mixed shrub border.

Set out in spring or fall in full sun or light shade (the further south, the more shade). They prefer a well-drained soil, but one that does not dry out; a summer mulch is beneficial.

Weigelas are remarkably free of serious problems. Powdery mildew will coat the leaves with its typical white powder, more unsightly than threatening; see "Leaves with powdery white coating" on page 237 for controls.

Wisteria

Wisteria. Vines.

Wisterias are vigorous vines with alternate, compound leaves. They climb, often to amazing heights; because of their ultimate size, they require strong supports. Their long flower clusters, often fragrant, appear in late spring, and a plant in full bloom is a delightful sight in the landscape.

Wisterias are among the finest and most impressive of ornamental vines. They grow best in full sun but will tolerate some shade. Set out in spring or fall in deeply prepared, moisture-retentive soil. Wisterias are often reluctant to bloom, sometimes taking many years. They seem to grow and flower best when given ample water; reluctant plants can sometimes be induced to bloom by severe root pruning, combined with pruning off some of the most vigorous shoots.

In warmer areas, where wisterias grow with amazing vigor, they are sometimes allowed to climb up on dead trees; be sure not to train them on a live one—it will quickly be strangled.

Problems

Leaves skeletonized or with large holes; branches may be webbed. Cause: Caterpillars. For control measures, see "Leaves skeletonized or with large holes; branches may be webbed" on page 236.

Leaves with notched edges. Cause: Black vine weevils. The wingless, 1/3″, brownish black adults feed on the foliage; spray leaves several times with pyrethrin for major infestations. The small, C-shaped, white grubs of these weevils feed on the roots and weaken the plant; drench the soil around the base of the plant with a solution of parasitic nematodes.

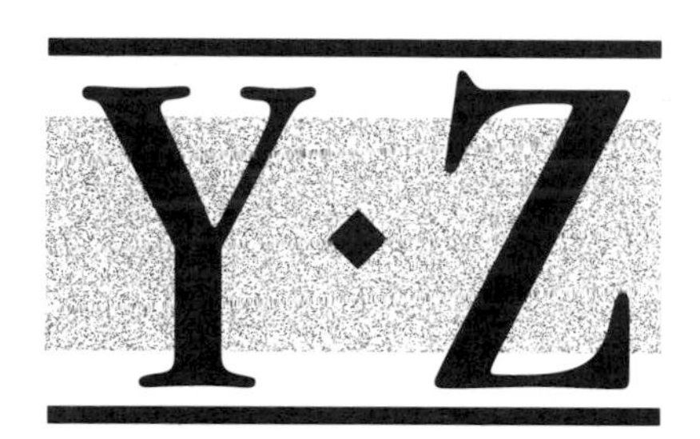

Yucca

Yucca, Adam's-needle. Perennials.

Clumps of stiff, sword-shaped, light green leaves arising from the ground give yuccas a strong presence in any landscape. Native in the southwestern United States and Mexico, where many species assume treelike form and heights up to 30′, only a few yuccas are hardy north of Zone 7. Adam's-needle (*Yucca filamentosa*) is a popular 3′ plant, hardy to Zone 4, with evergreen, 1½″ wide leaves and tall spikes of midsummer white flowers. Variegated cultivars are available.

Succulent yuccas grow well in conditions resembling the semi-desert of the Southwest: full sun and well-drained, sandy soil. Excess moisture, especially in the winter, encourages rots. New plants form at the base of mature yuccas and can be easily divided.

Insect pests such as yucca plant bugs and scales are more prevalent and likely to cause damage where yuccas are native. Aphids may infest Adam's-needle; see "Leaves, stems, and buds distorted, sticky; clusters of small insects" on page 177 for controls. Fungal leaf spots may appear and are best controlled with cultural practices: Limit excess water on foliage and remove severely infected plant parts.

Zinnia

Zinnia. Annuals.

You could fill an entire garden with zinnias and have a different type in every section. Plant height ranges from 6″ to 36″; flowers can be button-size to dinner plate-size, single or double, smooth or ruffled, solid, multicolored, or striped. Just about every color is represented except blue. Flowering is possible from spring until frost.

Zinnias are easy to raise from seed and can bloom within 2 months. Successive sowings are recommended to give continuous flowering. Indoors, sow seeds in individual pots 8 weeks before last frost. Move seedlings outdoors very carefully when soil is consistently warm. Direct-sowing is preferred, as zinnias resent root disturbance; in fact, some double-flowered cultivars may revert to single on transplanting. Direct-sow in spring and early summer when night temperatures stay over 50°F. Cover lightly and press soil down firmly. Germination takes 6 days. When seedlings are 3″ tall, thin to allow 4″-12″ between plants, depending on final size.

If you purchase seedling or potted zinnias, look for those that have not begun flowering or setting buds. Expect slow growth initially after transplanting. Make certain roots don't dry out in the process, water immediately, and give temporary shade in hot weather while plants adjust.

Zinnias need sun, ample water, good drainage, fertile soil, regular fertilizer, and good air circulation. They grow poorly in cool weather. Pinch initial buds to encourage side growth and flower formation. Remove spent flowers to prolong the blooming period. Water only from below; overhead watering weights down the already heavy flowers, causing stems to snap. Overhead watering also burns foliage and encourages mildew. Use a complete fertilizer once a month.

Problems

Leaves with powdery white patches. Cause: Powdery mildew. Zinnias are extremely prone to mildew if not given excellent air circulation. For more information on controlling powdery mildew, see "Leaves with powdery white patches" on page 19.

Plant wilts; leaves ragged. Cause: Stalk borers. Borers are long, thin, striped caterpillars that may have purple stripes. These larvae feed on leaves and within the wide zinnia stems. A small hole in the stalk marks their initial entrance. Stalk feeding can kill the plant.

Cut affected stems below the borer's hole; some plants may develop sideshoots that later flower. To save prize zinnias, try slitting affected stems and removing the borer, then binding stems together with green twine and keeping plants particularly well watered. Or inject BTK or parasitic nematodes into the stem with a syringe. A foliar application of BTK may be effective if borers feed on the leaves. Keep the garden weed-free to eliminate overwintering sites.

Leaves stippled with yellow; foliage webbed. Cause: Spider mites. For controls, see "Leaves stippled with yellow; foliage webbed" on page 18.

Seedlings or young plants cut off at soil level. Cause: Cutworms. See "Seedlings or young plants cut off at soil level" on page 20 for controls.

Blossoms and foliage disappear. Cause: Blister beetles. These metallic, dark-colored, ³⁄₄″ beetles have long antennae and long legs. They begin chewing on flowers and leaves in June. Blister beetles have body fluids that may cause painful blisters on skin contact. Handle only with gloves. Control serious infestations with pyrethrin or rotenone.

Seedlings die. Cause: Damping-off. See "Seedlings die" on page 20 for controls.

PART 2

INSECTS

Recognizing Your Friends—Eliminating Your Foes

Insects predate gardeners by eons. During the Paleozoic era 400 million years ago, huge cockroaches and dragonflies with 2-foot wingspans traveled the earth. Their diversity and range of adaptations to every climate and habitat have made insects the most varied, widespread group of all animals.

The pest insects that eat our treasured ornamentals and rob a share of the vegetable harvest are only a tiny fraction of the total insect population around us. The overwhelming majority of insects are harmless members of the natural community. Many are directly beneficial in their role as crop pollinators, predators on pests, and decomposers of plant material. They can be as common as houseflies, distributed globally, or as rare as the flea that lives on the skin of certain sea mammals or the midge that lives in the tiny pool of water in a pitcher plant.

It's important to keep this broader picture of the good side of insects in mind when fighting garden pests. Some control methods, especially chemical controls, kill beneficial insects as effectively as they kill the pests. The goal for the organic gardener is to work with nature as much as possible to let populations of pests and beneficials balance one another.

What Are Insects?

Insects are animals in the class Insecta, which is part of the larger group of animals known as arthropods, meaning creatures with jointed legs. They are cold-blooded and wear their skeletons on the outside like armor.

Spiders, millipedes, mites, crabs, and lobsters are also arthropods. Insects differ from these creatures because they have wings and only six legs. Their bodies are divided into three sections: the head; the thorax or midsection, where legs and wings are attached if they have them; and the abdomen or tail section, where the digestive and reproductive organs are located. On their heads, insects have a pair of antennae or "feelers," which are complex sensory organs. Insects also have at least one pair of eyes and often extra rudimentary eyes (called ocelli) as well. They breathe through a system of small, round openings, or spiracles, along the sides of their bodies. These open into small, branching tubes, known as trachea, that carry oxygen through their tissues.

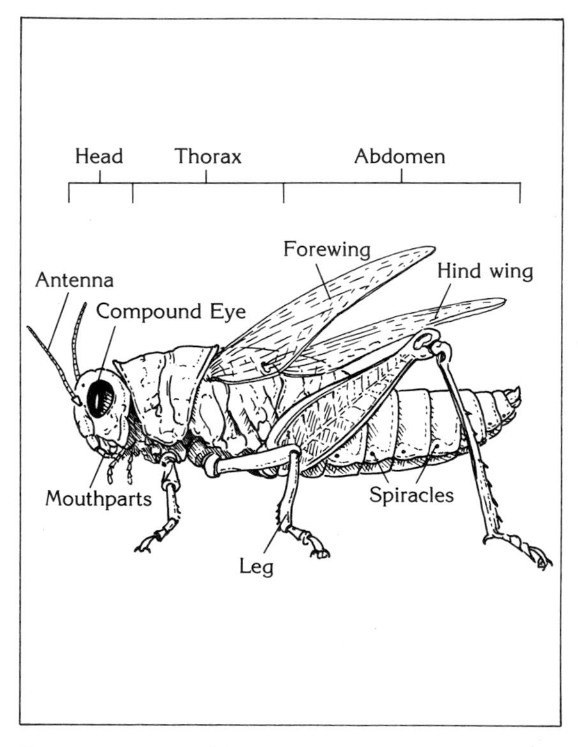

Insect anatomy. *While insects come in an amazing assortment of shapes and sizes, they all share the same basic body structure.*

Insect Life Cycles

Although there are insects with weird quirks in their life cycles that seem like something out of science fiction, the development of most insects follows one of two basic patterns of metamorphosis, or change of form.

Complete Metamorphosis

This pattern has an immobile stage, called a pupa, between the immature and adult forms of the insect. During the pupal stage, the tissues of the immature insect transform into an adult with an entirely different appearance. Butterflies, moths, wasps, beetles, and flies exhibit this type of metamorphosis.

The cycle begins with an egg, which hatches into a tiny, immature insect called a larva. All young insects are larvae, but we often use the terms caterpillar for moth or butterfly larvae, grub for beetle or wasp larvae, and maggot for fly larvae. As a larva grows, it periodically molts its skin to accommodate its enlarging body (one of the complications of having a skeleton on the outside).

When the larva reaches maximum size, usually after a number of molts, it contracts into a shorter, legless pupa. The pupa has a hardened skin to protect the developing adult inside. Some larvae spin a silken outer cocoon or chrysalis before pupating, for protection during the resting state. When the adult is ready to emerge, it splits open the pupal case and crawls out, still damp and soft. It slowly expands its wings. After its outer skeleton darkens and hardens, it is ready to fly away.

Larvae that undergo complete metamorphosis usually eat different food and live in different habitats than the adults of the species. This is important to remember if you want to attract beneficial insects. Generally, the adult form of beneficial species is the winged form that can migrate to your garden. These adults feed on pollen and nectar flowers. Be sure to provide a food source for them in order to reap the benefits of having their predatory offspring growing among your plants.

Incomplete Metamorphosis

Insects that develop gradually from immature stages to adults without pupating follow a pattern of incomplete metamorphosis. True bugs, including pests such as tarnished plant bugs and boxelder bugs as well as praying mantids, aphids, and whiteflies, exhibit incomplete metamorphosis.

The cycle starts with an egg, which hatches into a larva, usually called a nymph, that looks like a miniature, wingless version of the adult insect. The resemblance of nymphs to adults increases with each molt. They grow larger, their bodies lengthen, and small wing buds appear. With the final molt, they become adults with fully formed wings and reproductive organs. Adults and nymphs of species with this pattern usually eat the same kind of food.

Feeding Habits

Insects eat an incredible variety of foods: leaves, roots, plant sap, wood, other insects, other arthropods, blood of birds and mammals, decaying plant material, pollen, nectar, dung, particles of algae, and even fungi. Some, like cockroaches, can digest nearly anything they can get their mandibles (jaws) on, while others must find a particular species of plant to eat or they will die. The disappearance of many butterfly species is due to the loss of their particular host plants as wild areas are cultivated or paved.

Plant-eaters: Plant-eating insects usually are no friend to gardeners. They are adapted to chewing, sucking, or boring in leaves, stems, or roots. However, some species that eat weeds are beneficial. These insects usually must consume a large volume of plant material relative to their size to obtain enough nutrients to continue their development and reproduction.

Carnivores: Most of us are familiar with carnivorous insects such as mosquitoes, deer flies, and biting midges. These pests land on mammals or birds, suck blood for a short period of time, and then fly away. Others that feed on blood, such as lice and fleas, live on the skin of animals, hidden in their fur. Gardeners should also get to know the large group of carnivorous insects that feed on insects or other arthropods. These are divided into two main groups—predators and parasites.

Predatory insects such as lady beetles or ground beetles eat many other insects during their life cycles. Some have restricted tastes; for example, aphid midges feed only on aphids. Others, like praying mantids or assassin bugs, may be able to eat almost any species of insect they catch. These general predators often eat pollen as well and may suck plant juices when food is scarce or they need water.

Insects that parasitize other insects are called parasitoids. They lay eggs singly or in groups near, on, or inside the bodies of other insects. The parasitoid larvae develop as internal parasites. Parasitoids eventually kill the host, then pupate inside or crawl outside and pupate near the dead husk.

Scavengers: Dung beetles, carrion beetles, housefly larvae, and other insect species feed on decaying plant or animal material. These creatures perform a valuable task by breaking down these materials and hastening decomposition.

Omnivores: The ultimate survival strategy is to be able to eat nearly anything—a strategy favored by cockroaches, earwigs, and other pests. These species feed on all kinds of animal and vegetable materials, including soap, starch, and glue.

Fungus-feeders: Fungus gnat larvae, many kinds of soil-dwelling insects and mites, and a few species of lady beetle larvae actually eat fungi.

Beneficial Insects

As gardeners, we are most likely to notice the pest insects in our yards and gardens. However, entomologists estimate that more than 90 percent of all insects are beneficial. Wild and domestic bees and also some flies and moths pollinate crops; this is essential for the development of many fruits, vegetables, and crops grown for seed. Honeybees are also farmed to provide honey, and silkworms are farmed to yield silk from their cocoons for fine clothing.

Many species are decomposers: They recycle nutrients from organic materials and dung, and clean up the environment in the bargain. There are tremendous numbers of insects that have little direct effect on humans, but are essential food for fish, birds, and other animals.

Of most immediate interest to gardeners, however, are the thousands of species of predatory insects that attack pests. They are common worldwide and most numerous in gardens where pesticides are not used. Although they are often unseen, we reap immeasurable benefit from their presence. Sometimes the beneficials are difficult to distinguish from the pests. See "Insect Impostors" on page 260 for hints to help you see the difference between these two groups of insects.

Attracting and Conserving

The first and most important rule for the gardener who wants to encourage beneficial insects is to avoid using toxic sprays or dusts in the garden. Even botanical pesticides and insecticidal soap sprays kill beneficial insects as well as they kill pests. Use them only when absolutely necessary and only apply them on the plants being attacked.

To lure native beneficial insects into your garden, provide them with an attractive food supply. Since the adults of many beneficial insects feed only on pollen and nectar, the best way to attract them is to plant small-flowered plants, such as dill, fennel, parsley, and mint family plants. Members of the mustard family, including garden vegetables such as radishes or broccoli that have gone to flower, are also good choices. There are many small-flowered annual, biennial, and perennial flowers to suit any garden. You can interplant them among the vegetables as well as in borders and beds.

Once adults of beneficial species have arrived in your garden and have had a meal, the females will search for a good place to lay eggs, which later hatch into predatory larvae. Beneficial ground-dwelling insects, such as rove and ground beetles, find refuges in permanent walkways of sod, stone, or thick mulches. This provides them with a safe place to hide when plantings are being disturbed, cultivated, or harvested.

It is important to minimize dust and provide a water source in hot, dry areas to protect beneficial insects. They are easily killed by dehydration. Hedges, windbreaks, and even fences help keep down dust. An old birdbath filled with water and rocks or gravel (to provide safe landing places for tiny insects so they won't drown) will be used by many beneficial species. Hedgerows and permanent beds protect beneficial insects while there are disruptions in the garden and often provide them with an alternate food supply of nonpest insects living in the hedge.

Controlling Insect Pests

Pest problems concern every gardener. However, it's possible to have good harvests and beautiful gardens without exterminating every pest insect that enters your yard. The goal to strive for is to suppress pest populations to the point at which they don't interfere with our harvest or our enjoyment of ornamental gardens. Leaving a few pests in the garden attracts and sustains beneficial insects.

The concept of keeping pest populations at acceptable levels is an important part of organic pest management, a system of pest control that uses a variety of methods to reduce pest populations without resorting to synthetic chemicals. It is based on improving the health of the garden ecosystem—from maintaining fertile soils with high organic matter content to encouraging a complex, diverse plant community attractive to many species of insects and birds. Managing pests organically includes a range of strategies.

Cultural controls: These are steps the gardener takes while planting or cultivating crops to make the garden less hospitable to pests. It includes growing healthy plants on fertile soil to make them less attractive to insects (or able to recover faster), using pest-resistant cultivars, applying mulches to foil pests, and using sanitation methods such as removing garden trash or roots that harbor the overwintering stages of pests.

Physical controls: Using barriers, such as floating row cover, cutworm collars, or tree bands physically prevents insects from reaching plants. Cultivating soil in fall to kill overwintering pests, handpicking large pests, or knocking heavy insects such as Japanese beetles or plum curculios from foliage are also physical controls.

Biological controls: Using living organisms to control pests is called biological control. These controls include disease organisms such as *Bacillus thuringiensis* that infect insects as well as predatory or parasitic insects or mites. Many beneficial insects have been introduced from other countries to control imported pests and thousands of beneficial species are native to North America. Several dozen species are

(continued on page 262)

Insect Impostors

Hold your horses! Before you squash that bug or turn on that sprayer, take a close look. There are several species of common beneficial insects that are look-alikes of equally common pest species. Here are some that might stump you. (Hint: The good guys are always on the left.)

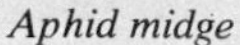
Aphid midge

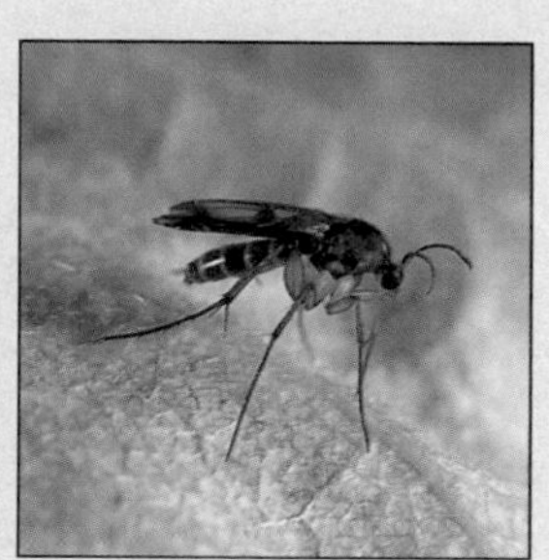

Fungus gnat

While aphid midges resemble fungus gnats, they are actually excellent predators of more than 60 species of aphids.

Rove beetle

Earwig

Before you stomp on that earwig, make sure it's not a rove beetle, which preys on many common garden pests.

Lady beetle

Mexican bean beetle

A lady beetle should be one of your best friends in the garden. But its cousin, the Mexican bean beetle, can be one of your worst enemies.

Spined soldier bug

Brown stink bug

Many kinds of stink bugs are pests, but spined soldier bugs are voracious predators of a variety of caterpillars and grubs.

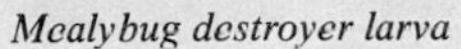

Mealybug destroyer larva

Mealybugs

It's easy to confuse mealybugs with the beneficial creatures that eat them—mealybug destroyer larvae.

Bigeyed bug

Chinch bug

Look for the big eyes! Bigeyed bugs are your friends, unlike chinch bugs, which munch on the roots of your lawn or corn crop.

Tachinid fly

Housefly

They may look like pesky houseflies, but don't use the swatter! Tachinid flies lay eggs on many types of caterpillar pests, and the tachinid larvae feed on and kill the caterpillars.

Hover fly

Baldfaced hornet

If it's a baldfaced hornet, you may fear its sting. But the only creatures that should be afraid of a hover fly are the aphids in your garden.

Coordinated Control

Combining different control methods that are effective against individual stages in a pest's life cycle can boost your success in keeping the pest from damaging plants. The codling moth, a serious pest of apples and other fruit trees, is a good example of a pest that is vulnerable to several types of controls.

In late winter, while trees are dormant, kill larvae overwintering in cocoons in the bark by scraping away loose bark and applying dormant oil sprays. Larvae overwintering in the soil litter are susceptible to attack by ground beetles, especially in orchards with cover crops, which protect the beetle populations.

In the spring, you can determine when adults begin to emerge from pupae by hanging pheromone traps among the trees to capture male moths. Pheromone lures hung in trees at this time will disrupt mating by confusing the males in their search for females. Once adults begin appearing in the traps, release parasitic *Trichogramma* wasps at 7–10 day intervals to attack the eggs that the adults will lay.

Most likely, some moths will still successfully mate and lay eggs, and some of those eggs will escape the parasites. Fortunately, you can control the larvae that do hatch before they enter the apples by spraying ryania or codling moth granulosis virus at 7–10 day intervals.

Once larvae have tunneled into fruit, they are relatively safe from control measures. However, after feeding, they will crawl down the trees to find a place to spin a cocoon. Intercept the migration by tying burlap tree bands around the trees. Check them daily and destroy the larvae.

You'll repeat the same control cycle for the second generation, and for a third generation in warmer areas. If you're diligent, you should have negligible damage from codling moths.

now sold by insectaries for release in gardens, orchards, and greenhouses.

Chemical controls: Sprays, dusts, or baits that repel or poison pests are types of chemical controls. You can make some types yourself from garden and kitchen ingredients. You can buy products including botanical pesticides such as rotenone, neem, or pyrethrin.

For some pests, using a single control method may be sufficient to prevent damage. Cutworm collars are reliable barriers against cutworms; once the spring risk to transplants has passed no further action need be taken.

For other pests, especially if they are numerous or very damaging, you'll need to combine several controls at one time or in succession throughout the season. See "Coordinated Control" above to find out how using several different control methods at various points throughout the life cycle of the codling moth gives more successful control.

Remember that some controls are not compatible with one another. For example, spraying botanical pesticides can harm beneficial insects, so don't spray when you have just released beneficial insects, and don't apply pesticides on plants that are attractive to beneficial insects.

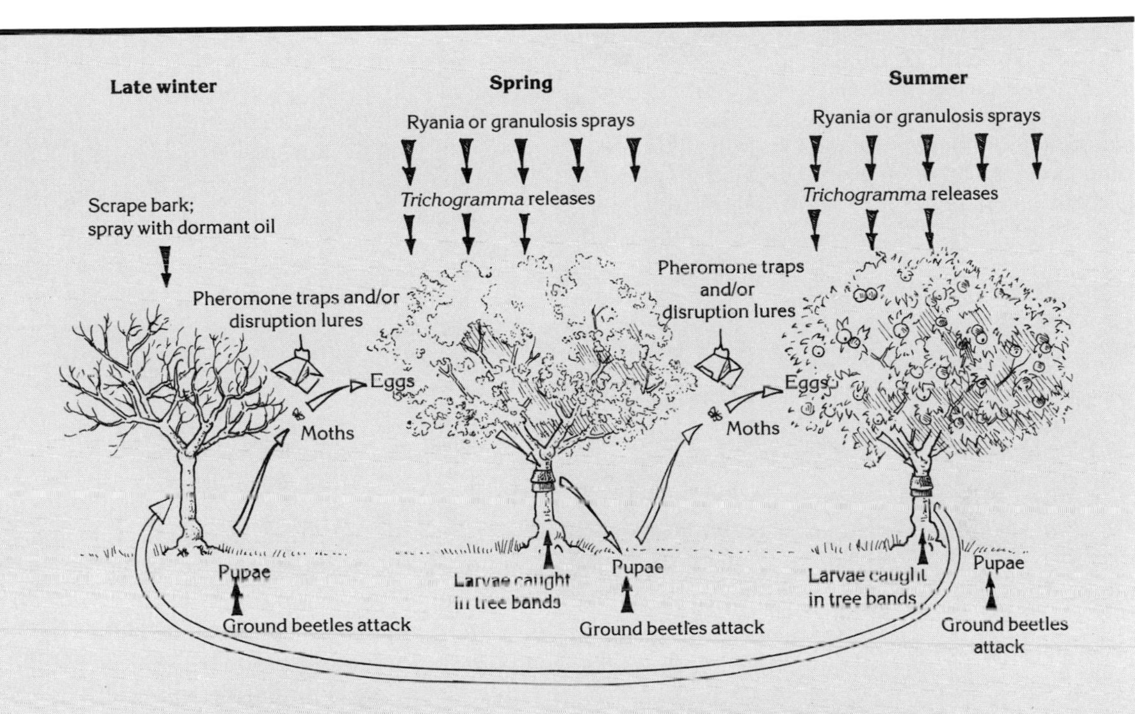

Timing insect-control methods. *White arrows follow the life cycle of the codling moth from the pupation of overwintering larvae in spring through 2 full generations to the end of summer. Black arrows indicate when to use various control methods to coincide with the vulnerable phase in the insect's life cycle.*

Planning Your Strategy

The first step in controlling a pest is identifying it. If you don't have any idea what pest has caused the damage you've found, start by referring to "Common Insect Damage Symptoms" on page 264. The listings of symptoms and possible culprits should help you determine what pest is causing the problem.

Some destructive insects look similar to beneficial species, so be sure your identification is accurate. You can check "Insect Impostors" on page 260 for a quick rundown on common beneficials that resemble pests.

If you aren't sure you've figured out the identity of your pest, get expert help from your local extension service, garden center, or university entomology department.

Remember that some damage symptoms can have many causes. The problem you see may be because of a disease organism or cultural problem. Or the damage may have been from insects that are no longer feeding. Once you've made a positive identification, look up the pest in the individual insect entries that follow. Plan a control strategy based on the suggested controls in these entries. For details on using a particular control method, see "Organic Controls" beginning on page 404.

Common Insect Damage Symptoms

If you have damaged plants and aren't sure what insect is causing the injury, use this table to narrow the list of possible culprits. Look for the description of damage symptoms in the left-hand column that matches your plant's symptoms. Next, find the appropriate category of host plants that suffer from those symptoms. Then look up the insects listed in the individual insect entries. The photographs and information in the entries will help you identify the pest and take appropriate steps to control it.

Symptoms	Host Plants	Insects
LEAF AND FOLIAGE DAMAGE		
Small plants cut off at soil line	Vegetables, flowers	Cutworms, slugs and snails
Large holes chewed in leaves; no excrement visible	Beans	Mexican bean beetles
	Cucumbers, corn, peanuts	Spotted cucumber beetles
	Flowers, vegetables	Japanese beetles, slugs and snails
	Potatoes	Colorado potato beetles
Large holes chewed in leaves, dark green excrement often visible	Cabbage family plants	Cabbage loopers, imported cabbageworms, diamondback moths
	Tomatoes	Tomato hornworms, tomato fruitworms
	Vegetables, fruits, ornamentals	Woollybear caterpillars
Small, round holes in leaves	Cabbage family plants, potatoes, spinach, flowers	Flea beetles
Puckered, twisted leaves; sticky honeydew present	Vegetables, fruits, ornamentals	Aphids
Puckered, twisted leaves; no sticky honeydew	Vegetables, fruits, ornamentals	Thrips, leafhoppers, plant bugs, buffalo treehoppers, spittlebugs/froghoppers, nematodes
Partially or fully defoliated plants	Asparagus	Asparagus beetles
	Conifers	Conifer sawflies, gypsy moths, spruce budworms
	Deciduous trees and shrubs	Gypsy moths, June/May beetles, cankerworms, tussock moths
	Flowers, vegetables	Armyworms, blister beetles, climbing cutworms, rose chafers, striped cucumber beetles
	Pears, cherries, plums, cotoneasters	Pear sawflies
	Squash family plants	Striped cucumber beetles

Symptoms	Host Plants	Insects
Chewed vines and leaves	Potatoes	Potato tuberworms
	Squash family plants	Pickleworms, squash vine borers
Shoot or branch tips wilt and die	Fruit and shade trees	Oriental fruit moths
	Squash family plants	Squash bugs
	Vegetables, fruits, flowers	Tarnished plant bugs and other plant bugs
Webbing on leaves, stems, and branch tips	Apples, roses, other deciduous trees and shrubs	Obliquebanded leafrollers, fruittree leafrollers, tent caterpillars, fall webworms
	Vegetables, strawberries	Garden webworms
Small, baglike cocoons with bits of leaves attached hanging from branches	Fruit trees, ornamental trees and shrubs	Bagworms
Mines between upper and lower leaf surfaces	Apples, roses, other deciduous trees and shrubs	Obliquebanded leafrollers (young larvae)
	Beets, chard, nightshade family plants	Leafminers
	Cabbage family plants	Diamondback moths (young larvae)
	Chrysanthemums and other ornamentals	Leafminers
	Elms, birches, alders	Leafminer sawflies
Yellow leaves	Citrus and other fruit trees	Soft scales, armored scales
	Lawn grasses	Chinch bugs
	Pears, quinces	Pear psyllas
Yellow and withered leaves	Evergreen trees and shrubs, strawberries, bramble fruits	Black vine weevil larvae
Russeted leaves	Apples, pears, tomatoes, ornamentals	Rust mites
Small, discolored spots on leaves	Vegetables, flowers	Garden fleahoppers
White, gray, or silvery speckled pattern on leaves	Vegetables, ornamentals, fruit and shade trees	Spider mites, thrips, lace bugs
Sticky honeydew on leaves	Pears, quinces	Pear psyllas
	Vegetables, fruits, ornamentals	Aphids, scales, mealybugs, whiteflies

(continued)

Common Insect Damage Symptoms — Continued

Symptoms	Host Plants	Insects
LEAF AND FOLIAGE DAMAGE—CONTINUED		
Galls on leaves	Oaks, roses	Gall wasps
	Maples	Gall mites
FRUIT DAMAGE		
Early-dropping fruit	Apples, plums, blueberries	Apple maggots, plum curculios
	Citrus and other tree fruits, grapes	Mealybugs
	Tree fruits, blueberries, currants	Fruit flies
Damage around pit in fruit	Almonds, walnuts	Navel orangeworms
	Cherries	Codling moths
Tunnels to core of fruit	Apples and other tree fruits	Codling moths, Oriental fruit moths, plum cucurlios
	Citrus, figs	Navel orangeworms
Large holes or damaged areas in fruit or ears	Corn	Corn earworms, European corn borers
	Squash family plants	Pickleworms
	Tomatoes	Tomato hornworms, tomato fruitworms
Distorted, scarred fruit	Vegetables, fruits	Tarnished plant bugs and other plant bugs, stink bugs, thrips
Russeted appearance of fruit	Apples, pears	Rust mites
Holes bored in seeds and pods	Beans	Bean weevils

Symptoms	Host Plants	Insects
STEM AND TRUNK DAMAGE		
Holes bored in trunk	Apple and other fruit trees, mountain ashes, hawthorns	Roundheaded appletree borers, peachtree borers
Holes bored in stems, buds, or shoots	Corn	European corn borers, southwestern corn borers
	Currants, gooseberries, raspberries, rhododendrons	Fruit borers
Galleries bored under bark	Elms	Elm bark beetles
	Fruit and shade trees	Flatheaded appletree borers, shothole borers
ROOT DAMAGE		
Holes tunneled in roots	Cabbage family plants	Cabbage maggots
	Carrot family	Carrot rust flies, carrot beetles, carrot weevils
	Gladiolus and other flower corms	Wireworms
	Irises	Iris borers
	Onions, leeks, garlic	Onion maggots
	Potatoes	Potato tuberworms, wireworms
	Strawberries, grapes, raspberries, other fruits	Strawberry root weevils
Knotted, lumpy roots	Beans, peas, other legumes	Beneficial bacteria in roots
	Tomatoes, lettuce, peppers, nonlegumes	Nematodes
Chewed, stunted, withered, or damaged roots	Lawn grasses	Japanese beetles, June/May beetles, rose chafers, chinch bugs

INSECT IDENTIFICATION GUIDE

Many of the pests that damage garden plants are widespread in the United States and Canada. Some are very damaging only in a particular region of North America. This guide provides information on the appearance, range, host plants, damage, life cycle, and control of more than 100 types of common garden pests. It includes some pests that are technically not insects, such as mites and slugs, but that cause similar types of damage and are vulnerable to many of the same control methods as insects.

The guide also offers information on the appearance, range, life cycle, and helpful effects of more than 20 beneficial types of insects, along with suggestions on how to attract them to your gardens. Note that beneficial insect names appear in green type; the names of the pest insects appear in black. So, you can tell at a glance which are the bad bugs and which are beneficial.

Aphid Midge *Aphidoletes aphidimyza*

Larva

Adult

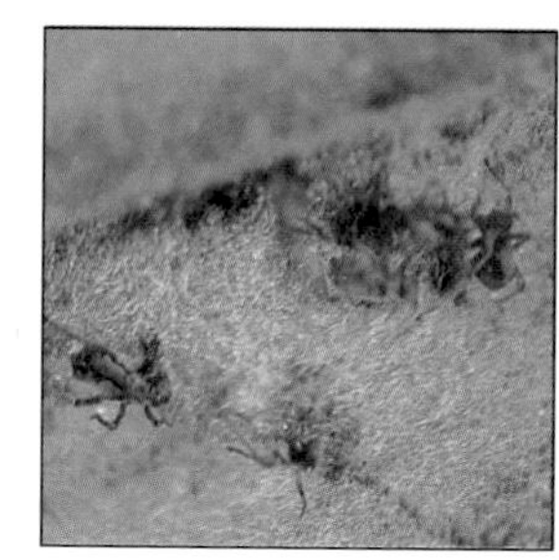

Aphid mummies

Description: Adults: delicate, long-legged, 1/16″ flies, active at night. Larvae: orange maggots, up to 1/8″. Eggs: minute orange ovals. Common throughout North America; sold commercially.

Beneficial Effect: Larvae paralyze aphids with toxic saliva, then suck their body fluids. Can attack more than 60 aphid species.

Life Cycle: Females lay eggs among aphids, eggs hatch in 2-3 days; larvae feed on aphids 3-5 days, then burrow into soil to pupate; adults emerge in 2 weeks. Overwinters as larvae in the soil.

How to Attract: Plant pollen and nectar plants; protect garden from winds; provide a water source; buy 200-300 cocoons for a small garden or greenhouse or use 3-5 cocoons per plant, 5-10 per fruit tree.

Aphids (Family Aphididae)

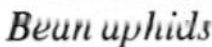
Bean aphids

Cabbage aphids

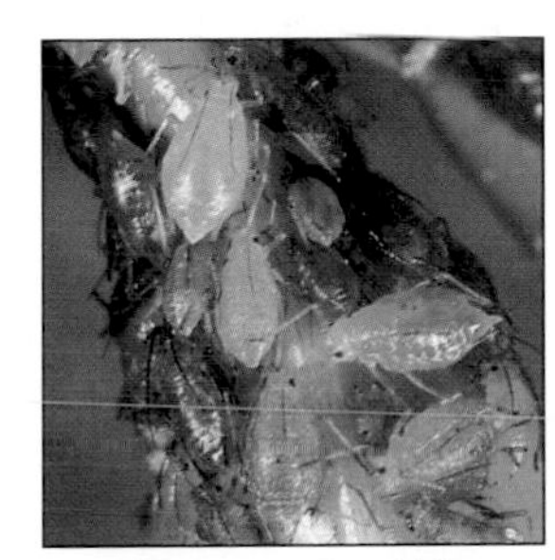
Green peach aphids

Melon aphid winged adult

Damage

Woolly apple aphids

Description: Adults: pear shaped, 1/32"–1/8" insects with 2 short tubes projecting backward from the abdomen; long antennae; green, pink, black, dusty gray, or with white fluffy coating; with or without wings. Nymphs: similar to adults. Colonies develop quickly; winged forms appear when they become crowded. Common throughout North America.

Damage: Nymphs and adults suck plant sap from most small fruits, vegetables, ornamentals, and fruit and shade trees. Their feeding causes leaf, bud, and flower distortions; severely infested leaves and flowers drop. Fruit that forms on infested branches are misshapen and stunted. Aphids secrete sticky honeydew that supports growth of sooty mold on leaves and fruit. Feeding can spread viral diseases.

Life Cycle: Eggs overwinter on woody stems, hatching in spring into stem females, which can give birth continuously to live nymphs without having to mate. Nymphs mature in 1–2 weeks. In fall, males and normal females are born; these mate to produce overwintering eggs. In greenhouses, some females continue to bear nymphs throughout the year. Some species feed on cereal crops or weeds for part of the year and on fruit trees at other times.

Control: For vegetable crops and small ornamentals, spray smaller plants frequently with a strong stream of water to knock aphids off; attract native predators and parasites by planting pollen and nectar plants; release purchased aphid midges, lady beetles, lacewings, or parasitic wasps; use homemade garlic, quassia, or tomato-leaf sprays; spray insecticidal soap; as a last resort, spray pyrethrin, sabadilla, nicotine, or rotenone. For fruit or shade trees, spray dormant oil to kill overwintering eggs, and plant flowering groundcovers in home orchards to attract predators and parasites.

Apple Maggot *Rhagoletis pomonella*

Adult

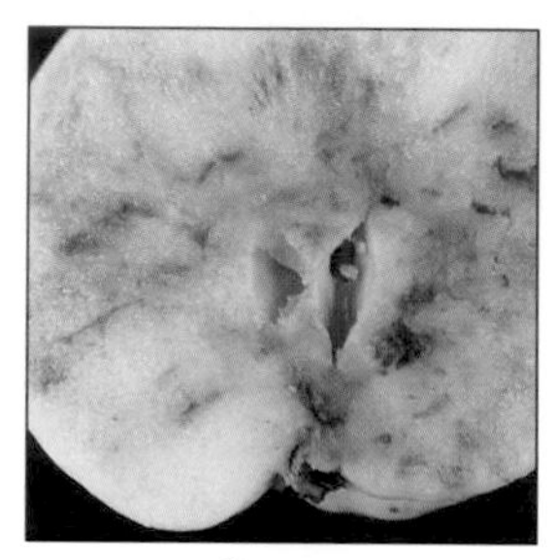
Damage

Description: Adults: 1/4″ flies with yellow legs and transparent wings patterned with dark, crosswise bands. Larvae: white, 1/4″ maggots. Found in eastern United States and Canada, also northern California.

Damage: Maggots tunnel through apples, blueberries, and plums. Fruit drops prematurely; early cultivars are most affected.

Life Cycle: Adults emerge from overwintering pupae mid-June to July and lay eggs in punctures in fruit skin; eggs hatch in 5-7 days, larvae tunnel in fruit until it drops, then leave to pupate in soil for winter. One generation per year. Some pupae remain dormant for several years.

Control: Collect and destroy dropped fruit daily until September, twice a month in fall; hang apple maggot traps in trees from mid-June until harvest (1 per dwarf tree, 6 per full-size tree); plant clover groundcover to attract beetles that prey on pupae; grow late-maturing cultivars.

Armyworms (Family Noctuidae)

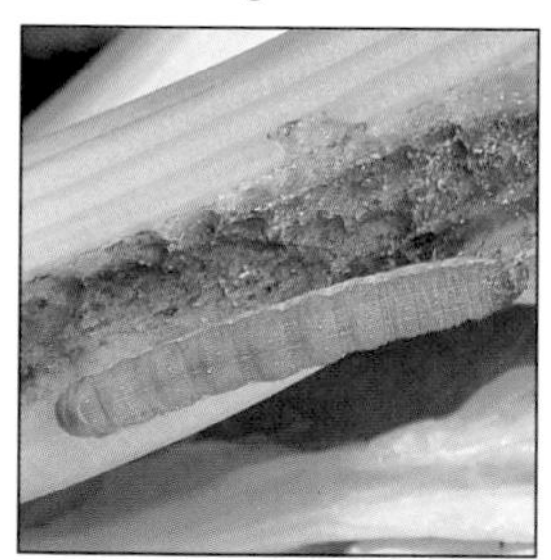
Beet armyworm

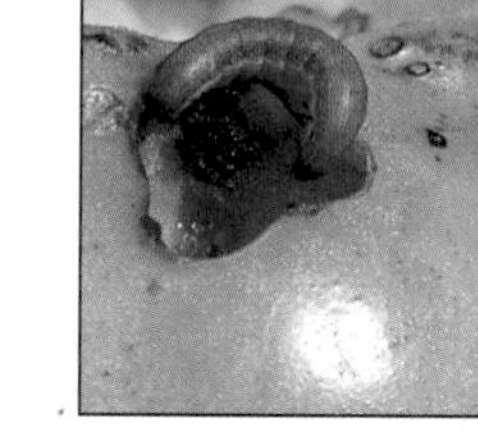
Beet armyworm

Fall armyworm

Description: Adults: pale, gray-brown moths with a white dot in center of forewing (1 1/2″-2″ wingspan), active only at night. Larvae: early stages smooth, pale green; older larvae reach 1 1/2″; greenish brown with white stripes on sides, dark or light stripes along backs. Eggs: greenish white, in masses on lower leaves. Found east of the Rockies and in southeastern Canada, also in New Mexico, Arizona, and California. Beet armyworm is common in southern United States.

Damage: Larvae feed on corn, field crops, and garden plants at night, hiding during the day in the center of corn plants or under stones or leaf litter. When food supply is depleted, they move en masse to a new location. Larvae can consume whole plants in 1 night. First generation larvae (June) usually cause the most damage.

Life Cycle: Larvae (sometimes pupae) overwinter in soil or litter around roots, resume feeding in spring, then pupate for 2 weeks. Two to 3 generations per year.

Control: Attract native parasitic wasps and flies; spray BTK to kill larvae; spray superior oil in July to kill eggs of second generation.

Asparagus Beetle *Crioceris asparagi*

Larva

Adult

Description: Adults: shiny, elongate, bluish black, 1/4" beetles with reddish brown thoraxes, 4 cream-colored spots and red borders on wing covers. Larvae: 1/3", plump, wrinkled, and gray with dark heads and legs. Eggs: shiny, black, glued on end to stems and young spears. Common throughout North America. The spotted asparagus beetle *(Crioceris duodecimpunctata)* causes similar damage, but is generally found east of the Mississippi River. Beetles are red-orange with 12 black spots on wing covers; larvae are orange.

Damage: Adults and larvae chew on green asparagus shoots, blemishing spears; also attack older stems and leaves.

Life Cycle: Hibernating adults emerge when first asparagus spears are ready to cut; they feed and lay eggs on spears. Eggs hatch in 1 week, larvae feed for 2 weeks, then burrow into the soil to pupate. Adults emerge in 10 days. Two or 3 generations per year.

Control: In fall, remove and burn old fronds and garden trash where beetles overwinter, or put it in sealed containers for disposal with household trash; in spring, cover spears with floating row cover until end of harvest; handpick beetles; spray pyrethrin or rotenone.

Assassin Bugs (Family Reduviidae)

Adult and prey

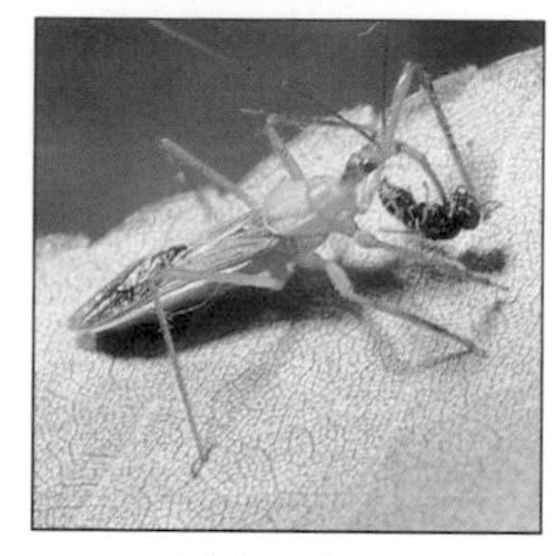

Adult and prey

Description: Adults: flattened, ¾″ bugs with long, narrow heads and stout, curving beaks, some with flared or sculptured thoraxes; may bite when handled; some species squeak. Nymphs: smaller, similar to adults, wingless, some brightly colored, others disguised by coating of dust or debris. Found throughout North America.

Beneficial Effect: General predators that help suppress populations of many insects, including flies and caterpillars.

Life Cycle: Adults lay eggs in crevices; nymphs develop until last molt and hibernate in a pre-adult stage, then develop to adults the following June.

How to Attract: Naturally present in most gardens. Avoid pesticide use.

Bagworm *Thyridopteryx ephemeraeformis*

Larvae in bags

Larva in bag

Description: Adults: males are black, clear-winged moths (1″ wingspan); females are wingless. Larvae: dark brown, ¾″–1″ caterpillars with white or yellow heads; feed inside cocoons. Eggs: light tan eggs laid inside bags. Found east of Rocky Mountains.

Damage: Larvae eat foliage of many trees and shrubs. Bags mar plant appearance. Use a knife to cut the silk from the twig; if you tear the bag away, you'll leave a coil of tightly wound silk ready to girdle the twig.

Life Cycle: Eggs hatch inside bags in spring. Larvae feed on leaves and build new bags with foliage of the host plant attached. Larvae enlarge bag as they grow, then pupate in September. Winged males emerge several days later and mate with wingless females in bags. Female moths lay eggs and die; eggs overwinter.

Control: Handpick and destroy bags; spray with BTK in early spring; set out pheromone traps in August to catch males.

Bean Weevil *Acanthoscelides obtectus*

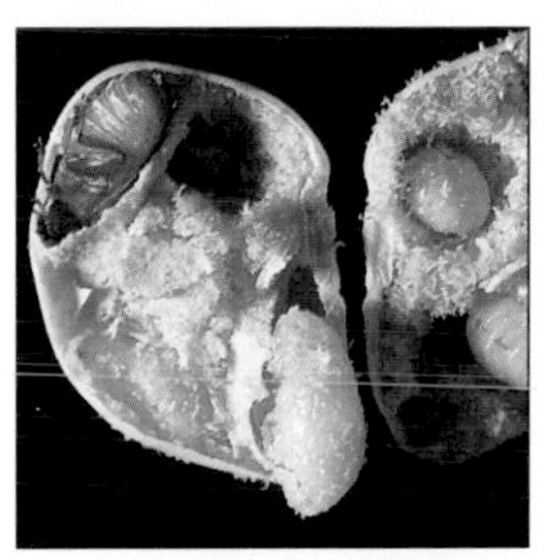

Larvae and pupae

Adult and damage

Adults and damage

Description: Adults: velvety gray or brown, 1/10″-1/8″ weevils with pale lines on their backs and red legs. Larvae: white, fat grubs. Found throughout North America.

Damage: Adults feed on leaves and pods of beans, peas, and lentils. Larvae bore through freshly harvested and stored seeds, ruining them.

Life Cycle: As plants bloom, overwintering adults emerge and feed, and lay eggs in pods. Larvae feed and pupate in seeds. In stored beans, eggs are laid on seeds and weevils brood year-round

Control: At harvest, treat seeds and stored legumes with diatomaceous earth; remove bean plants from the garden immediately after harvest to reduce overwintering populations.

Bigeyed Bugs *Geocoris* spp.

Adult

Description: Adults: fast-moving, 1/8″-1/4″ bugs with large eyes, minute black spots on heads and thoraxes. Nymphs: similar to adults, but wingless. May be mistaken for tarnished plant bugs. Common in western North America.

Beneficial Effect: Valuable predators of aphids, leafhoppers, plant bugs, spider mites, and small caterpillars in field crops and orchards.

Life Cycle: Females lay eggs on stems and leaf undersides; eggs hatch in 2 weeks; nymphs develop for 4-6 weeks and then molt. Adults overwinter in garden trash.

How to Attract: Soybeans, pigweed, and goldenrod are favored sites for the bugs to lay eggs. Interplant crops with soybeans; leave weedy plants in borders.

Black Vine Weevil *Otiorhynchus sulcatus*

Adult and damage

Description: Adults: oblong, brownish black, 1/3″ weevils. Larvae: white grubs with yellowish heads; up to 1/2″. Eggs: tiny white eggs laid in soil. Found in northern United States and southern Canada.

Damage: Larvae feed on roots, and adults chew on leaves, damaging many types of broad- and narrow-leaved evergreen trees and shrubs, strawberries, and bramble fruits.

Life Cycle: Larvae overwinter in soil, pupating in spring. Adults emerge in June. After 2 weeks, they lay eggs near the crowns of plants. Eggs hatch in 10 days; larvae burrow into soil and feed on roots. One generation per year.

Control: Shake weevils off plants at night onto a dropcloth and destroy them. Spray BTSD when adults are feeding on foliage. Apply parasitic nematodes to soil.

Blister Beetles (Family Meloidae)

Black blister beetle

Margined blister beetle

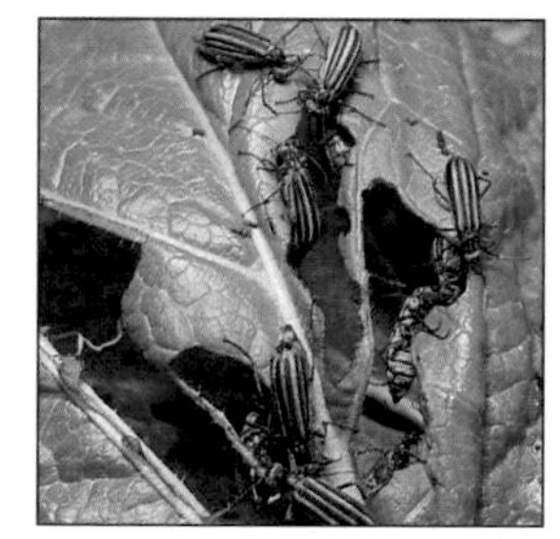
Striped blister beetles and damage

Description: Adults: metallic black, blue, purple, or brown, 3/4″ long beetles with soft, elongated bodies, narrow "necks," and long legs; beetles cling to plants when disturbed. Larvae: youngest are tiny, narrow, elongate grubs with large heads; later stages are progressively fatter with smaller heads, last stage are nearly legless. Found throughout North America.

Damage: Large numbers of adults feeding on flowers and foliage of many types of flowers, shrubs, and vegetables rapidly defoliate plants. Larvae of most species are beneficial because they prey on grasshopper eggs.

Life Cycle: Overwintering larvae pupate in spring, adults emerge and lay eggs in midsummer in grasshopper egg burrows; larvae feed on eggs for a month, then overwinter in the burrows for up to 2 years. Most species have 1 generation per year, coinciding with grasshopper life cycles.

Control: Except in areas where large adult populations do severe damage, tolerate adults to reap beneficial effects of larvae. To kill adults, knock them from plants into a pail of soapy water (wear gloves to avoid contact with crushed beetles, which cause skin burns); protect plants with floating row cover or screens in midsummer. For severe infestations, spray pyrethrin, sabadilla, or rotenone.

Boxelder Bug *Boisea trivittata*

Adult

Description: Adults: 1/2″ bugs with charcoal-colored wings with red veins. Nymphs: bright red, later marked with black. Found throughout North America.

Damage: Feeding causes deformities in flowers, leaves, shoots, or fruit of boxelder, ash, and maple trees. Occasionally attacks other shade or fruit trees. Damage is usually not serious. Swarms of adults congregating in fall on walls and trees in preparation for hibernation may be a nuisance.

Life Cycle: Females lay eggs in crevices, on foliage or seed pods; eggs hatch in 2 weeks; nymphs feed on foliage until final molt to adults in July. Second generation matures by fall. Adult females overwinter in buildings or sheltered areas.

Control: Usually not necessary. For severely annoying infestations, spray with pyrethrin or rotenone.

Braconid Wasps (Family Braconidae)

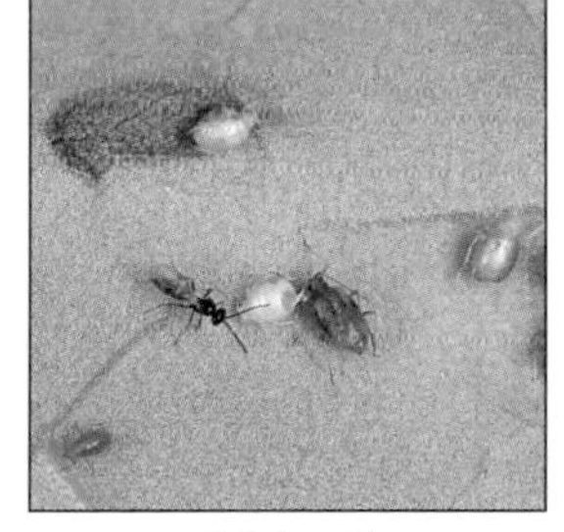

Adult

Wasp cocoons on tobacco hornworm

Adult with parasitized aphids

Description: Adults: slender, black or brown, 1/10″–1/2″ wasps with threadlike waists. Larvae: tiny cream-colored grubs that feed in or on other insects. Found throughout North America; some available commercially.

Beneficial Effect: Important native parasites of pests such as codling moths, elm bark beetles, cabbageworms, hornworms, corn borers, armyworms, aphids, and others. Some species attack flies and other insects.

Wasp larvae may develop inside the host insect. Or adult wasps may paralyze the host insect before laying eggs, and larvae then develop as external parasites.

Life Cycle: Females inject eggs into host insects, singly or in large numbers. When larvae complete development, they spin cocoons on or near the dead host, then pupate. Several generations occur per year.

How to Attract: Grow nectar plants with small flowers such as dill, parsley, and yarrow. Although some species are available by mail order, the best course for the homeowner is to attract native populations.

Buffalo Treehopper *Stictocephala bisonia*

Adult

Description: Adults: green or yellowish, wedge-shaped, 1/4″ bugs. Nymphs: green, humpbacked, with a spiny projection on their backs. Eggs: laid in crescent-shaped slits in bark. Found throughout North America.

Damage: Females puncture bark of fruit tree twigs to lay eggs. Nymphs damage tomatoes, potatoes, alfalfa, clover, and grasses by sucking plant sap.

Life Cycle: Eggs overwinter in tree bark and hatch in spring, nymphs drop to ground and feed on grasses and other plants for 6 weeks; after molting to adults, they return to trees to feed and lay eggs.

Control: Dormant oil sprays kill overwintering eggs; avoid alfalfa or clover cover crops where populations are damaging.

Bumblebees *Bombus* spp.

Adult

Description: Adults: plump, black-and-yellow, fuzzy, 1/2″–1″ bees; active even in cold weather. Larvae: fat, white grubs. Common throughout North America.

Beneficial Effect: Important wild pollinators; commercially reared colonies used as pollinators in greenhouses.

Life Cycle: Overwintering queen bee emerges from soil or leaf litter in May and makes nest on or below ground. The queen lays eggs in individual cells made of pollen; feeds developing larvae pollen and honey until they mature in 3–4 weeks to become workers. Workers collect pollen and nectar and rear subsequent 2 or 3 broods. In the fall, young queens develop and fly away to mate, and the colony breaks up.

How to Attract: Common in most gardens. Protect bees by minimizing pesticide usage; apply in evening when bees are not active.

Cabbage Looper *Trichoplusia ni*

Larva

Adult

Damage

Description: Adults: gray moths with a silver spot in the middle of each forewing (1½"-2" wingspan). Larvae: green, 1½" caterpillars with 2 white lines down their backs, 1 along each side; they move by looping their bodies. Eggs: light green, dome-shaped, on undersides of leaves. Common throughout most of United States and southern Canada.

Damage: Larvae chew large holes in leaves of cabbage family plants and many other vegetable crops. May destroy whole plants.

Life Cycle: Moths emerge from overwintering pupae in May and lay eggs on leaves; larvae feed 2-4 weeks, then pupate 10 days in cocoons attached to stems or leaves. Three to 4 generations per year.

Control: Handpick several times weekly; attract native parasitic wasps by planting pollen and nectar plants; till in crop residues before adults emerge in spring; spray larvae with BTK; spray with garlic oil, pyrethrin, or sabadilla.

Cabbage Maggot *Delia radicum (=Hylemya brassicae)*

Larvae on roots

Larva

Adult female

Description: Adults: gray, ¼" flies with long legs. Larvae: white, tapering, ¼" maggots in roots. Found throughout North America.

Damage: Maggots boring into roots of cabbage family plants ruin root crops and stunt or kill plants. Wounds allow disease organisms to enter roots. First sign of injury is usually wilting in midday.

Life Cycle: Adults emerge from overwintering pupae from late March onward. Females lay eggs in soil beside roots; larvae tunnel in roots 3-4 weeks, then pupate in soil for 2-3 weeks. Two to 4 generations per year.

Control: Cover seedlings with floating row cover, burying edges in soil; set out transplants through slits in tar-paper squares to prevent the females from laying eggs; burn or destroy roots of cabbage family plants when harvesting tops; apply parasitic nematodes to soil around roots. If populations are moderate, repel females by mounding wood ashes, diatomaceous earth, hot pepper, or powdered ginger around base of stems.

Cankerworms (Family Geometridae)

Larvae and damage

Description: Adults: males are light gray moths (more than 1″ wingspan); females are wingless with fuzzy, ½″ bodies. Larvae: slender, light green, yellow, or brown, ½″-1″ caterpillars with white stripes; they loop their bodies as they crawl. Eggs: gray-brown, round; laid in compact masses on plants. Found from Nova Scotia to North Carolina, west to Missouri, Montana, and Manitoba. Also found in Colorado, Utah, and California.

Damage: Larvae chew on young leaves and buds of apple trees and many deciduous shade trees and ornamental shrubs. They also feed on larger leaves, leaving only midribs and large veins. Heavily damaged trees look scorched.

Life Cycle: Adults emerge November to December and lay eggs on twigs and branches; eggs hatch in spring as first leaves open on trees. Larvae feed 3-4 weeks, then pupate in soil until early winter. One generation per year.

Control: Trap females in sticky tree bands as they climb trees to lay eggs; handpick and destroy egg masses on branches; spray with dormant oil to kill eggs; spray BTK to kill larvae.

Carrot Beetle *Ligyrus (=Bothynus) gibbosus*

Adult

Description: Adults: reddish brown or black, ½″ beetles with rows of fine punctures on wing covers. Larvae: bluish white, C-shaped, 1″ grubs with brown heads. Found throughout North America.

Damage: Adults feed on roots of carrot family crops, beets, corn, potatoes, sweet potatoes, and dahlias. Larvae feed on roots of grasses and cereal crops and on decaying matter. Damage worst in soils with high organic matter content.

Life Cycle: Adults overwinter in soil, emerging in spring to lay eggs in soil beside host plants. Eggs hatch in 1-3 weeks, larvae feed on roots until they pupate in late summer. One generation per year.

Control: Cultivate in fall to reduce overwintering populations; rotate crops.

Carrot Rust Fly *Psila rosae*

Adult

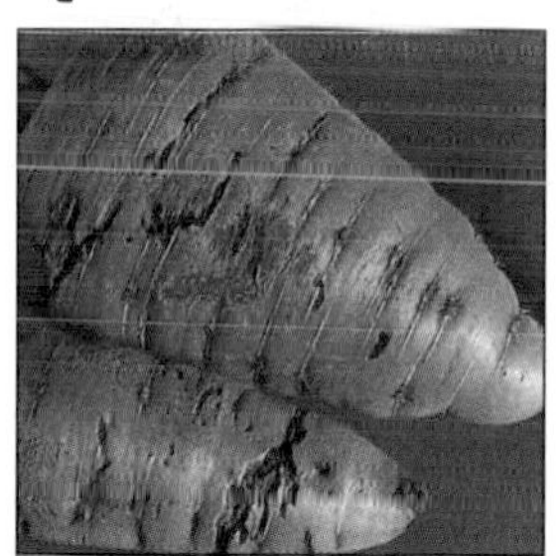

Damage

Description: Adults: shiny, metallic greenish black, ¼″ flies with yellow legs and head. Larvae: white, tapering, ⅓″ maggots. Found throughout North America.

Damage: Maggots eat root hairs and tunnel through roots of carrot family plants, stunting or killing plants, ruining root crops, and allowing disease organisms to enter. Maggots also feed on roots in storage.

Life Cycle: Adults emerge mid-April to May, laying eggs in soil near plants. Larvae burrow into roots for 3-4 weeks, then pupate. Two to 3 generations per year.

Control: Cover seedbeds with floating row cover, burying edges in soil, before seedlings emerge. Leave plants covered until harvest; apply parasitic nematodes to soil.

Carrot Weevil *Listronotus oregonensis*

Larva

Damage

Description: Adults: coppery brown, hard-shelled, 1/6″ weevils. Larvae: white, legless, C-shaped, 1/3″ grubs with brown heads. Found in New England and eastern United States.

Damage: Larvae tunnel through stems and roots of carrot family plants, stunting or killing plants. Infested carrot roots and celery stalks may be unfit to harvest.

Life Cycle: Overwintering adults emerge from grass or garden litter in May, lay eggs on plant stems; larvae bore into stems and downward into roots, then pupate in soil by late June. Second-generation adults emerge in July.

Control: Cover seedbeds with floating row cover; weevils do not fly, so plant susceptible crops in uninfested areas; drench soil with parasitic nematodes.

Centipedes/Millipedes

Garden centipede

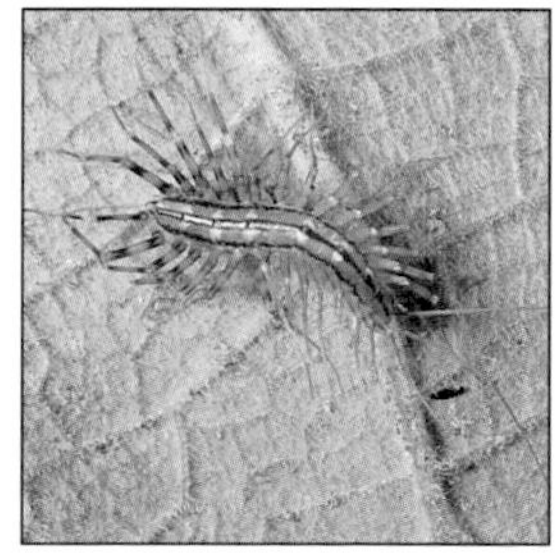

House centipede

Millipede

Description: Adults: slender, segmented creatures with many legs. Centipedes are 1″-5″, have fewer, longer legs (with only 1 set of legs per segment), and move quickly; millipedes are 1/2″-1 1/2″, move slowly, and have numerous short legs (with 2 sets of legs per segment). Some millipedes curl up when disturbed.

Beneficial Effect: Both centipedes and millipedes are generally beneficial in the garden. Centipedes generally feed on soil-dwelling mites, insects, and insect larvae.

Millipedes usually feed on decaying plant tissue; they are beneficial in compost piles because they help break down organic matter.

Damage: Centipedes occasionally feed on plants and earthworms. Millipedes may feed on plant roots, germinating seeds, and seedlings.

Life Cycle: Both lay eggs in soil. Eggs hatch into nymphs, which are similar to adults, but shorter, with fewer segments.

Control: Usually not needed; sprinkle wood ashes, diatomaceous earth, or cinders near rows of germinating seeds.

Chinch Bug *Blissus leucopterus*

Adult

Damage

Description: Adults: 1/6" bugs that have white forewings with a black triangular spot near margin. Nymphs: 1/8" insects; youngest are bright red with a white stripe across the back; older nymphs are dark with white spots on the middle. Most common in eastern half of United States and Canada.

Damage: Adults and nymphs suck sap from roots and stems of lawn grasses, corn, and cereal grain crops. Infested grass turns yellow, and patches may die off. Large infestations can devastate a grain crop or lawn. These pests usually congregate in open, sunny parts of the lawn. They also smell bad, especially when crushed, and you may be able to detect the odor when you walk across a severely infested lawn.

Life Cycle: Adults emerge from overwintering sites in sod along fence rows and hedges. Females lay eggs on grass roots, eggs hatch in 1-3 weeks, and nymphs chew on roots until molting to adults in late June. Two generations per year, 3 in southern areas.

Control: Avoid chinch bug problems by planting endophyte-containing grass cultivars. In small lawns, soak sod with soapy water (1 ounce liquid dish soap to 2 gallons water), then lay a flannel sheet over the grass to snare the bugs as they are driven out by the soapy solution; kill the bugs by washing them off the sheet in a bucket of soapy water. Encourage native predators—bigeyed bugs, minute pirate bugs, lacewings, lady beetles, and birds; bugs avoid shade, so shade base of crop plants by interplanting soybeans with corn or clover with grains.

Codling Moth *Cydia pomonella*

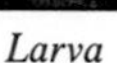

Larva

Adult

External damage

Internal damage

Description: Adults: gray-brown moths; forewings with fine, white lines and brown tips, hindwings brown with pale fringes (3/4″ wingspan). Larvae: pink or creamy white, 7/8″ caterpillars with brown heads. Eggs: flattened, white. Found throughout North America.

Damage: Larvae tunnel through apple, apricot, cherry, peach, pear, and plum fruit to center, ruining them.

Life Cycle: Overwintering larvae pupate in spring; adults emerge when apple trees bloom. Females lay eggs on fruit, leaves, or twigs; larvae burrow into fruit core, usually from blossom end, for 3-5 weeks, then leave fruit to pupate under tree bark or in ground litter. Two to 3 generations per year, 5-8 weeks apart.

Control: In early spring, scrape loose bark to remove overwintering cocoons and spray dormant oil; grow cover crops to attract native parasites and predators, especially ground beetles that eat pupae; use pheromone traps to determine main flight period for moths, then release parasitic *Trichogramma* wasps to attack eggs; trap larvae in tree bands, destroy daily; apply codling moth granulosis virus where available; in areas with severe infestations, spray ryania when 75 percent of petals have fallen, followed by 3 sprays at 1-2 week intervals.

Colorado Potato Beetle *Leptinotarsa decemlineata*

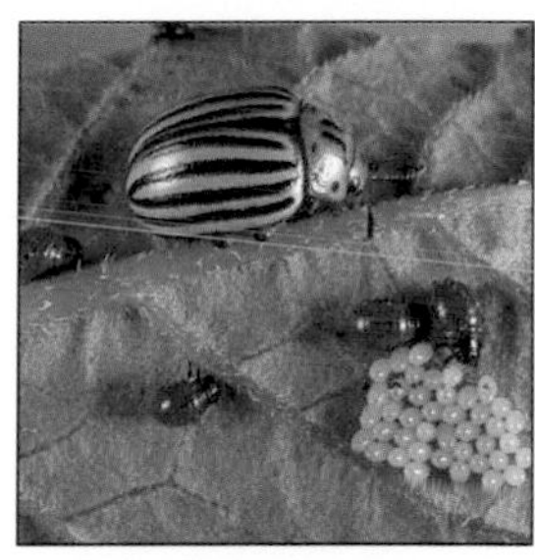

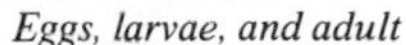

Eggs, larvae, and adult

Larva

Adult

Adults and larvae

Description: Adults: yellowish orange, 1/3" beetles with 10 lengthwise, black stripes on wing covers, black spots on thoraxes. Larvae: dark orange, humpbacked, 1/16"-1/2" grubs with a row of black spots along each side. Eggs: bright yellow ovals, standing on end in clusters on undersides of leaves. Found throughout North America.

Damage: Both adults and larvae chew leaves of potatoes, tomatoes, eggplants, and related plants, including petunias. Feeding can kill small plants and reduces yields of mature plants.

Life Cycle: Overwintering adults emerge from soil in spring to feed on young plants; after feeding, females lay up to 1,000 eggs during their lifespan of several months. Eggs hatch in 4-9 days; larvae feed 2-3 weeks, then pupate in soil. Adults emerge in 5-10 days. Two generations in most areas, 3 generations in southern states.

Control: When overwintering adults begin to emerge, shake adults from plants onto a dropcloth in the early morning. Dump beetles into soapy water. Attract native predators and parasites with pollen and nectar flowers; mulch plants with deep straw layer; cover plants with floating row cover until midseason; release 2-5 spined soldier bugs per square yard of plants; release parasitic wasp *Edovum puttleri* in southern areas to attack second-generation larvae; apply parasitic nematodes to soil to attack larvae as they prepare to pupate; apply double-strength sprays of BTSD on larvae; spray weekly with pyrethrin, rotenone, or neem.

Corn Earworm/Tomato Fruitworm

Helicoverpa (=Heliothis) zea

Larva and external damage

Larva and internal damage

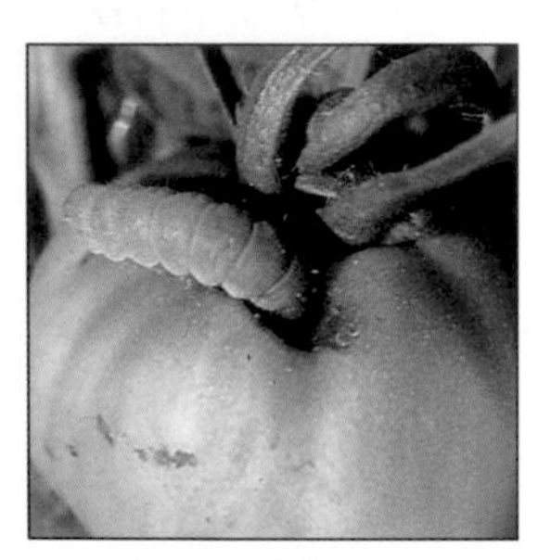

Larva and damage

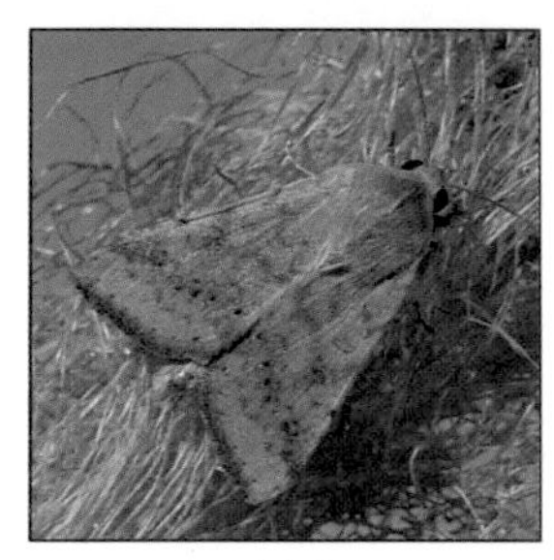

Adult

Description: Adults: tan moths (1½″-2″ wingspan). Larvae: 1″-2″ long, light yellow, green, pink, or brown; white and dark stripes along sides; yellow heads and black legs. Eggs: white, ribbed, and round. Found throughout North America; cannot overwinter in Canada, but migrates from United States in spring.

Damage: Larvae burrow into ripe tomatoes, eat buds, and chew large holes in leaves. In corn, larvae feed on fresh silks, then move down ears eating kernels, leaving trails of excrement. Early and late corn cultivars most affected. Larvae will also feed on a broad range of vegetable crops, fruits, and flowers.

Life Cycle: Adults emerge in early spring, migrating long distances to find food, if necessary. Females lay eggs on leaves or on tips of corn ears. Eggs hatch in 3 days, larvae feed 2-4 weeks, then pupate in soil. Adults emerge in 10-25 days. One to 4 generations per year.

Control: Plant corn cultivars with tight husks to prevent larvae from entering. After corn silks start to dry, spray BTK into tips of ears, or apply granular BTK; spray BTK on leaves and fruit of plants where fruitworms are feeding; attract native parasitic wasps and predatory bugs. Squirt parasitic nematodes into tips of corn ears or squirt mineral oil on the tips; open corn husks and dig out larvae in tip before they damage main ear; release lacewings or minute pirate bugs. Paint pyrethrin-and-molasses bait (3 parts spray solution to 1 part molasses) around base of plants to kill emerging adults; use pheromone traps to monitor appearance of moths; spray ryania or neem.

Cucumber Beetle, Spotted/Southern Corn Rootworm *Diabrotica undecimpunctata howardi*

Adult

Damage

Description: Adults: greenish yellow, 1/4″ beetles with 11 black spots on wing covers. Larvae: white, 1/2″ grubs with brown heads and brown patches on first and last segments. Found throughout the United States and southern Canada, east of the Rocky Mountains. Northern populations migrate north and south as seasons change, traveling up to 500 miles in a few days.

Damage: Larvae feed on roots of corn, often killing young plants; older plants are weakened and may topple easily. Adults feed on leaves and sometimes petals of squash family plants, other vegetable crops, and flowers. Both larvae and adults can transmit cucumber mosaic virus and bacterial wilt.

Life Cycle. Overwintering adults emerge from under crop residues in spring, lay eggs in soil close to plants. Eggs hatch and larvae feed in roots and crown of plants for 2–4 weeks, then pupate. One or 2 generations in northern areas, 3 in southern areas.

Control: Remove and destroy crop residues where adults overwinter; rotate garden crops with cover crops such as alfalfa; cover seedlings or plants with floating row cover, hand-pollinating covered squash family plants; apply parasitic nematodes to soil weekly to control larvae; control adults by spraying pyrethrin or rotenone.

Cucumber Beetle, Striped *Acalymma vittatum*

Adult and damage

Description: Adults: yellow, elongate, 1/4″ beetles with black heads and 3 wide black stripes on wing covers. Larvae: slender, white grubs. Found in United States west to Colorado and New Mexico; in Canada, west to Saskatchewan.

Damage: Adults feed on squash family plants, beans, corn, peas, and blossoms of many garden plants. The beetles swarm on seedlings, feeding on leaves and young shoots, often killing plants; they also attack stems and flowers of older plants and eat holes in fruit. Feeding can transmit wilt and mosaic viruses. Larvae feed on roots of squash family plants only, killing or stunting plants.

Life Cycle: Adults overwinter in dense grass or under leaves, emerging in April to early June. They eat weed pollen for 2 weeks, then move to crop plants, laying eggs in soil at base of plants. Eggs hatch in 10 days; larvae burrow into soil, feed on roots for 2-6 weeks, pupate in early August. Adults emerge in 2 weeks to feed on blossoms and maturing fruit. One to 2 generations per year.

Control: Cover seedlings or plants with floating row cover, and hand-pollinate covered squash family plants; pile deep straw mulch around plants to discourage beetles from moving between plants; apply parasitic nematodes to soil to control larvae; spray with sabadilla or rotenone when adults are seen feeding on pollen in flowers.

Cutworms (Family Noctuidae)

Army cutworm and damage

Black cutworm and damage

Variegated cutworm

Variegated cutworm adult

Description: Adults: brown or gray moths (1½″ wingspan). Larvae: fat, greasy-looking, gray or dull brown, 1″-2″ caterpillars with shiny heads. Found throughout North America.

Damage: At night, caterpillars feed on stems of vegetable and flower seedlings and transplants near the soil line, severing them or completely consuming small seedlings. During the day they rest below soil surface, curled beside plant stems.

Life Cycle: Some species overwinter as pupae; adults emerge and lay eggs on grass or soil surface from early May to early June. Eggs hatch in 5-7 days, larvae feed on grass and other plants for 3-5 weeks, then pupate in soil. Adults emerge late August to early September. Other species overwinter as eggs that hatch during first warm days and feed on early seedlings. One generation per year; a late second generation may damage crops in warm fall weather.

Control: Put collars made of paper, cardboard, or plastic around transplant stems at planting, pushing collars into soil until about half of the collar is below soil level. One week before setting out plants, scatter moist bran mixed with BTK and molasses over surface of beds; apply parasitic nematodes to soil; dig around base of damaged transplants in the morning and destroy larvae hiding below soil surface; set out transplants later in the season to avoid damage.

Damsel Bugs (Family Nabidae)

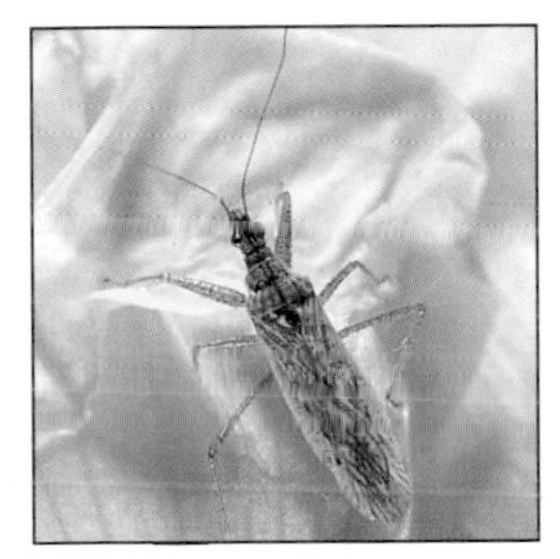

Adult

Description: Adults: elongate, gray or brown, fast-moving, 3/8″–1/2″ bugs. Nymphs: slender, wingless, smaller than adults. Found throughout North America.

Beneficial Effect: Important native predators of aphids, leafhoppers, plant bugs, thrips, and small caterpillars. Commonly found in unsprayed alfalfa fields.

Life Cycle: Females lay eggs in plant tissue, eggs hatch in 1 week; nymphs immediately begin feeding, often on prey larger than themselves. Nymphs develop 3-4 weeks until molt to adults. Adults overwinter. Two or more generations per season.

How to Attract: You can collect damsel bugs in alfalfa fields using a sweep net, and release them around the garden.

Diamondback Moth *Plutella xyllostella*

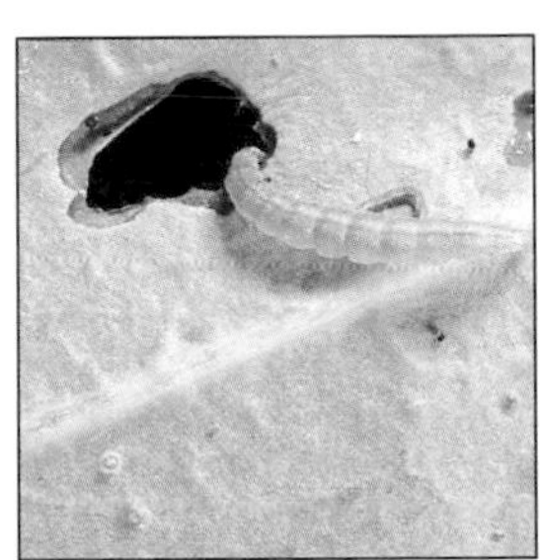

Larva and damage

Adult

Description: Adults: 1/2″ moths with light diamond pattern visible on the back when wings are folded; wing tips flare upward. Larvae: 5/16″ and pale green with light brown heads. Found throughout North America.

Damage: Youngest larvae mine tunnels in leaves of cabbage family plants and weeds.

Older larvae chew small, irregular holes in leaves, bore into cabbage heads, and chew curds of cauliflower and broccoli.

Life Cycle: Adults overwinter in mild climates. In spring females lay eggs on leaves; eggs hatch in a few days; larvae feed for 2 weeks, then pupate. Adults emerge in 7-10 days. Three to 6 generations per year.

Control: Cover crops with floating row cover; attract and conserve native parasites and predators, including birds; spray BTK or sabadilla to kill larvae.

Elm Bark Beetles

Scolytus multistriatus and *Hylurgopinus rufipes*

Smaller European elm bark beetle

Damage

Description: Adults: shiny, cylindrical, dark reddish brown, 1/10″ beetles with fine punctures in rows along wing covers; heads curved downward under broad thoraxes. Larvae: fat, C-shaped grubs with brown heads, living under bark. Found throughout United States except in the north-central states; also found in southern Ontario and New Brunswick.

Damage: Larvae and adults live in galleries engraved in a radiating pattern under the bark of elm trees. The tunneling does not cause serious damage to the trees. However, adults carry Dutch elm disease fungi on their bodies and infect healthy trees when they bore into bark.

Life Cycle: Overwintering adults emerge from holes in bark in spring; they feed in crotches of elm twigs, later moving to recently cut, dead, or dying elms. They carve galleries between the wood and inner bark and lay eggs. Each larva feeds in a separate branch of the gallery and pupates in a cell at the end; adults emerge in 10-14 days. Some species overwinter under the bark as larvae, others as adults. One to 3 generations per year.

Control: Maintain healthy trees; bury or burn all diseased or dying elms in the area to eliminate sources of disease; plant cultivars resistant to Dutch elm disease; trap adult beetles with pheromone traps; conserve braconid wasps (*Dendrosoter protuberans*), which were released in the eastern United States to parasitize the beetles.

European Corn Borer *Ostrinia nubilalis*

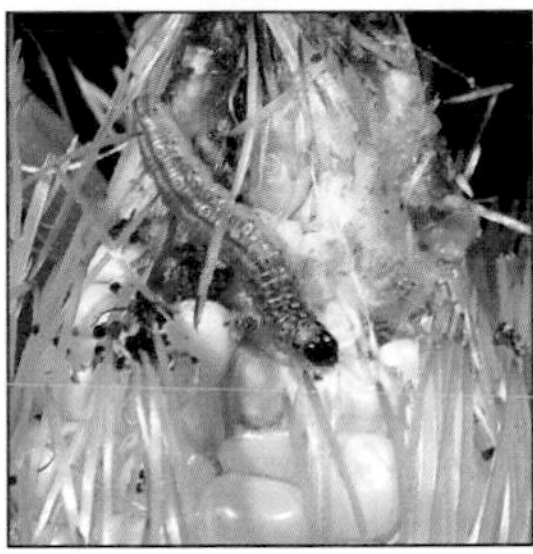

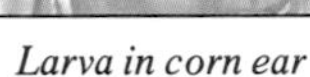

Larva in corn ear

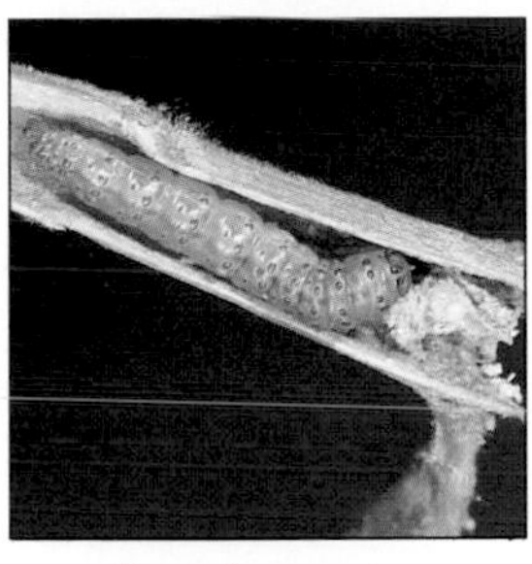

Larva in corn stem

Adult

Description: Adults: females pale yellowish brown with darker zigzag patterns across wings (1″ wingspan); males darker-colored. Larvae: beige with small brown spots, up to 1″. Eggs: white, overlapping, laid in masses of 15-20 on undersides of leaves. Found throughout northern and central United States and central and eastern Canada.

Damage: Young larvae feed on corn leaves and tassels and beneath husks. Older larvae burrow in corn stalks and ears; damaged stalks may break. Larvae also tunnel in stems or pods of beans, onions, peppers, potatoes, tomatoes, and other crops.

Life Cycle: Larvae overwinter in plant residue and pupate in early spring. Adults emerge in June; lay eggs late June to mid-July. Eggs hatch after 1 week; larvae feed for 3-4 weeks. One to 3 generations per year.

Control: Plant resistant corn cultivars; remove tassels from two-thirds of corn plants before they begin to shed pollen; spray BTK on leaf undersides and into tips of ears; apply granular BTK or mineral oil in tips of ears; rotate crops; release *Trichogramma* wasps for control in large fields; attract native parasites by allowing flowering weeds to grow between rows; pull out and destroy all infested crop residue immediately after harvest. For severe infestations, spray pyrethrin, ryania, or sabadilla when larvae begin feeding.

Flatheaded Appletree Borer *Chrysobothris femorata*

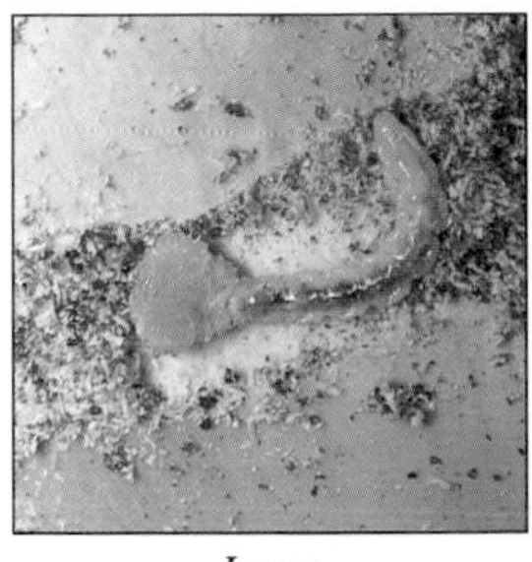

Larva

Adult

Description: Adults: flat, dark bronze, ½″ beetles. Larvae: white, legless grubs with brown, retracted heads; up to 1¼″. Found throughout United States and eastern Canada; similar species on Pacific Coast.

Damage: Adults feed on leaves of most

fruit and shade trees, and larvae tunnel into sapwood of young trees and under bark of older trees, forming galleries filled with castings. Attacked bark exudes gummy sap, turns dark, and dies; whole trees may be killed. Young trees or those in poor condition most susceptible.

Life Cycle: Grubs overwinter in chambers in wood, pupate in spring. Adults emerge May to July and lay eggs in cracks in bark. When eggs hatch, grubs tunnel under bark for rest of summer but are usually unable to complete development on vigorous trees. May take 2 years to complete life cycle.

Control: Maintain healthy, vigorous trees; avoid injury to bark; remove injured limbs as soon as damage occurs; protect trunks of young trees with white latex paint diluted with an equal amount of water, or wrap with paper or burlap to prevent attack.

Flea Beetles (Family Chrysomelidae)

Steelblue flea beetle

Damage

Description: Adults: black, brown, or bronze, 1/10″ beetles with well-developed hind legs; jump like fleas when disturbed. Larvae: thin, white, legless grubs with brown heads, up to 3/4″, living in soil. Found throughout North America.

Damage: Adults chew numerous small, round holes in leaves of most vegetable crops as well as many flowers and weeds. They are most damaging in early spring. Seedlings may be killed, larger plants usually survive. Larvae feed on plant roots. Adults may spread viral diseases as they feed.

Life Cycle: Overwintering adults emerge from soil in spring; they feed and lay eggs on plant roots, then die by early July. Eggs hatch in 1 week, larvae feed 2-3 weeks, then pupate in soil; adults emerge in 2-3 weeks. One to 4 generations per year.

Control: Delay planting to avoid peak populations; cover seedlings with row cover until adults die off. Flea beetles prefer full sun, so interplant crops to provide shade for susceptible plants; drench roots with insect parasitic nematodes to control larvae; spray with neem, pyrethrin, rotenone, or sabadilla.

Fruit Borers *Synanthedon* spp.

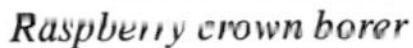
Raspberry crown borer

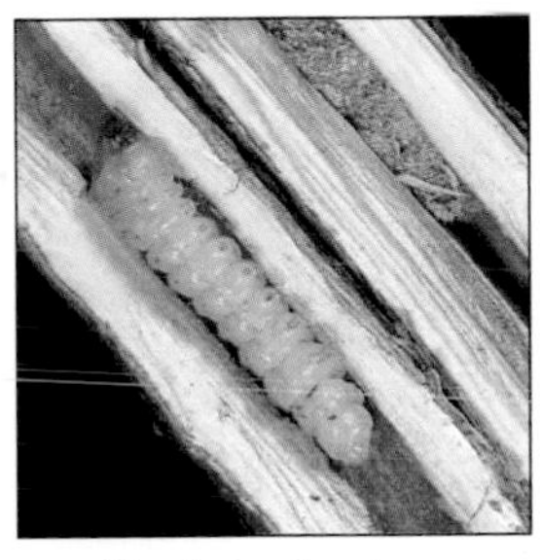
Rhododendron borer

Rhododendron borer adult

External damage

Description: Adults: wasplike moths have clear wings with darker markings and black or black-and-yellow striped, 1″ bodies. Larvae: pale yellow or white with dark heads, boring into canes or crowns. Found throughout North America.

Damage: Larvae bore into stems of currants, gooseberries, raspberries, and rhododendrons. Their tunneling can weaken canes, causing them to break easily, or can kill entire canes. Larvae boring into crowns girdle canes and destroy new shoots. These species are related to peachtree borers.

Life Cycle: Larvae overwinter in canes and pupate in early May; adults emerge in 2 weeks. Eggs are laid on canes and hatch in 10 days; larvae tunnel in canes all summer, remaining in canes for winter. Rhododendron borer larvae spend the winter in crowns of plants and work their way up into canes by July, pupating under the bark, several inches above the soil line; adults emerge in a month.

Control: Prune and burn all affected canes, remove infested plants; smash old stubs of plants with a mallet to kill pupae and larvae; try spraying superior oil in late May to kill eggs.

Fruit Flies *Rhagoletes* spp. and *Ceratitis capitata*

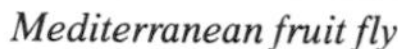
Mediterranean fruit fly

Walnut husk fly

Damage

Currant fruit fly

Description: Adults: 1/4" flies with yellow or white markings on their bodies, and transparent wings patterned with dark, crosswise bands. Larvae: white, 1/4" maggots feeding in fruit. Found throughout North America.

Damage: Larvae tunnel extensively through fruit of blueberry, currant, plum, cherry, and peach. Certain species also feed on walnut, citrus, and coffee. Infested fruit shrivels or drops early; earliest cultivars suffer the most damage. In walnut, main injury is shell staining.

Life Cycle: Adult flies emerge from pupae mid-June or later, lay eggs in punctures in fruit skin or around stems. Eggs hatch within 1 week, larvae tunnel in fruit until it drops, then leave fruit to pupate and overwinter in the soil. Most species have 1 generation per year; in very warm regions generations may continue all year.

Control: Pick up fallen fruit and destroy daily during summer, weekly in fall; hang yellow sticky traps baited with vials of 1 part ammonia and 1 part water or commercial fruit fly attractants in trees (1-2 traps per tree); encourage ground beetles and rove beetles, which feed on fly pupae, by planting groundcovers in orchards.

Fruittree Leafroller *Archips argyrospila*

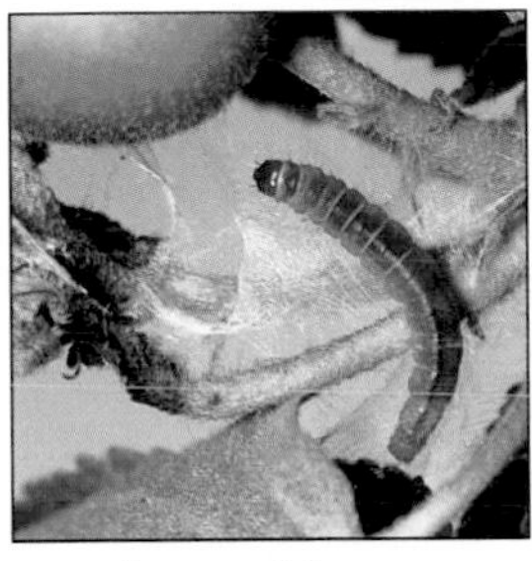
Larva and damage

Adult

Description: Adults: golden brown, mottled moths (3/4" wingspan). Larvae: green caterpillars with brown heads. Eggs: light yellowish brown with brown coating. Found throughout North America.

Damage: Larvae spin webs at branch tips of roses and most fruit and ornamental trees and feed on enclosed buds, leaves, and developing fruit.

Life Cycle: Eggs hatch in early spring. Larvae feed for 1 month, spin webs, and pupate within rolled leaves or in cocoons on bark. Adults emerge late June or July and lay overwintering eggs on bark.

Control: Scrape egg masses from branches in winter; apply dormant oil sprays to kill eggs; handpick caterpillars from young trees weekly; attract native parasitic wasps; apply BTK to larvae before they spin webs; spray pyrethrin or rotenone.

Gall Wasps (Family Cynipidae)

Gall

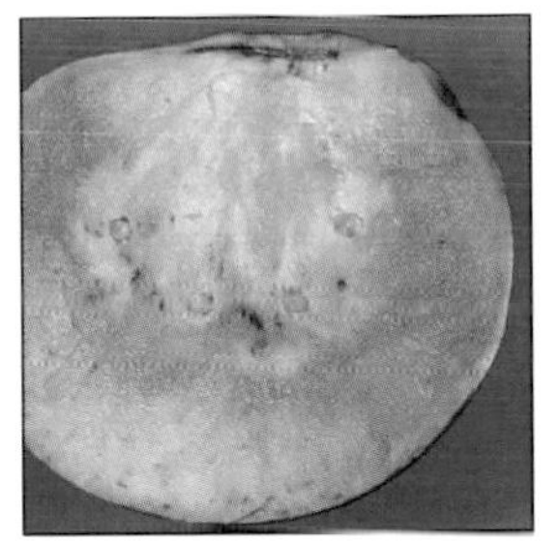
Gall and larvae

Galls

Description: Adults: brown or reddish, tiny wasps, rarely seen. Larvae: legless white grubs. Most common in western United States and Canada; some occur in the East.

Damage: Larvae feed on oaks, roses, thistles, and other plants. Plants respond by producing galls—enlarged masses of cells—of various shapes, attached to stems or leaves.

Life Cycle: Overwintering adults emerge from winter galls, usually on roots or fallen leaves; females lay eggs on host plants in early spring; feeding larvae stimulate gall formation, which serves as food and protects larvae. Adults emerge in summer and lay eggs that form overwintering galls.

Control: Usually not necessary; prune galls from roses or shrubs and destroy.

Grasshoppers (Family Acrididae)

Adult

Description: Adults: brown, yellow, or green, 1″-2″ insects with leathery forewings and enlarged hind legs; many have brightly colored underwings. Nymphs: similar to adults, but smaller. Found throughout North America.

Damage: Adults eat any kind of vegetation. In most areas of North America, economic damage occurs only in fields of grass or cereal crops. Little damage occurs to home gardens; on rare occasions swarms of grasshoppers devastate agricultural crops over large areas.

Life Cycle: In late summer females deposit elongate masses of eggs in soil; eggs hatch in spring; nymphs develop for 40-60 days until molting to adults. Adults feed until killed by cold weather. Swarms appear as a result of interaction of weather and biological influences.

Control: Usually controlled by natural enemies (blister beetle larvae, ground beetles, predatory flies, birds, parasitic nematodes, fungal dieases); cultivate fields in fall to kill overwintering eggs; aerial sprays of commercial protozoan disease (*Nosema locustae*) may be effective over large areas but is not useful on a home-garden scale.

Ground Beetles (Family Carabidae)

Larva and prey

Larva

Adult

Description: Adults: blue-black or brown, 3/4″-1″ beetles, usually iridescent; thorax well-defined, usually narrower than abdomen. Beetles hide under stones or other cover during the day. Larvae: dark brown or black grubs with 10 segments, tapering markedly toward the rear. Common throughout North America.

Beneficial Effect: There are more than 2,500 species of ground beetles. They prey on slugs, snails, cutworms, cabbage root maggots, and many other pests that have a soil-dwelling stage. Some species also pursue prey that live on plants or trees, such as Colorado potato beetle larvae, gypsy moths, and tent caterpillars. A single larva can eat more than 50 caterpillars; adults may live as long as 2-3 years and are fiercely voracious.

Life Cycle: Overwintering adults emerge from pupal cell and lay eggs in soil. Larvae feed on insects and slugs for 2-4 weeks, then pupate in soil. Adults remain in soil for the winter, emerging in spring.

How to Attract: Provide permanent beds and perennial plantings in garden to protect populations; plant white clover groundcover in orchards; make permanent stone, sod, or clover pathways throughout garden to provide refuges.

Gypsy Moth *Lymantria dispar*

Adult laying eggs

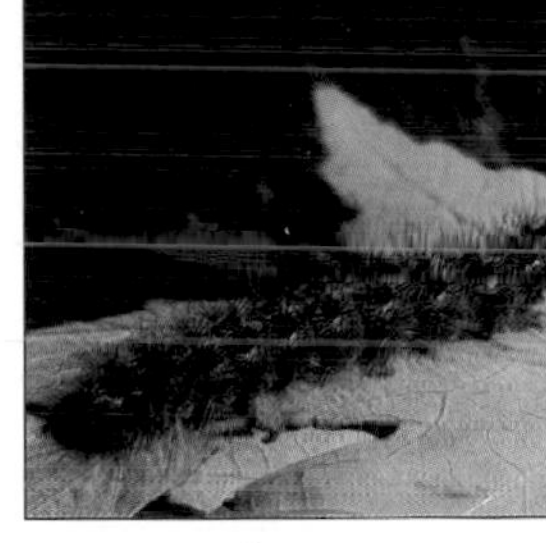

Larva

Adult

Description: Adults: females nearly white, 1″ moths with heavy bodies, unable to fly; males are smaller, darker, strong fliers. Larvae: up to 2½″, gray-brown caterpillars with 5 pairs of blue dots and 6 pairs of red dots on back, long hairs in tufts on body. Eggs: in masses under fuzzy yellow covering. Found in eastern and central United States and eastern Canada; isolated outbreaks in California and other western areas. Dispersal depends on human activity (trucking, camping, etc.) to move eggs, larvae, pupae to new areas.

Damage: Larvae feed on leaves of many trees and shrubs, including conifers; heavy infestations can defoliate trees. Repeated defoliations eventually kill deciduous trees; a single defoliation kills conifers.

Life Cycle: Overwintering eggs on tree trunks hatch in May, larvae feed in trees until mid-July, then pupate for several weeks. Adults emerge late July to early August. Females crawl up nearby trees or other objects to deposit egg masses that overwinter.

Control: Attract predators and parasites (especially tachinid flies, ground beetles, and parasitic wasps); prevent movement of pest by checking trailers, boats, camping gear, and so on for egg masses or larvae before leaving an infested region; use pheromone traps to catch males for monitoring or to prevent mating; spray BTK to kill larvae; wrap burlap tree bands around fruit and shade trees, check daily and destroy hiding larvae.

Honeybee *Apis mellifera*

Adult

Adult

Description: Adults: gold-and-black striped, 3/4″ bees with translucent wings. Larvae: white grubs in wax combs in hives. Found throughout North America.

Beneficial Effect: Extremely important pollinators of fruit, vegetables, and agricultural crops.

Life Cycle: Bees live in social colonies numbering up to 20,000. Queen bees lay eggs in wax cells in hives; workers feed and care for larvae, feeding those destined to become new queens a special diet of royal jelly. Males mate with new queens, who leave with swarms to start new colonies. Bees overwinter clustered in hives, living on stored honey.

How to Attract: Plant pollen and nectar plants; provide a water source in dry weather; avoid spraying fruit trees when flowers are in bloom; if you must apply insccticides, spray in evenings after bees return to hive.

Hover Flies/Flower Flies (Family Syrphidae)

Larva

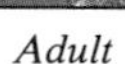

Adult

Adult

Description: Adults: yellow- or white-and-black striped, 1/2″-5/8″ flies, often seen hovering like hummingbirds over flowers. Larvae: gray or greenish, somewhat translucent, sluglike maggots. Eggs: white cylinders laid singly or in small groups near aphids. Various species common throughout North America.

Beneficial Effect: Larvae feed on many species of aphids. Common native predators in orchards.

Life Cycle: Females lay eggs among aphids; eggs hatch in 2-3 days; larvae feed on aphids for 3-4 weeks, then drop to the soil to pupate. Adults emerge after 2 weeks. Two to 4 generations per year.

How to Attract: Plant pollen and nectar flowers; allow flowering weeds such as wild carrot and yarrow to grow between crop plants.

Ichneumon Wasps (Family Ichneumonidae)

Adult

Adult

Description: Adults: slender, dark-colored, 1/10″–1½″ wasps with long antennae. Some species have threadlike ovipositors, as much as 3″ long (sometimes mistaken for stingers). Larvae: white, tapering grubs. Numerous species found throughout North America.

Beneficial Effect: Larvae develop as parasites of caterpillars, sawfly and beetle larvae, and other insects. Females also kill hosts by stinging them and feeding on body fluids. Ichneumon wasps are extremely important as native biological controls.

Life Cycle: Females lay eggs inside host eggs or larvae; wasp larvae develop inside hosts, eventually killing hosts and pupating in or on them. Many species overwinter as mature larvae in cocoons; in some species adult females overwinter. One to 3 generations per year usual; some with up to 10 generations.

How to Attract: Plant pollen and nectar flowers in gardens; grow flowering cover crops in orchards to attract females.

Imported Cabbageworm *Artogeia (=Pieris) rapae*

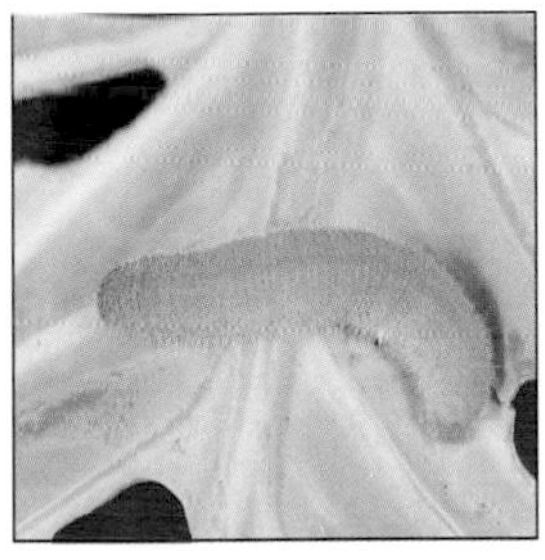
Larva

Adult

Description: Adults: common white butterflies; forewings with black tips and 2 or 3 spots (1½″–2″ wingspan). Larvae: up to 1¼″, velvety green caterpillars with a fine yellow stripe down the back. Eggs: yellow cones laid on undersides of leaves. Found throughout North America.

Damage: Larvae eat large, ragged holes in leaves and heads of cabbage family plants, soiling leaves with dark green droppings.

Life Cycle: Adults emerge from overwintering pupae in early spring to lay eggs. Larvae feed for 2-3 weeks, then pupate in debris on soil surface; adults emerge in 1-2 weeks. Three to 5 overlapping generations per year; all ages of larvae present all season.

Control: Cover plants with floating row cover; handpick larvae in light infestations; use yellow sticky traps to catch female butterflies; spray with BTK at 1-2 week intervals. As a last resort, spray sabadilla.

Iris Borer *Macronoctua onusta*

Larvae and damage

Description: Adults: moths with dark brown forewings, yellowish hind wings (2″ wingspan). Larvae: fat, pinkish borers with brown heads and a light stripe down the back, rows of black dots on sides; up to 2″ long. Found in eastern United States west to Iowa, and in Quebec and eastern Ontario.

Damage: Young larvae tunnel within leaves, leaf sheaths, and buds of iris, moving down into crowns and rhizomes as they develop. Soft rots usually follow larval damage.

Life Cycle: Eggs overwintering on old leaves hatch in late April or early May. Larvae enter leaves, feed for several weeks, then pupate in soil near rhizomes; adults emerge in late summer and lay eggs.

Control: Remove and destroy dead iris leaves and stems in late fall to eliminate overwintering eggs; dig infested rhizomes, remove larvae and pupae, dip in sulfur fungicide or other antibiotic before replanting if soft rot present.

Japanese Beetle *Popillia japonica*

Larva

Adult

Adults and damage

Description: Adults: chunky, metallic blue-green, 1/2″ beetles with bronze wing covers, long legs, and fine hairs covering body. Larvae: fat, dirty white grubs with brown heads; up to 3/4″; found in sod. Found in all states east of the Mississippi River.

Damage: Adults eat flowers and skeletonize leaves of a broad range of plants; plants may be completely defoliated. Larvae feed on roots of lawn grasses and garden plants.

Life Cycle: Overwintering larvae deep in the soil move toward the surface in spring to feed on roots, pupating in early summer. Adults emerge, feed on plants, and lay eggs in late summer; eggs hatch into larvae that overwinter in soil. One generation occurs every 1-2 years.

Control: In early morning, shake beetles from plants onto dropcloths, then drown them in soapy water; cover plants with floating row cover; apply milky disease spores or parasitic nematodes to sod to kill larvae; attract native species of parasitic wasps and flies; organize a community-wide trapping program to reduce adult beetle population; spray plants attacked by beetles with rotenone.

June/May Beetles *Phyllophaga* spp.

Larva

Adult

Description: Adults: blocky, 3/4" beetles with tip of abdomen showing behind wings; most are shiny brown or black, some with stripes on back or fine hairs on body. Larvae: fat, C-shaped, white grubs with dark heads. Found throughout North America.

Damage: Larvae feed on roots of corn, potatoes, grasses, vegetable transplants, and strawberries. Adults feed on leaves of many species of trees and shrubs.

Life Cycle: Females lay eggs in balls of earth in soil; eggs hatch in 2-3 weeks. Grubs feed on decaying vegetation the first summer, hibernate in the soil and feed on plant roots the second summer. After hibernating again, they feed until June of the third summer, then pupate 2-3 weeks. Adults remain in pupal cells in the soil until spring of the fourth year, when they emerge to feed and lay eggs. Largest broods appear in 3-year cycle; some species with 1- or 4-year life cycles.

Control: Populations usually suppressed by native predators and parasites; where infestations are severe, apply milky disease spores or parasitic nematodes to the soil to control grubs.

Lace Bugs (Family Tingidae)

Adult

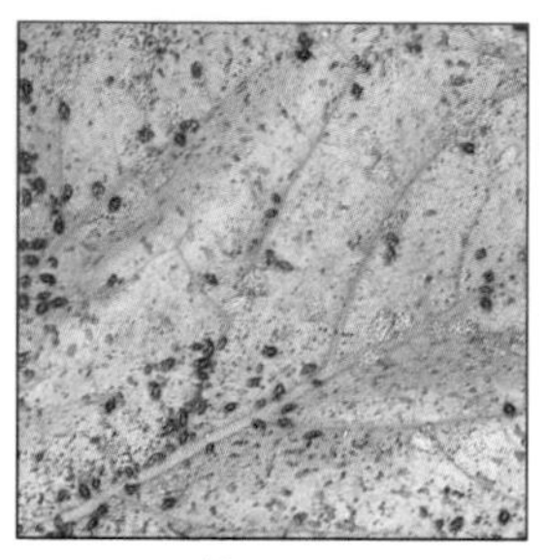
Damage

Description: Adults: oval or rectangular, 1/8″ bugs with lacy pattern and wide, flattened extensions on thoraxes. Nymphs: smaller, darker, covered with spines. Eggs: inserted in leaves along midribs on undersides with conelike caps projecting from leaves. Found throughout North America.

Damage: Adults and nymphs suck juices from undersides of leaves of flowers, trees, and vegetable plants, spotting leaves with excrement and leaving speckled white or gray, blotchy appearance on leaf surfaces.

Life Cycle: Most species overwinter in egg stage, some as adults under bark of trees. Eggs hatch into nymphs that feed for several weeks until molting to adults. Three or more generations per year.

Control: Spray superior oil (not on chrysanthemum flowers); spray pyrethrin or rotenone as a last resort.

Lacewings *Chrysoperla* (=*Chrysopa*) spp.

Common green lacewing larvae and eggs

Common green lacewing larva and prey

Common green lacewing

Description: Adults: fragile, green or brown, 1/2″-3/4″ insects with small heads, large eyes, and netted, transparent wings. Larvae: spindle-shaped, mottled yellow or brown. Eggs: laid on tips of fine stalks. Found throughout North America; sold commercially.

Beneficial Effect: Common general predators in gardens and orchards.

Life Cycle: Adults or pupae overwinter; adults emerge in spring to lay eggs. Eggs hatch in 4-7 days; larvae feed for about 3 weeks, then pupate for 5-7 days. Three to 4 generations per year.

How to Attract: Plant pollen and nectar flowers; allow some flowering weeds to grow between rows; provide water source; scatter purchased eggs widely throughout garden.

Lady Beetles (Family Coccinellidae)

Convergent lady beetle larva

Convergent lady beetle larvae pupating

Convergent lady beetle and prey

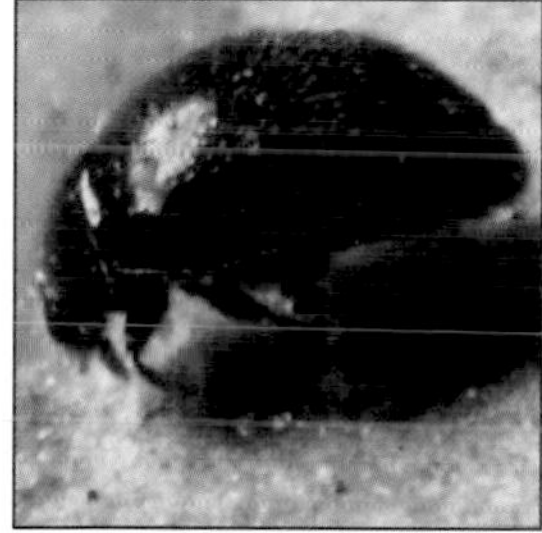

Spider mite destroyer

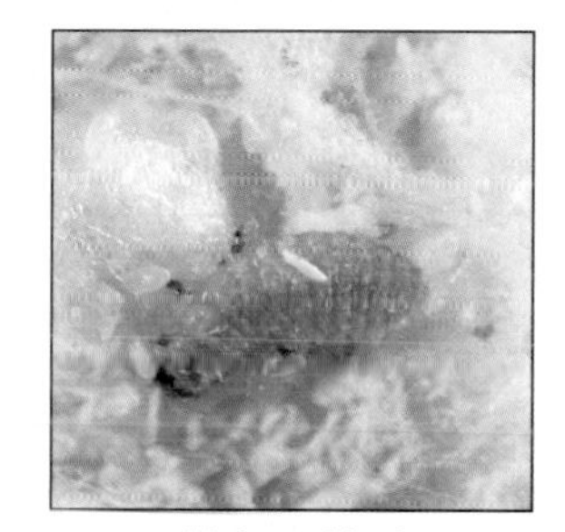

Twicestabbed lady beetle larva

Twicestabbed lady beetle

Description: Adults: shiny, round, 1/16"–3/8" beetles with short legs and antennae. Common species are pale yellow to dark reddish orange with or without black spots; some species solid black or black with red spots. Larvae: spindle-shaped, alligator-like, usually with short spines or knoblike projections on body. Eggs: white or yellow ovals, usually deposited in upright clusters. Numerous species common throughout North America; convergent lady beetle (*Hippodamia convergens*) sold commercially.

Beneficial Effect: Adults and larvae of many species feed on aphids and soft-bodied pests; some species prefer mealybugs, spider mites, or soft scales. Native lady beetles are important biological controls for aphids in gardens and orchards. In western regions, *Stethorus* spp. provide excellent control of spider mites in raspberries and other crops. *Chilocorus* spp. are voracious predators of soft scales. A few species, such as Mexican bean beetles, are plant pests.

Life Cycle: In spring, overwintering adults seek food, then lay eggs among aphids or other prey. Eggs hatch in 3–5 days, larvae feed 2–3 weeks, then pupate. Adults emerge in 7–10 days. In fall, local species overwinter as adults in leaf litter; *H. convergens* migrates to Sierra Nevada Mountains to overwinter in large groups.

How to Attract: Plant pollen and nectar flowers; leave weeds such as dandelions, wild carrot, and yarrow between crop plants; purchased *H. convergens* is effective in greenhouses with screened vents but is not advisable for release in gardens, because they will fly away.

Leafhoppers (Family Cicadellidae)

Beet leafhopper

Potato leafhopper

Damage

Damage

Redbanded leafhopper

Description: Adults: wedge-shaped, slender, green or brown, 1/10″–1/2″ insects; a forward point above the head is very pronounced in some species. Some have brightly colored bands on wings; all jump rapidly into flight when disturbed. Nymphs: pale, wingless, similar to adults; hop rapidly when disturbed. Found throughout North America.

Damage: Adults and nymphs suck juices from stems and undersides of leaves of most fruit and vegetable crops, also some flowers and weeds. Their toxic saliva distorts and stunts plants and causes tipburn and yellowed, curled leaves with white spots on undersides. Fruit may be spotted with drops of excrement and honeydew. Pests may spread viral diseases as they feed.

Life Cycle: Overwintering adults start laying eggs in spring when leaves begin to appear on trees. Some species do not survive winter in northern United States and in Canada; they migrate from the south every summer. Females lay eggs in leaves and stems; eggs hatch in 10-14 days, nymphs develop for several weeks. Most species have 2-5 generations per year, overwintering as adults or eggs.

Control: Wash nymphs from plants with stiff sprays of water; attract natural enemies (predatory flies and bugs and parasitic wasps); spray with insecticidal soap, pyrethrin, rotenone, or sabadilla.

Leafminers (Family Agromyzidae)

Adult

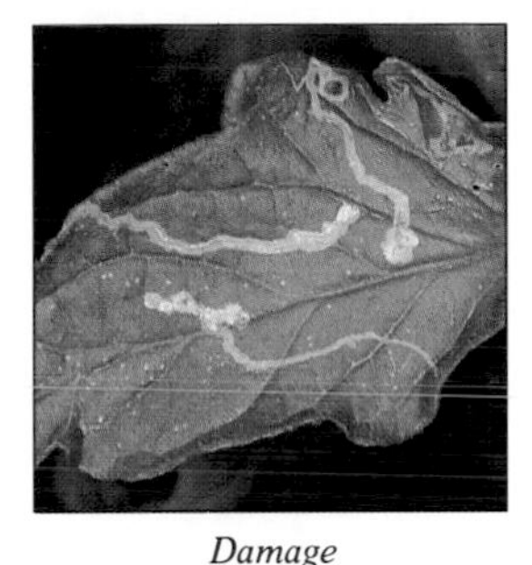
Damage

Damage

Description: Adults: black or black-and-yellow, 1/10" flies; rarely seen. Larvae: pale green, stubby, translucent, 1/8" maggots, found in tunnels in leaves. Eggs: white, cylindrical, laid in clusters on undersides of leaves. Several species found throughout North America.

Damage: Larvae tunnel within leaves of many vegetable crops and ornamentals. They feed on leaf tissue and make round or winding, hollow mines, often destroying seedlings. On larger vegetables, more of a nuisance than serious problem; damaging on ornamentals because mines are unsightly.

Life Cycle: Adults emerge from overwintering cocoons in early spring and lay eggs on leaves. Larvae mine leaves for 1-3 weeks, then drop to soil to pupate 2-4 weeks. Two to 3 generations per year, more in greenhouses.

Control: Cover seedlings with floating row cover; pick and destroy mined leaves and remove egg clusters; remove nearby dock or lamb's-quarters, which are hosts for beet leafminers; attract native parasitic wasps by planting nectar plants; spray with nicotine tea; spray with neem or avermectins.

Mealybug Destroyer *Cryptolaemus montrouzieri*

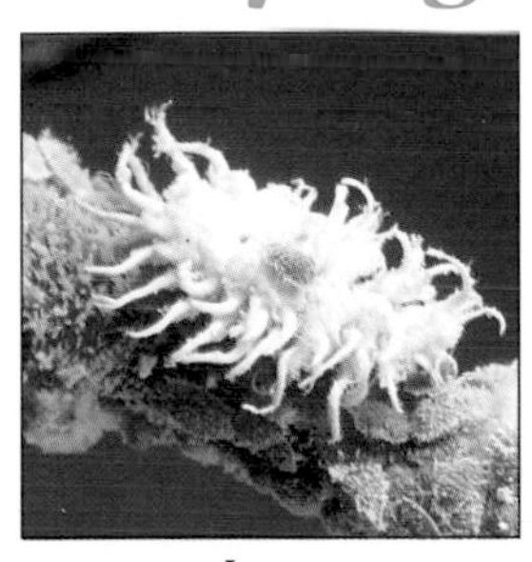
Larva

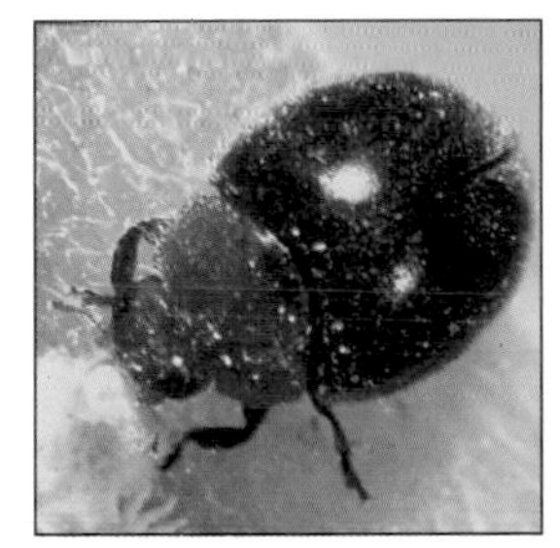
Adult

Description: Adults: oval, 1/3" beetles with black wing covers, head and tip of abdomen is coral-colored. Larvae: cream-colored, segmented, covered with long, waxy hair. Eggs: yellow ovals. Native to California and western coastal region of United States; sold commercially.

Beneficial Effect: Both adults and larvae prey on aboveground species of mealybugs on citrus, grapes, and ornamentals. Good controls for greenhouses and on houseplants.

Life Cycle: Females lay eggs among mealybug egg masses; eggs hatch in 8-10 days, larvae eat mealybug eggs and young mealybugs for 3 weeks, then pupate on plants. Adults emerge in 2-3 weeks. Adults overwinter in mild coastal climates; generations continue year-round indoors.

How to Attract: Conserve native populations by avoiding pesticides; release 250-500 in small orchards or vineyards; release 2-5 per infested plant in interior plantings, twice yearly; on houseplants, confine 10-20 adults per plant for 4-5 weeks by draping sheer curtain material over plant and tying it around the pot.

Mealybugs (Family Pseudococcidae)

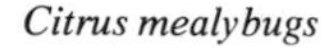

Citrus mealybugs

Longtailed mealybugs

Description: Adults: females are soft, oval, 1/10″ insects with distinctly segmented, pinkish bodies covered by white waxy fluff; males are tiny, 2-winged insects; rarely seen. Nymphs: similar to adults, but smaller. Most species found in southern United States, longtailed mealybug throughout North America; all species found in greenhouses.

Damage: Adults and nymphs suck plant juices from many types of fruit crops, avocados, potatoes, ornamentals, and tropical foliage plants. They feed on all parts of a plant, particularly new growth, causing leaves to wither and yellow and fruit to drop prematurely. Honeydew excreted on leaves supports the growth of sooty mold fungus.

Life Cycle: Females lay eggs in a fluffy white mass; eggs hatch in 10 days and crawlers wander away to find feeding sites where they develop for 1-2 months or longer. Several generations per year.

Control: Rinse plants with stiff streams of water to dislodge mealybugs; spray with insecticidal soap; release mealybug destroyers (*Cryptolaemus montrouzieri*) in citrus or grape plantings or on indoor plants; for citrus mealybugs release parasitic wasp *Leptomastix dactylopii;* attract native parasitic wasps, which usually keep populations in check outdoors.

Mexican Bean Beetle *Epilachna varivestis*

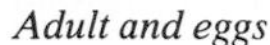

Adult and eggs

Larva

Larvae and damage

Adult and larvae

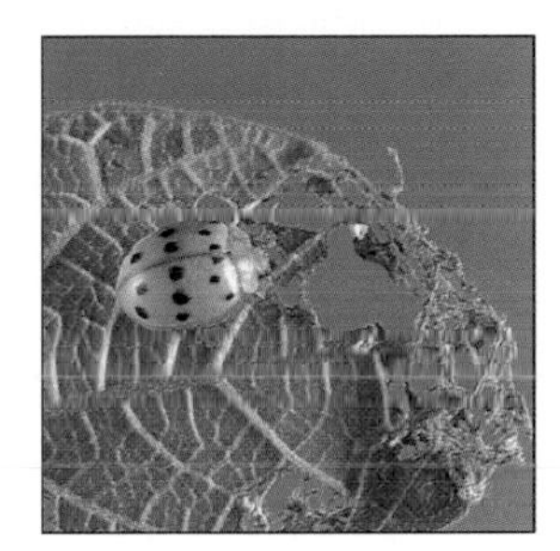

Adult and damage

Description: Adults: oval, yellowish brown, $1/4''$ beetles with 16 black spots on wing covers. Larvae: fat, yellowish orange, $5/16''$ grubs with long, branching spines. Eggs: yellow ovals laid on end on undersides of leaves. Found in most states east of the Mississippi River; also Texas, Arizona, Utah, Colorado, and Nebraska.

Damage: Both larvae and adults skeletonize leaves of cowpeas, lima beans, snap beans, and soybeans. They feed from the undersides of leaves, leaving characteristic lacy damage; severely defoliated plants may be killed. Beetles are most abundant in weedless fields.

Life Cycle: Adults overwinter in leaf litter in nearby fields; in spring, females lay eggs on beans. Eggs hatch in 5-14 days, larvae feed 2-4 weeks, then pupate on leaves. One to 3 generations per year.

Control: Plant early season bush beans to avoid main beetle generations; plant soybeans as trap crops, destroy them when infested with larvae; handpick larvae and adults daily in small bean patches; cover plants with floating row cover until plants are large enough to withstand damage; attract native predators and parasites by leaving flowering weeds between rows or by interplanting flowers and herbs; dig in crop residues as soon as plants are harvested; release spined soldier bugs (*Podisus maculiventris*) to control early generation; release parasitic wasps *Pediobius foveolatus* when weather warms; spray weekly with pyrethrin, sabadilla, rotenone, or neem.

Minute Pirate Bug *Orius tristicolor*

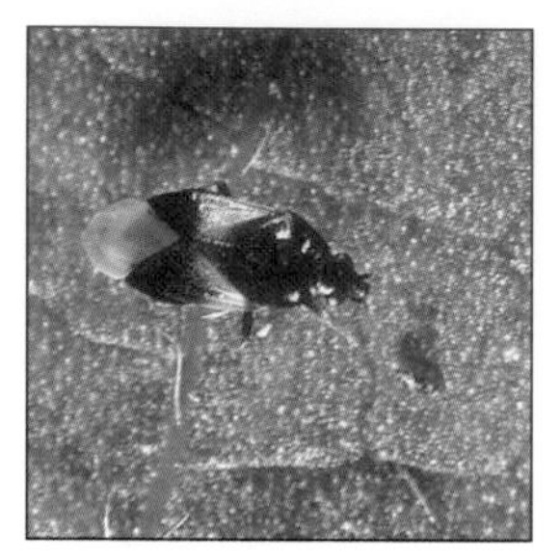

Adult

Description: Adults: quick-moving, black-and-white-patterned, 1/4″ bugs. Nymphs: shiny, wingless, changing from yellow through orange to mahogany brown as they grow. Found throughout North America; sold commercially.

Beneficial Effect: Voracious predators of thrips, spider mites, small caterpillars, leafhopper nymphs, other small insects, and insect eggs.

Life Cycle: Females lay eggs in plant stems or leaves; eggs hatch in 3-5 days and nymphs feed on insects in flowers and undersides of leaves for 2-3 weeks. Adult females overwinter in crevices of bark, weeds, and plant residues. Two to 4 generations per year.

How to Attract: Plant pollen and nectar plants, especially goldenrod, daisies, yarrow, alfalfa, and stinging nettle; in greenhouses, release at the rate of 1 pirate bug per 1-5 plants.

Mites, Gall (Family Eriophyidae)

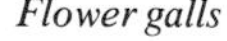

Flower galls

Maple velvet gall

Description: Adults: nearly invisible, elongate, cylindrical, pale yellow or tan mites with 2 pairs of legs at the front end. Nymphs: similar to adults, but smaller. Found throughout North America.

Damage: Mites burrow from undersides into leaf tissue of pears, currants, and many ornamentals. Leaves react by forming raised blisters, puckers, or galls along leaf margins. Some blisters, especially on maples, are bright red, others are yellow or brown.

Life Cycle: Numerous overlapping generations all season. Mites overwinter in crevices in bark, crawling onto new growth in the spring.

Control: Spray dormant oil or lime-sulfur on dormant plants.

Mites, Predatory (Family Phytoseiidae)

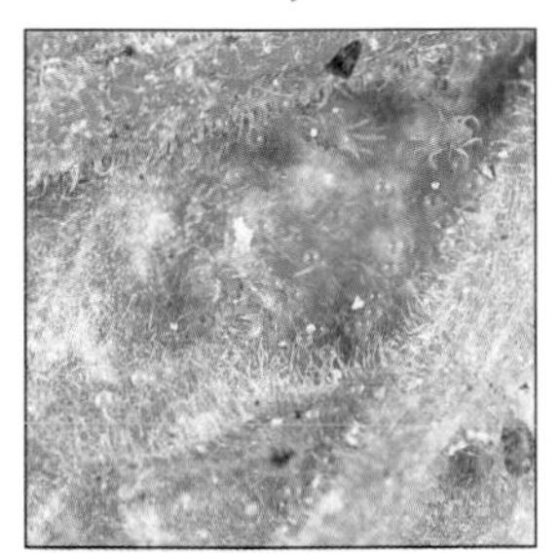

Adults and prey

Description: Adults: minute, beige to reddish tan, fast-moving mites. Nymphs: similar to adults, but smaller. Found throughout North America; several species sold commercially.

Beneficial Effect: These attack spider mites, especially European red mites and citrus red mites; some feed on pollen, thrips, or other types of mites.

Life Cycle: Overwintering females emerge from crevices in bark or soil litter and lay eggs on leaves among prey; nymphs hatch in 3-4 days, molting several times until they reach the adult stage in 5-10 days. Numerous overlapping generations.

How to Attract: Avoid pesticide use; sustain native species by sprinkling pollen (especially from ice plants, cattails, or dandelions) on plants. For apples and strawberries, release *Metaseiulus occidentalis* to control European red mite and other spider mites; in greenhouses, release *Phytoseiulus persimilis* or other species to control spider mites; and release *Amblyseius cucumeris* to control small thrips on peppers and cucumbers.

Mites, Rust (Family Eriophyidae)

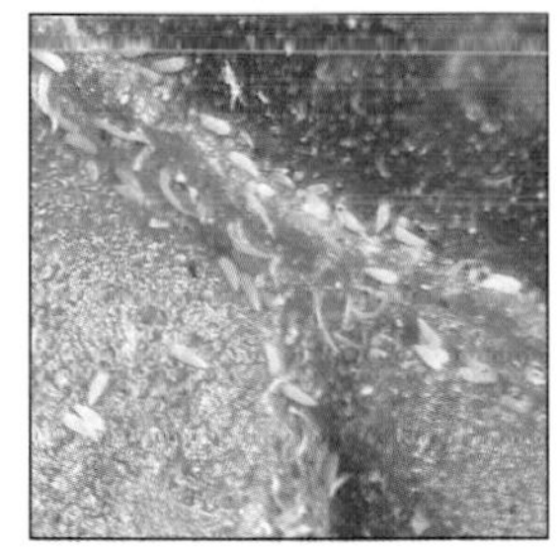

Tomato russet mites

Citrus rust mite damage

Description: Adults: nearly invisible, elongate, cylindrical, pale yellow or tan mites, with 2 pairs of legs at the head end. Nymphs: similar to adults, but smaller.

Damage: Mites burrow from undersides into leaf tissue of apples, pears, tomatoes, and ornamental trees and shrubs. Infested leaves and fruit have russeted appearance.

Life Cycle: Mites overwinter at base of buds or in cracks in bark, moving to developing flowers in spring. Numerous overlapping generations all season; populations usually decline in hot weather. By late August, most species move to overwintering sites.

Control: Spray dormant oil with lime-sulfur on dormant trees; spray foliage with sulfur fungicide.

Mites, Spider (Family Tetranychidae)

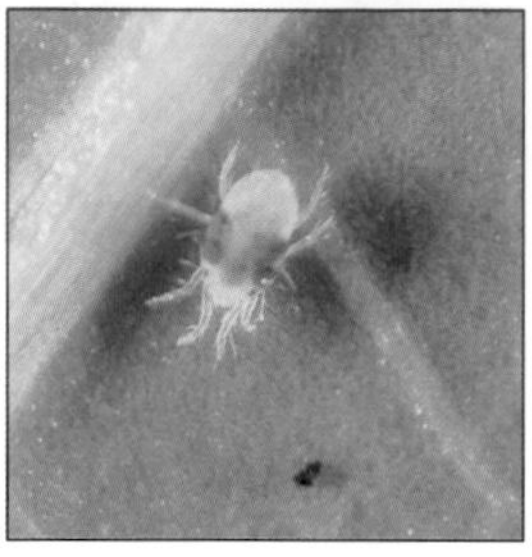

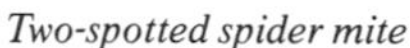
Two-spotted spider mite

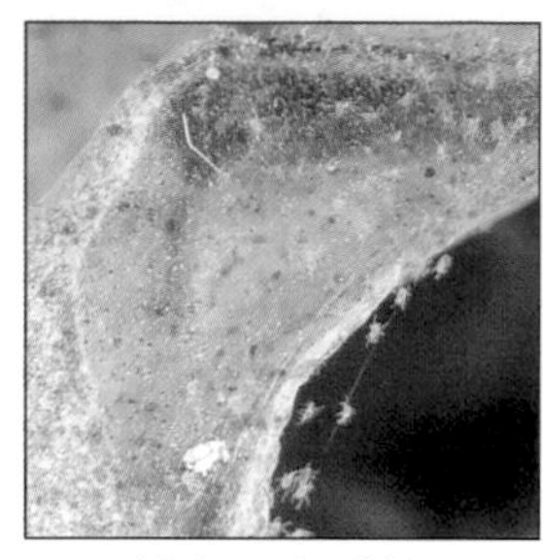
Adults and webbing

Damage

Description: Adults: minute, 8-legged, 1/50″ mites with fine hairs on body, reddish, pale green or yellow; most, but not all, species spin fine webs. Nymphs: similar to adults, but smaller; early stages with 6 legs. Found throughout North America.

Damage: Adults and nymphs suck juice from cells on undersides of leaves of many food crops, ornamentals, and fruit trees. Plants are weakened, leaves may drop and fruit may be stunted. Early damage appears as yellow specked areas, with or without webbing, on leaf undersides; later, leaves are bronzed or turn yellow or white with brown edges, webs may cover leaves and growing tips.

Life Cycle: Eggs or adults overwinter in bark crevices or garden debris, emerging in early spring. Eggs hatch in 2-3 days, nymphs develop to adults in 7-10 days. Many overlapping generations every season; reproduction continues all year in greenhouses.

Control: Spray fruit trees with dormant oil to kill overwintering eggs; in garden or greenhouse rinse plants with water and mist daily to suppress reproduction of mites; release predatory mites *Metaseiulus occidentalis* on fruit trees, *Phytoseiulus persimilis* or similar species on vegetables, strawberries, and flowers; spray insecticidal soap, pyrethrin, or neem; as a last resort, spray avermectins or rotenone.

Navel Orangeworm *Amyelois transitella*

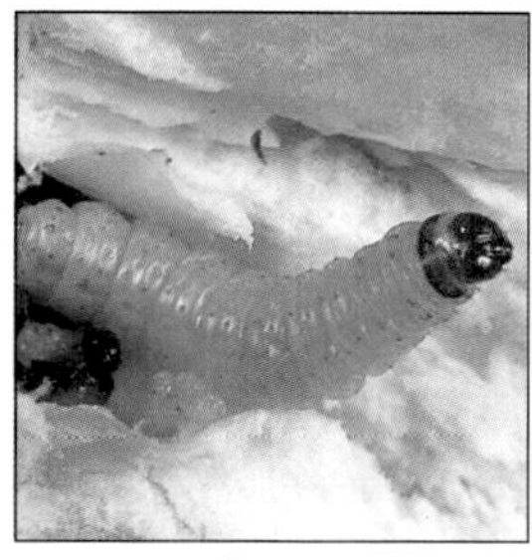
Larva

Adult

Description: Adults: light gray moths with dark mottling and a dark border on hind wings (3/4″ wingspan). Larvae: reddish orange, brown-headed caterpillars with 2 crescent-shaped marks on the second segment behind the head. Found in southwest United States, Oklahoma, and Georgia.

Damage: Larvae are key pests in

almonds, walnuts, citrus, figs, and other fruits. Larvae bore into cracks in nut husks or into damaged fruit, spinning webs inside and pupating.

Life Cycle: Adults lay eggs on mummified nuts, dropped nuts, or on the blossom end of injured fruit; larvae bore into nuts or fruit and spin webs inside; larvae feed and pupate inside the webs and then emerge the next spring. One generation per year.

Control: Pick up dropped fruit or nuts and knock all mummified nuts out of trees in winter before adults emerge in spring; harvest nuts early to avoid damage; larvae in fallen nuts survive best on bare ground; therefore, grow cover crops under trees. On fruit, spray BTK to kill larvae before they bore into fruit.

Nematodes

Root knot nematode

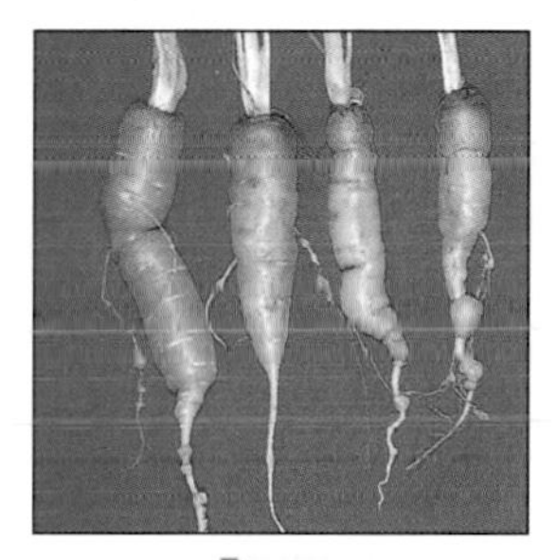

Damage

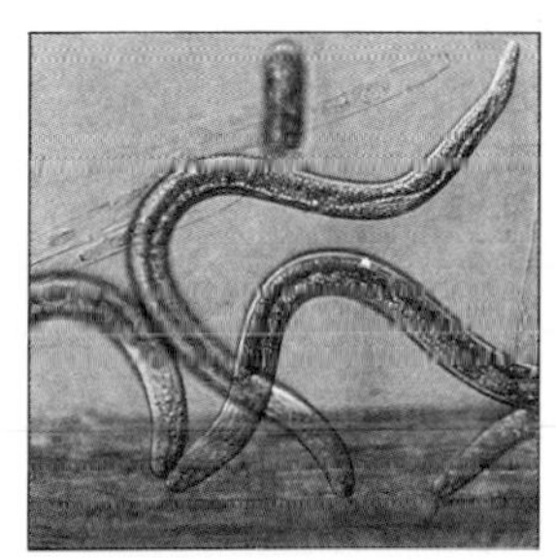

Lesion nematode

Damage

Description: Slender, translucent, unsegmented worms; most pest species are microscopic plant parasites only about 1/50″ long; species that parasitize insects are larger—1/25″ to several inches long; beneficial, soil-dwelling species that break down organic matter are easily visible—about 1/4″ long. Found throughout North America.

Damage: Pest nematodes cause root knots or galls, injured root tips, excessive root branching, leaf galls, lesions of dying tissue, and twisted, distorted leaves. Plants most commonly attacked at the roots include tomatoes, potatoes, peppers, lettuce, corn, and other vegetables; plants that sustain leaf and stem injury include chrysanthemums, onions, rye, and alfalfa.

Beneficial Effects: Some parasitic nematode species are sold as biological controls for root weevils, crown and stem borers, corn rootworms, and other lawn and garden pests. Beneficial nematodes also decompose organic material and are common in compost heaps.

Life Cycle: Most species have a mobile larval stage that moves through the soil or on a film of water to infect the host plant or insect; larvae molt several times to reach the adult stage; adults lay eggs in masses. Life cycle for many pest species takes 3-4 weeks.

Control: Nematodes move slowly in soil and take several years to build up to damaging numbers. Control pest nematodes by crop rotation with nonsusceptible crops, by planting a nematode-suppressing cover crop, such as marigolds, or by soil solarization. To solarize soil and destroy nematodes, cover moist soil with clear plastic mulch for the summer months; this procedure heats soil sufficiently to kill pest nematodes.

Northern Corn Rootworm *Diabrotica longicornis*

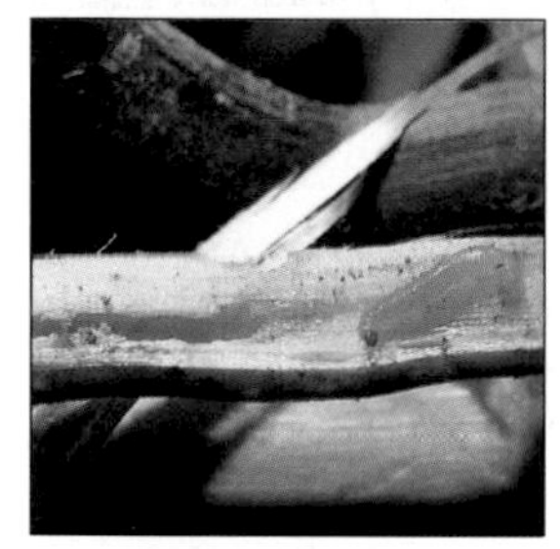

Larva and damage

Adult

Description: Adults: yellowish green, 1/4″ beetles, sometimes with brown heads and thoraxes. Larvae: slender, wrinkled, white, 1/2″ grubs with light brown heads. Eggs: laid on soil near corn roots. Found in north-central United States.

Damage: Larvae severely damage corn roots; adults feed on pollen and corn silk, damaging ears and interfering with pollination; larvae also spread bacterial wilt.

Life Cycle: Females lay eggs around corn roots in late summer and fall; eggs hatch the following spring; larvae migrate to newly growing corn roots and feed until early summer, then pupate in a soilborne cell. Adults emerge in July and August. One generation per year in most areas.

Control: Since larvae cannot move far through soil, rotate crops to prevent pest buildup; encourage predatory ground and rove beetles with permanent borders; cultivate corn patches well after harvest and before seeding to kill eggs and larvae.

Obliquebanded Leafroller *Choristoneura rosaceana*

Larva

Adult

Description: Adults: reddish brown moths with light and dark V pattern on their backs when wings are folded (3/4" wingspan). Larvae: pale green, black-headed caterpillars. Eggs: greenish, laid in overlapping masses on undersides of leaves. Found throughout North America.

Damage: Young larvae mine in leaves of apples, strawberries, roses, and other ornamentals in early spring. They later spin webs around branch tips and feed on enclosed buds, leaves, and developing fruit. Second generation larvae most damaging to fruit.

Life Cycle: Larvae overwinter in tough silken cocoons on tree bark; resume feeding in spring, pupate, and emerge as adults in June. Eggs are laid on leaves; larvae mine in leaves, then move to branch tips. One generation per year; a second generation occurs in warmer regions

Control: Spray BTK, pyrethrin, or rotenone when larvae are young, before they spin webs; handpick webbed leaves and buds, destroy; on small trees, check for egg masses and destroy.

Onion Maggot *Delia antiqua*

Adult

Damage

Description: Adults: gray, bristly, humpbacked, 1/4" flies with large wings; about half the size of houseflies. Larvae: white, bluntended, 1/4" maggots. Eggs: laid in soil near plants. Found throughout the northern half of the United States as well as Canada.

Damage: Maggots burrow into developing onions or leeks, killing young plants and hollowing out or stunting older plants; rot diseases enter bulbs injured by maggot feeding; 1 maggot can kill over a dozen seedlings during its development. This pest rarely infests onions in the South. In the North, infestations are worst in cool, wet weather, sometimes killing up to 80 percent of a spring crop.

Life Cycle: Flies overwinter as brown pupae, resembling grains of wheat, in soil or garden trash; adult flies emerge from pupae from mid-May to the end of June and lay eggs at the bases of onion or leek plants; eggs hatch in a week; maggots burrow into onions and feed 2-3 weeks, then pupate in soil nearby; adults emerge in 1-2 weeks. Commonly 2 generations per growing season; a third generation may attack onions just before harvest and cause storage rot.

Control: Cover seedlings with floating row cover; sprinkle rows liberally with ground cayenne pepper, ginger, dill, or chili powder; plant cull onions around the borders and down the rows of seedling onions to act as a trap crop; pull and destroy cull trap crops 2 weeks after they sprout; bury, burn, or destroy all unwanted onions at the end of harvest; plant onion sets late to avoid the first generation of flies.

Oriental Fruit Moth *Grapholitha molesta*

Larva and internal damage

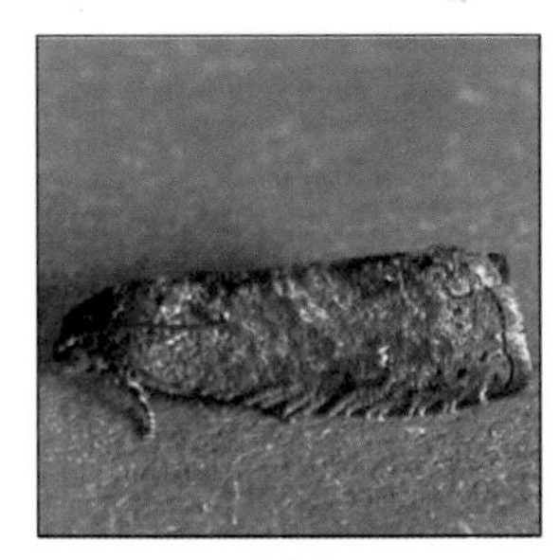

Adult

Twig damage

Description: Adults: small, dark gray moths with dark brown mottled forewings (½″ wingspan). Larvae: white to pinkish gray, ½″ caterpillars with brown heads. Eggs: flat, white, laid on twigs or leaf undersides. Found in eastern United States, Pacific Northwest, and Ontario.

Damage: In spring, young larvae bore into green twigs of peach, almond, cherry, apple, pear, or other fruit trees causing twig wilting and dieback; second generation larvae bore into developing fruit, leaving masses of gummy castings; later generations enter the stem end of maturing fruit leaving no external signs of damage; perfect-looking fruit injured through stem entry will usually break down in storage.

Life Cycle: Larvae overwinter in silken cocoons on bark or weeds or in soil around trees, pupating in early spring; adults emerge from early May to mid-June; females lay eggs, which hatch in 10-14 days; first generation larvae bore into tender stems and twigs, feed for 2-3 weeks, then pupate. Second generation adults appear in mid-July; second generation larvae bore into young fruit and don't feed on twigs; a third generation of larvae arrive by the end of August in northern United States; these bore into the stem ends of mature fruits and feed on the pits. Three to 4 generations in the North, 6-7 in southern states.

Control: Where possible, plant early-bearing peach and apricot cultivars that are harvested before midsummer; to destroy overwintering larvae, cultivate soil 4 inches deep around trees in early spring; attract native parasitic wasps and flies with flowering cover crops planted around trees; disrupt mating with pheromone patches applied to lower limbs of trees (1 patch per 4 trees); spray superior oil to kill eggs and larvae.

Peachtree Borers *Synanthedon exitiosa* and *S. pictipes*

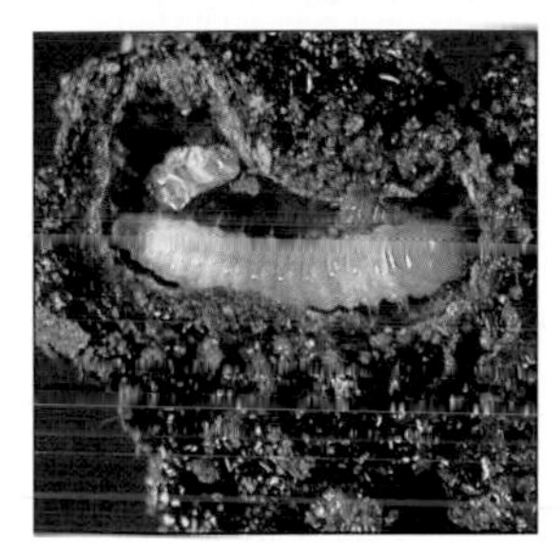

Larva

Adult female

Description: Adults: blue-black, 1¼" moths with yellow or orange bands across the body and narrow translucent wings; adults resemble wasps, and, unlike most moths, are active during the day. Larvae: white caterpillars with dark brown heads; ⅛"-1" long, depending on age. Found throughout North America.

Damage: Larvae bore beneath bark of peach trees at the base as well as into main roots near the surface. Also may attack plum, prune, cherry, apricot, and nectarine trees. Trees are often girdled. Burrow entrance holes exude gum mixed with sawdustlike material. Young or weak trees may be seriously damaged or killed; older trees are less affected.

Life Cycle: Larvae overwinter in tree trunk burrows; in the spring they spin brown silken cocoons at the surface of the burrow or in soil; cocoons may be covered with dirt and gum from the tree; first adult moths emerge in July, and adult emergence continues into early fall; in late summer females begin laying eggs on tree trunks or in cracks in soil within a few inches of trunks; eggs hatch in 10 days, and new larvae burrow into tree trunks to feed and overwinter until next year. One generation per year; some with 2-year life cycle.

Control: Since adult borer moths are attracted to injured or diseased trees, maintain vigorous trees and avoid mechanical injury to trunks; beginning in late summer and into fall, inspect tree trunks from a foot or so above ground level to a few inches below ground level, digging away soil to expose the trunk area below the ground surface; kill borers in exposed burrows by inserting a fine, flexible wire; in the fall and spring cultivate soil around the base of the trunk to expose and destroy larvae and pupae; in the spring encircle the trunk with a mound of tobacco dust 4"-6" wide; attract native parasitic wasps and predators.

Pear Psylla *Cacopsylla pyricola*

Nymph

Adult

Damage

Description: Adults: red to green, 1/10″ long insects, resembling tiny cicadas with wings folded rooflike over the back. Nymphs: oval, green to brown, wingless, 1/80″ insects. Found in eastern United States and Canada, Pacific Northwest, and California.

Damage: A major pest of pears and quinces, psyllas suck plant juices, causing leaves to yellow from the toxic saliva; honeydew secretions support growth of sooty mold. Feeding spreads pear decline virus.

Life Cycle: Overwintering adults emerge in spring from bark and leaf litter to lay eggs on fruit spurs and buds. Nymphs spend early stages protected by honeydew; later stages are more active. Three to 5 generations per year.

Control: Spray superior oil in the spring; plant cover crops to attract native predatory insects; spray insecticidal soap.

Pear Sawfly *Caliroa cerasi*

Larva

Adult

Damage

Description: Adults: shiny, black-and-yellow, 1/5″ insects, resembling houseflies. Larvae: young resemble green-black, 1/2″ slugs; older larvae are yellow and segmented. Found throughout North America.

Damage: Larvae skeletonize upper leaf surfaces of pears, cherries, plums, and cotoneasters, leaving scorched areas. Young trees may be defoliated and eventually killed. Second generation larvae cause the most damage. Related species on roses.

Life Cycle: Larvae overwinter in cocoons in soil and pupate in spring; adults emerge in late May; eggs hatch in a week; larvae feed 3-4 weeks, then pupate in soil. Second generation adults appear in late July, with larvae hatching in mid-August.

Control: Spray trees with strong streams of water to remove larvae; spray rotenone or pyrethrin.

Pickleworm *Diaphania nitidalis*

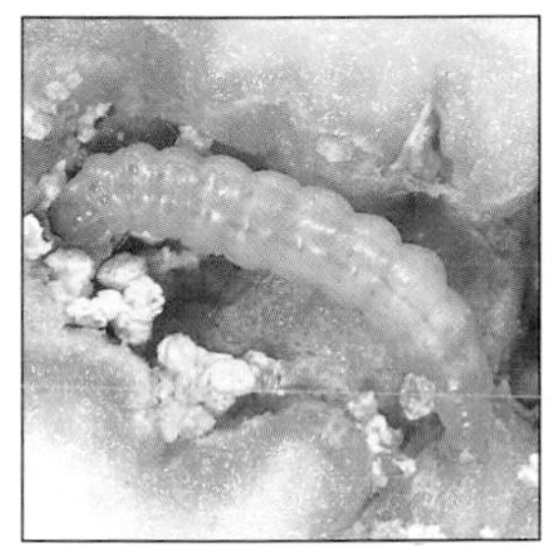
Larva and damage

Adult

Description: Adults: moths with brown-banded wing edges (1¼″ wingspan) and hair tufts on the tip of the abdomen. Larvae: pale green caterpillars with black spots when young. Found mostly in southeastern United States.

Damage: Larvae chew blossoms and vines and bore into fruit of cucurbit plants, especially cucumbers, cantaloupes, and squash. A serious pest of summer squash.

Life Cycle: Pupae overwinter in weeds or rolled leaves of cucurbits; adults emerge in spring and lay eggs on vines; eggs hatch in a few days; larvae feed for 2 weeks, then pupate in cocoons inside rolled leaves. Up to 5 generations per year.

Control: Plant early maturing cultivars; plant and destroy a summer squash trap crop to protect cucumber and melon; destroy or till under all residues from cucurbit vines right after harvest.

Plum Curculio *Conotrachelus nenuphar*

Adult

Adults

Description: Adults: dark, brownish gray, ¼″ beetles with warty, hard wing covers, prominent snout, and white hairs on body. Larvae: plump, white, ⅓″ grubs with brown heads. Eggs: round, white, laid individually under a crescent-shaped cut in the fruit skin. Found in eastern North America.

Damage: A major plum and apple pest in many areas, adult curculios feed on petals, buds, and young fruit; females deposit a single egg just under fruit skin, leaving a crescent-shaped scar at each egg-laying site; newly hatched larvae feed inside the fruit, causing it to drop, rot, or develop deformed growth. Other susceptible fruits include peaches, cherries, and apricots.

Life Cycle: Adult beetles overwinter under fallen leaves, stones, logs, or other garden debris, flying to trees just as blossoms open; adults feed and lay eggs, which hatch in 5-10 days; larvae feed in fruit 2-3 weeks; when fruit drops, they exit and pupate in the soil. Second generation adults emerge in late July to late October, feed on ripe or fallen fruit until fall, then move to shelter to hibernate. Up to 2 generations per year.

Control: Knock beetles out of trees onto a dropcloth by sharply tapping branches with a padded stick; gather and destroy beetles; for this control to be effective, you must do it twice a day throughout the growing season. Every other day pick up and destroy all fallen fruit, especially early drops; keep chickens around fruit trees to feed on dropped fruit. In areas where severe infestations occur, check developing fruit for egg scars twice a week; when first fruit scars appear, apply a botanical pesticide containing pyrethrin, ryania, and rotenone, such as Triple Plus; repeat in 7-10 days. Do not use a botanical pesticide before petals drop—it kills beneficial pollinators.

Potato Tuberworm *Phthorimaea operculella*

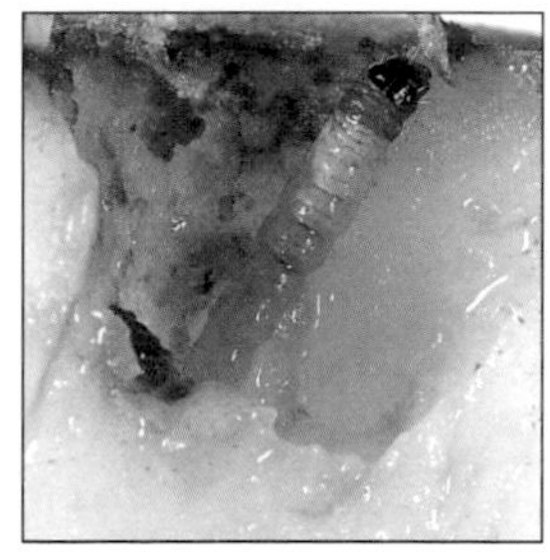

Larva

Adult

Description: Adults: mottled gray moths with narrow, fringed wings (½″ wingspan). Larvae: slender, pinkish white, ½″ borers with dark heads. Found mostly in southern United States.

Damage: Early summer larvae feed in leaves and stems of potatoes, tomatoes, and eggplants; later-generation larvae feed inside potato tubers, allowing rot to enter.

Life Cycle: Larvae or pupae overwinter in soil or potato culls left outdoors; in spring adults emerge and lay eggs on plants; eggs hatch in a few days; larvae feed 2-3 weeks, then pupate for a week; up to 6 generations per year in warm regions.

Control: Plant potatoes early; harvest early; use heavy mulch to prevent cracked, dry soil, which exposes potato tubers to attack; at harvest, destroy cull potatoes and infested vines before nightfall; adult moths lay eggs on culls left out overnight.

Praying Mantid *Mantis religiosa*

Adult with egg case

Adult

Description: Adults: large, elongated, green or brown insects with prominent eyes; up to 4″ long. Nymphs: similar to adults, but smaller and without wings. Eggs: laid in a grayish frothy case of 50–400 eggs glued to stems or twigs. Found in southern and eastern United States, north into Ontario.

Beneficial Effect: Mantids catch and devour both pests and beneficial species. They eat virtually any insects they catch, including each other.

Life Cycle: Eggs are glued in a gray, frothy, sticky mass to plants where they harden and remain for the winter; adults hatch in the spring; 1 generation per year.

How to Attract: To protect native species, don't release purchased mantids, and avoid pesticides; provide sites for overwintering eggs by keeping permanent plantings around the garden.

Rose Chafer *Macrodactylus subspinosus*

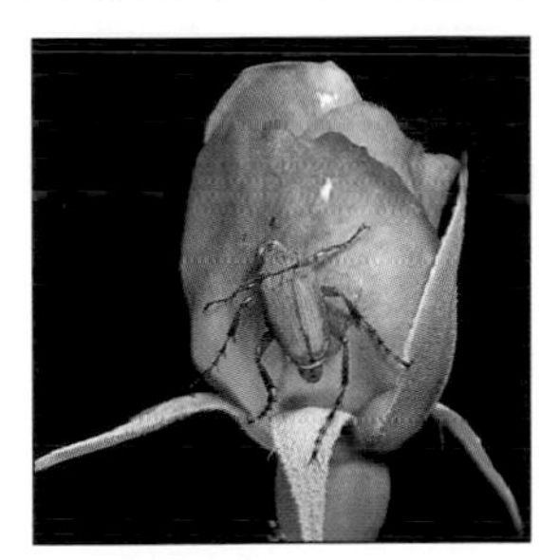
Adult

Description: Adults: reddish brown, 1/3″ beetles with black undersides and wing covers cloaked in thick, yellowish hairs. Larvae: small white grubs. Found throughout North America.

Damage: Larvae feed on roots of grass and weeds; adults chew on the flowers, leaves, and fruit of grapes, roses, tree fruits, brambles, strawberries, peonies, irises, dahlias, hollyhocks, and vegetables. Serious damage only with heavy infestation.

Life Cycle: Larvae overwinter in soil, pupate in spring, and emerge as adults in late May to early June; adults lay eggs in soil until early July; eggs hatch in about 2 weeks; grubs feed on roots until fall.

Control: Control is not usually necessary; for severe infestations, spray pyrethrin or rotenone.

Roundheaded Appletree Borer *Saperda candida*

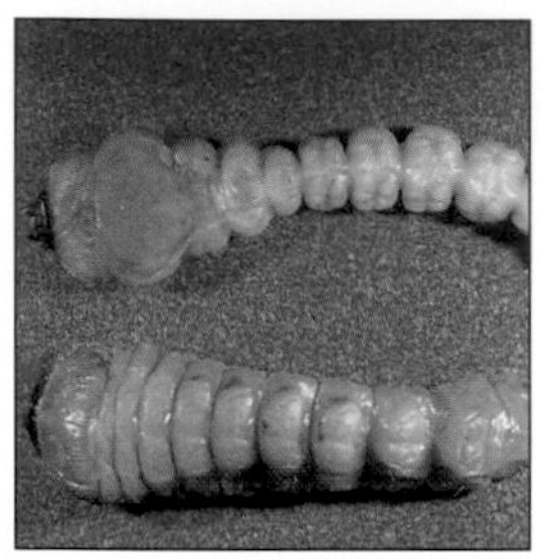

Roundheaded appletree borer larva (bottom) *and flatheaded appletree borer larva* (top)

Description: Adults: yellow or reddish brown, 3/4″ beetles with white stripes. Larvae: creamy white, dark-headed grubs. Found in eastern Canada and United States west to Nevada.

Damage: Larvae bore into trunks of apple trees near ground level, girdling the tree or penetrating into heartwood. Other hosts include pears, plums, cherries, crab apples, mountain ashes, and hawthorns.

Life Cycle: Adults emerge in June and lay eggs in bark just above soil line; larvae burrow into sapwood, overwinter in soil, and tunnel into heartwood the second year. In the third spring, larvae pupate in tree tunnels and emerge as adults several weeks later. Two- to 4-year life cycle.

Control: Remove nearby wild host trees; impale larvae in holes with flexible wire; inject parasitic nematodes in borer holes.

Rove Beetles (Family Staphylinidae)

Adult

Description: Adults: brown or black, slender, 1/10″-1″ long beetles with stubby wings covering only part of the body. Larvae: resembling adults, but wingless. Over 3,100 species native to North America.

Beneficial Effect: Valuable for controlling aphids, springtails, mites, nematodes, flies, and cabbage maggots. Also help decompose organic matter.

Life Cycle: Most species overwinter as adults, becoming active in the spring and laying eggs in the soil; larvae molt 3 times as they feed, then pupate in soil.

How to Attract: Maintain permanent beds and plantings in garden to protect overwintering adults; interplant with cover crops or mulch planting beds; make stone or plank walks in garden to provide shelter.

Sawflies, Conifer *Neodiprion* spp.

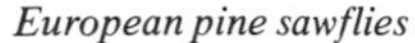

European pine sawflies

European pine sawfly larvae and damage

Redheaded pine sawfly larvae and damage

Description: Adults: stout, 1/4″–1/2″ insects with membranous wings. Larvae: similar to caterpillars, gray with light stripes or yellow with black dots. Found in eastern United States and Canada.

Damage: Larvae strip needles from the upper branches of pines, spruces, and hemlocks. Repeated infestations kill trees.

Life Cycle: Eggs overwinter in branch slits, hatch into needle-devouring larvae in early spring, then pupate in ground litter; adults emerge in fall and lay eggs. Some species lay eggs in spring and overwinter in cocoons. One generation per year.

Control: Shrews, viral diseases, predators, and parasites generally suppress populations. Spread dropcloths under trees to collect, and destroy larvae as they drop to the ground. Apply superior oil to trees. (Note: Oil on blue spruce damages foliage.)

Sawflies, Leafminer *Fenusa* spp.

Birch leafminer adult

Elm leafminer damage.

Description: Adults: black, stout-bodied, 1/5″ insects with transparent wings. Larvae: flattened, white, legless maggots with brown heads. Found in northern and eastern United States, west to Great Lakes states. Related species in Canada.

Damage: Larvae mine leaves of elm, birch, or alder, feeding between upper and lower leaf surfaces and leaving brownish, wrinkled blisters in leaves; damaged trees may be attacked by borers and other pests.

Life Cycle: Larvae overwinter in cocoons in the soil, pupating in the spring; adults emerge in mid-May and lay eggs in leaves; larvae feed in leaves until ready to pupate. Up to 4 generations per year.

Control: Maintain healthy, vigorous trees.

Scales, Armored (Family Coccidae)

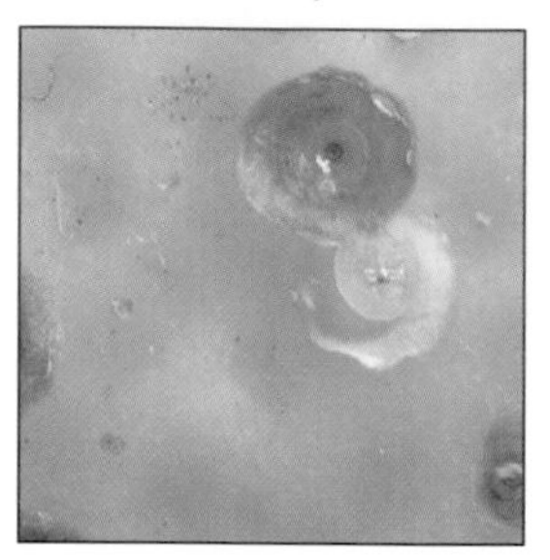

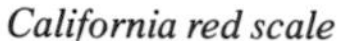
California red scale

California red scale

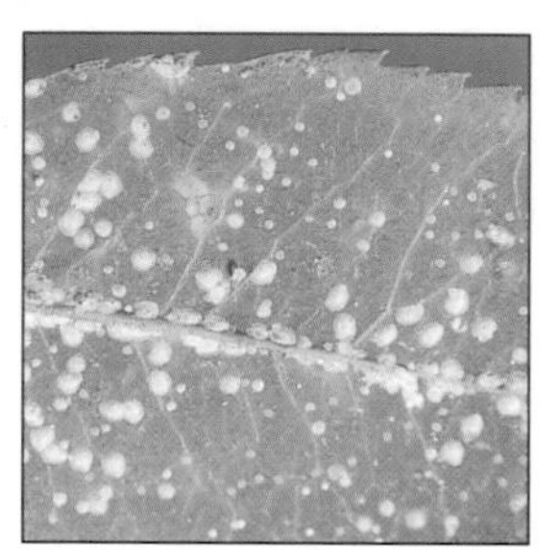

Oleander scale

Oystershell scale

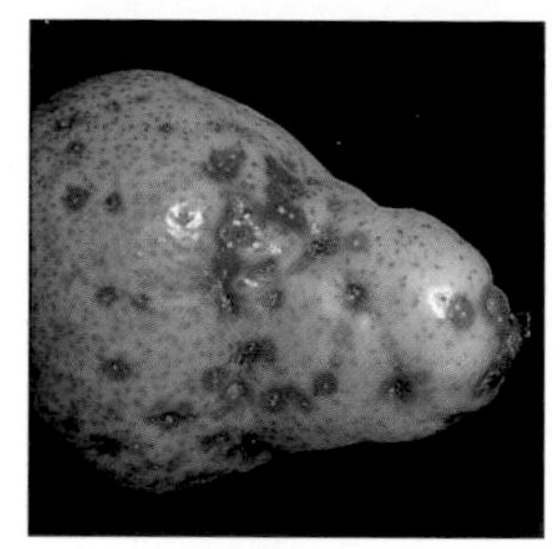

San Jose scale

Description: Adults: circular or oval, hard bumps less than 1/10″ in diameter and sometimes bearing a dimple in the center. Colors vary, including gray, yellow, white, and reddish or purplish brown. All secrete an armor of wax in an oyster-shell or circular pattern. Nymphs: early stages are mobile crawlers; later stages are legless and sedentary. Most species are found in southern United States, some in northern United States and southern Canada.

Damage: Scales weaken plants by sucking plant juices. In the South they attack citrus, roses, palms, and tropical ornamentals; northern species attack fruit trees, shade trees, grapes, currants, raspberries, and shrubs. Scales also cause injury by injecting toxic saliva into plant tissues. Some species secrete honeydew, which attracts ants and encourages the growth of sooty mold, a fungus that feeds on the honeydew. Severe infestations may kill trees.

Life Cycle: Females lay eggs or give birth to live nymphs; these wander for a few hours or days, then settle and molt to an immobile adult form, a process that takes a month or more. Most scales overwinter as nymphs or eggs hidden in tree bark. One to 2 generations in northern regions; in southern regions, up to 6 generations per year.

Control: Difficult to control with pesticides because they are protected by a waxy covering. Dormant oil sprays smother overwintering eggs and provide good control; superior oil sprays kill eggs and nymphs. Release predatory beetles *Chilocorus nigritus* or *Lindorus lophanthae.* Release parasitic wasps *Aphytis melinus* against California red scale and oleander scale.

Scales, Soft (Family Coccidae)

Black scale

Sooty mold due to scale

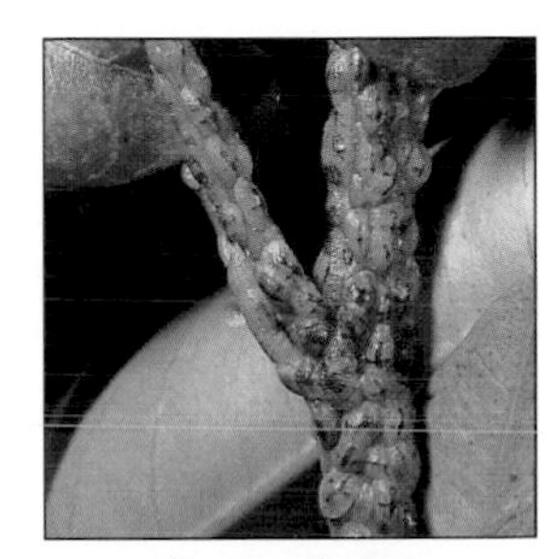
Brown soft scale

Cottony cushion scale

Cottony maple scale

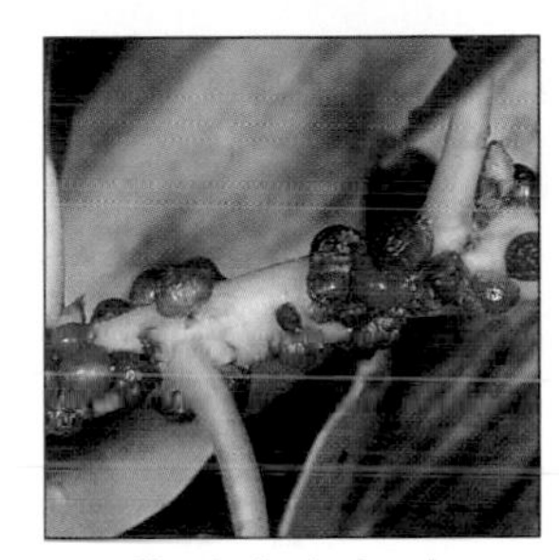
Hemispherical scale

Description: Adults: females are oval or round, soft, legless bumps, 1/10"–1/5" in diameter and without wings or appendages; males are minute, yellow-winged insects. Larvae: mobile crawlers resembling tiny mealybugs. Found throughout North America.

Damage: All stages suck plant sap, weakening plants, especially citrus, fruit trees, ornamental shrubs, trees, and houseplants. In severe infestations, leaves yellow and drop and the plant dies. Most scales secrete large quantities of honeydew onto foliage and fruit. Sooty mold, a black fungus, feeds on the honeydew.

Life Cycle: Females of some species lay as many 2,000 eggs; others give birth to several nymphs per day. Mobile nymphs move around for a short time, then settle at one spot. Females molt to a legless, immobile form. Males molt to tiny, flylike insects. One or 2 generations per year outdoors; up to 6 generations on indoor plants.

Control: Prune and dispose of infested branches and twigs. Attract native predatory beetles, such as soldier beetles and lady beetles as well as parasitic wasps. Remove scales from twigs with a soft brush or your fingernail, or from leaves with a soft cloth and soapy water; rinse well. Release predatory beetles *Chilocorus nigritus* or *Lindorus lophanthae.* Release parasitic wasp *Metaphycus helvolus* to control soft brown scale. Spray dormant oil on fruit and ornamental trees. Spray superior oil (not on citrus after July). As a last resort, spray pyrethrin or rotenone.

Shothole Borer *Scolytus rugulosus*

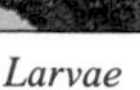

Larvae

Damage

Damage

Description: Adults: brown or black, 1/10″ beetles with red-tipped wing covers. Larvae: white grubs with reddish brown heads. Found throughout North America.

Damage: Adults and larvae bore and tunnel under the bark of peach, plum, cherry, apple, and pear; unhealthy trees are most often attacked.

Life Cycle: Adults bore through bark and deposit eggs in galleries. Resulting larvae bore at right angles to parent gallery and pupate under the bark. Emerging adults bore new escape holes through bark. Two to 3 generations per year.

Control: Maintain healthy trees; destroy infested branches or trees in winter when larvae are still in wood; destroy borers by inserting a flexible wire into burrows; protect young trunks with whitewash or latex paint diluted with an equal amount of water.

Slugs/Snails

Slug

Brown garden snail

Brown garden snails

Description: Adults: soft-bodied, gray, tan, green, black, yellow, or spotted, wormlike animals that are members of the same zoological group as clams, mussels, and scallops. Slugs have no shells, while snails have coiled shells. Measuring 1/8″–8″, both slugs and snails leave characteristic trails of mucus wherever they crawl. Eggs: clear, oval or round, laid in jellylike masses under stones or debris. Garden slugs and snails occur throughout North America.

Damage: Both slugs and snails rasp large holes in foliage, stems, and bulbs. They feast on any tender plant or shrub and may demolish seedlings. Most damaging in wet years or regions.

Life Cycle: Adults lay eggs in moist soil or under rocks. Eggs hatch in 2–4 weeks and young grow 5 months to 2 years before reaching maturity.

Control: Wrap copper strips around trunks of trees or shrubs, or use copper flashing as edging for garden beds. Trap under flowerpots or boards. Attract with pieces of raw potato or cabbage leaves set out in the garden; collect and destroy every morning. Trap in shallow pans of beer buried with the container lip flush to soil surface. To encourage predatory beetles, maintain permanent walkways of clover, sod, or stone mulch. Protect seedlings with wide bands of cinders, wood ashes, or diatomaceous earth, renewed frequently.

Soldier Beetles (Family Cantharidae)

Downy leatherwinged soldier beetle

Description: Adults: elongated, slender, nearly flat, ⅓"–½" beetles, often with downy, leathery wing covers. Larvae: flattened, elongated, covered with hairs, and usually dark-colored. Found throughout North America.

Beneficial Effect: Both larvae and adults prey on cucumber beetles, corn rootworms, aphids, grasshopper eggs, caterpillars, and beetle larvae.

Life Cycle: Females lay eggs in soil; newly hatched larvae remain inactive for a short period before developing; larvae overwinter in soil and pupate in the spring. Up to 2 generations per year.

How to Attract: To attract adult soldier beetles, plant goldenrod, milkweed, hydrangeas, or catnip. To protect pupating beetles, maintain some permanent plantings where soil is not disturbed.

Southwestern Corn Borer *Diatraea grandiosella*

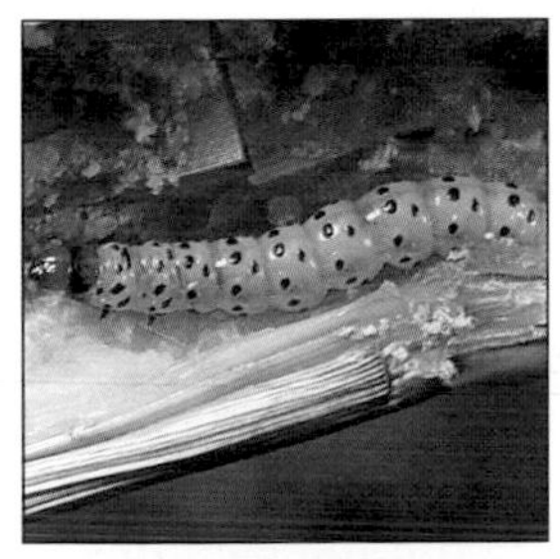

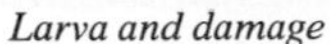

Larva and damage

Adult

Description: Adults: pale beige, lightly marked moths (1¼" wingspan); males darker. Larvae: white with brown heads, spotted in summer. Found throughout the South.

Damage: Larvae bore into leaf whorls, stalks, and roots of corn, sorghum, and some grasses, causing stem breakage.

Life Cycle: Larvae overwinter in plant roots and pupate in early June; moths emerge a week later and lay eggs; larvae bore into leaf whorls and stalks, then pupate and emerge in August. Second generation larvae bore into stalks, then roots.

Control: Plant corn early; plant resistant cultivars; cut stalks at soil level and remove right after harvest; cultivate deeply in fall.

Sowbugs/Pillbugs

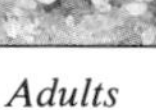

Adults

Adult

Adults

Description: Adults: slate gray or brown, 1/4"–5/8" crustaceans with jointed armor and 7 pairs of tiny legs; some curl up in a ball when disturbed. They are not insects, but are related to crayfish. Nymphs: same as adults, but smaller. Found throughout North America.

Damage: Usually none to established plants; however, in high numbers damage to seedlings can be severe. Feed on decaying organic material and young seedlings.

Life Cycle: Eggs are laid in damp locations; young are similar to adults.

Control: Drain wet areas; sprinkle diatomaceous earth around foundations where bugs congregate; trap under stones or boards, then destroy; make paper traps painted with a sticky coating, such as Tanglefoot, then folded tentlike, with sticky side down.

Spined Soldier Bug *Podisus maculiventris*

Adult and prey

Description: Adults: shield-shaped, black-speckled, 1/2" bugs with sharp points on shoulders. Nymphs: similar to adults, but wingless. Native to North America.

Beneficial Effect: Important native predators attack caterpillars and grubs, including tent caterpillars, fall armyworms, sawfly larvae, and Mexican bean beetle larvae.

Life Cycle: Overwintering adults emerge in spring; females lay eggs on leaves; nymphs drink water or plant juices for a short period, then become predators; nymphs develop to adults in 6–8 weeks; adults live 5–8 weeks. One to 2 generations per year.

How to Attract: Maintain permanent beds of perennials to provide shelter.

Spittlebug/Froghopper *Philaenus spumarius*

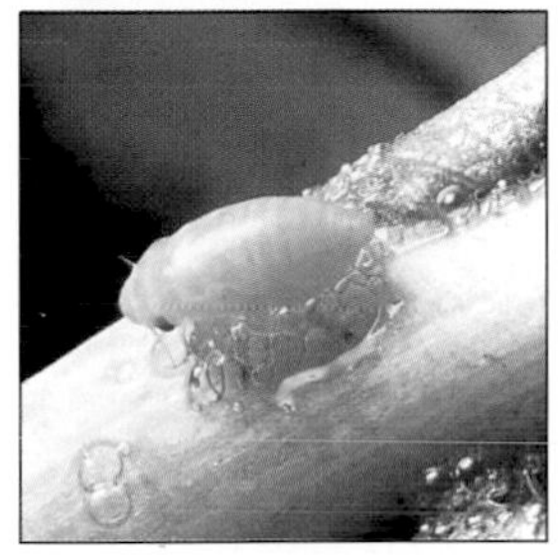

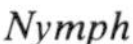

Nymph

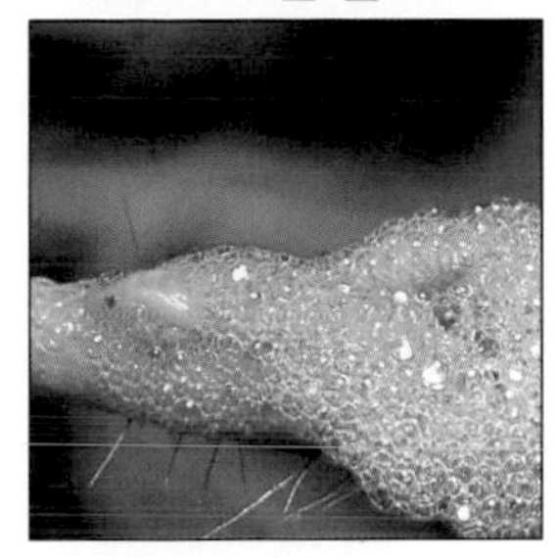

Nymph in spittle mass

Meadow spittlebug nymph and spittle mass

Description: Adults: oval, frog-faced, 1/4″–1/2″ insects; tan, brown, or black. Nymphs: yellow to yellowish green, similar to adults, but wingless; inside a foamy mass of "spittle." Eggs: white to beige, laid in rows. Found throughout North America.

Damage: Adults and nymphs suck plant juices of strawberries, legumes, forage crops, and nursery plants, causing stunted, dwarfed, weakened plants with reduced yields. Adults migrate from hayfields to nearby crops when hay is cut.

Life Cycle: Overwintering eggs hatch in mid April; nymphs develop in spittle masses for 6–7 weeks. Adults feed over the summer, laying eggs in field stubble by fall.

Control: Not usually a serious problem in a home garden; if many nymphs apparent in summer, then turn in stubble of susceptible crops in the fall to kill eggs.

Spruce Budworm *Choristoneura fumiferana*

Adult and larva

Damage

Description: Adults: grayish brown moths (3/4″–1 1/4″ wingspan); hind wings dark brown with white fringe. Larvae: dark brown caterpillars with lighter sides, dark spines, and 2 rows of white dots along their backs. Found throughout north-central and eastern United States; in Canada to the Yukon; related species in western North America.

Damage: Larvae mine in needles, buds, cones, or twigs of balsam fir, spruce, Douglas fir, hemlock, larch, and pine. One of the most damaging forest pests in North America; also damaging in nurseries and ornamental trees. Budworms are the most serious conifer pests in North America, destroying billions of board feet of fir and spruce each year. After 3-5 years of heavy infestation, entire stands of conifers may be killed; surviving trees are weakened and susceptible to bark beetles.

Life Cycle: Moths lay eggs from late June to early August; eggs hatch in 8-12 days. Young larvae disperse throughout tree, then spin cocoons and hibernate. The following spring larvae emerge and mine in old needles, then in young buds. As new growth expands, they spin webs around tips of twigs and feed inside, pupating by late June. Adults emerge in 10 days. One generation per year.

Control: Spray BTK as soon as tiny larvae appear in late summer and again in early spring when they resume feeding; avoid using balsam fir in ornamental plantings. Since outbreaks of this pest are most common in overmature stands of spruce and balsam firs, avoid planting these trees as ornamental specimens if you live in an area where spruce and fir grow in great abundance.

Squash Bug *Anasa tristis*

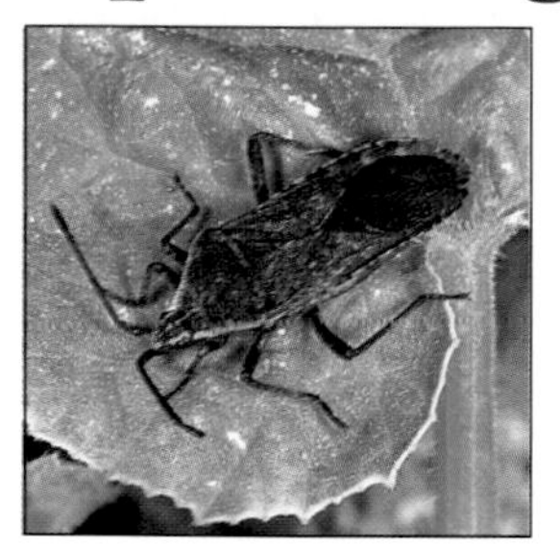

Adult

Description: Adults: brownish black, flat-backed, 1/2″ bugs covered with fine, dark hairs. They give off an unpleasant smell in defense. Nymphs: whitish green or gray young, similar in shape to adults, with darker thorax and abdomen as they mature; usually covered with a grainy white powder. Eggs: shiny yellow to brown ellipses, in groups on underside of leaves. Found throughout North America.

Damage: Both adults and nymphs suck plant juices of all cucurbit crops, especially squash or pumpkins, causing leaves and shoots to blacken and die back; attacked plants fail to produce fruit. Winter squash are most severely affected.

Life Cycle: Unmated adult insects overwinter under garden litter, vines, or boards to emerge, mate, and lay eggs in spring; nymphs take all summer to develop, molting 5 times before maturity.

Control: Maintain vigorous plant growth; handpick all stages from undersides of leaves; support vines off the ground on trellises; attract native parasitic flies with pollen and nectar plants; cover plants with floating row cover (hand-pollinate flowers); spray rotenone or sabadilla.

Squash Vine Borer *Melittia cucurbitae*

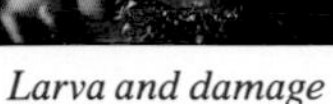
Larva and damage

Adult

Damage

Description: Adults: narrow-winged, olive-brown, 1″–1½″ moths, with fringed hind legs, clear hind wings, and red abdomens with black rings. Larvae: white with brown heads. Found throughout the United States and Canada east of the Rocky Mountains and south to Mexico.

Damage: Larvae bore into vines of squash, pumpkins, cucumbers, melons, and gourds. They chew the inner tissue near the base, causing vines to wilt suddenly; girdled vines rot and die.

Life Cycle: Larvae or pupae overwinter in the soil. Adults emerge in spring and lay eggs on stems and leaf stalks near the base of the plant. Newly hatched larvae bore into vine stems, causing sudden wilting and death of stems. Larvae feed for up to 6 weeks, then pupate in the soil.

Control: Early in the growing season, cover vines with floating row cover; uncover later for pollinators or hand-pollinate; spray base of plants with rotenone or pyrethrin repeatedly to kill young larvae before they enter vines; to save attacked vines, slit infested stems and remove borers, and heap soil over vines to induce rooting.

Stink Bugs (Family Pentatomidae)

Brown stink bug

Harlequin bug eggs

Harlequin bug

Description: Adults: shield-shaped, green, tan, brown, or gray, ½″ bugs; most species smooth, but a few spiny or rough-textured. Nymphs: oval-shaped, wingless; similar to adults. Eggs: barrel-shaped, often with a fringe of spines at one end, laid in clusters. Found throughout North America.

Damage: Adults and nymphs suck plant sap from leaves, flowers, buds, fruit, and seeds of cabbage family crops, squash, beans, peas, corn, tomatoes, and peaches. Feeding punctures in fruit cause scarring and dimpling known as cat-facing. The harlequin bug, a species with bright red and black markings, is the most important pest of cabbage family crops in the South.

Life Cycle: Adults overwinter in weeds in waste areas; females lay 300-500 eggs each when weather warms; eggs hatch in a week, and nymphs develop to adults in about 5 weeks. Two or more generations per year.

Control: Control weeds in susceptible crops; remove or mow weedy areas adjacent to garden beds; attract native parasitic wasps and flies by planting small-flowered plants. As a last resort, dust with sabadilla.

Strawberry Root Weevil *Otiorhynchus ovatus*

Adult

Description: Adults: black, shiny, hard-shelled, flightless, 1/4″ weevils. Larvae: white, legless, brown-headed grubs. Found throughout North America.

Damage: Larvae bore into crowns and roots of strawberries and into roots of raspberries, grapes, apples, peaches, arborvitae, and pine and spruce seedlings. Adults feed on leaves, needles, and fruits, clipping half-circles from edges of leaves.

Life Cycle: Larvae overwinter in soil and pupate in spring; a few adults overwinter in debris; in spring newly emerged adults feed for 2-3 weeks, then lay eggs throughout the summer; new larvae burrow into soil to feed on roots, then move deeper into soil until spring.

Control: Cover plants with floating row cover; apply insect parasitic nematodes to soil to control larvae; spray rotenone to kill leaf-feeding adults.

Tachinid Flies (Family Tachinidae)

Adult

Description: Adults: robust, gray, brown, or black, 1/3″-1/2″ insects resembling overgrown, bristly houseflies; some with mottled bodies but without bright colors. Larvae: maggots that feed inside host insects. Found throughout North America.

Beneficial Effect: Excellent predators of many caterpillar pests, including cutworms, armyworms, tent caterpillars, cabbage loopers, and gypsy moth larvae; some also attack sawflies, squash bugs, and stink bugs.

Life Cycle: Females lay eggs on newly hatched larvae or on leaves on which caterpillars are feeding; caterpillars then ingest eggs and larvae hatch inside the host; larvae feed and devour the host from within; in some cases females place live young on a caterpillar's skin and the maggots burrow into the host. As maggots develop inside, they kill the host, then pupate inside or in soil nearby. Up to 2 generations per year.

How to Attract: Adult flies feed on the nectar of flowers of dill, parsley, sweet clover, and other herbs, so allow weeds to flower throughout the garden; don't destroy caterpillars with white eggs stuck to their backs—these will develop into more tachinid flies.

Tarnished Plant Bug *Lygus lineolaris*

Tarnished plant bug nymph

Tarnished plant bug

Four-lined plant bug

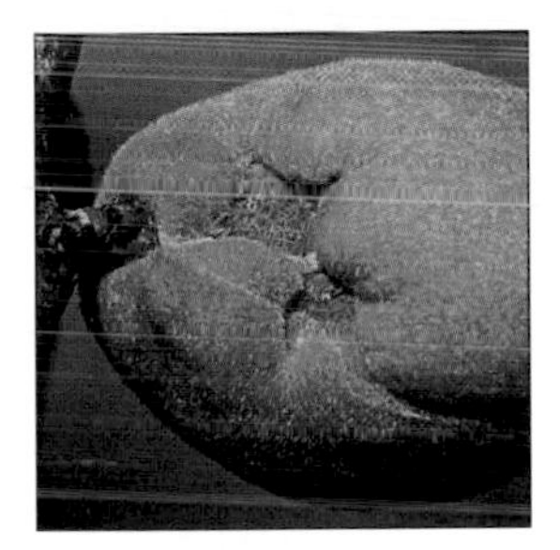

Damage

Description: Adults: oval, light green to brown, mottled, ¼″ bugs; each forewing with a black-tipped yellow triangle. Nymphs: yellow-green, wingless; similar to adults. Several other plant bug species (Family Miridae) also damage garden plants. Found throughout North America.

Damage: Adults and nymphs suck plant juices of most flowers, fruits, and vegetables, causing shoot and fruit distortion, bud drop, wilting, stunting, and dieback.

Life Cycle: Adults overwinter under bark or leaf litter, emerge in early spring to lay eggs in leaf tissue; eggs hatch in 10 days; nymphs feed 3-4 weeks, then molt to adult. Up to 5 generations per year.

Control: Cover plants with floating row cover; attract native predators (bigeyed bugs, damsel bugs, pirate bugs) with groundcovers and pollen plants; try releases of minute pirate bugs; spray rotenone or sabadilla as last resort.

Tent Caterpillars *Malacosoma* spp.

Eastern tent caterpillar egg mass

Eastern tent caterpillar

Eastern tent caterpillar

Tent

Eastern tent caterpillar and tent

Description: Adults: yellowish tan to brown moths with 2 narrow, diagonal stripes across wings (1″-1½″ wingspan). Larvae: black, hairy, 2″-2½″ caterpillars with a white stripe or rows of dots along the back and irregular, brownish blue or red marks along sides; most spin large "tents" of silk webbing in branch crotches of trees. Eggs: laid on twigs in masses, covered with hardened foamy layer. Eggs resemble a dark, shiny belt encircling a twig. Found throughout North America.

Damage: Larvae feed on leaves of most deciduous trees and shrubs, especially apples, aspens, and wild cherries. Trees may be fully defoliated in years of high caterpillar populations. Trees usually leaf out again later in the summer but growth may be stunted for several years.

Life Cycle: Moths lay eggs on twigs in midsummer; eggs overwinter and hatch in early spring; caterpillars move to nearest branch crotch and spin a silk tent for protection during rain or at night, and leave it to feed during the day. After feeding 5-8 weeks, they pupate in white cocoons attached to tree trunks or leaf litter; adult moths emerge in 10 days. One generation per year.

Control: Prune infested branches and burn them, or remove tents filled with caterpillars from branches by winding them onto a broomstick with nails projecting from it; in winter, remove egg masses from bare branches; attract native parasitic flies and wasps by growing small-flowered herbs, such as catnip, and wildflowers, such as Queen Anne's lace. Do not destroy wandering caterpillars with white eggs or cocoons attached to their backs; they are hosts for native parasites. Spray BTK weekly while larvae are small; try releases of spined soldier bugs.

Thrips (Family Thripidae)

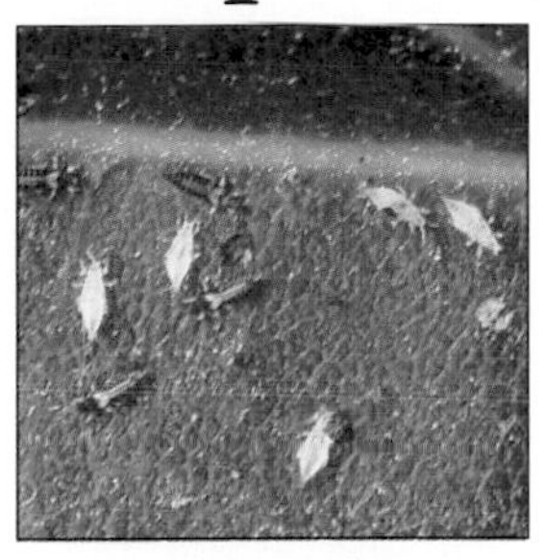
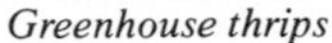
Greenhouse thrips

Western flower thrips

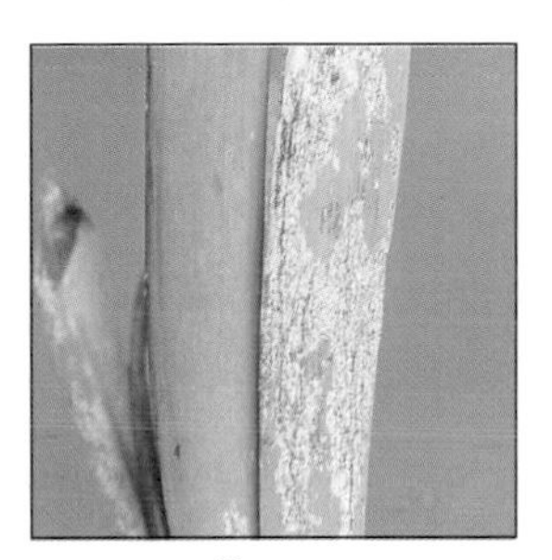
Damage

Damage

Description: Adults: slender, 1/50″–1/25″ long insects; colors range from yellowish to brown or black; these fast-moving insects leap or fly away on narrow, fringed wings when disturbed; individual insects difficult to see without a magnifying glass. Nymphs: similar to small adults; light green or yellow, some with red eyes. Found throughout North America.

Damage: Adults and nymphs suck contents of plant cells from a variety of garden plants, flowers, fruits, and shade trees. Their feeding leaves silvery speckling or streaks on leaves; severe infestations stunt plants and damage flowers and developing fruit; some species spread tomato spotted wilt virus.

Life Cycle: Adults overwinter in sod, plant debris, or cracks in bark, becoming active in early spring. Eggs are laid in plant tissue and hatch in 3-5 days; nymphs feed for 1-3 weeks, then rest in soil or on leaves until they molt to adult form in 1-2 weeks. Up to 15 generations per year outdoors; may breed continuously in greenhouses.

Controls: Spray dormant oil on fruit trees; encourage native predators, such as pirate bugs, lacewings, and lady beetles, for onion or western flower thrips, release the predatory mite *Amblyseius cucumeris* or minute pirate bugs (*Orius tristicolor*); hang blue or yellow sticky traps to catch adults; spray insecticidal soap, pyrethrin, or neem; as a last resort, spray ryania or dust undersides of leaves with sabadilla or diatomaceous earth.

Tiger Beetles (Family Cicindelidae)

Six-spotted tiger beetle

Description: Adults: long-legged, ½″-¾″ beetles with bright colors and patterns on their bodies; often attracted to lights at night. Larvae: segmented, S-shaped larvae, each with a pronounced hump covered with strong hooks on the fifth segment. Found throughout North America.

Beneficial Effect: Both adults and larvae feed on a variety of insects; generally beneficial for controlling pests.

Life Cycle: Females lay single eggs in soil burrows; larvae prey on insects that fall into burrows; larvae develop for several years before adulthood. One generation takes 2-3 years.

How to Attract: Maintain permanent plantings as refuges; do not use insect light traps.

Tomato Hornworm *Manduca quinquemaculata*

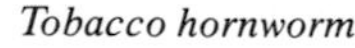

Tobacco hornworm

Tomato hornworm adult

Description: Adults: large, gray moths (4″-5″ wingspan). Larvae: green caterpillars up to 4½″ long with a black horn on tail, 8 diagonal white marks along the sides. The tobacco hornworm (*Manduca sexta*), a related species, has a red horn and 7 white marks on the sides. Found throughout North America.

Damage: Larvae of both species consume leaves, stems, and fruit of nightshade family plants. Feeding can kill young plants.

Life Cycle: In June and July, moths emerge from soilborne pupae; adults lay eggs on undersides of leaves; eggs hatch in a week; larvae feed for a month, then pupate in soil until the following summer.

Control: Handpick caterpillars from foliage; attract native parasitic wasps; spray BTK while caterpillars are still small.

Tussock Moths *Orgyia* spp.

Western tussock moth larva

Western tussock moth winged adult male and wingless adult female

Description: Adults: males are ash gray moths ($1\frac{1}{2}$″ wingspan); flightless females have stubby wings. Larvae: pale yellow caterpillars with red heads; bodies covered with tufts of light and dark hairs. Eggs: covered in white, stiff lather. Found throughout eastern United States and Canada, west to Colorado and British Columbia.

Damage: Larvae feed on leaves of deciduous trees and shrubs, but cause little permanent damage unless numerous.

Life Cycle: Overwintering eggs hatch in spring; larvae develop for 4–6 weeks, then pupate in cocoons on nearby trees; adults emerge in 2 weeks, mate, and lay eggs on old cocoons. Up to 3 generations per year.

Control: Scrape egg masses from tree trunks or branches; spray BTK to control larvae; attract birds and parasitic and predatory insects.

Vinegar Fly/Fruit Fly *Drosophila melanogaster*

Adult

Description: Adults: yellowish, clear-winged, $\frac{1}{10}$″ flies, also called fruit flies. Larvae: small white maggots. Found throughout North America.

Damage: Flies are a nuisance around commercial canning and packing houses, also in kitchens; maggots feed on microorganisms in decaying fruit or plants.

Life Cycle: Females lay up to 2,000 eggs each in overripe or fermenting fruit; maggots feed for several days then pupate for several days. Life cycle takes 10 days.

Control: The best control is sanitation, removing garbage and fermenting fruit; grow crack-resistant tomato cultivars to prevent fruit from being infested in the field.

Webworm, Fall *Hyphantria cunea*

Larvae and nest

Larvae and nest

Description: Adults: satiny white moths with black dots on forewings (1½″ wingspread). Larvae: beige caterpillars covered with dense yellow to brown hairs and long white hairs on sides. Found throughout United States and in southern Canada.

Damage: Larvae chew leaves and spin large, conspicuous webs over ends of branches of many deciduous trees and shrubs.

Life Cycle: Adults emerge from overwintering pupae in late spring and lay eggs; groups of larvae cover foliage with webbing and feed inside it through midsummer; larvae leave webs to pupate in soil debris. Up to 2 generations per year.

Control: Prune and destroy branches with webs; spray BTK on leaves around web when larvae are small or when last stage larvae wander outside of web; attract native parasitic wasps.

Webworm, Garden *Achyra rantalis*

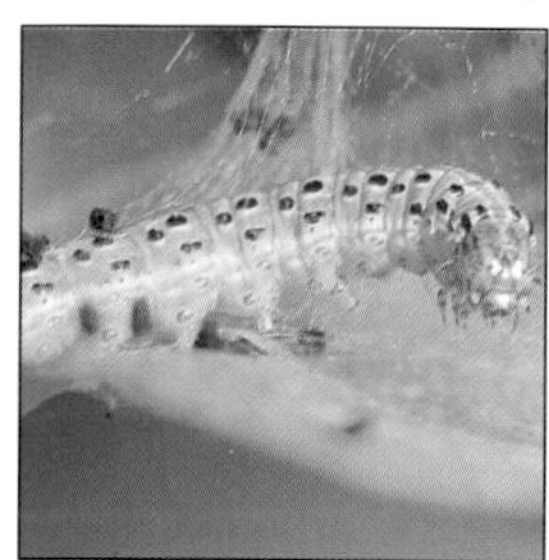
Larva and webbing

Adult

Description: Adults: brown moths with pale gold marks on wings (1″ wingspan). Larvae: pale green to nearly black caterpillars with a dark or light stripe down the back and 3 dark spots on the side of each segment. Found throughout North America.

Damage: Caterpillars spin silk webbing around leaves of vegetables, strawberries, and some weeds, devouring enclosed leaves.

Life Cycle: Pupae overwinter in soilborne cocoons; moths emerge in late spring to midsummer and lay eggs, which hatch in a week. Caterpillars feed for a month, then pupate. Up to 3 generations per year.

Control: Pick caterpillars from plants and drop into a pail of soapy water; destroy webbing; spray BTK on leaves when larvae are small, before webbing is extensive; spray pyrethrin or rotenone as a last resort.

Whiteflies (Family Aleyrodidae)

Citrus whiteflies

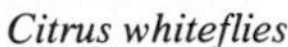

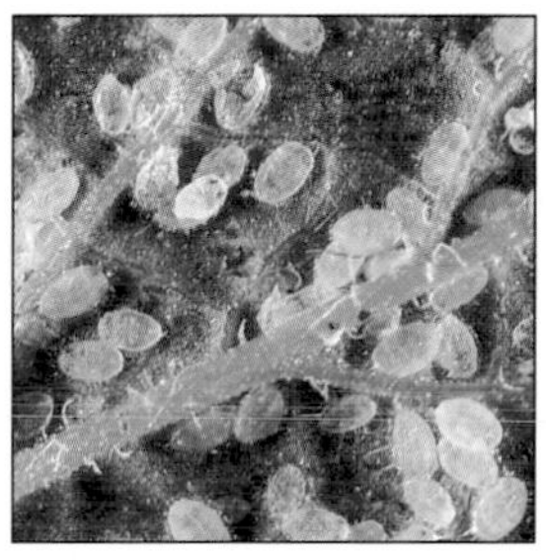

Greenhouse whitefly larvae

Greenhouse whiteflies

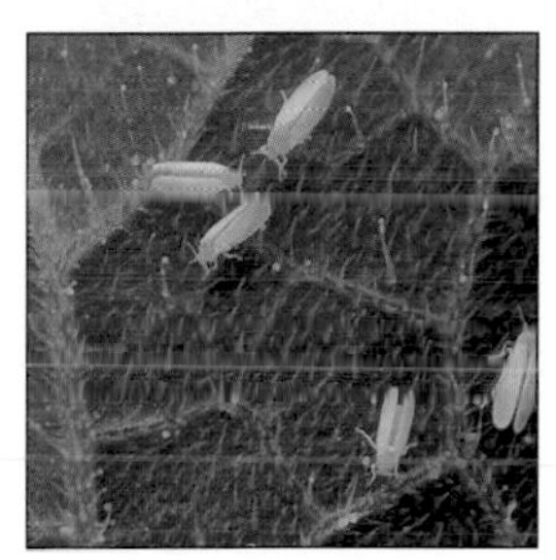

Sweet potato whiteflies

Description: Adults: minute sucking insects with powdery white wings; whiteflies rest in huge numbers on leaf undersides and fly out in clouds when disturbed. Larvae: flattened, legless, translucent, $^{1}/_{30}$" scales on leaf undersides. Eggs: gray or yellow cones the size of a pinpoint. Commonly found in greenhouses throughout North America; also found outdoors in warm regions of California, Florida, the Gulf states, and area on the West Coast.

Damage: Nymphs and adults suck plant juices from citrus, greenhouse foliage plants, ornamentals, and vegetables. Their feeding weakens plants; they also secrete a sticky, sugary substance called honeydew. Sooty mold, a black fungus, grows on the honeydew-coated leaves and fruit. Whitefly feeding can also spread viral diseases.

Life Cycle: Females lay eggs on undersides of leaves; these hatch in 2 days into tiny, mobile scales; while continuing to feed on plant juices, scales molt to a legless stage in a few days. After several growth stages, nymphs rest in a sort of pupal stage before emerging as adults. Most whitefly species require 20-30 days for a complete life cycle at room temperature, less in the summer. Numerous overlapping generations per year, continuing all winter in greenhouses and warm climates.

Control: Catch adults on yellow sticky traps; vacuum adults from leaves; indoors, release *Encarsia formosa* parasitic wasps to control greenhouse whitefly; outdoors, attract native parasitic wasps and predatory beetles; spray with insecticidal soap, kinoprene (Enstar), or garlic oil; as a last resort, spray with pyrethrin or rotenone.

Wireworms *Limonius* spp.

Larvae

Adult

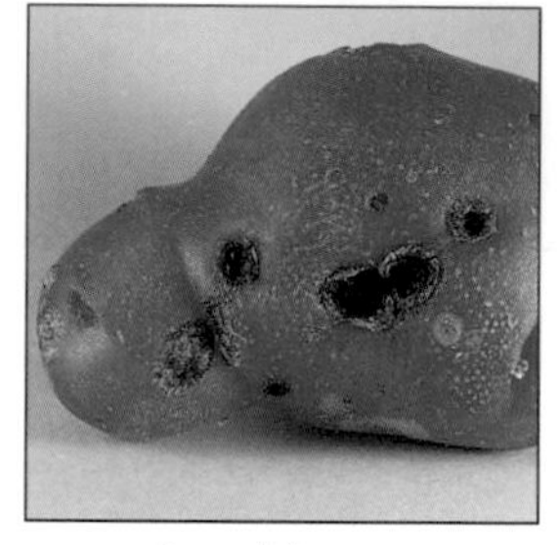
Larval damage

Description: Adults: hard-shelled, elongated, dark-colored, 1/3″-3/4″ beetles with lengthwise grooves on wing covers; often called click beetles, they make a clicking sound as they flip from their backs onto their feet. Larvae: yellow to reddish brown, jointed, wirelike, 1″-1½″ long worms. Found throughout North America.

Damage: Larvae bore into newly planted seeds or into plant roots, tubers, and bulbs, preventing germination or stunting and killing plants. Plants attacked include gladiolus and other flower corms, small grains, and most vegetable crops. Wireworms are most common in soil recently in sod. Adult beetles feed on leaves and flowers, but cause little damage.

Life Cycle: Adults lay eggs on roots in early spring; larvae hatch in 3-10 days; larvae spend 2-6 years feeding on surface roots in the spring and fall and move deeper into the soil to overwinter. Mature larvae pupate in late summer and overwinter as adults. One generation in 2-6 years.

Control: Cultivate thoroughly every week for 4-6 weeks in fall to expose and destroy larvae; delay planting tubers and corms until soil is very warm, keeping soil bare until planting; allow chickens to run on infested ground to eat larvae; bury raw potato pieces 4″-6″ deep to attract larvae, check every 1-2 days and destroy wireworms; apply parasitic nematodes to soil.

Woollybears (Family Arctiidae)

Saltmarsh caterpillar

Saltmarsh caterpillar adult

Description: Adults: white or yellowish moths with a small dark spot on each wing (1½″ wingspan). Larvae: densely hairy, 1″-2″ caterpillars; hair may be yellow, brown, or black at either end of caterpillar and rust brown around the middle. Often seen in fall;

rolls into a ball when touched. Found throughout North America.

Damage: Caterpillars feed on tender stems, leaves, or flowers of garden plants and ornamentals, chewing ragged holes. Populations usually not seriously damaging.

Life Cycle: Pupae overwinter in woolly cocoons among leaves, trash, or clods of soil. Moths emerge to lay eggs in spring. Two or more generations per year.

Control: Control usually not necessary. Spray BTK when caterpillars are feeding on plants.

Yellow Jackets *Vespula* spp.

Baldfaced hornet

Western yellow jacket

Description: Adults: ½″–¾″ wasps with a yellow and black striped abdomen and 2 pairs of membranous wings. Larvae: white grubs that develop inside cells in paper nests. Found throughout North America.

Beneficial Effect: Good general predators on flies, caterpillars, and other pests. Adult wasps generally feed on sugary solutions, such as flower nectar or juices of ripe fruit. Can be pests at picnics.

Life Cycle: Queen wasps overwinter under bark or in protected burrows, emerging in spring to build a small nucleus of paper cells in the ground, in an old log, or attached to a tree limb. Queens lay eggs and capture prey for the first brood of larvae, which hatch into workers; the feeding task is then taken over by succeeding broods of worker wasps. The colony expands until late summer; then worker wasps die off before winter, and a few young, newly mated females leave the nest to overwinter in protected spots as next year's queens. The original nest is deserted.

How to Attract: Not usually necessary to attract them or to control them. Since some species do sting when disturbed, remove nests from areas frequented by people; it is not necessary to remove nests located out of harm's way.

PART 3

DISEASES

Identifying the Causes—Implementing the Cures

The best approach to disease problems is to prevent them. Many good organic gardening practices are natural disease-preventive measures. Keeping soil healthy, keeping plants properly watered, cleaning up the garden, and rotating crops in the vegetable garden are on the list of practices that help discourage disease.

Dealing with plant diseases can be one of the most challenging aspects of pest control. Unlike insects, which are usually easy to spot, most organisms that cause disease are too small to see without a magnifying glass or microscope. Disease symptoms can be variable and subtle, so they're tricky to diagnose and sometimes easy to miss until the problem is quite severe. Once you diagnose a problem, there is not a broad range of organically acceptable controls. And in some cases, if you haven't noticed symptoms in time, there's no remedy, chemical or organic, for plant diseases. Your only option is to pull up the diseased plant and destroy it.

What Is Disease?

What do we mean when we say a plant is diseased? If you accidentally skin the bark off the base of a tree with your lawn mower, is that tree diseased? If spring frosts slightly burn the leaves on broccoli transplants, are those plants diseased? Are pea plants infected with beneficial root-nodule bacteria diseased?

Even plant pathologists—scientists who study plant diseases—don't agree on a single definition of disease. A working definition accepted by most is that disease is an irritation that disturbs a plant's normal functions (such as water uptake or cell division). Some scientists further restrict the above definition to conditions or organisms that cause continuous irritation, making the above-mentioned "lawn-mower blight" not a disease (unless you injure your plant every time you mow!).

In this chapter, we'll divide plant diseases into two broad categories. Infectious diseases are those diseases that can be transmitted from one plant to another. Noninfectious diseases—which we call plant disorders—are problems that cannot be transmitted.

How Diseases Affect Plants

The effect of a disease or disorder on a plant can range from a hardly noticeable decrease in yield to sudden wilting and death. All diseases interfere with normal plant growth, but the ways in which different diseases cause damage varies. Diseases caused by fungi and bacteria often weaken plants by literally eating food the plant has made for itself. Disease-causing organisms also harm plants by injecting them with toxins and by plugging up water and nutrient-conducting vessels. A disease can alter the hormone balance within a plant, resulting in galls that upset movement of food and water within a plant. Disorders such as nutrient deficiencies can slow growth, and prevent proper development of flowers or fruit.

However, not all conditions that meet the technical definition of disease are detrimental to plants. Nodules on the roots of peas and beans are caused by bacteria, and they do interrupt some of the normal plant functions, so they are a "disease." Their net effect, however, is beneficial because these bacteria take nitrogen from the air and convert it into a form that plants can use. Similarly, the roots of almost all plants are infected with mycorrhizal fungi. Although they sap some food from plants, these beneficial fungi help their host plants garner nutrients from the soil by increasing the effective surface area of the root system.

The first step you can take toward vanquishing diseases from your garden is to learn

Plant Diseases Make History

When humans first began cultivating plants, they unknowingly also increased the likelihood of plant diseases. As agriculture spread, farmers took plants out of their diverse natural environments and began growing similar plants close together; this made it easier for disease-causing microorganisms to spread from one plant to the next. Today, over 50,000 diseases are known to afflict cultivated plants.

Early Concepts of Disease

Though the causes of diseases were a mystery, plant diseases themselves were not overlooked by the earliest gardeners. Diseases figure into early religious writings and practices, including the Bible: "I smote you with blight and mildew; I laid waste your gardens and vineyards." (Amos 4:9).

In ancient times, many societies interpreted crop failure as a sign of retaliation from some higher power. Wheat rust, for example, was such a problem for the Romans that they created a god of rust. Each spring, before the rust appeared, the Romans held a festival to appease the god and attempt to save their wheat crop.

Although they did not develop any practical cures, there were some careful observers of plant diseases among the ancients. Theophrastus, a Greek philosopher and naturalist in the third century B.C., recorded which plants were most susceptible to diseases as well as the influence of soil and weather on disease.

The invention of the compound microscope in the early 1500s set the stage for an understanding of the causes for plant diseases. Almost a century later, the Dutch naturalist Antonie van Leeuwenhoek was the first person to actually see bacteria through the lens of a microscope.

Late Blight Blasts Ireland

It was a large-scale tragedy—the Irish potato famine—that provided the impetus for more research into the causes of plant diseases. In 1845 and 1846, cool, rainy conditions provided ideal conditions for the spread of the disease we now know as late blight. This blight swept through the potato crop, destroying the major food source of the Irish people. A million Irish peasants starved and more than a million additional people fled the country, many to America.

This great tragedy stirred a young scientist, Heinrich Anton DeBary, to review many of the previous experiments concerned with the causes for plant diseases. In 1853, at the age of 22, he published a book that conclusively proved that fungi were a cause and not a result of disease.

The Downy Mildew Disaster

Other diseases further spurred study of the biology and control of plant diseases. In the 1870s, downy mildew was accidentally introduced from its native home of North America into France. Downy mildew caused severe damage to French vineyards in just a few years.

During this time, Dr. Pierre Millardet, a French scientist studying vine diseases, was walking along a road in the Bordeaux region one day, and he happened to notice that some grape vines bordering the road were free of mildew. Upon inquiry, he found that the owner of the vineyard had made a solution of bluestone (copper sulfate) and lime and spattered it with a broom on those vines bordering the road, hoping that the ominous blue-green color of the grapes would discourage pilferers. Within three years, Millardet perfected the compound, which came to be known as bordeaux mix and is still in use today.

about the various types of diseases and how they develop or spread. The three most common kinds of organisms that cause infectious plant diseases are fungi, bacteria, and viruses. By learning about these organisms—how they infect plants and how they reproduce—you can plan a gardening strategy that will minimize disease. For example, it helps to know that warm, wet conditions encourage the spread of most fungal diseases and that insect feeding is one of the most important factors in the spread of viral diseases. You'll learn more about the types of infectious diseases and the environmental conditions that favor their development under "Infectious Diseases" below.

Understanding how environment and nutrient balance affect plant growth can help you prevent plant disorders. Learning about the different light requirements of plants will help you choose the right planting sites to encourage best plant performance. Realizing that a plant label can tighten around a growing stem to the point that it cuts off the flow of water and nutrients through the plant will help motivate you to remember to remove labels from newly purchased plants. You'll find more information about causes of plant disorders and how you can avoid them on page 355.

Infectious Diseases

When we talk about disease in relation to animals or humans, we talk about their being spread by germs. Germs are the tiny organisms such as viruses and bacteria that can cause disease. These types of organisms also cause plant diseases.

A more scientific term for disease-causing organisms is the word *pathogen*. In addition to fungi and bacteria, viruses, nematodes, and parasitic plants are plant pathogens. These organisms run the spectrum in food preferences. Some are nourished mostly from dead organic materials, and occasionally a living plant. Others can grow and multiply only when they have infected a living plant. Plant pathogens do not attack humans or other animals. An exception is certain viruses that multiply within the insects that carry them.

How Disease Develops

If you place a Colorado potato beetle on a potato leaf, and the beetle doesn't eat anything, you don't have a pest problem. Similarly, if you put rust fungus particles on a snapdragon leaf, and they don't infect the leaf, you don't have a disease problem. Disease only occurs when the proper environmental conditions exist to allow the pathogen to penetrate and grow into the host plant. You can picture this interrelationship between organism, environment, and plant as a triangle, like the one shown below.

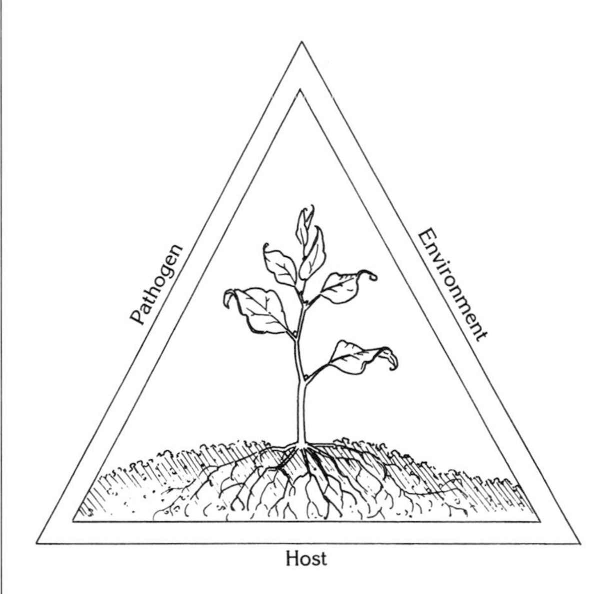

The development of disease depends on an interaction of several factors, including the host plant, the pathogen, and the environment. Each of these factors has to be present for disease to occur.

Before any disease symptoms show up on your plant, three conditions must be met.

1. The pathogen must be present. The stage of the pathogen that can infect a plant is technically known as inoculum. Inoculum may be a virus particle, a bacterial cell, a fungal spore or hypha, a nematode, or a seed or a piece of a parasitic plant.

2. There must be a susceptible plant. For example, a tomato plant is susceptible to damping-off disease when it is a small seedling, but is resistant to this disease when it is growing luxuriantly in midsummer. Club root fungi in the soil can attack cabbage roots, but cannot infect tomato roots.

3. Environmental conditions must be favorable. The most important environmental factors affecting disease development are moisture, temperature, light, and soil fertility. In the case of many diseases, a susceptible plant and the inoculum are present throughout the growing season. It is the daily changes in the environment that determine the extent of disease. A summer thundershower or a gusty, drying wind can change conditions almost within minutes, perhaps providing the right environment for disease to develop.

The Disease Cycle

The steps that a disease goes through over time is known as the disease cycle. The survival and spread of a disease depends on its success in completing all stages of its cycle.

Inoculation. This is the first step in the development of any infectious disease. Inoculation occurs when the disease-causing factor (the inoculum) comes into contact with the plant. There are many ways by which pathogens can spread to your plants, including wind, rain, insect feeding, contaminated tools, and in infected seeds or transplants. For more information on how pathogens disperse, see "Reproduction, dissemination, and overwintering" on page 344.

Penetration and infection. If the environment is right, infection begins soon after the inoculum comes into contact with a susceptible plant. Bacteria, viruses, and mycoplasmas can infect plants only through wounds or natural openings. Wounds are holes in the plant caused by pruning, careless use of equipment, animal feeding, storm damage, or rough handling by the gardener. Natural openings include leaf pores, called stomates, and openings near the ends of veins at the margins of leaves, called hydathodes. The scar that remains when a leaf falls may also provide entry for disease. Flowers provide natural openings for pathogen entry, and the pathogen may even be brought to the flower on an insect seeking nectar. The bark and branches of woody plants are pocked with small pores developed for gas exchange (called lenticels) that also provide entry for pathogens.

Fungi, nematodes, and parasitic plants may enter host plants either through natural openings or directly through intact surfaces. As spores of fungi and seeds of parasitic plants germinate, they develop small shoots and thin, elongating strands that can push right into the plant. Then, as they grow, the fungi secrete enzymes or toxins that dissolve or kill plant cells in their path. Parasitic plants send tentacle-like structures into the stem of the host plant. Nematodes penetrate cells with their sharply pointed mouthparts.

Just because a pathogen penetrates a host does not mean that the host will become diseased. Sometimes pathogens enter hosts that are not susceptible to the disease; in this case, the pathogen will die and no disease symptoms develop. However, once the pathogen successfully penetrates a susceptible host, the pathogen can establish contact with a cell and begin getting nourishment from the host plant. This is known as infection.

Incubation and invasion. The time between infection and the appearance of visible symptoms is called the incubation period. The incubation period may last days, months, or even years (in the case of some viral diseases). During this time, the pathogen will be growing, multiplying, and spreading, yet the plant will not show any symptoms of a disease problem.

Disease organisms may infect only a part of the plant, or may spread through the entire plant. Viruses and bacteria are systemic, living within the cells. In contrast, the fungus that causes sooty mold never penetrates the plant at all. (This fungus feeds on insect honeydew dripped on the surfaces of leaves and fruit.) Powdery mildew fungi grow mostly on the outside of the plant, except for small structures called penetration pegs that grow into the plant to extract food. In a disease such as apple scab, the fungus grows just beneath the outer layer of plant cells. The fungus that causes peach leaf curl lives within the plant but between plant cells, and the club root fungus is an example of a fungus that invades the plant cells.

Reproduction, dissemination, and overwintering. As the disease organisms spread through the host and obtain nutrients, they increase in size, number, or both. The way pathogens reproduce varies. Fungi produce spores, bacteria split apart, viruses are replicated by the cells they infect, nematodes lay eggs, and parasitic plants produce seeds.

To ensure their survival, pathogens have devised various ways of traveling to new host plants. Wind or water can spread inoculum over great distances. Inoculum also may hitchhike on insects. Many viruses, for example, are carried from one plant to the next on the mouthparts of aphids. Birds and animals may carry inoculum on their feet. Humans can spread inoculum as they work in the garden, touching plants with their hands and tools. A soilborne fungus such as *Fusarium* may be carried from infected soil to healthy soil on a trowel. Nematodes also transmit viruses as they feed and can carry bacteria and fungal spores as they move from plant to plant.

Plant Defenses

Considering the many methods pathogens have evolved to ensure their survival, you may wonder how you can have any healthy plants at all in your garden! Fortunately, plants are by no means passive players in the disease process.

Structural defenses. Plants have natural features that help ward off attack by a pathogen. A thick, waxy layer on the leaf surface can stop mechanical penetration by fungi; an impenetrable cell wall may impede progress of a fungus within a plant. A plant may be resistant to bacterial attack because it has small pores that bacteria cannot get through easily. Similarly, hairs on leaves or fruit of many plants may shed water, increasing resistance to diseases that need water in order to develop.

Chemical defenses. Some plants contain biochemicals that help them resist diseases. Cells commonly release various substances, some of which may be toxic to nearby fungi or bacteria. For example, wilt-resistant pea cultivars exude from their roots a chemical that is toxic to wilt-producing fungi. Plant cells may also contain protective biochemicals that can slow or stop the growth of a pathogenic fungus or bacterium once it penetrates the plant. Part of the Verticillium resistance of potato is due to such protective biochemicals within the cells of resistant cultivars.

Some protective biochemicals are not normally found in healthy plants, but are produced after a plant has been attacked by a disease-causing organism. For example, one way fungi advance into plant tissue is by secreting enzymes that dissolve cell walls. If a particular plant can produce biochemicals that inactivate that enzyme, the attack will be stopped.

In other cases, a plant might produce biochemicals directly toxic to the attacking pathogen. For instance, specific biochemicals have been found that protect certain cultivars of sweet potato from black rot. Some plants can transform a toxin produced by a pathogen into a nontoxic product. Tomato cultivars resistant to Fusarium wilt can metabolize the toxin produced by the *Fusarium* fungus into a harmless product.

Significant Symptoms

Infectious diseases are often classified by the type of symptom they cause. If you can identify the symptoms, you may be able to successfully control the disease, even if you don't know the specific pathogen causing the infection. The appearance of some common disease terms are explained below. If these symptoms seem to match ones on your plants, turn to "Disease Symptom Guide" on page 370, where you'll find additional descriptions and photographs of particular diseases as well as information on how to prevent and control them.

Blights: When plants suffer from blight, leaves or branches suddenly wither, stop growing, and die. Later, plant parts may rot. Common blights include fire blight, Alternaria blight, and bacterial blights.

Cankers: Cankers usually form on woody stems and may be cracks, sunken areas, or raised areas of dead or abnormal tissue. Sometimes cankers ooze conspicuously. Cankers can girdle shoots or trunks, causing everything above the canker to wilt and die.

Galls: Galls are swollen masses of abnormal tissue. They can be caused by fungi and bacteria as well as certain insects. If you cut open a gall and there is no sign of an insect inside, suspect disease.

Leaf blisters and curls: Blisters are yellow bumps on the upper surfaces of the leaves with gray depressions on the lower surfaces. On plants suffering from leaf curl diseases, the new leaves are pale or reddish and the midrib doesn't grow properly. The leaves become puckered and curled as they expand.

Mildews: There are two common types of mildews: downy mildew and powdery mildew. The primary symptom of downy mildew is a white to purple, downy growth, usually on the undersides of leaves and along stems, which turns black with age. Powdery mildew first appears as a white to grayish powdery growth, usually on the upper surfaces of leaves.

Rots: Rots are diseases that decay roots, stems, wood, flowers, and fruit. Some diseases cause leaves to rot, but those symptoms tend to be described as leaf spots and blights. Rots can be soft and squishy or hard and dry.

Rusts: Rusts are a specific type of fungal disease. Many of them require two different plant species as hosts to complete their life cycles. Typical rust symptoms include a powdery tan to rust-colored coating or soft tentacles. Cedar-apple rust and white pine blister rust are two common rust problems that can appear in home landscapes.

Wilts: Plants wilt when they don't get enough water. When fungi or bacteria attack or clog a plant's water-conducting system, they can cause permanent wilting, often followed by the death of all or part of the plant. Wilt symptoms may resemble those of blights. Wilting may also be from a cultural problem, such as improper watering.

Fungi

Of all the plant pathogens, the ones you'll deal with most frequently are the fungi. All plants are susceptible to attack by some type of fungus. There are more than 100,000 species of fungi, about 8,000 of which cause plant diseases.

What They Are

From the gardener's perspective, fungi are generally beneficial. Fungi decompose dead plants and animals, recycling nutrients back into the soil. Fungi also help aggregate soil particles into clumps, creating pore spaces that allow the soil to hold both air and water for good plant growth. Many beneficial fungi suppress the development of other fungi that cause plant diseases. Yeasts that ferment malt into beer or grapes into wine are fungi, as are mushrooms that we eat.

If you examine the roots of almost any plant, you'll find that they are infected with a beneficial type of fungus called mycorrhizal fungus. Infection actually helps the plants by improving the uptake of nutrients, especially phosphorus. The fungi also influence a plant's ability to tolerate drought and to ward off microorganisms that attack roots. Soil sterilization and certain pesticides will discourage mycorrhizal associations.

However, fungi are responsible for some common garden plant diseases, including powdery mildew, damping-off, late blight, apple scab, and corn smut.

Fungi can often be seen with the naked eye, as is the case with mushrooms, molds, and mildews. Fungi are multi-celled and have threadlike bodies called hyphae that spread over plants. The hyphae sometimes grow mostly on plant surfaces and sometimes penetrate plant cells. With powdery mildew infection, hyphae form a white coating on the plant surface and occasionally send feeding pegs into plant cells. In other diseases, fungal hyphae grow right into the plant, secreting enzymes that dissolve cell walls as they proceed.

Fungi also form spores, which are tiny, seedlike structures ranging from 1 micron (0.001 millimeter) to 1 millimeter in size. Spores are more tolerant of unfavorable conditions, such as winter cold or summer heat, than actively growing hyphae, so spores are the overwintering form of most fungi.

Most fungi produce two or three different types of spores during their life cycles; fungi that cause rust diseases may produce five different types. Spores may be produced on fruiting bodies that range in size from microscopic to the size of a basketball. Mushrooms and truffles are also fruiting bodies.

What You See

Fungal diseases result in a spectrum of plant symptoms on roots, stems, leaves, and flowers. Fungal diseases fall into one of two general categories: those from fungi that live in the soil, attacking roots or crowns of plants; and those from fungi whose spores are dispersed in the air, attacking aboveground parts of plants.

The list below identifies some of the most common symptoms caused by fungal diseases.

- **Damping-off** can kill seedlings before they even break through the soil, but it also strikes seedlings just an inch or so tall. The fungi rot the stem right at the soil line and, overnight, infected seedlings topple over.
- **Root rots** generally attack older plants, killing the tiny rootlets and appearing above ground as stunting and wilting.
- **Fungal wilts** damage a wide range of plants, plugging up the plant's water-conducting vessels and causing leaves to wilt and die.
- **Club root** commonly infects cabbage family plants, causing large swellings on roots and stunted or dead plants.
- **Blights** include early and late blight, which attack tomatoes and their relatives. The fungi can damage or kill leaves and cause rot

in the fruit or tubers. Other fungal blights, such as juniper blight, attack woody plants.

■ **Mildews** include downy and powdery mildew. Infection results in spots or white patches on leaves, shoots, and other plant parts. Downy mildew can kill plants rapidly; powdery mildew commonly causes poor growth and lower yield, but seldom kills the plant.

■ **Rusts** produce orange or white spots, usually on leaves and stems, weakening plants and reducing crop yields.

■ **Leaf spot** symptoms are caused by a wide range of fungi, including Alternaria, Septoria, and anthracnose.

Other fungal diseases cause swellings on plant parts, such as black knot of plum and cherry, or sunken areas in stems, as with Cytospora canker. On fruit, fungi can cause hard, black patches (scab), soft spots (rot), or fuzzy gray mold (Botrytis rot).

How They Spread

Spore production gives fungi the ability to travel great distances. The spores are easily picked up and carried by water or animals—including gardeners! Spores are light enough to waft up into plants from the ground, as apple scab spores do when they drift from dead leaves lying on the ground up into apple trees in spring. Spores of a fungus such as cedar-apple rust can travel miles from cedars to infect apples, and spores of a disease such as wheat rust can hitchhike hundreds of miles on the atmospheric jet stream.

How to Prevent

One of the best ways to prevent fungal problems is to select plants that are resistant to the fungal diseases common in your area. However, even if you plant a susceptible cultivar or species, there are some steps you can take to help lessen the chance of fungal infection.

Once spores alight on a susceptible plant, they will germinate and produce hyphae if conditions are favorable. Most spores germinate best when they are surrounded with water and temperatures are warm. There's not much you can do about temperature (and some spores, such as those causing late blight of potatoes, actually germinate best at cool temperatures). However, you can help minimize the times when plant leaves are wet by choosing a well-drained, upland planting site that is basked in sunlight and bathed in gentle breezes. Pruning keeps plants open, allowing air and sun to quickly dry branches and leaves following rains or morning dew. Crop rotation can starve out soilborne pathogens.

Although they are not usually thought of as fungicides, antitranspirants and superior oil sprays may help protect plants from fungal infections. Antitranspirants, such as Wilt-Pruf, are normally sprayed on trees and shrubs to protect them from winter damage. Oil sprays are commonly used for controlling insects. Both of these products form a coating on leaves that seems to prevent spore germination and penetration. For more information on using antitranspirants, see page 441; see page 478 for details on applying superior oil.

How to Control

Even though preventive measures can go a long way in reducing the incidence of fungal diseases, you will probably have occasional need for some control measure. For airborne fungi, which produce symptoms on aboveground plant parts, picking off the infected part can help reduce the spread of the disease. If a plant is seriously infected, remove the entire plant. Place the infected pieces or plants in the center of a hot compost pile, or place in sealed containers and dispose of them with household trash.

For serious outbreaks, you may choose a homemade or commercial organic fungicide. Home remedies include baking soda spray for black spot on roses and garlic spray for a range of fungal problems.

Sulfur is probably the most commonly used organic fungicide, although plain sulfur is more a protective measure than a control. Sulfur doesn't kill fungal spores, but it does prevent them from germinating on the plant surface. Another useful control is lime-sulfur, which can kill recently germinated disease spores. Copper-based fungicides, such as copper sulfate and bordeaux mix, also inhibit the germination and growth of fungal spores.

Soilborne fungi are the most difficult to control. If possible, dig up and dispose of the infected soil. Or try solarizing your soil by laying a sheet of clear plastic over the moistened area and letting the sun "cook" the soil for a few weeks. For more details on soil solarization, see page 424.

Bacteria

Bacteria are found almost everywhere on Earth, even in such inhospitable habitats as deserts, hot springs, and highly acidic waste products from mining operations. These single-celled organisms can cause many serious plant diseases, including soft rot, crown gall, fire blight, and bacterial wilt.

What They Are

Bacterial cells are large enough to be visible through the common light microscope; still, 25,000 cells laid end-to-end would make up only 1 inch. The cells are of various shapes, including spheres, rods, spirals, and filaments. Those that cause plant diseases are mostly rod-shaped.

Bacterial cells divide by a process called fission. Each cell pinches itself in half, then the halves separate, resulting in two cells. Under ideal conditions, a single cell can divide every 30 minutes. If the resulting cells from each division keep on dividing, this would result in 8,388,608 bacterial cells in only 12

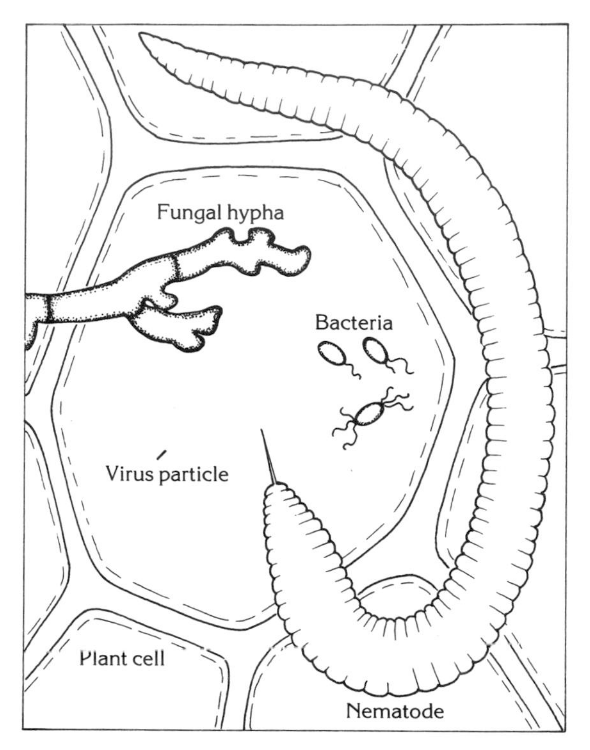

Plant pathogens come in a range of sizes. Large pathogens, including nematodes and many fungi, live mainly outside plant cells and insert specialized feeding structures into cells. Other pathogens, such as virus particles, are small enough to live entirely within cells.

hours! We are not knee-deep in bacteria because conditions are rarely ideal for continued growth of bacteria. As bacterial populations grow, they become overcrowded, use up their food supply, and wallow in their own waste products. Growth then stops or slows dramatically. Bacteria also are susceptible to infection by viral diseases.

Most bacteria are beneficial, increasing the fertility of the soil as they take nitrogen from the air and make it available to plants, and recycling nutrients in dead plants and animals. A single teaspoon of good garden

Name That Disease

Plant pathologists classify most disease organisms with a scientific name—called a binomial—consisting of two parts. The first word of the name is the genus, the second word is the species. Usually, these words are derived from Latin words. In most cases gardeners don't use scientific names when describing diseases. However, in some cases, knowing the scientific name of a disease can be helpful.

For example, both apples and cucumbers can suffer from powdery mildew. *Podosphaera leucotricha* is the organism that causes powdery mildew on apple, while *Sphaerotheca fuliginea* is the one that causes powdery mildew on cucumber. Armed with the knowledge that the fungus that causes powdery mildew on apples is different from the one that causes powdery mildew on cucumbers, you know that you need not worry about mildew spreading from your apple tree to your cucumber plants, or vice versa.

Common names of diseases are often based on disease symptoms: Apple scab, tomato leaf spot, fire blight, and damping-off are examples. Sometimes the names don't describe the disease with complete accuracy. For example, *blight* technically refers to rapid browning and death of leaves, flowers, or stems. In the case of chestnut blight, however, death is not so quick. Other common names of diseases include the scientific name of the pathogen, as in the case of Verticillium wilt and Fusarium wilt.

A virus usually has a common name that includes the plant it was first identified on, along with the symptom produced. Thus, cucumber mosaic virus was first identified on cucumber and produces a patchy yellow mottling of the leaves. But cucumber mosaic virus also attacks many other plants, including beans, celery, petunias, and delphiniums, and it doesn't always produce the distinctive mosaic symptom. Scientists are working on a more precise classification system for viruses, based on their chemical makeups, their shapes, and their modes of transmission.

soil is teeming with millions of bacteria. Friendly bacteria inhabit the digestive tracks of animals (including humans), aiding digestion, and are used in making such foods as yogurt and cheese. Special strains of bacteria even have been used to help clean up oil spills.

About 200 different bacteria are responsible for plant diseases. Warmth and moisture are most conducive to bacterial growth, so bacterial diseases generally are worse in warm, humid climates. Hence, pears, which are subject to the bacterial disease fire blight, are not extensively grown on a commercial scale in the hot, humid Southeast.

What You See

One distinctive symptom of some bacterial diseases is the sticky, gummy material secreted by active bacterial cells. If the leaves on your cucumber plant are wilting and you suspect bacterial wilt disease, cut the stem or leaf stalk with a sharp knife. If you see threads of slime when you pull the stem or stalk apart, this confirms bacterial wilt disease. Active fire blight cankers are covered with a similar bacterial slime. In many cases the bacterial slime has an unpleasant odor.

The following list explains some other common bacterial disease symptoms.

■ **Wilts** grow within a plant, causing bacteria to clog the plant's water-conducting vessels and the leaves to droop. If your cucumbers have ever been infected with bacterial wilt, you have seen a dramatic demonstration of this symptom—almost overnight, first one, then successive leaves collapse.

■ **Leaf spots** can appear on leaves when attacking bacteria kill plant cells.

■ **Soft rot** develops when bacteria infect fleshy fruit, tubers, or roots, secreting enzymes and perhaps toxins that break down the cells. These infections begin as small, water-soaked spots that turn mushy and smelly.

■ **Galls** form when plants respond to bacterial attack by growing an excess of new cells. Crown gall, for example, produces swollen knots of plant tissue on roots and stems.

■ **Cankers** are sunken areas produced by dead plant cells. Cankers often ooze a slimy or gummy substance.

How They Spread

Bacterial cells are spread around your garden by splashing rainwater, running water, insects, and animals, or on tools and diseased plants that you move from one place to another. Once bacterial cells are close to a susceptible plant, they can enter through wounds or natural openings. Wounds through which bacteria penetrate plants may be caused by insects, such as the cucumber beetles that transmit bacterial wilt, or by nematodes, which help spread bacterial wilt of tomato. You may inadvertently pick up and spread bacterial disease with your pruning shears. Wounds from hail damage also provide easy entry for bacteria.

Natural openings through which bacteria may enter plants include the small holes in leaves through which water vapor, carbon dioxide, and oxygen pass in and out. Other entrance points include the tiny glands at the base of flowers (known as nectaries) and the pores in tree bark (known as lenticels). Generally, bacteria most easily infect and cause greatest damage to young, succulent shoots, perhaps because their natural barriers are less well developed than those of mature wood.

How to Prevent

Good cultural practices are essential to prevent bacterial diseases. Always start with disease-free seeds and plants. To allow woody plants enough time to toughen up their new growth before fall, don't stimulate them in late summer by fertilizing, watering, or pruning. Trees with soft or succulent growth may suffer cold damage such as bark cracking in winter, providing entrance for bacteria. To reduce the chances of spreading disease inoculum, avoid working among wet plants. Crop rotation helps control soilborne bacterial disease, such as crown gall. If you know a particular bacterial disease is prevalent in your area, look for disease-resistant species or cultivars.

The best way to avoid soft rot on stored produce, such as carrots and potatoes, is careful harvesting to minimize wounding. During storage, keep moisture levels just high enough to prevent shriveling.

How to Control

Once they get started, bacterial diseases are generally difficult to control. Pick or prune off infected plant parts as soon as you spot them. Be sure to dip pruning shears in a 10 percent bleach solution (1 part bleach to 9 parts water) between cuts. Bury the infected pieces in the center of a hot compost pile or dispose of them with household trash. Certain sprays, such as copper compounds, are partially effective, but will not control these diseases when used alone. Bacteriophages (viruses that cause disease in bacteria) may one day find more widespread use against bacterial diseases.

Mystifying Mycoplasmas

Some diseases once thought to be caused by viruses, such as aster yellows, are now known to be caused by mycoplasma-like organisms (MLO). Mycoplasmas are extremely small, somewhere in size between common bacteria and viruses. Although generally grouped with bacteria, mycoplasmas differ from bacteria because they lack a cell wall. Each mycoplasmal cell is enclosed in a soft membrane. But mycoplasmas and MLOs, like bacteria, respond to antibiotics. They are also similar to viruses in that they are transmitted by insects.

Viruses

Among plant pathogens causing infectious diseases, the smallest are viruses and related viruslike organisms. Viruses attack every form of life on Earth, including humans (causing such diseases as smallpox, measles, and mumps) and other animals, trees, herbaceous plants, algae, fungi, and bacteria.

What They Are

Virus particles are only about 0.1 to 0.01 microns in size (1 micron=0.001 millimeter). They can only be seen with the aid of an electron microscope. The typical virus can hardly be considered alive, consisting not of a cell but merely of nucleic acid (RNA or DNA) surrounded by a protein sheath. Viroids are even simpler—each is just a strand of RNA.

Viruses and viroids are inactive outside of living cells. Once inside a live cell, though, these pathogens use the cell's "machinery" to multiply themselves, upsetting the cell's metabolism and causing disease. A single infected plant cell may become home to over a million virus particles.

While they generally are considered a problem, viruses can sometimes give a plant desirable properties. For example, solid-color tulips, when infected with certain viruses, become mottled with bold blotches of color (these types are sold as Rembrandt tulips). Another virus infects an apple cultivar known as 'Malling 9', which is used as a rootstock to dwarf apple trees. When special propagation techniques were used to rid some 'Malling 9' plants of the virus, resulting trees no longer were as dwarfed.

What You See

Of all potential problems you encounter in your garden, viral diseases may be the most difficult to identify. Symptoms of a specific viral disease can vary from one plant to the next, and also may vary depending on plant age or growing conditions. A plant can harbor a virus but not show any symptoms, or show symptoms only when cool weather slows growth. Virus symptoms may be hardly noticeable, causing a slight reduction in yield or growth, or may slowly become more pronounced, causing a gradual decline ending in death of the plant.

You may be able to identify a viral disease that produces characteristic symptoms and is common in your area. Green and yellow mottling on cucumber leaves, for example, very likely indicates cucumber mosaic. For positive identification where doubt exists, scientists rely on tests of plant sap, or they graft a bud from a suspected plant onto another plant known to produce characteristic symptoms of the particular virus in question.

Some general types of symptoms that characterize viral diseases include:

■ **Mosaic** causes normally green leaves or fruit to become mottled with patches of light green, yellow, or white areas. On flowers, mosaics can result in color breaks, and the flowers may be disfigured as well.

■ **Rosetting** is the term used to describe the short, bushy growth caused by some viruses. Peach rosette is an example of a virus that causes stunting by telescoping down the distance from one leaf to the next along the stem. Leaves or side branches are so close together that they grow in a rosette.

■ **Ring spot viruses,** including those that cause peony or dahlia ring spot disease, show up as pale, yellow spots on the leaves.

■ **Leaf curling** or deformed leaves is another common virus-produced symptom.

How They Spread

Viral diseases are not spread by wind or water. The virus particles must be brought in contact with plants and then either rubbed against or injected into the plant so they enter the sap.

Many insects, including aphids, leafhoppers, and whiteflies, carry viral diseases from infected to healthy plants. Mites, nematodes, and fungi also transmit viral diseases, as do parasitic plants. Any of these organisms cause plant damage by themselves, but they pose an even greater threat when carrying a virus. Plants tolerate a certain amount of aphid feeding, for example, with no apparent harm. However, if those aphids inject a virus into one of your plants, the plant becomes permanently diseased.

Even you can unknowingly spread viral diseases to your plants. Smokers are likely to pick up tobacco mosaic virus on their fingers, and from there, transmit it to other susceptible plants. The virus can infect hundreds of different plants, including tomatoes, peppers, eggplants, petunias, apples, and grapes, and can survive for decades in dried tobacco leaves.

You also may transmit viruses as you propagate plants. If you graft part of a virus-infected plant—even a single bud—onto a healthy tree, viruses from that scion may infect the whole tree. Other methods of vegetative propagation, such as division or cuttings, will usually also produce infected plants if the parent plant was carrying a virus. Occasionally, viruses move from one plant to another by naturally occurring bark or root grafts. Viruses are only rarely carried in seeds or pollen.

How to Prevent

The best way to deal with potential viral infections is to avoid them in the first place. Since viruses usually spread within a whole plant, it is useless to prune off infected leaves or stems. Also, there are no sprays that cure viral diseases. If possible, do not handle plants when they are wet. Wet plants bruise more easily, providing entrance for viruses.

If you propagate plants yourself, make sure you begin with disease-free plants. When buying virus-prone plants, such as strawberries and raspberries, choose ones that are certified to be virus-free. The raspberry plants your neighbor offers you may cost you nothing, but if they are a few years old, they also probably have viral infection. To avoid spreading tobacco mosaic virus, wash your hands after handling any tobacco product, and do not smoke or chew tobacco when working with plants.

Many viruses also attack weeds, so good weed control can help reduce problems. When possible, remove nearby wild or abandoned plants known to be susceptible to viruses that also attack your plants. Ragweed, nightshade, and milkweed, for example, are hosts of cucumber mosaic virus.

Sometimes it is possible to avert viruses by controlling insects that spread the disease. But good control of the insect is necessary, and you must keep the insect from the plant or kill it before it has a chance to inject virus particles into the sap. You can protect your plants by covering them with floating row cover. Reflective mulches of aluminum foil have also been successful in repelling virus-transmitting aphids from plants.

If you expect a virus problem, you may be able to plant a virus-resistant cultivar. For example, many peas and beans have been bred for resistance to bean mosaic, and breeders have developed tomatoes that resist tobacco mosaic virus. In some cases, a cultivar may be susceptible to a virus, but not prone to getting it because it is resistant to insects that transmit the virus. Such is the case with 'Royalty' purple raspberry, which, though prone to mosaic, is resistant to aphids that spread the disease.

How to Control

When the presence of a viral disease is suspected or confirmed, ruthlessly pull out and destroy unhealthy plants to prevent the disease from spreading to those that are healthy.

In the future, control may be achieved by deliberately infecting plants with certain viruses. A plant infected with one strain of a virus usually is protected from infection by another strain of the same virus. Ongoing research involves purposely inoculating a plant with a virus strain that produces only mild disease symptoms to prevent infection by more damaging strains.

Nematodes

We share this planet with thousands of species of nematodes. While most nematodes are not harmful, there are a number that parasitize and cause disease in humans and other animals. Also, unfortunately, there are many that attack and feed on living plants.

What They Are

While they are often described as wormlike, nematodes are not closely related to true worms. They are multi-cellular animals with smooth, unsegmented bodies. The nematode species that feed on plants are so tiny that you need a microscope to see them. They are often long and slender, although some species appear pear-shaped during their life cycles.

Feeding habits of nematodes vary, depending on the species. Some feed on the outside of the plant, while others burrow into plant tissue. While soil-dwelling nematodes are the most common culprits, some species damage stems, foliage, and flowers.

No matter where they feed, these tiny creatures can seriously damage your plants. Plant-parasitic nematodes have sharply pointed mouthparts that they use to puncture cell walls. The real damage occurs when the nematode injects saliva into the cell through its mouthparts, and then sucks out the cell contents. The plant responds with swellings, distorted growth, and dead areas. Nematodes can also carry viruses and inject them into plants. The feeding wounds they make also provide an easy entrance point for bacteria and fungi.

There are many species of beneficial nematodes that live in the soil. They may feed on decaying material, insects, or other nematodes. For more information on these beneficial nematodes, see page 309.

What You See

Unlike most other disease-causing organisms, plant-parasitic nematodes seldom produce any characteristic symptoms. Most of the symptoms that do appear are vague and often resemble those caused by other factors, such as viruses, nutrient deficiencies, or air pollution. Nematodes feeding above ground may cause leaves, stems, and flowers to be twisted and distorted.

If nematodes are feeding on the roots, the plant may be yellowed, wilted, or stunted, often with a reduced crop yield. If you suspect nematodes on the roots, carefully lift one of the infected plants and wash off the roots for easier inspection. If nematodes are causing damage, you may see small galls or lesions, injured root tips, root rot, or excessive root branching. For a positive diagnosis, contact your local extension office for information on where you can get your soil tested.

How They Spread

Whether they feed above or below ground, most nematodes spend at least part of their life cycle in the soil. While they can't move very far under their own power, they can swim freely in water, and they move more quickly in moist soil. They are also spread by anything that can carry particles of infested soil, including tools, boots, animals, and infected plants.

How to Prevent

There are several cultural techniques you can use to reduce the chances of nematode problems. Rotate crops that are not prone to the same types of nematodes to keep pest populations low. Use clean tools and other good sanitation practices to minimize the spread of nematodes. Dig plenty of organic matter into your soil to promote populations of beneficial fungi that feed on nematodes. Look for plant species and cultivars that are resistant to nematode damage.

How to Control

If you suspect you have a nematode problem, consider solarizing the infested area by covering the wet soil with a sheet of clear plastic. The heat that builds up during a few hot weeks in the summer will kill many of the nematodes. For more details on soil solarization, see page 424.

Another control technique is to plant a cover crop of marigolds in infested soil to lower nematode populations. Apparently, some species of nematodes are lured to the roots of these plants, but once they are there, they are unable to lay eggs. Turning under the marigolds may also suppress nematode feeding for several years. Drenching the soil with neem may also be effective.

Parasitic Plants

While they are probably the least common problem you'll face, parasitic plants can attack the plants growing in your garden.

What They Are

Unlike most plants, parasitic plants seldom produce their own food through photosynthesis. Instead, these plants attach themselves to host plants and withdraw water and nutrients from the hosts. The two most common parasitic plants are mistletoe and dodder.

What You See

Plants attacked by parasitic plants show a variety of symptoms and signs. Here are some common characteristics.

- **Dwarf mistletoes** generally produce small tufts of short yellowish or greenish stems. They attack many kinds of conifers, such as pines, causing cankers or swellings on stems. The plants may be stunted, deformed, or killed; branches often break off at the cankers.
- **True mistletoes** are leafy. They primarily attack deciduous trees. The clusters of green stems and leaves are most obvious in winter, when the host plant's leaves have dropped.
- **Dodder** is a twining plant that has a threadlike, leafless, orange or yellow stem. Weakened by the parasite, infested plantings are quickly smothered by the rampantly growing vines.

How They Spread

Mistletoes produce sticky berries, which are either carried by birds or dropped from the mistletoe plant. Dodder spreads by seeds, which can be brought in with crop seeds or spread by animals or equipment.

How to Prevent

The best way to avoid problems with parasitic plants is to buy clean seeds and plants from reputable companies.

How to Control

For mistletoe, the only control measure is to remove the parasite. Either cut off the mistletoe stems or remove whole infested branches or plants.

Once you spot dodder on crops, clean any equipment you use in the area thoroughly before moving to an uninfected spot. Persistent hand-weeding is the only organic control.

Disorders

Various environmental and cultural problems are also considered diseases since they upset the plant's normal function. Because such diseases cannot be transmitted from one plant to the next, they can be called noninfectious diseases or physiological disorders.

Just because they cannot spread, however, does not make plant disorders any less serious than infectious diseases. Fortunately, disorders are often easy to avoid and to remedy with good garden management. It's important to be aware of the various factors that affect plant growth, so you can try to keep them balanced. It's also important to recognize the symptoms of disorders so you can treat them effectively, instead of mistaking them for an infectious disease.

Environmental Problems

Environmental problems are caused by a lack or excess of something that a plant needs to grow. If, for example, you try to grow a plant that likes cool temperatures, shade, and moist soil on a hot, dry site, you will obviously have difficulty. In other cases, though, the source of the problem is not so easy to determine.

What They Are

Some of the factors that can affect plant health are water, nutrients, temperature, light, oxygen, and air pollution. An imbalance in any of these factors can interfere with normal plant growth.

What You See

Sometimes it is easy to diagnose environmental problems. If the soil is dry, your wilted tomato plant is most likely suffering from drought, not from Verticillium wilt. Other problems, though, produce much more subtle symptoms. For example, continual exposure to even low levels of air pollution can reduce yields and plant vigor, making plants more susceptible to attack by pests or pathogens. Here are some common environmental problems and their associated symptoms.

- **Excessive water** often causes greenish yellow leaves, or plant wilting due to root rot. Some plants develop a condition called edema, when tiny white or brown blisters appear on stems or lower leaf surfaces.
- **Drought,** or too little water, can cause wilting, along with leaf scorch (browned leaf edges), early fruit or leaf drop, stem dieback, and plant death.
- **Irregular watering** or a sudden change in water status, such as a heavy soaking after a dry period, can cause fruit and root crops to crack and lower leaves to yellow and drop.
- **Nutrient deficiency,** if minor, may have barely noticeable symptoms. As a deficiency becomes more severe, the crop yield may decrease and the plant may show other symptoms. Symptoms common to several kinds of deficiencies include abnormal leaf color, curled leaves, dead growing tips, or smaller-than-normal leaves.
- **Nutrient excess** of some nutrients can cause symptoms similar to those of nutrient

deficiencies. In other cases, the effects of a nutrient excess are indirect. Too much nitrogen, for example, will produce lush, healthy-looking plants that produce hardly any fruit.

■ **Cold temperatures** can kill tender buds, growing tips, leaves, stems, flowers, or fruit. Roots may die, and trunks often crack or form cankers.

■ **Excessively high temperatures,** usually coupled with strong, direct sunlight, cause browned and blistered stems, leaves, or fruit. Young plants may die.

■ **Too much light** or strong sunlight may burn the leaves of shade-loving plants, causing brown patches or dead leaves. Heavy pruning may expose previously shaded tree limbs to the bright sun, producing brown patches on trunks and branches. Plants with purplish or yellow leaves often fade or burn in direct sunlight.

■ **Lack of light** or too much shade may cause pale leaves, spindly yellow stems, or death of the plant. Leaves of variegated plants may turn evenly green if they don't get enough sun.

■ **Lack of oxygen** is most commonly from overwatering. Water fills up the soil pores that normally hold air, so no oxygen is available to the roots. This causes the roots to die, which reduces the root area available for water uptake and causes the plant to wilt.

■ **Ozone pollution** causes mottling or yellowing of leaves, especially on the upper leaf surfaces.

■ **Peroxyacetyl nitrate (PAN) pollution** is common in urban areas. Damage due to exposure to PAN appears as silvery white or brown spots on the undersides of leaves.

■ **Sulfur dioxide pollution** may cause leaves to yellow or brown in between the veins. Sulfur dioxide also combines with moisture in the air to form acid rain. While the exact effect of acid rain on plants isn't known, it can lower the soil pH and cause nutrient imbalance symptoms.

How to Prevent

Putting the right plant into the right spot will go a long way in preventing environmental problems. Learn about a plant before you buy it, and make sure you have an appropriate site for it. For example, no matter how much care and love you give them, roses will seldom perform well in a heavily shaded yard. Figure out what growing conditions your yard has to offer, and then choose from the range of plants that are best adapted to that site. If air pollution is a problem, look for resistant species or cultivars.

Good garden maintenance also helps to keep plants healthy. During dry spells, water plants deeply and evenly. Observe your plants closely and test the soil every few years. Most organic fertilizers, and especially compost, supply a wide spectrum of plant nutrients.

How to Control

Once you identify the cause, "curing" an environmental problem may be fairly simple. If the soil is dry, soak the ground slowly and deeply to restore soil moisture. A 5- to 10-minute blast from a hose may make the top of the soil look wet, but it's probably still dry a few inches below the surface. A soil that is too wet, depriving plant roots of oxygen, is more difficult to fix. The best solution is to remove damaged plants and replace them with plants that are better adapted to wet soil.

If you think your plants have a nutrient imbalance, consider having the leaves analyzed. These tests may be available from your local extension agent or a private laboratory. If symptoms are from a nutrient imbalance, there may not be sufficient time to add soil nutrients and for plants to take them up before the growing season ends and the plant dies. In such cases you can spray your plant with a nutrient solution, which will be taken in through the plants' leaves. Organic fertilizers such as fish emulsion and seaweed extract can be applied in this manner, but make sure to follow directions to avoid burning your plants' leaves with too concentrated a solution. You can also use well-diluted compost tea in this way. For directions on making this tea see page 427.

Temperature extremes and air pollution can't be controlled. On woody plants, pruning off damaged growth and providing extra fertilizer and water may help the plant recover from damage. If too much sun or shade seems to be the problem, try moving the affected plant to a different light level, if possible.

Cultural Problems

Cultural problems are the things people and other animals do that injure plants. With a little care and common sense on your part, most of these problems are easy to avoid.

What They Are

People can damage plants in a variety of ways. Almost anything that you do in the garden, if you do it the wrong way or at the wrong time, can cause problems for your plants. Some insect controls, such as soaps and oil sprays, can damage plants if you apply them at the wrong time.

Other types of damage include the infamous "lawnmoweritis" and string trimmer injury. When these machines bump into trees, they cause unsightly wounds that are perfect entrance points for disease organisms and boring insects. Trees on construction sites are highly prone to being hit by equipment or to suffer root damage due to soil compaction. Tight plant labels and staking wires may cut into the bark of your tree and interrupt the flow of water and nutrients.

Poor cultural practices can make plants more susceptible to environmental problems. Fertilizing at the wrong time of year, for example, can encourage plants to produce tender growth that is easily damaged by cold temperatures. Cultivating too close to plants may damage their root systems, reducing their water uptake.

Animals also cause their share of plant damage. Birds such as sapsuckers can peck holes into bark, creating entry sites for diseases and insects. Tender plant crowns, roots, and bark make tasty winter fare for hungry deer, mice, and rabbits.

What You See

Cultural problems produce a wide range of symptoms, so it may be difficult to determine the exact cause. Observe the plant closely, though, and the problem may become apparent. Look for wounds in the stems, stakes or plant labels that are too tight, or teeth marks caused by animals.

Keeping complete garden records may also help you make a diagnosis. If a plant's leaves suddenly become distorted or damaged, looking at your records may help you realize that you applied some form of pest control under the wrong conditions. For example, you might have applied bordeaux mix when the weather was too cool and damp, which would account for the discolored patches on your apples.

How to Prevent

In general, the best way to avoid cultural problems is to be a perfect gardener. Then you would never bump your tree with a lawn mower, fertilize too late in the season, or cultivate too deeply. Obviously, few gardeners achieve this level of expertise. The garden is a constantly changing environment, reflecting the varying weather conditions and the changes in the plants themselves. If, however, you keep records of what you do in the garden and how the plants respond, you'll gradually learn what works best for your plants and conditions.

There are also some active steps that you can take to avoid cultural problems. To reduce the chances of hitting a tree with a mower or trimmer, plant a groundcover or apply mulch around the base of the tree so you won't have to trim close to the trunk. During the winter, though, pull the mulch away from the base of the tree; otherwise, the mulch will provide a great place for mice to live and feed on the bark and roots.

Remember to remove labels and to check wrapped or staked trees frequently, loosening the ties, if necessary. Whenever possible, protect trees from construction damage by roping off a large area around them, so trucks won't compact the soil around the roots or hit the tree.

How to Control

Cultural problems are often difficult to "cure." You won't know that something went wrong until the damage shows up. For example, if you cultivate too close to plants, you won't know that the roots were severely damaged until the plants start to wilt.

In most cases plants can outgrow damage caused by a cultural problem if it is not too severe. With some special care, such as careful fertilization and extra water, your plant will often recover and continue to grow. If you cause a wound on a tree, you can smooth out the edges to help it close faster; for instructions, see the illustration "Treating tree wounds" on page 235.

Disease Prevention and Control

Plant diseases tend to be minor problems in organic gardens. Techniques such as interplanting, following good sanitation practices, and choosing resistant cultivars tend to minimize the incidence and spread of disease organisms. While there aren't many organically acceptable chemical controls for disease, observant gardeners can catch problems early and treat them with less-toxic controls.

There are some plants that seem to be disease-prone, no matter how carefully you grow them. For example, many gardeners have problems with powdery mildew on garden phlox *(Phlox paniculata).* In many cases those plants just aren't well-adapted to the climate or growing conditions in a given area. The answer is to not plant those plants or to search for disease-resistant substitutes. For example, *P. maculata* is a good mildew-resistant alternative to garden phlox.

Preventing disease problems is largely a matter of common sense. Before buying a plant, learn about its potential problems and avoid those plants that are very disease-susceptible. Also, choose plants that are well-adapted to your growing conditions, and care for them properly to keep them vigorous and healthy. Be aware of the problems that tend to attack each plant and take the appropriate precautions to avoid a disease outbreak.

Most important, observe plants frequently. Early detection is the key to easy and effective disease control. If you catch a disease early on, a simple control such as handpicking the infected part may eliminate the problem. A few days of undisturbed development, however, may be all that disease needs to threaten the life of your plants. And as it is weakened by the disease, the infected plant becomes more susceptible to other problems. Plants with root rot, for instance, are more likely to suffer from winter cold injury and drought damage than healthy plants. And plants damaged by air pollution become more susceptible to insect attack. By keeping a watchful eye on your plants, you can spot diseases before they get out of hand.

Identifying Disease Problems

Suppose you wander out into your garden on a summer morning and you notice some spots on the tomato leaves. What should you do? Nothing—until you determine what caused the spots.

To determine what is causing the injury, you need to look closely at the plant for more information. While some diseases may only cause a single symptom, such as leaf spots, many produce more than one indication of their presence. Finding these other clues can help you identify the problem quickly and accurately. And by being aware of common symptoms and signs of diseases, you'll be able to spot problems early on, before they get out of hand.

Putting Together a Diagnosis

Diagnosing a plant problem is easier if you have a systematic approach. Follow these steps.

1. Identify the plant that is afflicted. If you use reference books to key out symptoms and signs, you may be on a wild goose chase if you haven't figured out what plant is affected. Also, some control products, such as horticultural oil, can injure plant foliage of certain plant species. If you haven't identified the plant, you won't know whether you're using the appropriate control.

2. Observe the symptoms and signs. When making a diagnosis, don't be too quick to assume the problem is a disease. For example, if your geranium leaves are covered with yellow spots, they may be suffering from a bacterial or fungal disease. But if you observe the leaves closely and find webbing and tiny black specks on the leaves as well, the plants are infested with spider mites. See "Pest Patrol Checklist" on page 9 for a detailed rundown of how to examine your plants.

3. If you know what diseases are very common in your area, decide whether the symptoms and signs you see are typical of one of those common problems. For example, you may know that powdery mildew on lilacs, scab on apples, black spot on roses, and leaf spot on tomatoes generally appear every year in your area. Other diseases may appear only sporadically in your area.

4. If the problem isn't one of those ever-present diseases, you'll have to do some research. You'll find information on common diseases in the "Disease Symptom Guide" beginning on page 370. You can also check to see if your plant is listed in the "Problem-Solving Plant Guide" beginning on page 11.

5. Once you're familiar with the symptoms caused by different disease organisms, you may not need to make a pinpoint diagnosis in order to know what to do. For example, if you see that plants in one part of a perennial bed are dying off and have blackened, soggy roots, you won't need to identify the specific organism causing the problem. You'll know that you should improve drainage in that part of the bed to make the environment less favorable for fungi and bacteria that cause root rot.

6. If you do want to make a specific diagnosis, keep in mind that disease symptoms can change as the disease progresses and that secondary symptoms can mask the original problem. Primary symptoms are those symptoms produced at the point where infection occurred. These are usually the most obvious clues for identifying a disease. Secondary symptoms are produced elsewhere on the plant, away from the original infection site. Tomatoes, for instance, commonly suffer from leaf spot disease. The spotted leaves are the primary symptom of disease, but as the disease progresses, the leaves may yellow and drop, exposing the fruit to intense sunlight. The resulting sunscald on the fruit is a secondary symptom.

With most root diseases, you probably will notice secondary symptoms first. If some of your cabbage plants are wilting (a secondary symptom), dig one up and inspect its roots. Are the roots white and well-formed, or are they a gnarled mass (the primary symptom), indicating club root disease? Wilting strawberry plants may have damaged roots, caused by black root rot (producing darkened roots) or red stele (with no side roots and a red core evident when a root is slit lengthwise).

7. If you have a serious disease problem in your garden that you cannot diagnose, you may want to turn to private consultants or to government sources, such as the Cooperative Extension Service, for help. Your local extension agent may immediately recognize the problem. If not, he or she can serve as a contact with specialists from your state university.

All the information you've gathered—symptoms, signs, weather, soil, disease distribution, and so on—will be useful in helping others pinpoint your plants' problem. The more detailed and the more accurate your information, the more reliable the diagnosis will be.

You also may be requested to send a sample of the diseased plant or plant part. Choose representative diseased samples: Don't send leaves that have been dead for a long time or those that are so affected that they are unrecognizable. Succulent plant parts such as leaves or young shoots shipped in sealed plastic bags tend to rot. Instead, wrap them in several layers of newspaper, which also will prevent crushing. Dry or woody plant material ships well in plastic bags. If possible, send the entire plant, and also include a specimen of a healthy plant or part of a plant of the same species and cultivar. Pack specimens for shipping in a sturdy container, such as a cardboard box or mailing tube, to prevent crushing.

Plant Detective Kit

When you're diagnosing plant problems, it's helpful to assemble as much information as you can. You can compare samples and your notes to reference texts or show them to your extension agent if you can't key out the problem yourself. When you go out to the garden, keep handy a small magnifying glass and small bags or vials for specimens. Also carry a notebook to record what symptoms occur and when. These notes may help in the diagnosis of the problem at hand and also serve as a quick reference in the future.

A rain gauge and a thermometer that registers minimum and maximum temperatures can also provide data that will be useful in making a diagnosis. Position the rain gauge and thermometer at a site similar to that experienced by your plants. To get the most accurate temperature reading, shield the thermometer from the direct rays of the sun; the north side of a pole is a good location.

General weather conditions can play a major role in determining which diseases affect your plants. Prolonged rainy, cool weather, for example, promotes late blight of potatoes. Powdery mildew, on the other hand, is favored by dry weather, especially when days are hot and nights are cool. Remember that weather not only influences the development of infectious diseases, but can itself lead to diseased conditions. Cold injury, for example, causes water-soaked splotches on young leaves or cracks in stems.

Keeping Disease Out

One of the best ways to avoid the worry of having to diagnose and treat a disease problem is to keep disease-causing organisms out of your yard and garden. If you're aware of how disease organisms can enter your yard or garden, you'll know how to keep them out.

Choose clean seeds. Several kinds of disease organisms, including those that cause anthracnose and bacterial blight, can overwinter on seeds. Buy seeds from reputable companies to minimize your chances of getting infected seed. If you choose to save home-grown seed, save seed only from healthy plants and pods, and store the seeds in a cool, dry place.

Select clean plants. Obviously, buying disease-free plants is a key part of excluding disease from your garden. Check plants carefully before you buy them. If you order plants through a catalog, inspect them as soon as you receive them in the mail. Look for stunted growth, cracked stems, discolored leaves, or other unusual signs. Inspect the roots and crowns of bare-root plants. Don't buy those with disease symptoms, such as wartlike swellings indicative of crown gall. Select bulbs and tubers that appear healthy and undamaged. Avoid any that are moldy, soft, or bruised. If you have doubts about the health of a new plant, grow it in isolation from related plants until you are sure it is healthy.

Some diseases, such as viral diseases that commonly attack small fruits, may not show symptoms. Buy from reputable nurseries to get "clean" plants. Commercial nurseries maintain stock plants in special structures and sell certified virus-free plants propagated from those protected plants.

Quarantine new arrivals. Even plants that look healthy can harbor unseen pests and disease pathogens. If you suspect that a new plant might be carrying a pest or disease, keep it separated from your other plants for a while. Observe the plant closely for a few days or weeks, depending on how careful you want to be. If any problems do occur, you can easily control the problem or destroy the plant.

Also think about quarantines if you plan to import or export plants from the United States. In 1912, the Federal Plant Quarantine Act was passed to prevent the introduction of foreign pathogens into the United States. Federal plant quarantine laws limit the importation of plants, plant products (even seeds), and soil from other countries. Some states have their own, more restrictive, exclusions. Before you bring any plant into this country, check with the U.S. Department of Agriculture (USDA) and your state department of agriculture to see if any restrictions apply. Write to USDA/APHIS/PP&Q, Federal Center Building, Hyattsville, MD 20782, for information.

Discouraging Disease

Avoiding disease outbreaks in the garden requires a two-pronged approach: You'll want to optimize growing conditions for each plant while minimizing the growth of pathogens.

Providing Good Growing Conditions

A plant growing under good conditions is more likely to resist disease and to survive in spite of disease than one growing under poor conditions. Learn the needs of your individual plants in terms of light, water, fertility, and soil pH, and provide the combination that is best for each plant.

Fit plant to place. While you cannot do anything to change your local climate, you can amend the soil, provide water and fertility, and select microclimates around your property to suit the needs of individual plants. Many garden plants are exotics, native to places where conditions are different from your garden. Cabbages originated along the chalky coast of England, tomatoes are indigenous to subtropical regions of western South America, and tulips are native to the dry steppes of Turkestan. Yet these three plants, along with other plants from various corners of the world, are commonly grown together on the same plot of land in gardens across the country. Take time to learn about the light, soil, and other site requirements of plants, especially exotics, and try to find the best match of plant and site on your property.

Time garden tasks. Optimizing growing conditions for a plant is not only a question of what to do, but of when to do it. Corn or bean seeds planted before the soil has warmed sufficiently do not germinate rapidly and are susceptible to damping-off. Tomatoes set out too early in the spring may succumb to frost or show phosphorus deficiency until their roots become active.

Trees and shrubs that are fertilized too late in summer, or overfertilized with nitrogen at any time, do not harden off with the approach of cold weather in autumn. Pruning or overwatering too late in the summer can also cause a late flush of succulent growth. In the winter, insufficiently hardened plants may experience cold damage and subsequent disease problems, such as Cytospora canker, also known as Valsa canker, of peach trees.

To optimize growing conditions for your plants, consider all your cultural practices together. For example, if you plan to prune off a large amount of winterkilled wood from a tree, don't fertilize that tree before you prune. Pruning stimulates new growth, and the combination of pruning and fertilization that season will result in overly lush growth.

Minimizing Disease Development

Making your garden an unfavorable place for pathogens is an important way to reduce the chances of disease outbreaks. Infectious diseases are not a problem unless the environment is suitable for the pathogen to infect, grow, and multiply in or on susceptible plants.

Water wisely. Water is especially important in the development of bacterial and fungal diseases. Many soilborne fungal diseases become a problem in soils that are too wet. The fungi thrive in waterlogged conditions and roots are damaged by saturation, rendering them more susceptible to infection. If you have wet soils, put in a drainage system, dig ditches to carry away excess water, or plant in raised beds to reduce problems with soilborne diseases, such as root rots and damping-off. Avoid overwatering, and when you do water, keep excess water from flooding around the trunks of woody plants.

How you water can be as important as how much you water. One advantage of drip irrigation over sprinklers is that it doesn't wet foliage. Thus, relative humidity is lower and fungal spores are less apt to germinate because there is no water on the leaves. There is also no splashing water to spread fungi and bacteria within or between plants. If you do water your plants by sprinkling, water early in the day, when the warming sun will dry the leaves quickly. (Leaves also dry quickly with afternoon watering, but much water may be wasted through evaporation.)

Most fungi and bacteria that attack the branches, fruit, and flowers of plants thrive in high humidity. Except for plants that need humid conditions, grow your outdoor plants where gentle breezes rapidly dry leaves and remove humid air hovering near them. Inside your home, especially in winter, the air is too dry for most aboveground diseases to develop. However, a home greenhouse can have the humid conditions that set the stage for such diseases as Botrytis on greenhouse geraniums. Use fans and vents to moderate humidity levels in greenhouses.

Allow for air circulation. Good air circulation will go a long way in discouraging disease development. Overcrowded plants do not dry rapidly after rain or watering; the resulting high humidity can encourage disease. Don't sow seeds too densely, and thin emerged plants to an adequate spacing. Thin some stems from large clumps of multistemmed perennials, such as phlox and bee balm, to allow more air flow through the plants. Also, avoid dense weed growth that can cause the air to stagnate around your plants.

Proper pruning will also encourage good air circulation. Train young trees and shrubs to an open framework so air can easily circulate around the branches. Use mostly thinning cuts—pruning branches back either to ground level or to a branch collar—to avoid bushy regrowth after pruning.

Adjust soil pH. Soil acidity can have a dramatic effect on plant diseases. You can control certain diseases merely by adjusting soil pH. The fungus that causes club root, a serious disease of cabbage family crops, can survive for years in the soil, even in the absence of a host plant. However, it thrives only in acidic soils; you can eliminate club root problems by adjusting soil pH to at least 7.2.

Other diseases can be checked by making your soil more acidic. Scab is most destructive in soils with pH higher than 5.7. To avoid scab, do not plant potatoes in soil that has been recently limed, and adjust soil pH, if necessary, to below 5.7. Cotton root rot, which is prevalent in the Southwest and attacks more than 1,700 plant species, favors a soil pH higher than 8.0. One part of the multifaceted approach needed to control this disease is to acidify the soil.

Balance soil fertility. Plants suffering from nutrient deficiencies are weakened and are likely candidates for disease problems. In some cases, nutrient excesses can also encourage disease. High levels of nitrogen, for example, promote succulent plant growth, increasing susceptibility of pears to fire blight as well as increasing the incidence of viruses, rust, powdery mildew, and Verticillium wilt on some plants.

Rather than trying to adjust specific nutrients to ward off disease, the best approach is to strive for balanced fertility. Work toward this by maintaining an abundant supply of organic matter, such as compost, in your soil and by applying additional fertilizers according to the needs of your plants, as indicated by plant growth or soil tests. For more information on managing soils, see "Cultural Controls" on page 410.

Resort to resistance. Planting disease-resistant species and cultivars also reduces the chances of disease problems. These special plants can occur naturally, appearing in wild populations or among cultivated plants. Breeders also induce mutations in plants, hoping that the mutations prove desirable. Or, using the techniques of genetic engineering, researchers can insert specific genes into desired plant cells or fuse the contents of different plant cells. See "Resistance" on page 412 for a complete discussion of types of resistant plants and how to use them in your garden.

Growing resistant plants does have its limitations in controlling disease. First of all, plants may not have been developed or found that are resistant to certain diseases. Also, disease-resistant cultivars may not have all the other attributes you want in your plants. In the case of tomatoes, for example, does the fruit taste good? Do you want a "meaty" tomato for canning, or a juicy one for fresh eating? There's no point in growing a plant that doesn't fit your needs, just because it is disease-resistant.

Be sure you need the type of resistance that the cultivar you select offers. For instance, it's not necessary to plant a Verticillium-resistant tomato cultivar if your soil is not infested with *Verticillium* fungi.

Stopping the Spread of Disease

At some point, despite your best efforts, disease problems will crop up in your garden. Don't despair—there are many easy things you can do to avoid a widespread disease outbreak.

Breaking the Cycle

In most parts of North America, gardens are still and seemingly lifeless during the winter. However, many disease organisms are present beneath that quiet surface, overwintering in dead plant tissue or in the soil. If you remove or destroy the pathogens' overwintering sites, you can lessen the chances of new disease outbreaks in the following growing season. For example, the bacteria that cause fire blight of pear survive the winter in sunken lesions, called cankers, on the branches. By inspecting your pear trees carefully, pruning off all cankers, and destroying the diseased prunings, you can eliminate bacteria that would have become active in spring and spread disease. Apple scab fungi, on the other hand, survive on dead apple leaves until a new growing season begins; by gathering up and destroying these old leaves, you remove a source of infection for the coming season.

Crop rotation also helps interrupt pathogen life cycles, especially for those diseases that only attack certain types of hosts. For more details on how crop rotation can control disease, see "Rotate your crops" below.

Breaking the life cycles of some other disease organisms is less straightforward. For example, the spores that will cause next season's bacterial wilt of cucumbers survive in the digestive tracts of striped cucumber beetles. To limit the spread of the disease, you must diligently control the beetles.

Destroying Diseased Plants

Infected plant parts may themselves become a source for further infection. These secondary cycles, as they are called, often are important in intensifying the disease. Purposely destroying diseased plants or those suspected of harboring disease is a way to keep disease from spreading to healthy plants.

Removing and destroying diseased plants is often the only safe and practical way to deal with a disease outbreak. This process, also called rogueing out, is most commonly used on annuals, such as vegetables and bedding plants; in some cases, such as root rot, you may have to remove whole trees. While you do lose part of your crop or floral display during the current season, you will then be aware of the problem and can take steps to prevent it in the future. For example, you could plant disease-resistant cultivars, improve soil drainage to discourage root rot organisms, or try preventive sprays of an organic fungicide that keeps the disease organism from infecting plants.

Removing wild disease-prone plants growing in or around your garden is another way to reduce the spread of disease. Eliminating wild cherries, for example, should reduce populations of black knot fungi, which also infect cultivated plums. Many common weeds, including ground cherries, catnip, milkweed, and pokeweed, provide overwintering sites for cucumber mosaic virus. Eliminating these hosts will reduce the chances of your cucumber and squash plants being infected by the virus.

Using Good Garden Management

Tilling the soil, seed-sowing, weeding, and fertilizing are examples of cultural practices associated with growing and caring for plants. You can discourage or eliminate many diseases by giving special attention to cultural practices. Any cultural practice you employ for disease control must, of course, fit into your scheme of gardening. On the other hand, you may want to change some of the ways in which you grow a plant to control a specific disease.

Keep the garden clean. One simple but effective method of disease control is to keep your garden area free of debris that could harbor pathogens. As you walk around your garden, carry a bag so you'll have a safe place to put any diseased material as soon as you spot it. As you harvest fruits and vegetables, remove and destroy any damaged ones that you see. Gather up old plants from the garden at the end of the season, along with any dead, dying, or diseased leaves.

Prune for disease control. One of many reasons for pruning plants is as a sanitation measure to control diseases. As you look over your plants, cut off any leaves or stems that you suspect are diseased. This not only prevents disease from spreading within the plant, but also checks the spread of disease to other plants. Pruning is useful in controlling such diseases as fire blight on pear and bud and twig blight of rhododendron. See "Pruning for Pest Control" on page 432 for specific instructions on how to prune diseased plant materials.

Destroy diseased material. You have several options for disposing of diseased plant material; the best method varies according to the type of disease organism involved.

Viruses are very difficult to destroy. If you suspect that your plant is infected with a virus, bury the affected parts away from cultivated areas of garden, or place them in a sealed container and dispose of them with your household trash.

You can dispose of most other types of diseased materials in your compost pile. By composting them, you conserve organic matter and nutrients, which are recycled to nourish your plants. A newly made hot compost pile should heat up to at least 160°F, a temperature that will kill most pathogenic nematodes, fungi, and bacteria. Make sure you put the diseased material near the center of the pile, where it will be exposed to the highest temperatures. If you don't have a hot compost pile, burn or bury the diseased material, or put it in a sealed container for disposal with your household trash.

Tilling is another way to clean up old plant material at season's end. Chopping and mixing plants into the soil subjects them to rapid attack by soil microorganisms. The old plant material then becomes humus, rather than remaining as a potential source of inoculum for the next season.

Keep tools clean. Sanitation also involves cleaning tools and pots that come into contact with plants. Soil clinging to trowels and shovels can carry disease organisms such as *Verticillium* and damping-off fungi. Clean these tools after use, scraping off the soil (but not over garden beds) and then wiping them clean. Sweep potting benches clean after each use. Disease-causing organisms can also survive in old pots and seedling flats, so clean these containers thoroughly before use, then disinfect them by dipping them in a 10 percent bleach solution (1 part bleach to 9 parts water). Do not allow bleach solution to come into contact with plants.

Rotate your crops. Avoid the buildup of disease organisms by growing your crops in different parts of the garden each year. This practice, called crop rotation, is one of the best ways to reduce the chances of a disease outbreak. The premise is simple: Many garden pests survive the winter and can reinfect a crop if it is grown in the same spot year after year. However, if the suitable host plant is absent, the pest starves.

Most pathogens tend to attack all the members of a given plant family. That's why it's helpful to learn which vegetable crops belong in the same family. You might be surprised to learn that plants that seem very different, such as carrots and parsley, actually belong to the same family (Umbelliferae). Other common families include the nightshade family (tomatoes, eggplants, and peppers), the cabbage family (with cabbage, broccoli, cauliflower, and brussels sprouts), and the legume family (including beans and peas). To learn how to plan an effective crop rotation program, see "Crop Rotation" on page 415.

Crop rotation is more effective in controlling some diseases than others. Soilborne diseases are not very mobile, so you can control them more easily by crop rotation than you can aboveground diseases, which can send spores wafting across from neighboring gardens. Crop rotation also is most useful in starving out a pathogen that attacks only one or a few species of plants. If a disease has many host species (such as the Verticillium wilt fungus, which can attack nearly 300 different species), crop rotation isn't a practical control.

As a cultural practice, crop rotation does more than control infectious diseases. It also helps control certain insect pests and is valuable for avoiding nutrient disorders. Growing different types of plants each year at a given location prevents the soil from being selectively depleted of the nutrients utilized by one type of plant. Instead, a spectrum of nutrients is removed from the soil over time, which generally tends to maintain balanced fertility.

Control disease carriers. Because pathogens are so small, they cannot move very far under their own power. Many nematodes, for example, only move a few inches from the site where they hatched. However, disease inoculum can also spread by other methods, including insect vectors, splashing or running water, wind, and bits of soil carried on boots and tools.

Many diseases can spread from one plant to the next only by means of a vector. A vector is an animal (usually an insect) that carries disease from one plant to the next. In such cases, if you can control the vector, the disease will not occur. Thus, one approach to controlling Dutch elm disease is to control the elm bark beetle, which spreads the disease from infected to healthy elms.

Insect vectors, such as elm bark beetles,

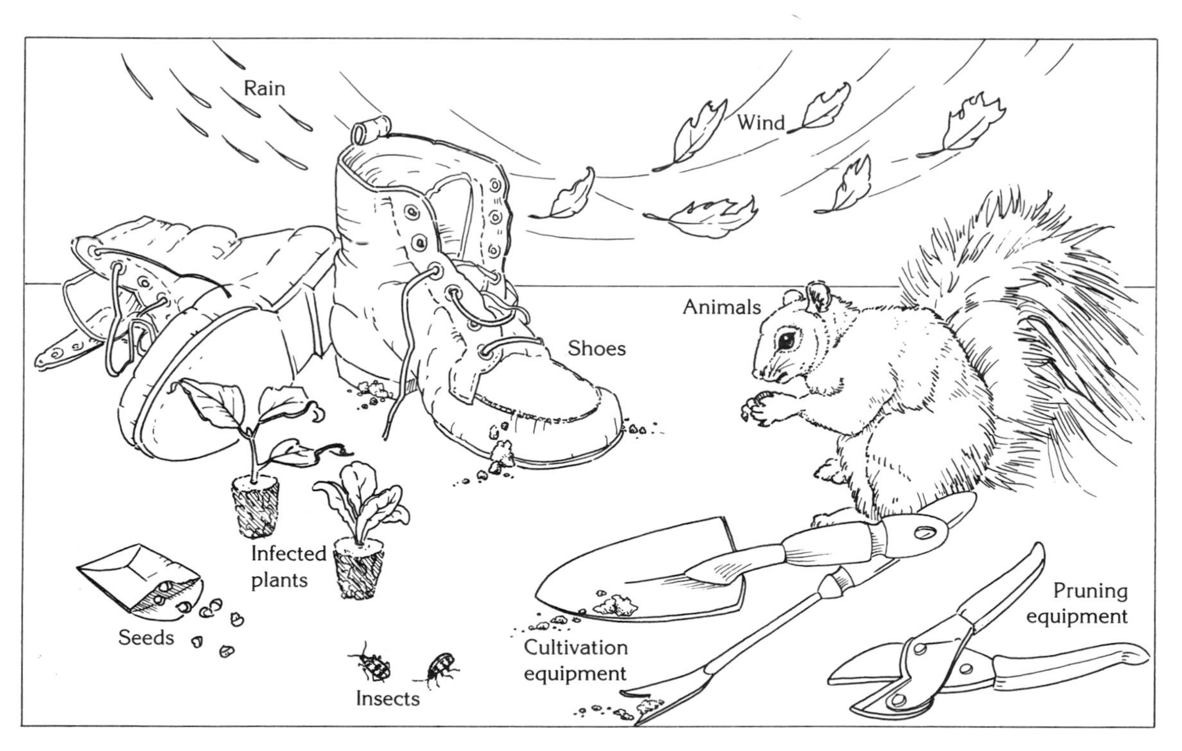

Diseases spread by a wide variety of means. Some carriers, such as wind and rain, are beyond the gardener's control. In many cases, though, you can directly reduce the spread of disease by excluding insects and animals, keeping gardening equipment clean, and avoiding infected seeds and transplants.

aphids that transmit virus diseases, and cucumber beetles that spread bacterial wilt, require more thorough control if the objective is to prevent disease rather than merely to prevent insect damage. While feeding by a few aphids won't produce much visible damage, a few feeding punctures from a virus-carrying aphid are enough to inoculate that plant. The virus will eventually spread throughout the plant, reducing yield and perhaps killing the plant.

Plant nurseries that maintain virus-free stock grow virus-susceptible plants in cages covered with fine mesh screen to keep insect vectors from feeding. You can keep insects away from valuable plants in your garden by covering them with floating row cover. If the plants are not self-pollinating, you must eventually uncover them to allow for insect pollination. By then, though, the plants will be well-established and more resistant to infection.

Laying down reflective mulch, such as aluminum foil, around plants is an effective aphid deterrent. The reflection disorients the pests sufficiently to keep plants free from aphids and the diseases they transmit. Chemical insect controls may be useful in avoiding some insect-transmitted diseases, but in other cases do not control the insect sufficiently to prevent inoculation.

Remember that humans also can be vectors of disease. You can carry crown gall from an infected plant to a healthy plant on a shovel you use for planting, or fire blight from one branch to the next on infected pruning shears. Your tomato plants could become infected with tobacco mosaic virus if you handle the plants after touching cigarettes or other tobacco products infected with tobacco mosaic virus.

Use temperature for disease control. You can control certain diseases by exposing plants or soil to high or low temperatures. In many cases, though, too much of a good thing can be fatal, so be sure you know the pros and cons of each method before treating your whole crop.

The most common use of temperature for disease control is cold storage. In general, storing harvested crops at low temperatures slows the growth of possible pathogens. The optimum storage temperature, however, varies from crop to crop. Apples and many other harvested fruit, root, and leafy crops store best at temperatures as close to freezing as possible. The flavor of bananas and avocados, however, are ruined at such low temperatures. Similarly, winter squash, pumpkins, and sweet potatoes keep best at 50°-60°F.

Using high temperatures to control disease on living plants is a tricky proposition, because there is always the chance of damaging or killing the plant in the process. Heat has been used to rid plants of systemic virus diseases, but such treatments are probably unfeasible for backyard gardeners. Plants may be dipped in hot water for a few minutes up to many hours, the temperature and the timing depending on the plant and the virus. Or, plants may be grown at 95°-110°F for a period of weeks or months. These treatments may damage the plants, but the ones that survive are used commercially to provide disease-free material for propagation.

Heat-treatment of seeds and bulbs is more common, because these structures are more tolerant of high temperatures. Hot-water dips for seeds were developed at the end of the 1800s and are still used today for some commercial seeds. Temperature and timing are critical and vary depending on the type of plant. See page 422 for full instructions on this technique.

Treating soil with high temperatures is often very effective for controlling soilborne diseases, such as root rots and damping-off. You can heat potting soil in the oven and use the sun's heat to raise temperatures in garden soil, a process called solarization. You'll find directions for heat-treating soil in "Pathogen-Free Potting Soil" on page 413.

Attacking Disease Problems

An important decision you must make is whether or not controlling a specific disease problem in your garden is worth the effort, time, expense, and possible hazards. Spraying lime-sulfur to control mildew, for example, may burn your plants' leaves in hot weather. Over time, your experience will help you project how severe a problem will become later if you don't take steps to control it.

Controlling certain diseases may not be worthwhile toward the end of the growing season. For example, potatoes grown in the North can be attacked by late blight in September and suffer no reduction in yield, because the leaves are ceasing to function anyway. This same severity of blight infection in June, however, could result in no potatoes at all.

In other cases, damage from certain diseases is mostly cosmetic. Lilacs covered with powdery mildew in late summer may look unsightly, but the disease usually does the plants little harm. Cosmetic damage affects you rather than the plant, so you must weigh your tolerance for the unsightliness against the costs in time, money, and risk associated with various controls.

The nature of a particular disease can help you predict whether control is necessary; hence the importance of disease identification. Noninfectious diseases do not spread from one plant to the next. For instance, leaf yellowing due to iron deficiency on an azalea planted in alkaline soil will not spread to nearby healthy azaleas planted in acidic soil. While you should take steps to revitalize the affected plants, you won't have to do anything to protect the healthy plants.

There are probably some diseases that are so common in your area that you will know you can expect them (although seasonal weather variations will determine its severity each year). Apple scab reliably appears on susceptible cultivars of apples in the Northeast, and powdery mildew just as reliably appears on many plants on the West Coast each summer. It's good practice to assume you should take preventive measures against these diseases every season.

Choosing a Control

When disease threatens or strikes, you usually have a few options available for controlling the disease. In most cases, it is advisable not to limit yourself to a single approach for best results.

Keep a close eye on your plants when you implement disease-control measures, so that you can note their effectiveness. Also take note of weather conditions and cultural practices, such as fertilization and watering, because these factors can influence the effectiveness of many control measures. Don't trust your memory for all these details! Keep written records of disease incidence, control measures, and growing conditions.

Biological controls. One approach to controlling many infectious diseases is to use "friendly" microorganisms to fight off disease-producing organisms. This is known as biological control.

One approach in biological control is to make the pathogens sick. Bacteriophages, which are viruses that infect bacteria, have been used experimentally to infect and weaken the bacteria that cause bacterial wilt of tomatoes and fire blight on pears.

Beneficial fungi also could be used to attack or inhibit the growth of fungi that cause plant disease. Some practical success has been made in this direction, not by inoculating soils with these beneficial fungi, but by providing soil amendments that foster their growth. Generally, if you grow plants in well-drained soils and amend the soil annually with good-quality compost or other organic materials, your soil will have high populations of beneficial bacteria and fungi.

In some cases, inoculating a plant with one microorganism protects the plant from attack by another microorganism. The microorganisms used for this technique either don't cause disease or cause only mild symptoms in the treated plant. For example, chestnut blight has been successfully controlled in Europe by deliberately inoculating trees with blight fungus that has been weakened by a virus. Trees infected with the weakened fungus are resistant to infection by the healthy fungus.

One commercial product that makes use of "friendly" microorganisms to fight disease is Galltrol-A. Dipping tree roots in Galltrol-A, which contains the nonpathogenic bacterium *Agrobacterium radiobacter,* can control crown gall disease.

Research has shown that, at least indirectly, adding compost to soil also can help in the fight against disease. By raising the level of organic matter, compost encourages large populations of beneficial, disease-suppressing organisms in the soil. In fact, turning any kind of organic matter, such as shredded leaves or a cover crop, into the soil can reduce the populations of nematodes and other pathogens.

Chemical controls. Contrary to popular belief, most chemical controls for plant diseases cannot "cure" an infected plant. Rather, chemical fungicides and bactericides act to provide protection against further disease infection. Most of these products form a coating around plant parts, inhibiting the germination of fungal spores, or killing germinated spores or bacteria before they invade the plant. This means that early treatment is critical to successfully prevent further infection. Seriously infected plants may not benefit from a chemical application.

Presently, most chemicals used to control plant diseases are fungicides. A fungicide may act either to cure or to protect a plant from disease. Most fungicides used on growing plants are protectants rather than curatives. Fungicides also may be used as disinfectants to kill fungal hyphae or spores on plant tubers, seeds, bulbs, or other parts used in propagation.

Sulfur, lime-sulfur, and bordeaux mix are acceptable for use as fungicides in organic gardens. They can be applied in different formulations. These compounds are caustic and have the potential to harm nontarget organisms such as insects, animals, humans, beneficial fungi, even the plant that you are trying to protect. Use them with caution and strictly according to directions on the label.

Sulfur does not kill fungal spores, but instead forms a protective layer that inhibits spore germination. A severe limitation to the use of sulfur is the foliar damage it causes in hot weather. Plants such as tomatoes, grapes, and melons are especially sensitive to sulfur damage under a wide range of conditions.

Lime-sulfur prevents spore growth and can also kill recently germinated spores. Lime-sulfur is useful against such diseases as apple scab, brown rot on peaches, and powdery mildews. This compound is more toxic to plants than plain sulfur, so use it with caution. Lime-sulfur is often used as a dormant spray, when there is less chance of damaging the plant.

Copper-based compounds inhibit the growth of fungal spores and also are somewhat effective against bacteria. Copper sulfate can be toxic to plants, so it is often combined with water and lime to produce a compound called bordeaux mix. The lime is reduces the chances of the copper burning the plant leaves. Even so, bordeaux can cause damage such as russeting of apples, delayed fruit set of tomatoes, and yellowing and dropping of rose leaves. This damage is most likely during cool, overcast weather when the spray dries slowly on the plant.

A few sprays that you can make yourself are effective against some diseases. Baking soda sprays can prevent damage from black spot on roses. And some gardeners use garlic sprays to treat a range of disease problems.

To learn more about specific chemical control options, see "Chemical Controls" beginning on page 463.

DISEASE SYMPTOM GUIDE

Scientists who study plant diseases separate the visible characteristics of disease into two categories: symptoms and signs. A *symptom* is a plant's response to a disease-causing organism or condition. Two common disease symptoms are changes in plant color and wilting. Disease symptoms usually result from death of cells, inhibited cell development, or overstimulation of cell development.

A *sign* of an infectious disease is the disease-causing organism itself or its products (such as spores). Examples of signs produced by fungi include the white powdery spores of mildew, the black film of fungal strands on leaves with sooty mold, and the galls on cedar trees produced by cedar-apple rust. One common sign of bacteria is slimy ooze that often has a foul odor.

Diseased plants may show both symptoms and signs. Lilacs infected with powdery mildew may have curled leaves (a symptom), but even more obvious is the powdery white coating of fungal spores (a sign).

While it's not critical for you to know the technical difference between a symptom and a sign, it is helpful to know the terms used to describe common symptoms and signs of disease. If you suspect your plant is diseased, study its symptoms, and then review the symptom descriptions listed on the following pages.

Once you match your plant's symptoms to one of the symptom descriptions and/or photographs, read the corresponding discussion to learn more about the disease, including the type of problem, other common symptoms, plants affected, and prevention and control measures.

The disease guide includes photographs and discussions of common plant diseases and disorders. They are organized by symptoms, according to the plant parts affected: leaves; flowers and fruit; stems and roots; or whole plants. The symptom categories, and the pages on which you'll find descriptions of diseases that cause those symptoms, are as follows:

Leaf Symptoms

Flower and Fruit Symptoms

Stem and Root Symptoms

Whole Plant Symptoms

Leaf Symptoms: Leaves with Spots

The death of leaf cells can result in well-defined, circular spots. Spots can be of various colors and may change colors as symptoms progress. In some cases, as in cherry leaf spot, the dead cells eventually fall out, leaving holes.

Anthracnose

Symptoms on dogwood

Symptoms on maple

Type of Problem: Fungal.

Symptoms: On leaves, anthracnose diseases generally appear first as small, irregular yellow or brown spots that darken as they age. These spots may also expand and join to cover the leaves. On vegetables, anthracnose diseases can affect any part of the plant. See "Anthracnose" on page 386 for a discussion of anthracnose symptoms on fruit. On trees, infection can begin before the leaves appear, killing the tips of young twigs. More often, anthracnose fungi strike the young leaves, producing brown spots and patches. Defoliation may occur, forcing the tree to produce a new set of leaves in the summer.

Plants Affected: Many kinds of woody and herbaceous plants. Vegetables such as beans, cucumbers, melons, peppers, and tomatoes are particularly susceptible. Anthracnose-prone trees include dogwoods, maples, and sycamores.

Prevention and Control: Avoid anthracnose on vegetables by selecting resistant cultivars (when available), buying healthy transplants, planting in well-drained soil, and not touching plants when they're wet. Remove and destroy infected plants. Spraying with a copper-based fungicide may provide some protection.

On trees, prune out the dead wood and the water sprouts. Avoid drought stress by watering trees during dry spells and keeping the root zone mulched. Gather up and destroy infected leaves. A dormant spray of bordeaux mix may provide some control.

Apple Scab

Symptoms on apple

Type of Problem: Fungal.

Symptoms: Apple scab symptoms first appear on leaves as olive green spots that gradually turn black. These spots may expand and run together, forming large blotches. Leaves may drop prematurely. The leaves may be deformed or smaller than normal. Brown or black spots may also appear on the fruit; for more details, see "Apple Scab" on page 387.

Plants Affected: Apples and crab apples.

Prevention and Control: Plant resistant cultivars. Rake up and dispose of fallen leaves, which carry overwintering spores. Apple scab is most prevalent in areas with cool, moist summers because the spores spread rapidly in wet weather. In spring, regular copper, sulfur, or lime-sulfur sprays may prevent apple scab fungi from infecting young leaves and fruit, reducing damage and the chance of apple scab recurring later in the season.

Bacterial Spot

Symptoms on zinnia

Type of Problem: Bacterial.

Symptoms: Depending on the plant they attack, these bacteria will produce round, angular, or elongated discolorations on leaves. The spots are tiny at first but may spread and join to cover whole leaves. The spots are usually brown and are sometimes surrounded with a yellow ring referred to as a halo. The damaged tissue often drops out of the leaves, leaving small holes. Severely infected leaves may fall early. The bacteria can also attack fruit, causing sunken spots; raised, scabby spots; or cracking.

Plants Affected: Many kinds of woody and herbaceous plants.

Prevention and Control: Bacterial spot pathogens overwinter in infected plant parts and seeds, in the soil, or on contaminated tools and pots. Reduce the chances of disease by using clean seeds, rotating crops, and practicing good garden sanitation. Plant resistant cultivars. Preventive copper sprays can be helpful if bacterial spots have been a past problem. If possible, remove and destroy severely infected plants.

Black Spot

Symptoms on rose

Type of Problem: Fungal.

Symptoms: This disease appears as circular black spots on infected leaves. The spots usually have fringed or indistinct margins and are often surrounded by a ring of yellow tissue. Severely infected leaves may fall early. Black spot fungus can also infect stems, causing purplish or black blisters on young canes.

Plants Affected: Roses.

Prevention and Control: Black spot fungus thrives in moist conditions. Avoid wetting plant leaves when watering. Prune plants to increase air circulation. Plant resistant species or cultivars. If you expect black spot to be a problem based on past infections, spray plants weekly with sulfur or fungicidal soap. Once symptoms appear, black spot is very difficult to control. Remove and destroy infected leaves and canes. A 0.5 percent solution of baking soda (1 teaspoon baking soda in 1 quart water) may help to control the disease. Spray infected plants thoroughly.

Botrytis Blight

Symptoms on peony

Type of Problem: Fungal.

Symptoms: Besides affecting fruit and flowers, *Botrytis* fungi can also damage leaves; infected leaves develop water-soaked spots that later turn brown or dry. For information about Botrytis on flowers, see "Botrytis Blight" on page 385; for Botrytis on fruit, see "Botrytis Fruit Rot" on page 389.

Plants Affected: A wide range of woody and herbaceous plants, including cabbage, onions, peonies, and strawberries.

Prevention and Control: After they die down, cut or pull off and destroy the tops of herbaceous plants to remove a potential source of inoculum for the following season. Provide good air circulation through pruning and site selection. Remove and destroy infected parts.

Cedar-Apple Rust

Symptoms on apple

Type of Problem: Fungal.

Symptoms: On apples, rust symptoms commonly appear in spring. Tiny yellow spots, which later expand and turn orange, form on upper leaf surfaces and on fruit. Brown spots may appear on the undersides of leaves. For details on how cedar-apple rust affects cedar trees, see "Cedar-Apple Rust" on page 398.

Plants Affected: Apples and crab apples. Similar rust diseases affect hawthorns (cedar-hawthorn rust) and quinces (cedar-quince rust).

Prevention and Control: Cedar-apple rust completes its life cycle only if fungal spores can travel between cedar (*Juniperus* spp.) and apple trees. Fungi growing in cedars send spores to infect apple trees. However, infections on the apple tree do not spread within the tree; the fungus can only can send spores back to infect cedar.

Prevention is the best control. Rust fungi need moisture, so promote drying through pruning and site selection to limit disease problems. Plant apple trees only if cedars are at least 4 miles away; this will reduce the chances of the disease spreading. If you want to grow both cedars and apple trees, plant rust-resistant species or cultivars of these plants. Many fungicides (including sulfur and lime-sulfur) that are effective against other fungal diseases are not very effective against rust diseases; preventive sprays of copper-based fungicides may provide some protection.

Cherry Leaf Spot

Symptoms on cherry

Type of Problem: Fungal.

Symptoms: The first noticeable symptoms are tiny purple spots on the upper leaf surfaces. Corresponding whitish spots on the undersides of leaves may appear. The centers of these spots often dry and fall out, giving the leaves a shothole appearance. Entire leaves may turn yellow and drop early. Fruit, as well as leaf and fruit stems, can also show symptoms.

Plants Affected: Cherries and, less often, plums.

Prevention and Control: Plant resistant cultivars. Clean up fallen leaves in autumn to remove overwintering fungi. Preventive sulfur sprays help to reduce disease severity.

Downy Mildew

Symptoms on grape

Type of Problem: Fungal.

Symptoms: Downy mildew infections begin as angular yellow spots on the upper leaf surfaces; these spots eventually turn brown. Corresponding white, tan, or gray, cottony spots form on the undersides of the leaves. Downy mildew can also attack young shoots and fruit, forming a white coating.

Plants Affected: A wide range of woody and herbaceous plants. This disease is a serious problem on grapes.

Prevention and Control: Downy mildew thrives during cool, moist weather. Control downy mildew on your plants by promoting drying (through pruning and site selection) and growing resistant cultivars. Plant disease-free seeds and bulbs. Remove and destroy badly infected leaves. Sprays of bordeaux mix or other copper-based fungicides may reduce the spread of the disease.

Early Blight

Symptoms on tomato

Type of Problem: Fungal.

Symptoms: Early blight symptoms appear first on lower leaves as brown spots with concentric rings; these spots eventually spread to cover the leaves. Affected leaves drop early, exposing fruit to sunscald. Spots and cankers may also appear on stems.

Plants Affected: Tomatoes and potatoes.

Prevention and Control: Clean up plant debris to remove overwintering sites. Use disease-free seeds and seed potatoes. Rotate crops and plant resistant cultivars. Preventive copper-based fungicide sprays may help reduce the spread of early blight. Remove and destroy severely infected plants.

Late Blight

Symptoms on potato

Type of Problem: Fungal.

Symptoms: On leaves, late blight begins as tiny brown spots, which develop into greenish gray or brown areas that can expand to cover whole leaves. These spots may be surrounded by a ring of yellow tissue on the upper surfaces of leaves and a ring of whitish fungal growth on the leaf undersides. Brownish black areas may form on stems. Fruit rots and shrivels quickly; tubers may show a reddish brown dry rot.

Plants Affected: Potatoes and tomatoes.

Prevention and Control: The fungus that causes potato late blight overwinters on diseased tubers. Harvesting all tubers and disposing of those that are infected limit the disease the following season. Rotate crops and select resistant cultivars. Plant only certified disease-free seed potatoes.

Leaf Blister

Symptoms on oak

Type of Problem: Fungal.

Symptoms: Swollen, yellow or brownish blisters appear on upper surfaces as leaves develop in the spring. The spots may expand and run together to cover leaves; seriously damaged leaves may fall early.

Plants Affected: Oaks.

Prevention and Control: Mild, moist weather conditions promote the development of this disease. Oak leaf blister usually only attacks young, developing leaves in the spring; older leaves, later in the season, are not affected. Control is usually not necessary. If your trees were severely infected the preceding season, apply a dormant spray of lime-sulfur or bordeaux mix before the buds open in spring.

Needlecast

Symptoms on pine

Symptoms on pine

Type of Problem: Fungal.

Symptoms: The symptoms of this disease appear on developing needles. Mottled yellow spots appear first, changing to reddish or orangish brown. Severely damaged needles may fall by midsummer.

Plants Affected: Many kinds of needle-leaved plants.

Prevention and Control: Site plants where they will get good air circulation. Clean up fallen needles; prune off damaged tips. If plants have been infected in previous seasons, spray with bordeaux mix when new shoots are half-grown; repeat 2 weeks later.

Rust

Symptoms on hollyhock

Symptoms on blackberry

Type of Problem: Fungal.

Symptoms: Yellow or white spots form on upper leaf surfaces. Orange or yellow spots or streaks appear on the undersides of leaves. Spots are fungal structures that release spores.

Plants Affected: A wide range of woody and herbaceous plants.

Prevention and Control: Provide good air circulation and avoid wetting leaves when watering. Remove and destroy seriously affected parts. Starting early in the season, dust plants with sulfur to prevent infection or to keep mild infections from spreading. For bramble fruits, immediately destroy any infected plants and replant resistant cultivars.

Septoria Leaf Spot

Symptoms on tomato

Type of Problem: Fungal.

Symptoms: Leaf damage starts as small yellow spots that gradually turn brown; they are often surrounded by a ring of yellow or brownish black tissue. Whole leaves may turn yellow and drop, exposing fruit to sun, which may result in sunscald. This disease usually starts on lower leaves and progresses upward.

Plants Affected: A wide range of herbaceous plants.

Prevention and Control: Remove and destroy infected leaves. Clean up plant debris in fall. Use disease-free seed. Crop rotation and the use of resistant cultivars will reduce the chances of Septoria leaf spot.

Leaf Symptoms: Leaves Yellow or Discolored

Leaf discoloration can be from cell death, or it can be from an interruption of plant biochemical processes as a result of nutrient deficiencies or environmental pollutants.

Foliar Nematodes

Symptoms on chrysanthemum

Type of Problem: Nematodes.

Symptoms: Leaves show yellow patches that later turn brown or black; these blotches may enlarge to cover whole leaves, causing early leaf drop. Symptoms start near the bottom of plants and work upward. Infected shoots are stunted, and flowers may be deformed.

Plants Affected: A wide range of herbaceous plants. Plants that are particularly susceptible include chrysanthemums, asters, dahlias, phlox, primroses, and strawberries.

Prevention and Control: Avoid planting susceptible species. Foliar nematodes move up stems in a film of water. Choose a site with good air circulation, and thin stems for quick drying. Clean up debris to eliminate overwintering sites. Destroy infected plant parts.

Iron Deficiency or Overly High pH

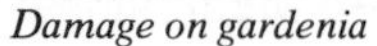

Damage on gardenia *Damage on azalea*

Type of Problem: Environmental.

Symptoms: The youngest leaves, those near the tips of shoots, turn yellow except for the veins, which remain green.

Plants Affected: Acid-loving plants, including blueberries, oaks, hollies, azaleas, and rhododendrons.

Prevention and Control: Symptoms of iron deficiency appear when the soil is not sufficiently acidic. Symptoms commonly occur on susceptible plants growing near buildings because lime that leaches out of the concrete foundations raises soil pH. In most cases, iron is present in the soil in adequate amounts, but plant roots cannot absorb it if the pH is not in the appropriate range for that plant. The long-term solution to this problem is to plant in acidic soil or to make the soil acidic by adding sulfur or peat moss. Planting in raised beds amended with peat moss can also provide good growing conditions for acid-loving plants. Mulch acid-loving plants with evergreen needles. The way to quickly save an ailing plant is to spray a commercial chelated iron product on its leaves. Spraying leaves with seaweed extract will also help to provide the necessary nutrients.

Leaf Scorch

Damage on maple

Type of Problem: Environmental.

Symptoms: Leaf scorch appears as yellowing and browning of leaves beginning along the margins and tips. Associated symptoms may include wilting, rolling of leaves, stunted growth, and death of the plant.

Plants Affected: A wide range of woody and herbaceous plants.

Prevention and Control: The 2 major causes of leaf scorch are drought and reflected heat. Avoid drought injury to plants with timely watering. On the average, most garden plants need the equivalent of about 1 inch (2 gallons per square foot) of water per week for every foot depth of rooting. Avoid frequent, light watering, which will promote shallow rooting. Reflected heat from surrounding pavement is mostly a problem for street trees; there is no control. Avoid planting highly susceptible trees, such as horse chestnuts and maples, in these sites.

Mosaic

Symptoms on squash

Type of Problem: Viral.

Symptoms: Mosaic-infected leaves are mottled with yellow, white, and light and dark green spots or streaks. Fruit may show similar symptoms. Plants are often stunted (see photograph on page 403).

Plants Affected: A wide range of woody and herbaceous plants.

Prevention and Control: Plant resistant cultivars when available. Mosaic viruses are mostly spread by insect pests, especially aphids and leafhoppers. If possible, deny these carriers access to your crop by covering the plants with floating row cover. For more information on preventing and controlling aphids and leafhoppers, see pages 269 and 302. Once plants are infected, there are no controls; remove and destroy infected plants.

Nitrogen Deficiency

Damage on pepper

Type of Problem: Environmental.

Symptoms: Nitrogen deficiency appears as a uniform yellowing of the oldest leaves (those nearest the base of the stem). A less-obvious symptom is stunted and spindly growth.

Plants Affected: All plants.

Prevention and Control: Yearly applications of supplemental nitrogen are most important in the vegetable garden. Other plantings can get the nitrogen they need from regular applications of compost or aged manure. Organic nitrogen sources include soybean meal, dried blood, composted manure, and compost. In the vegetable garden, you can provide nitrogen by growing a leguminous cover crop every other season. If your plant shows deficiency symptoms, spray the leaves with fish emulsion.

Ozone Damage

Damage on nasturtiums

Type of Problem: Environmental.

Symptoms: Ozone damage appears as white or tan stippling or flecking on leaves. High concentrations of ozone in the atmosphere may cause early fall color and leaf drop. This damage can be difficult to diagnose because symptoms mimic many other disease conditions. Check for similar damage on other ozone-susceptible plants growing in the same area.

Plants Affected: A wide range of woody and ornamental plants. Plants such as beans, spinach, tomatoes, blackberries, sweet gums, pines, and tulip poplars are particularly sensitive to ozone.

Prevention and Control: Ozone levels in the air may reach damaging levels during the hot, calm days of mid- to late summer. Automobile exhaust is a major source of ozone pollution. There is no cure for ozone damage. Avoid it in following years by planting less-susceptible plants, including beets, lettuce, and strawberries.

PAN Damage

Damage on romaine lettuce

Type of Problem: Environmental.

Symptoms: Exposure to PAN (peroxyacyl nitrate) causes a silvery glaze on the lower surfaces of leaves, resembling damage due to frost, sunscald, mites, thrips, or leafhoppers. Young, rapidly growing tissue is most sensitive to PAN injury.

Plants Affected: A wide range of woody and herbaceous plants. Plants that are particularly susceptible include petunias, beans, lettuce, peppers, and tomatoes.

Prevention and Control: PAN is a major component of engine exhaust and is a problem in smoggy areas. If you live where PAN damage is a recurring problem, grow plants that tolerate these pollutants, such as sugar maples, arborvitae, winged euonymus, English ivy, snapdragons, cabbage, cucumbers, and squash.

Phosphorus Deficiency

Damage on corn

Type of Problem: Environmental.

Symptoms: A bluish or purplish cast to leaves or stems is the most common symptom of phosphorus deficiency; some plants develop purple spots. Phosphorus-deficient plants also do not flower and fruit as well as healthy plants do.

Plants Affected: All plants.

Prevention and Control: Phosphorus deficiency symptoms develop fairly frequently in transplants set out in early spring. Unlike some other nutrients, phosphorus does not move through the soil; roots must grow to reach it. Until the soil warms sufficiently to stimulate root growth, plants may not be able to take up enough phosphorus.

Organic matter contains both phosphorus and potassium, so enrich your soil with plenty of compost, leaf mold, and other organic materials. Other sources of phosphorus include bonemeal and rock phosphate.

Potassium Deficiency

Damage on tomato

Type of Problem: Environmental.

Symptoms: Symptoms usually appear on older leaves first as yellowing and browning of the leaf margins. Dead areas on edges may drop, giving the leaves a ragged appearance. If deficiency is severe, young leaves will show symptoms as well; consider having leaf tissue analyzed to confirm.

Plants Affected: All plants.

Prevention and Control: Soil tests can alert you to deficiency problems before they become serious. Regular applications of compost or other organic fertilizers will help to maintain an even potassium supply. Sources of potassium include kelp meal, granite dust, greensand, and wood ashes. Use wood ashes sparingly, as they can raise the soil pH.

Sulfur Dioxide Injury

Damage on blackberry

Type of Problem: Environmental.

Symptoms: Mild cases of sulfur dioxide damage show up as general leaf yellowing.

More severe damage involves yellowing or browning of the tissues between leaf veins.

Plants Affected: A wide range of woody and herbaceous plants.

Prevention and Control: Sulfur dioxide is an air pollutant, primarily a result of industrial processes. Where these pollutants are a problem, grow tolerant trees such as ginkgos, junipers, sycamores, and arborvitae. Vegetables that tolerate sulfur dioxide include cucumbers, corn, and onions.

Verticillium Wilt

Symptoms on tomato

Symptoms on strawberry

Type of Problem: Fungal.

Symptoms: Infection by wilt fungi cause leaves to yellow and leaf stems to droop, giving plants a wilted appearance. The yellow leaf patches turn brown and may spread to cover whole leaves. Leaves often fall early, and plants will die. Symptoms usually first appear on the lower or outer parts of plants. The interior of the stem near the base may be discolored (see photograph of stem damage on page 396).

Plants Affected: A wide range of woody and herbaceous plants. Some particularly susceptible plants include tomatoes, peppers, melons, asters, chrysanthemums, peaches, cherries, strawberries, and maples.

Prevention and Control: Verticillium and Fusarium wilt fungi cause similar symptoms. Verticillium wilt is more common in cool weather in temperate areas of the country.

There is no cure for Verticillium wilt. The most effective control is the use of wilt-resistant cultivars, when available. Crop rotation is not very effective because so many species are susceptible to Verticillium wilt. Soil solarization may help to control it.

Winter Injury

Damage on cherry laurel

Type of Problem: Environmental.

Symptoms: Symptoms of cold injury can be similar to sunscald symptoms: blotchy, water-soaked areas on leaves. Shoot tips often die back (see stem damage photograph on page 396).

Plants Affected: A wide range of woody and herbaceous plants.

Prevention and Control: In winter, evergreens can suffer from drying when their roots cannot get sufficient water from frozen ground. Water plants thoroughly in late fall. Spray leaves with an antitranspirant to reduce water loss. If winter damage has been a problem in the past, move plants to a more sheltered spot or erect a barrier (such as a piece of burlap) to protect the plants from drying winds. Prune out damaged wood in spring.

Cold, or in some cases cool temperatures, also damages herbaceous plants. If you plant corn or beans too early, these heat-loving plants suffer during extended cool weather, even though the temperature never dips below freezing. Sub-freezing weather in the spring can damage tender seedlings as well as new shoots and flowers of woody plants. Acclimate indoor seedlings to outdoor temperatures gradually to help the plants adapt to the cold. Protect garden seedlings and transplants with cloches or row cover.

Yellows

Symptoms on marigolds

Type of Problem: Mycoplasma-like organisms.

Symptoms: This disease produces a gradual yellowing of leaves. Plants often appear dwarfed. Plant parts, including roots and flowers, may be deformed. On trees, leaves turn yellow, then brown, and may drop early. Symptoms appear over the whole crown of the tree. Plants may die in a single growing season.

Plants Affected: A wide range of woody and herbaceous plants. Aster yellows affects many plants, including carrots, lettuce, tomatoes, China asters, and gladiolus. Elm yellows (also known as elm phloem necrosis) attacks several species of elms.

Prevention and Control: Yellows diseases are commonly transmitted by leafhoppers, so controlling the pests will reduce the chances of disease (see page 302 for controls). Remove weeds that provide overwintering sites for the pathogen, including thistles, Queen Anne's lace, dandelions, and wild chicory. Remove and destroy infected plants.

Leaf Symptoms: Leaves with White, Gray, or Black Patches

Discolored patches on leaves may be signs of the pathogen. For example, the white powdery covering of powdery mildew is actually the fungus spreading across leaf surfaces.

Powdery Mildew

Symptoms on pumpkin

Symptoms on zinnia

Type of Problem: Fungal.

Symptoms: Plants suffering from powdery mildew look as if they have been dusted with flour. Powdery mildew fungi mostly attacks new leaves, causing distorted growth. It can also attack fruit; see "Powdery Mildew" on pages 387 and 388.

Plants Affected: A wide range of woody and herbaceous plants. Some plants that are particularly susceptible include lilacs, phlox, bee balm, squash, roses, and zinnias.

Prevention and Control: Powdery mildew thrives in hot weather, especially with cool nights. Unlike most fungal diseases, it is actually less of a problem in rainy weather.

Control by providing good air circulation and by growing resistant cultivars. Spraying with sulfur or lime-sulfur may reduce the spread of this disease. A 0.5 percent solution of baking soda (1 teaspoon baking soda in 1 quart water) may help to control the disease. Spray infected plants thoroughly.

Salt Injury

Damage on ivy

Type of Problem: Environmental.

Symptoms: The plant responds to excess soil salt just as it would to drought: stunting, wilting, drying out of leaves, even death. White, crusty material—salt—may build up on leaves of outdoor plants exposed to salt spray or road de-icing salts. A white crust on the surface of the potting mix in which you grow houseplants can indicate salt buildup due to poor drainage or overfertilization.

Plants Affected: A wide range of woody and herbaceous plants.

Prevention and Control: How you prevent salt damage depends on how salt gets to the plants. In the North, this salt may come from sodium chloride used for road de-icing. Plants in coastal areas are often exposed to blowing sea spray. Some soils of the West are naturally high in salts, as are some irrigation waters in that region. Soluble salts from animal urine can also damage plant roots. Excess fertilizer, even manure, can cause salt buildup wherever drainage is poor or rainfall is insufficient to leach excess salt out of the soil.

Remove excess salts from soils by watering heavily. Use sand or sawdust rather than de-icing salts to improve traction on icy sidewalks. Improve soil drainage, if necessary, by digging open trenches or burying perforated plastic drainage pipe within the soil to carry away excess water. In areas of the West where drainage is poor due to excess sodium in the soil, apply gypsum to loosen the soil structure. Choose salt-tolerant plant species and cultivars.

Sooty Mold

Symptoms on leaves

Type of Problem: Fungal.

Symptoms: Leaves and stems are speckled or coated with a thin black film, which can be wiped off to expose healthy green leaf surfaces. Plants may also feel sticky.

Plants Affected: A wide range of woody and herbaceous plants.

Prevention and Control: This fungus grows on the sticky excretions (known as honeydew) produced by insects such as aphids, scales, and mealybugs. The fungus does not directly injure the plant, but the black coating is unsightly and may interfere with photosynthesis and reduce plant growth.

On small plants, you can wipe the coating off with a damp cloth. The best treatment is to control the insects producing the honeydew (see pages 269, 304, and 320–21 for aphid, mealybug, and scale controls, respectively). If the plants do not have an insect problem, the honeydew may be from an overhanging plant.

Leaf Symptoms: Leaves Curled or Distorted

Disease can alter the normal pattern of leaf growth by inhibiting or stimulating cell development in unusual ways. Leaves may twist or form growths called galls.

Curly Top

Symptoms on tomato

Type of Problem: Viral.

Symptoms: Leaves of infected plants twist and curl upward, becoming stiff and leathery. They eventually turn yellow and then brown. Leaf stems bend downward. The plant may appear stunted, and fruit production stops.

Plants Affected: A wide range of herbaceous plants. Beets, tomatoes, beans, melons, and spinach are particularly susceptible.

Prevention and Control: Plant resistant cultivars. Remove surrounding weeds, such as thistles and plantain, that provide disease overwintering sites. Leafhoppers carry the curly top virus, so keep these pests away by protecting plants with floating row cover. See page 302 for more leafhopper controls. Remove and destroy affected plants.

Leaf Gall

Symptoms on azalea

Type of Problem: Fungal.

Symptoms: Reddish or yellowish leaf spots often appear first. Infected leaves develop light green galls that later turn white and then brown. Flowers may also be damaged.

Plants Affected: Fungal leaf galls are most common on azaleas and rhododendrons. Camellias may also get leaf galls.

Prevention and Control: Pick off and destroy infected leaves as soon as you spot the galls.

Peach Leaf Curl

Symptoms on peach

Type of Problem: Fungal.

Symptoms: Infected plants develop yellowish or reddish blisters on leaves, which become curled and distorted. The blisters eventually turn powdery gray. Entire leaves may turn yellow and fall early. Fruit can be deformed and may drop early. New growth can be stunted; infected shoot tips may die back.

Plants Affected: Peaches and nectarines.

Prevention and Control: Plant resistant cultivars. Remove and destroy infected leaves. Peach leaf curl is usually worst during cool, wet springs. If this disease has been a problem in past years, apply a dormant spray of lime-sulfur or bordeaux mix.

Leaf Symptoms: Leaves Wilted

Leaves on diseased plants can become limp from the death of cells that move water and nutrients in the stems. Wilting is usually accompanied by other signs or symptoms.

Bacterial Wilt

Symptoms on cucumber

Type of Problem: Bacterial.

Symptoms: Leaves appear limp and wilted. Infected stems wilt and collapse quickly. All affected parts are soft at first, but turn hard and dry. When you pull apart a cut stem, you may see long, sticky strands of whitish bacterial ooze. Spots may occur on fruit.

Plants Affected: A wide range of herbaceous plants. Cucumbers, melons, and squash are very susceptible. Similar wilts affect tomatoes and beans.

Prevention and Control: Plant resistant cultivars and use disease-free seed. Control cucumber beetles (see page 285) and grasshoppers (see page 294), which transmit the disease as they feed; protect plants with floating row cover. Remove and destroy infected plants.

Fusarium Wilt

Symptoms on tomato

Type of Problem: Fungal.

Symptoms: Wilt fungi cause leaves to yellow and leaf stems to droop, giving plants a wilted appearance. The yellow leaf patches turn brown and may spread to cover whole leaves. Leaves often fall early, and the plants will die. Symptoms usually first appear on the lower or outer parts of plants. In some cases, the symptoms are most apparent on only 1 side of a plant. If you cut the stem near the base, you may notice a brown discoloration in the interior.

Plants Affected: A wide range of woody and herbaceous plants. Common hosts include tomatoes, peas, peppers, melons, dahlias, and mimosa trees.

Prevention and Control: Fusarium and Verticillium wilt fungi cause similar symptoms. Fusarium wilt thrives in warmer areas.

Plant resistant cultivars. Crop rotation is of limited value in starving these pests; *Fusarium* fungi can survive in the soil a number of years in the absence of a susceptible plant. Remove and destroy infected plants. Soil solarization may reduce the incidence of this disease.

Verticillium Wilt

Symptoms on tomato

Symptoms on strawberry

Type of Problem: Fungal.

Symptoms: Wilt fungi cause leaves to yellow and leaf stems to droop, giving plants a wilted appearance. The yellow leaf patches turn brown, and may spread to cover whole leaves. Leaves often fall early, and the plants will die. Symptoms usually first appear on the lower or outer parts of the plant. The interior of the stem near the base may be discolored (see photograph of stem damage on page 396).

Plants Affected: A wide range of woody and herbaceous plants. Some particularly susceptible plants include tomatoes, peppers, melons, asters, chrysanthemums, peaches, cherries, strawberries, and maples.

Prevention and Control: Verticillium and Fusarium wilt fungi cause similar symptoms. Verticillium wilt is more common in cool weather in temperate areas of the country.

There is no cure for Verticillium wilt. The most effective control is the use of wilt-resistant cultivars, when available. Crop rotation is not very effective because so many species are susceptible to Verticillium wilt. Soil solarization may help to control the disease.

Waterlogged Soil

Damage on yew

Type of Problem: Environmental.

Symptoms: Because waterlogging inhibits root function, it causes essentially the same symptoms as droughty conditions do—wilting. Other common symptoms include yellowed leaves and sudden leaf drop.

Plants Affected: A wide range of woody and herbaceous plants.

Prevention and Control: When the soil is flooded with water, pores that previously held air become filled with water. But root cells need oxygen in order to function, and the cells may die if they are deprived of air long enough. Besides causing wilting, waterlogged soils provide ideal conditions for bacteria to attack the damaged roots.

Waterlogging results if you apply too much water to the soil. Also, certain soils are naturally prone to waterlogging. Improve soil drainage by adding organic matter or making raised beds. Water plants evenly, according to their needs. Severely damaged plants may not recover and should be removed.

Flower and Fruit Symptoms: Flowers Discolored

Flower discoloration may be caused by cell death or from signs of the disease organism appearing on flower parts.

Botrytis Blight

Symptoms on rose

Type of Problem: Fungal.

Symptoms: Botrytis blight generally begins on flowers, producing a white, gray, or tan, fluffy growth. The fungus then spreads to the flower stalk, weakening the stalk and causing the flowers to droop. Affected plant parts eventually turn brown and dry. For information about Botrytis on leaves, see "Botrytis Blight" on page 373; for Botrytis infection on fruit, see "Botrytis Fruit Rot" on page 389.

Plants Affected: A wide range of woody and herbaceous plants. The blooms of such flowers as roses, begonias, peonies, chrysanthemums, dahlias, and geraniums are particularly susceptible.

Prevention and Control: Provide good air circulation through pruning and site selection. Remove and destroy affected parts.

Brown Rot

Symptoms on cherry

Type of Problem: Fungal.

Symptoms: Infected flowers appear wilted and browned. Eventually they are covered with light brown spore masses, which then attack developing fruit. Small cankers appear near branch tips. For more on fruit damage, see "Brown Rot" on page 390.

Plants Affected: Peaches, cherries, plums, and other stone fruits.

Prevention and Control: Plant resistant cultivars. Prune trees to provide for good air circulation. Prune out and destroy damaged shoots. Pick off and clean up rotted and shriveled fruit. Spray with sulfur just before blossoms open and again after blossoming to protect the fruit. Another spray, just before harvest, will protect fruit from brown rot during storage.

Flower Blight of Camellia

Symptoms on camellia

Type of Problem: Fungal.

Symptoms: Flower blight fungi produce small brown spots on petals. These spots enlarge and run together, turning whole flowers brown.

Plants Affected: Camellias.

Prevention and Control: Avoid bringing the fungi into your garden by only purchasing bare-root plants; also, pick off and destroy any flower buds before planting. If disease strikes, remove and destroy all infected flowers and buds, including those that have fallen from the plants. Remove the existing mulch and replace with a fresh, 3″ thick layer. Use a preventive spray of bordeaux mix the following spring.

Flower and Fruit Symptoms: Fruit with Spots

Disease organisms can also invade fruit, killing cells in the fleshy fruit tissues. Fruit with mild symptoms may still be harvestable, but as symptoms progress, fruit may be ruined.

Anthracnose

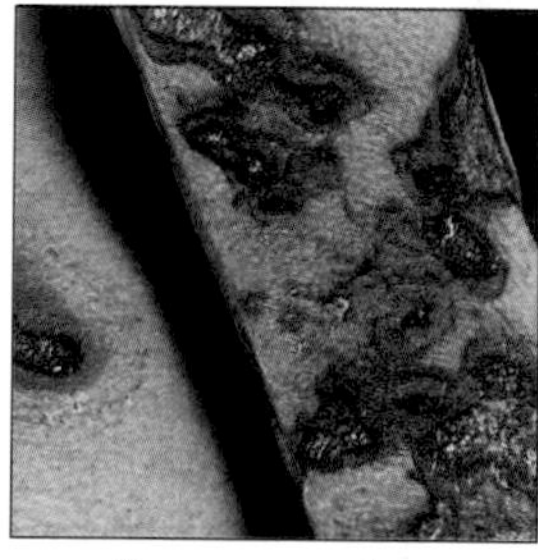

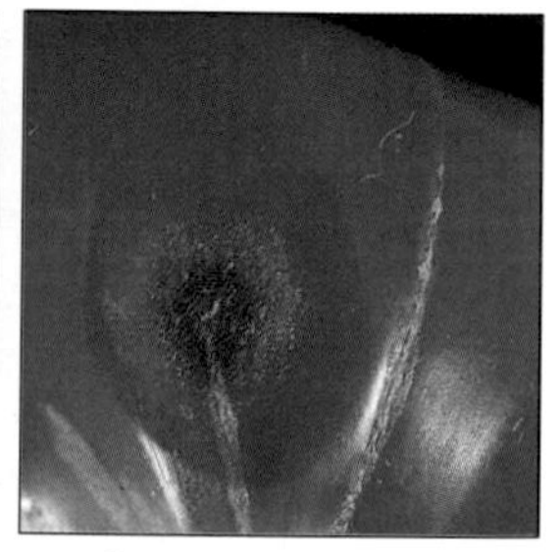

Symptoms on bean *Symptoms on tomato*

Type of Problem: Fungal.

Symptoms: Various anthracnose diseases affect fruit, producing small, dark, sunken spots. As the disease progresses, the spots may spread. Pinkish spore masses appear in the center of the spots in moist weather. Fruit eventually rots. Other plant parts are also affected by anthracnose; see "Anthracnose" on page 371 for more information.

Plants Affected: A wide range of woody and herbaceous plants. Tomatoes as well as cucumbers and melons are often affected; similar fungi attack beans.

Prevention and Control: Plant resistant cultivars. Using disease-free seeds and crop rotation may help prevent anthracnose diseases in vegetables. Copper sprays may provide some protection. Remove and destroy severely infected plants.

Apple Scab

Symptoms on apple

Type of Problem: Fungal.

Symptoms: Fruit infected with apple scab shows green or velvety brown spots, which later turn into raised, brown, corky areas. Leaves are usually also infected; see "Apple Scab" on page 371 for more details.

Plants Affected: Apples and crab apples.

Prevention and Control: Plant resistant cultivars. Rake up and dispose of fallen leaves. Apple scab is most prevalent in areas with cool, moist summers because the spores spread rapidly in wet weather. In spring, regular copper, sulfur, or lime-sulfur sprays may prevent apple scab fungi from infecting young leaves and fruit, reducing damage and the chance of apple scab recurring later in the season.

Flower and Fruit Symptoms: Fruit with Off-Color Patches

Odd-colored patches on fruit may be from cultural or environmental problems, or may be signs of a pathogen.

Blossom End Rot

Symptoms on tomato

Type of Problem: Environmental.

Symptoms: A water-soaked spot on the end of fruit gradually enlarges and turns brown or black, with a leathery appearance. The end of the fruit will appear flattened.

Plants Affected: Tomatoes and peppers.

Prevention and Control: Blossom end rot is commonly from calcium deficiency. This often occurs when hot, dry weather or rapid growth (due to excess nitrogen, for example) draws extra water and nutrients to the leaves, starving fruit for this nutrient. Root damage can also interfere with calcium uptake.

To avoid blossom end rot, keep the soil evenly moist by watering regularly and putting down a thick layer of organic mulch. Cultivate carefully to avoid damaging roots.

Powdery Mildew

Symptoms on apple

Type of Problem: Fungal.

Symptoms: Fruit infected with powdery mildew can crack or have raised webbing (called russeting) on their surfaces. The fruit may also show a white powdery coating; see "Powdery Mildew" on this page. This disease can also attack leaves; see "Powdery Mildew" on page 381.

Plants Affected: A wide range of woody and herbaceous plants. The fruit of apples and grapes are particularly susceptible.

Prevention and Control: Powdery mildew fungi thrive in hot weather, especially in conjunction with cool nights. Unlike most other fungal diseases, it is actually less of a problem in rainy weather.

Plant resistant cultivars: Control powdery mildew on your plants by providing good air circulation and growing resistant cultivars. Spraying with sulfur or lime-sulfur may reduce the spread of this disease. A 0.5 percent solution of baking soda (1 teaspoon baking soda in 1 quart water) may help to control the disease. Spray infected plants thoroughly.

Sunscald

Damage on tomato

Type of Problem: Environmental.

Symptoms: Sunscald appears on fruit as a water-soaked, blistered spot that eventually dries out and turns brown. Sunscald can also affect leaves, starting as pale green areas that later turn brown and dry.

Plants Affected: A wide range of woody and herbaceous plants. Fruit most commonly damaged includes tomatoes, peppers, and apples.

Prevention and Control: Bright light or excessively high temperatures can cause sunscald. On tomatoes this problem commonly occurs when a disease such as leaf spot causes leaves to fall, suddenly exposing once-shaded fruit to full sunlight. Plant cultivars that are resistant to major foliage diseases. Once damage occurs, there is no control.

Flower and Fruit Symptoms: Fruit with Powdery White Coating

The powdery mildew fungus can be on the surfaces of fruit as well as on leaves.

Powdery Mildew

Symptoms on apple

Symptoms on grape

Type of Problem: Fungal.

Symptoms: The fruit may color slowly or not at all and may show a powdery white coating. Other fruit symptoms include russeting of the skin; see "Powdery Mildew" on page 387 for details. For information about this disease on leaves, see "Powdery Mildew" on page 381.

Plants Affected: A wide range of woody and herbaceous plants. The fruit of apples and grapes are particularly susceptible.

Prevention and Control: Powdery mildew fungi thrive in hot weather, especially in conjunction with cool nights. Unlike most other fungal diseases, it is actually less of a problem in rainy weather.

Plant resistant cultivars: Control powdery mildew on your plants by providing good air circulation and growing resistant cultivars. Spraying with sulfur or lime-sulfur may reduce the spread of this disease. A 0.5 percent solution of baking soda (1 teaspoon baking soda in 1 quart water) may help to control the disease. Spray infected plants thoroughly.

Flower and Fruit Symptoms: Fruit Rotted or Deformed

When fruit cells die, they can become soft, and the fruit may rot. If cell development is inhibited or stimulated, fruit may have unusual shapes.

Black Rot

Symptoms on grape

Type of Problem: Fungal.

Symptoms: Purplish brown spots appear on green fruit. Infected grapes shrivel up, turn black, and cling to the bunch. Leaves and shoots can also be affected, showing circular, reddish brown spots.

Plants Affected: Grapes.

Prevention and Control: Plant resistant cultivars. Pick off and destroy infected fruit. Remove existing mulch or cover it with a fresh layer. Apply preventive sprays of bordeaux mix or another copper-based fungicide just before and immediately after bloom.

Botrytis Fruit Rot

Symptoms on strawberry

Type of Problem: Fungal.

Symptoms: If you have ever picked strawberries in moist spring weather, you have probably seen some berries covered with a fuzzy gray or tan mold. When you touch an infected berry, it sends up a small cloud of spores. Other fruits, such as apples, may instead show an internal rot. For information about Botrytis on leaves, see "Botrytis Blight" on page 373; for Botrytis on flowers, see "Botrytis Blight" on page 385.

Plants Affected: A wide range of woody and herbaceous plants. Fruits of strawberries, grapes, and brambles are most commonly affected.

Prevention and Control: Cool temperatures and high humidity promote Botrytis fruit rot. Removing infected fruit, whether living or dead, reduces the spread of the disease. Plants overstimulated with nitrogen fertilizer are most susceptible to gray mold, as are plants around which air cannot circulate to quickly dry leaves, stems, and fruit.

Brown Rot

Symptoms on peach

Symptoms on plum

Symptoms on plum (with mummies)

Type of Problem: Fungal.

Symptoms: Infected fruit shows small brown spots that often enlarge to cover the surface. These patches produce masses of grayish brown spores. The fruit eventually rots and shrivels up (forming a mummy); the damaged fruit may drop or persist in the tree. Small cankers may form on branch tips. Early in the season, flowers may be infected; for more details, see "Brown Rot" on page 386.

Plants Affected: Peaches, cherries, plums, and other stone fruits.

Prevention and Control: Plant resistant cultivars. Prune out to remove damaged shoots and to provide good air circulation. Pick off and clean up rotted and shriveled fruit. Spray with sulfur after blossoming to protect the fruit. Spray again just before harvest to protect fruit from brown rot in storage.

Corn Smut

Symptoms on corn

Type of Problem: Fungal.

Symptoms: Smut appears as swollen, grayish white galls on any part of the plant, especially ears and tassels. These galls continue to swell, and eventually release large quantities of powdery, dark brown spores.

Plants Affected: Corn.

Prevention and Control: Plant resistant cultivars. Rotate crops. Clean up garden debris in the fall. Remove infected plant parts as soon as you spot them.

Frost Damage

Damage on strawberry

Type of Problem: Environmental.

Symptoms: Flowers affected by frost have black centers, unlike the yellow centers of undamaged flowers. The tips of damaged fruit may be deformed. For stem symptoms, see "Frost Damage" on the opposite page.

Plants Affected: Strawberries.

Prevention and Control: Once damage occurs, there is no control. If you expect frost, cover plants with a blanket overnight.

Stem and Root Symptoms: Stems with Cracks or Holes

Openings in stems include cankers caused by disease organisms. Stem cracks or cuts may also be due to damage from frost or power equipment.

Cytospora Canker

Symptoms on peach

Type of Problem: Fungal.

Symptoms: Cytospora canker (also known as Valsa or Leucostoma canker) causes yellowing, wilting, and dieback of new shoots. Inner bark on infected twigs may show black or reddish brown discoloration. Gummy cankers form on trunks and branches and increase in size until they girdle and kill the affected part. For other stem symptoms, see "Cytospora Canker" on page 393.

Plants Affected: A wide range of woody plants. Stone fruits (such as peaches and plums), apples, pears, spruces, maples, poplars, and willows are among the most susceptible plants.

Prevention and Control: Plant resistant cultivars when available. Vigorously growing trees are less susceptible to this disease. Prune out infected branches during dry weather; disinfect your pruners between cuts. Avoid making unnecessary wounds in the bark, which can provide an entry for the fungus. On stone fruits, it is particularly important to avoid winter damage, which can be caused by fertilizing the trees late in the season.

Frost Damage

Damage on honey locust

Type of Problem: Environmental.

Symptoms: In the dead of winter, a combination of heat and cold can cause sunken areas or long cracks to form in the bark of woody plants. Damage is most common on the southwest side of a trunk. The last light of day warms the bark, then the bark rapidly cools as the sun drops below the horizon. For symptoms on fruit, see "Frost Damage" on the opposite page.

Plants Affected: A wide range of woody and herbaceous plants.

Prevention and Control: Young trees that have thin bark (such as sugar maples) or trunks that are unshaded by side branches are particularly susceptible. Prevent damage to tree trunks by wrapping them with tree wrap or by painting the trunks with white latex paint diluted with an equal amount of water to reflect heat and keep the bark temperature uniform.

Lawn Mower Damage

Damage to tree

Type of Problem: Cultural.

Symptoms: When a lawn mower hits a tree, it can cut a thin horizontal groove into the bark. Repeated damage may cause branch dieback or kill the tree. Lawn mowers can also damage surface roots. These wounds provide entrance points for diseases and insects. String trimmers also pose a hazard; careless use can cause the tool to strip the bark off the base of the tree, girdling and killing the plant.

Plants Affected: Any woody plant.

Prevention and Control: Use tools carefully around plants. Trim around trees and shrubs by hand, put plastic guards around trunks, or surround base of plants with mulch or groundcovers to eliminate the need to trim.

Lightning Damage

Damage on cottonwood

Type of Problem: Cultural.

Symptoms: The most common symptom is a large scar running down the trunk of the tree to the ground. In some cases, though, the only symptom you'll see is branch dieback caused by the root injury. Or, in extreme cases, branches or the whole tree may explode.

Plants Affected: Any woody plant.

Prevention and Control: Help trees recover by removing damaged bark or branches. Water during dry spells, and fertilize to promote vigorous growth. If you have an especially valuable tree, you might want to protect it with a lightning rod.

Sapsucker Damage

Damage on pine

Type of Problem: Cultural.

Symptoms: Tree trunks and limbs are covered with even rows of closely spaced holes.

Plants Affected: Any woody plant.

Prevention and Control: Several birds in the woodpecker family feed on tree sap. These birds, known as sapsuckers, prefer trees with a high sugar content, such as sugar maples, birches, beeches, and apples. Sections of bark may fall off of severely damaged trees. The holes also provide entrance points for disease organisms and wood-boring insects. Damage is difficult to control; wrapping the trunk with tree wrap may help prevent further feeding.

Stem and Root Symptoms: Stems Discolored

Discolored or blighted areas on stems may be due to death of cells or may be signs of the pathogen.

Cane Blight

Symptoms on raspberry

Type of Problem: Fungal.

Symptoms: This disease commonly produces brownish purple spots on canes. Infected canes may wilt and die.

Plants Affected: Black raspberries are most commonly affected, although other bramble fruits may also be attacked.

Prevention and Control: Choose a planting site with plenty of sunshine and good air circulation. Thin canes to avoid overcrowding. Remove diseased canes at ground level and destroy. The following year, spray with lime-sulfur as leaf buds begin to open in spring.

Cytospora Canker

Symptoms on peach

Type of Problem: Fungal.

Symptoms: Cytospora canker (also known as Valsa or Leucostoma canker) causes yellowing, wilting, and dieback of new shoots. Inner bark on infected twigs may show black or reddish brown discoloration. Gummy cankers form on trunks and branches and increase in size until they girdle and kill the affected part. Note: Bacterial canker may cause symptoms similar to those shown in this photo for Cytospora canker; controls for this disease are the same as those for Cytospora canker. For other stem symptoms, see "Cytospora Canker" on page 391.

Plants Affected: A wide range of woody plants. Stone fruits (such as peaches and plums), apples, pears, spruces, maples, poplars, and willows are among the most susceptible plants.

Prevention and Control: Plant resistant cultivars when available. Vigorously growing trees are less susceptible to this disease. Prune out infected branches during dry weather; disinfect your pruners between cuts. Avoid making unnecessary wounds in the bark, which can provide an entry for the fungus.

Slime Flux

Symptoms on tree

Type of Problem: Bacterial.

Symptoms: Slime flux (also known as bacterial wetwood) is indicated by the slimy liquid that oozes out of cracks and wounds in the bark, causing streaks on the trunk. The liquid may have a foul odor. In severe cases, shoot tips may wilt and die back.

Plants Affected: A wide range of woody plants. Elms, maples, and poplars are particularly susceptible.

Prevention and Control: In most cases, the damage caused by slime flux is cosmetic. The oozing cankers may be very unsightly. No control is available.

Stem and Root Symptoms: Stem Tips Stunted

When development of a single stem is inhibited, the distances between leaves or branches telescopes down so the leaves or branches grow close together. The result is what appears to be a whorl of leaves or broomlike growth of branches.

Peach Rosette

Symptoms on peach

Type of Problem: Viral.

Symptoms: This virus often produces stunting by causing trees to grow shoots that have abnormally short distances between the leaf nodes. Leaves may be discolored. The tree usually dies within a few months.

Plants Affected: Peaches and, less often, plums.

Prevention and Control: Similar symptoms are caused by a zinc deficiency. Try spraying the leaves with kelp extract. If the infected plant shows no response, remove and destroy it.

Stem and Root Symptoms: Stem Tips Die Back

Stem tips may die back as a result of disease problems, such as fire blight, or from physical damage to the plant.

Construction Damage

Damage to tree

Type of Problem: Cultural.

Symptoms: Branch tips die back over the crown of the tree; overall growth is poor.

Plants Affected: All woody plants.

Prevention and Control: Prevent damage by fencing off the area to keep equipment from hitting the plant or compacting the soil. Once the tree is damaged, there is little you can do; cut off any damaged bark and smooth off the wound edges so they can close properly. Fertilize, if necessary, and water during dry spells to help the plant recover.

Dutch Elm Disease

Symptoms on elm

Type of Problem: Fungal.

Symptoms: Dutch elm disease (DED) causes leaves to wilt, yellow, and drop early. Usually, branches show symptoms and die back one at a time. Sometimes the whole tree will wilt and die suddenly.

Plants Affected: Elm trees, particularly American elm.

Prevention and Control: DED spores are carried by elm bark beetles or transmitted through natural root grafts. Prevent insect attacks (see page 288) and subsequent disease infection by keeping trees healthy and vigorous. Once the disease infects a tree, there is no effective organic control. Remove and immediately destroy all infected elms. Remove the stump, if possible, or peel the bark off to below the normal soil line to deter elm bark beetles from feeding there. (The beetles can spread the disease as they move from diseased to healthy trees.)

Fire Blight

Symptoms on apple

Type of Problem: Bacterial.

Symptoms: Flowers usually show symptoms first, browning and shriveling. Leaves turn brown or black. Dead leaves remain on twigs. Symptoms progress from the tips of shoots toward the roots. Shoot tips turn black, wilt, and curl downward. Cankers form on branches. Fruit turns black and may cling to the tree.

Plants Affected: Many plants in the rose family, especially pears, apples, and quinces.

Prevention and Control: Plant resistant cultivars. Do not prune susceptible woody plants too severely or overfeed them, because both encourage succulent, disease-susceptible growth. Prune out infected branches, along with 6″-12″ of healthy tissue; disinfect pruners in between cuts by dipping them in a 10 percent bleach solution (1 part bleach to 9 parts water). Spray bordeaux mix during dormancy.

Oak Wilt

Symptoms on oak

Type of Problem: Fungal.

Symptoms: While symptoms vary according to the tree species affected, oak wilt commonly causes leaves to brown, wilt, and drop. Plants usually die within a year.

Plants Affected: Oaks.

Prevention and Control: Prune oaks only when they are dormant to reduce the chances of the fungus entering through the wounds. Once oak wilt begins, there is no control; remove and destroy infected trees. Dig a narrow, 36″-40″ deep trench between infected and healthy trees to break the natural root grafts through which the fungus can spread. Backfill the trench immediately to keep the healthy roots from drying out.

Twig Blight

Symptoms on juniper

Type of Problem: Fungal.

Symptoms: Infection begins on young leaves, causing tiny yellow spots. Branch tips turn reddish brown and die back. You may see a grayish band at the base of the dead shoot.

Plants Affected: Many needle-leaved evergreens, including junipers and cypresses.

Prevention and Control: Plant resistant species and cultivars. Prune off and destroy infected shoots on a dry day.

Verticillium Wilt

Symptoms on maple

Type of Problem: Fungal.

Symptoms: Infection by *Verticillium* fungi causes leaves to yellow and leaf stems to droop, giving plants a wilted appearance (see photographs on page 384). The yellow leaf patches turn brown and may spread to cover whole leaves. Leaves often fall early, and the plants will die. Symptoms usually first appear on the lower or outer parts of plants. The interior of the stem near the base may be discolored.

Plants Affected: A wide range of woody and herbaceous plants. Some particularly susceptible plants include tomatoes, peppers, melons, asters, chrysanthemums, peaches, cherries, strawberries, and maples.

Prevention and Control: Verticillium and Fusarium wilt fungi cause similar symptoms. Verticillium wilt is more common in cool weather in temperate areas of the country.

There is no cure for Verticillium wilt. Remove infected woody plants and replant with resistant cultivars, when available. Crop rotation is not very effective because so many species are susceptible. Soil solarization may help to control Verticillium wilt.

Winter Injury

Damage on yew

Type of Problem: Environmental.

Symptoms: Symptoms of cold injury can be very similar to the symptoms of sunscald: blotchy, water-soaked areas on leaves. Shoot tips often die back. For other information on leaf damage, see "Winter Injury" on page 380.

Plants Affected: A wide range of woody and herbaceous plants.

Prevention and Control: In winter, evergreens can suffer from drying when their roots cannot get sufficient water from frozen ground. Water plants thoroughly in late fall. Spray leaves with an antitranspirant to reduce water loss. If damage has been a problem in the past, move plants to a sheltered spot or erect a barrier (such as a piece of burlap) to protect them from drying winds. Prune out damaged wood in spring; remove severely affected plants.

Stem and Root Symptoms: Stems Entangled with Orange "Strings"

Dodder is a parasitic plant that entwines around host plants. The parasite draws food and water from its hosts, eventually weakening and smothering the infected plants.

Dodder

Dodder on ivy

Type of Problem: Parasitic plant.

Symptoms: Found throughout North America, dodder grows as a tangle of orange or yellowish threads that winds around stems and other parts of host plants. The parasite draws nutrients from the host plants, weakening them. Dodder's rampant growth can rapidly smother plantings.

Plants Affected: Many ornamentals as well as vegetables such as potatoes and onions.

Prevention and Control: Dodder does not have roots (except in early seedling stage), leaves, or chlorophyll. Beginning early in the season, dodder produces tiny flowers. Seeds are spread by animals and tools. The best way to get rid of dodder is to ruthlessly destroy it repeatedly throughout the season. Keep an eye out for seedlings that may develop the next season and weed them out as soon as you spot them. Don't add dodder to the compost pile, or you may be spreading seeds as you spread the finished compost.

Stem and Root Symptoms: Stems with Clusters of Evergreen Leaves

Mistletoe is a parasitic plant that is better known as a seasonal decoration than as a cause of plant disease.

Mistletoe

Mistletoe on oak

Mistletoe

Mistletoe berries

Type of Problem: Parasitic plant.

Symptoms: The mistletoe of Christmas, called leafy mistletoe, attacks mostly hardwoods in the southern parts of the country. Clusters of evergreen growth appear on the limbs of these deciduous trees. Leafy mistletoe is a weak parasite, so it does little harm to the trees besides creating a strange appearance in winter.

Another type, known as dwarf mistletoe, attacks conifers throughout the world, but most seriously threatens trees in this country on the West Coast. It weakens, deforms, and even kills trees. The parasite looks like a tuft of branches, varying in color from yellowish to brownish green.

Plants Affected: A wide range of woody plants.

Prevention and Control: Cut off and destroy infected branches as soon as you notice them to avoid the spread of this parasite.

Stem and Root Symptoms: Stems with Swollen Growths

Galls and other swellings on stems and roots are due to overstimulated cell development. Root swellings can interfere with water and mineral uptake, resulting in yellowing and wilting of the aboveground portions of the plants.

Black Knot

Symptoms on plum

Type of Problem: Fungal.

Symptoms: Black knot appears as unsightly swellings on twigs and branches. These swellings are usually black, but they may appear velvety green in early spring. Tips of infected branches often die back. Severe infections can kill whole limbs, and the tree may be stunted.

Plants Affected: Cherries and plums are most commonly affected.

Prevention and Control: In fall or late winter, prune off infected limbs, 6″–12″ below the knots; disinfect pruners in between cuts with a 10 percent bleach solution (1 part bleach to 9 parts water). Destroy the prunings. Remove any wild plum or cherry trees nearby. For persistent infections, apply 2 sprays of lime-sulfur, 7 days apart, before the buds begin to grow in spring.

Cedar-Apple Rust

Symptoms on cedar

Type of Problem: Fungal.

Symptoms: Hard, brown swellings appear on branch tips. These galls do not seriously damage cedar trees, but they can mar the plants' appearance. Warm, moist weather in spring causes these galls to swell dramatically, and they produce gelatinous horns that release rust-colored spores. The spores then infect apple trees. For more details about this disease on apples, see "Cedar-Apple Rust" on page 373.

Plants Affected: Eastern red cedars and other species of juniper.

Prevention and Control: Cedar-apple rust completes its life cycle only where the fungal spores can travel back and forth between cedar and apple trees. Spores from cedar trees send spores to infect apple trees, but infections on the apple tree do not spread within the tree; they only can send the disease back to infect cedar.

Prevention is the best control. Rust diseases thrive in moist conditions, so anything you can do to promote leaf drying will limit disease problems. Plant apple trees only if cedars are at least 4 miles away; this will reduce the chances of the disease spreading. If you want to grow both cedars and apple trees, plant rust-resistant species or cultivars of both plants. Prune off and destroy galls before late winter.

Crown Gall

Symptoms on rose

Type of Problem: Bacterial.

Symptoms: Above ground, the plant is stunted with yellowing leaves. Just beneath the soil line of an infected plant (or sometimes just above the soil line), you will see the tuberous swellings that upset the mineral- and water-conducting vessels, causing the aboveground symptoms.

Plants Affected: A wide range of woody and herbaceous plants. Plants commonly attacked include stone fruits, grapes, brambles, euonymus, chrysanthemums, and roses.

Prevention and Control: This common bacterium enters plants through wounds, such as those that occur during transplanting. The bacteria may be transmitted from one plant to another on a trowel or shovel, which can wound a plant during digging. The bacteria may also survive on dead plants for years in infested soil, waiting for a suitable host plant.

Inspect nursery plants carefully before you buy to avoid infected plants. Protect healthy plants by dipping their roots in a solution of *Agrobacterium radiobacter* (sold as Galltrol-A) before planting. To control mild infections, prune off diseased growth; disinfect your pruners in between cuts with a 10 percent bleach solution (1 part bleach to 9 parts water). Remove and destroy severely infected plants. Avoid replanting the area with susceptible plants.

Stem and Root Symptoms: Roots with Swollen Growths

Various diseases can cause roots to swell, but root growths may also be caused by the presence of beneficial nitrogen-fixing bacteria.

Club Root

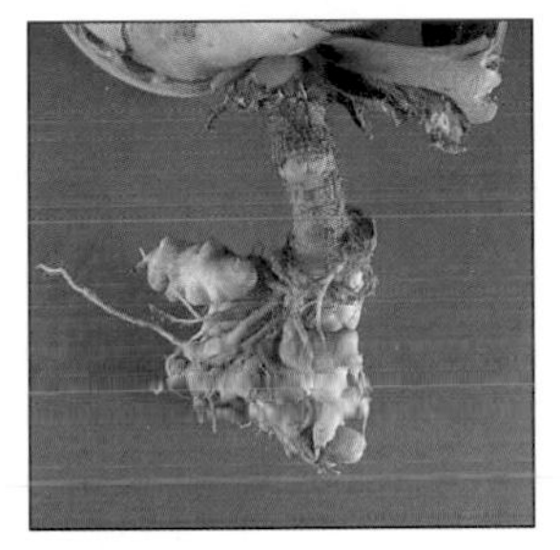

Symptoms on cabbage

Type of Problem: Fungal.

Symptoms: Above ground, the plant may appear stunted, wilted, and yellowed. Below ground, infection causes the roots to swell into a gnarled mass that cannot supply water and nutrients to the plant.

Plants Affected: Cabbage family plants, such as cabbage, cauliflower, broccoli, and brussels sprouts.

Prevention and Control: Once it is in the soil, the causal fungus can infect wild cabbage family plants (such as wild mustard and shepherd's purse) or survive for years even in the absence of a host plant. This fungus, however, thrives only in acidic soils, so you can completely check the disease by adjusting the soil pH to 7.2 or slightly above. Soil solarization can also reduce the incidence of club root. Remove and destroy seriously infected plants.

Nitrogen-Fixing Nodules

Nitrogen-fixing nodules on bean

Type of Problem: Bacterial (beneficial).

Symptoms: Small, hard nodules form on the roots of host plants; plant appears healthy.

Plants Affected: Legumes, such as peas.

Prevention and Control: The relationship of legumes and these bacteria (*Rhizobium* spp.) is generally beneficial to both. Plants supply bacteria with nutrients; bacteria convert nitrogen from the air into a form that plants can use. To promote growth of these beneficial organisms, buy the bacteria in powder form (called inoculant) and dust the seeds of appropriate crops before planting. Certain species work better with particular plants, so buy the right inoculant for your crop.

Root Knot Nematodes

Symptoms on tomato

Type of Problem: Nematodes.

Symptoms: The symptoms produced by root knot nematodes are similar to those of other diseases and disorders that interfere with proper root function. Above ground, plants may be yellowed and wilted; flowers and fruit are stunted or nonexistent. Below ground, however, you'll find numerous clubbed galls on the roots caused by the nematodes feeding inside.

Plants Affected: A wide range of woody and herbaceous plants. These nematodes commonly attack fruit trees, corn, lettuce, tomatoes, and potatoes.

Prevention and Control: Plant resistant cultivars, when available. Rotate crops. Avoid spreading soil from infested to noninfested areas. Soil solarization helps to reduce nematode populations. Or try a cover crop of marigolds, and turn them into the soil at the end of the season. Incorporate plenty of organic matter into the soil to promote natural nematode-controlling microorganisms. Drenching the soil with neem may also help.

Stem and Root Symptoms: Roots Die Back

Aboveground symptoms such as stunting and wilting may be the result of the death of cells in diseased roots.

Red Stele

Symptoms on strawberry

Type of Problem: Fungal.

Symptoms: Symptoms include stunted growth and wilting and death of older leaves or whole plants. Infected plants bear little or no fruit. Young feeder roots die; older, infected roots have dead tips and a reddish brown discoloration in the core of the roots.

Plants Affected: Strawberries.

Prevention and Control: Remove and destroy infected and surrounding plants. If red stele has been a problem in the past, prepare a new site for your strawberry bed and plant resistant cultivars.

Root Rot

Symptoms on birch

Symptoms on oak

Type of Problem: Fungal.

Symptoms: Symptoms include leaves that are small, yellow or brown, and wilted. Decline may be gradual, over years, or rapid.

Other symptoms, which indicate Armillaria root rot, may appear at the base of the plant. These include white mats of fungi and brownish black fungal strands on the roots or between the bark and the wood. Honey-colored mushrooms often grow around the base of dead or dying plants. Severely damaged plants may fall over during a storm. For more information, see "Armillaria Root Rot" on page 403.

If you don't see any signs of fungal growth near the base of the plant, suspect other types of root rot. To check for root damage, lift the plant out of the soil and carefully wash off the soil. Look for roots that are damaged with brown or black tips.

Plants Affected: A wide range of woody and herbaceous plants.

Prevention and Control: These common soilborne diseases attack fully grown plants as well as seedlings. Root rots are caused by a few different types of fungi, each most prevalent in certain areas or under certain conditions.

Plant resistant cultivars when available. Reduce the incidence of root rot with good cultural practices: Plant in well-drained soil, do not overwater, and direct surface water away from the crown of plants. Incorporating compost into the soil may also help to prevent some types of root rot. If a plant is diseased, you may be able to save it by pulling the soil back away from the crown, pruning off diseased roots, and allowing the remaining exposed roots to air-dry. Also, disinfect tools after use with a 10 percent bleach solution (1 part bleach to 9 parts water) so that you do not spread disease from infested to healthy soil. Remove and destroy seriously infected plants.

Stem and Root Symptoms: Tubers Discolored or Rotted

Disease can also cause cells of storage organs such as tubers to die, either in the ground or in storage.

Bacterial Soft Rot

Symptoms on potato

Type of Problem: Bacterial.

Symptoms: In the garden or in storage, the fruit or storage roots first develop small, water-soaked spots. These areas enlarge, becoming soft, sunken, and discolored. You may notice a foul odor. Above ground, plants may appear yellowed, stunted, and wilted.

Plants Affected: A wide range of herbaceous plants. Vegetables with fleshy fruit or succulent stems are quite susceptible.

Prevention and Control: To avoid soft rots, handle fruits and vegetables (especially those you plan to store) carefully, both during and after harvest. Soft rot bacteria enter wounds produced by rough handling as well as those caused when fruits, tubers, and roots are frozen or damaged by insects. Soft rots usually need a high moisture level to develop, so store produce in a cool, dry place. If you have had trouble with soft rot in past years, rotate crops to reduce the spread of the disease in the garden.

Scab

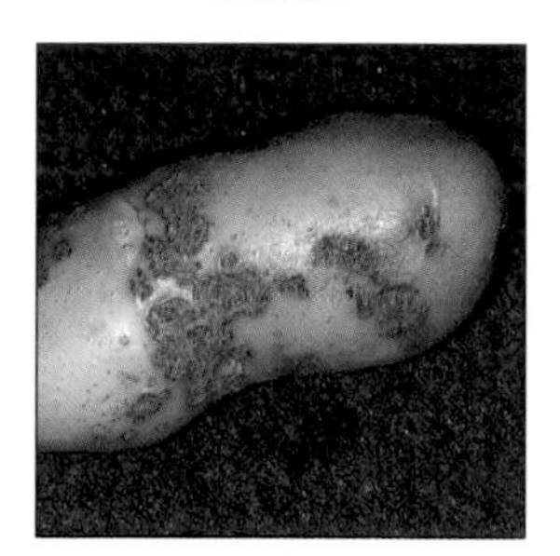

Symptoms on potato

Type of Problem: Fungal.

Symptoms: Scab begins as small brown spots on the tubers or roots. These areas enlarge and run together, producing russeting and rough, corky areas. These spots are usually just on or below the surface, so the crop is still edible, although visually unappealing.

Plants Affected: Potatoes, beets, radishes, and other root crops.

Prevention and Control: Scab fungi thrive in light, sandy soils with a neutral or alkaline pH. Lowering the soil pH to below 5.2 will reduce the chance of disease. Start with certified disease-free seed potatoes. Rotate crops and plant resistant cultivars. Keep the soil evenly moist.

Whole Plant Symptoms: Seedlings Die

Damping-off fungi can infect and kill a wide range of plants at the seedling stage.

Damping-Off

Symptoms on seedlings

Type of Problem: Fungal.

Symptoms: Damping-off fungi can kill seedlings before they even break through the soil, but a more dramatic demonstration of this disease occurs when it strikes seedlings just an inch or so tall. The fungi rot the stems right at the soil line and, overnight, infected seedlings topple over.

Plants Affected: All plants.

Prevention and Control: Plants growing in containers are more susceptible to damping-off than are those growing outdoors in the garden. Good cultural conditions usually control it; to prevent it from attacking your seedlings, grow them in well-drained soil with plenty of light. Do not allow them to crowd each other, or the stagnant air that results will promote the growth of damping-off fungi.

A thin layer of dry material, such as sand or perlite, sprinkled on the soil surface keeps seedling stems dry at the soil line, where damping-off often strikes. Even better is a layer of fine sphagnum moss, which not only keeps the surface dry but also reduces the chance of fungal growth. After a few weeks of growth, seedling stems toughen and no longer are as susceptible to attack by damping-off fungi. If you've had past problems with damping-off, use sterile potting mix or pasteurize your own potting soil (see "Pathogen-Free Potting Soil" on page 413) when growing plants in containers.

Whole Plant Symptoms: Plants Stunted

Stunted growth is due to inhibited cell development. Many viruses cause stunting, but stunted growth also may be from drought or insufficient nitrogen.

Mosaic

Symptoms on bean

Type of Problem: Viral.

Symptoms: Mosaic-infected leaves are mottled with yellow, white, and light and dark green spots or streaks (see photograph on page 377). Fruit may show similar symptoms. Plants are often stunted.

Plants Affected: A wide range of woody and herbaceous plants.

Prevention and Control: Plant resistant cultivars. Mosaic viruses are mostly spread by insect pests, especially aphids and leafhoppers. If possible, deny these carriers access by covering your plants with floating row cover. For more control information on aphids and leafhoppers, see pages 269 and 302, respectively. Once plants are infected, there are no controls; remove and destroy infected plants.

Whole Plant Symptoms: Plants Fall Over

If disease kills off a large portion of the root system, it can't support the aboveground part of the plant; the plant may collapse.

Armillaria Root Rot

Armillaria root rot

Type of Problem: Fungal.

Symptoms: Common symptoms of root rot include leaves that are small, yellow or brown, and wilted. Decline may be gradual, over a period of years, or rapid. Other symptoms that may appear include white mats of fungi and brownish black fungal strands on the roots or between the bark and the wood. Honey-colored mushrooms often grow around the base of dead or dying plants. Severely damaged plants may fall over during a storm. For more on root rot, see "Root Rot" on page 401.

Plants Affected: A wide range of woody and herbaceous plants.

Prevention and Control: Armillaria root rot is prevalent throughout the country, especially at sites that were recently oak forests. Avoid Armillaria root rot by planting in sites not harboring the fungus or by planting disease-resistant plants. Unfortunately, many plant species are susceptible; some resistant plants include pears, white fir, sweet gums, Oregon grape holly, and pines. Once plants fall over, there is obviously no way to restore them; remove the plant and what is left of the roots.

PART 4

ORGANIC CONTROLS

Using Remedies Safely

Just as interest in organic gardening methods has steadily increased in the past two decades, so has the range of control options for organic gardens. These range from new, highly sophisticated products that use insect hormones called pheromones to prevent pest insects from mating to more effective formulations of standard products such as horticultural oils.

In this section, you'll find information on control practices to incorporate into your gardening routine to combat widespread problems, such as aphids and powdery mildew, along with specific products or methods that affect a single pest, such as surrounding seedlings with collars to prevent cutworm damage. While all the controls described are considered organically acceptable, they vary tremendously in their convenience and environmental impact. Always begin with the safest, least-toxic method before moving on to more toxic—and usually broader spectrum—controls. Even plant-derived pesticides such as rotenone and ryania can pollute water, kill honeybees, or make you sick if you mishandle them. However, few, if any, people, pets, wildlife, or beneficial insects will suffer ill effects when exposed to crop rotation, companion planting, row cover, or other control methods that prevent pest damage without having to actively kill the pests involved.

The control methods described in this section come from a variety of sources. Some are gardening practices that have been handed down as folklore through the ages, while others are based on the latest scientific research. Many of them are simply matters of common sense—keep your plants healthy and they'll be better able to fend off problems; use barriers to exclude pests from your plants and your crops won't be eaten. Regardless of the methods you choose to use to protect your plants, use care when applying them. You'll notice that special precautions for use of some chemical controls are included in the descriptions that follow. Pay heed to the warning information included in these listings, and when using commercial products, always apply them according to the label specifications.

Types of Controls

The control methods described in this section are divided into four categories according to the way in which they act on pests and also according to their level of potential harm to nontarget organisms, such as honeybees, fish, and even humans. The order of presentation reflects the order in which you should choose controls when dealing with the particular pest problems in your yard and garden.

- **Cultural controls.** These are gardening practices that help prevent pest problems. You can integrate these controls into your overall gardening routine. There are few, if any, harmful side effects on beneficial insects or other organisms.
- **Physical controls.** These controls exclude or remove pests from plants. Handpicking insect pests is the most basic form of physical control.

■ **Biological controls.** These methods use living organisms to kill garden pests. They include releasing and encouraging beneficial insects and applying bacterial insecticides.

■ **Chemical controls.** These include homemade pesticidal sprays, sulfur and copper compounds, and botanical poisons. They have varying levels of toxicity and are only to be used for severe problems when other types of control methods have proven inadequate.

Specific control methods do not always fit neatly into one of these four categories. As growers and researchers learn more about organic control methods, they often uncover information that causes them to rethink the way in which traditional remedies are viewed. For example, chemical controls generally are substances that kill organisms. However, sulfur, generally classified as a chemical control, may actually prevent fungal diseases by coating plants' leaves so that pathogens are unable to infect a crop. This is technically a physical method of protection.

Certain botanically derived insecticides classified as chemical controls actually act as growth or feeding regulators on pests. This means their mode of action is more biological than chemical.

Many cultural control practices are thought to work on a variety of levels. Companion planting is generally thought of as a general cultural practice that lessens pest problems. In specific cases, the companion plants may attract beneficial insects, and so in a sense, are a biological control. Other plant companions may serve as a trap crop for pests—a physical method of pest management. Often it is the oldest methods that are least understood in terms of how and why they work.

Use the information about each control category to form a mental checklist for dealing with pest and disease problems. Consider the advantages and disadvantages of each approach and how it fits into your garden plan. Most organic methods of pest and disease control rely on the gardener's ability to implement them completely—starting with awareness of pest emergence times through application of controls to subsequent monitoring of remaining pests. Always be sure to try all possible low-impact controls before considering more hazardous methods.

Planning a Control Strategy

How do you choose which control methods to use against a particular pest, such as Colorado potato beetles on potatoes or scab on apples? A good place to start is the "Problem-Solving Plant Guide" beginning on page 11. The Potato entry will suggest some control methods for the beetle, the Apple entry describes ways to prevent or lessen scab problems. You can then refer to the descriptions of various products and methods that follow for information on how to make or use specific controls.

For insects and related pests, you'll also find information to help you identify and understand the pest, its life cycle, and how to manage it in the "Insect Identification Guide" beginning on page 268. The "Disease Symptom Guide" beginning on page 370 provides similar information on plant diseases and disorders. Don't forget the possibility that the damage you find in your garden could be caused by a large animal pest. For some tips on stopping animal feeding, see "Stopping Animal Pests" on page 408. After you read the appropriate information in these sections, you should be able to craft the safest and most effective strategy for your garden.

Stopping Animal Pests

Whether you garden in the suburbs or in a rural area, you may find that animals and birds cause you more headaches than insect pests do. The trickiest part of coping with animal pests can be identifying the pest. Since many animals feed at dawn or dusk, you'll need to notice such signs as feeding patterns, tracks, tunnels, or excrement to figure out what culprit is invading your garden. For example, gnawed strawberries may be the work of birds, mice, or slugs. Controls that work for one pest may do nothing to stop another, so it's important to figure out exactly what pest is to blame before you act.

Animal deterrents fall into one of four categories.

- **Fences and barriers.** These hinge on the assumption that there's probably some way to keep the pest out of your yard.
- **Traps.** If you use live traps, you must be willing to transport the captured pest to a new location. Other types of traps kill the pest to stop the damage.
- **Repellents.** These work on the principle that most animals won't eat something that tastes or smells bad.
- **Scare tactics.** These employ such gadgets as inflatable owls, plastic snakes, hawk-shaped kites, and elaborate scarecrows.

At least one of these approaches should give you a fighting chance against whatever animal is dining in your garden. After you've discovered just which animal you're trying to discourage, the following tactics should help you out.

Deer

An electric fence, and not necessarily a high one, is the most effective way to keep deer out. One design suitable for small areas uses two fences, an inner chicken-wire fence 4′ high and a single-wire electric fence only 2½′ off the ground and located 3′ outside the chicken wire. Deer find it hard to jump the chicken-wire fence with the electrified wire in the way. If you prefer a conventional woven wire fence instead of going electric, choose one at least 8′ high. A second inner fence about 3′ high increases the effectiveness of a nonelectric fence because double obstacles confuse deer. If deer are nibbling just a few shrubs, consider enclosing these in woven wire fence cages. For a slight deer problem, this is an inexpensive, effective solution.

Repellents deter deer as long as the pressures of starvation or overpopulation don't force them to eat anything in sight. Buy soap bars in bulk and hang them from strings in trees. Or nail each bar to a 4′ stake and drive the stakes at 15′ intervals around your property. Soap fragrance is more pervasive when the soap is wet, so a good strategy is to mist soap bars with water early in the evening just before deer begin to feed. Some gardeners report that human hair hung in mesh bags or old stockings is an effective repellent. You can probably collect all the hair you need from a barber.

One repellent that seems to work is made from eggs and water. Mix one egg per quart of water, multiplying the recipe as many times as you need to get the right amount of spray. With rainfall or heavy dew, some repellent will wash away. You can prolong the time it remains on plants by mixing it with a small amount of antidesiccant, such as Wilt-Pruf or VaporGard. Other homemade repellents include water solutions of blood meal, hot sauce, or garlic oil. Many commercial repellents are currently available, including BGR (Big Game Repellent), made from dried eggs, Hot Sauce, made from ground hot peppers, and Hinder, a soap-based repellent.

Rabbits

The best way to keep rabbits out of a garden is to erect a chicken-wire fence with mesh no larger than 1″. If you have existing picket or woven-wire fence, simply attach a 2′ wide strip of chicken wire to the bottom of the fence. Rabbits also sometimes burrow under a fence, so you may need to dig a 6″ deep trench and sink the chicken wire down into the soil to completely keep them out. If your soil is rocky, pile a 1′ wide border of small stones around the periphery of your fence to discourage burrowing. To protect young trees and shrubs from rabbits gnawing bark in the winter, erect cylinders made of ¼″ hardware cloth; the cages should be 1½′–2′ high (higher if you live in an area with deep snowfall) and should be sunk 2″–3″ below the soil surface. This method is also effective in protecting trees from bark feeding by mice and voles.

Most deer repellents are also effective against rabbits. Some gardeners report that used cat box filler sprinkled on the lawn around ornamentals deters rabbits. Since effectiveness wears off after about a week, you'll need to replace the litter often, especially after rain.

Moles

Moles eat grubs and earthworms, not plants. Nevertheless, their extensive tunnels ruin lawns and provide easy entry for mice and voles, which do dine on plant roots and bulbs. If your yard is full of mole runs, chances are your soil is full of beetle grubs, the favorite food of foraging moles. The most humane way to discourage moles is to eliminate soil grubs by applying milky disease spores (see more information about this control on page 460). When the grubs die, the moles will move on to better feeding grounds. But this approach requires patience; it may take several years after initial application of milky disease spores for the grubs to disappear.

If you'd like to dispatch an active mole immediately, you must locate a permanent tunnel leading from the underground nest. Find a long, straight tunnel and press it down in a few places. If you find the tunnel reopened the next day, you've found a permanent passage. You can try flooding the mole out by filling this passage with water, but you must be prepared to kill it as soon as it comes to the surface in order for the method to be effective. You can also place any of the various commercial mole traps along a permanent passage and eventually eradicate your pest. Deciding whether getting rid of mole tunneling warrants killing these animals is a matter of personal choice.

Birds

Birds are both the gardener's friends and foes. While they eat insect pests, they also consume entire fruits or vegetables or will pick at your produce until it is damaged enough to be unappealing. Most bird controls involve making the area you wish to protect either less appealing to birds or more difficult for them to feed in.

The most effective control to protect bush and vine fruits and small fruit trees is to cover them with lightweight plastic netting; cover row crops with floating row cover. You can also use a variety of commercial or homemade devices to frighten birds away from your crops. Tactics that keep birds guessing include moving a scarecrow to a different spot every few days and tying pie pans, pinwheels, or strips of plastic or foil to its arms. Other bird-scaring devices include plastic snakes, bird-scaring balloons, and inflatable owls and hawks. These are most effective when mounted on a fence at the garden's edge and relocated every few days.

CULTURAL CONTROLS

Everything you do to encourage healthy plant growth can be considered a form of cultural pest control. As such, cultural controls are not usually applied in response to a specific problem, but instead are an integral part of an organic approach to gardening. From the moment you sketch out a planting scheme, turn up a shovelful of soil, or buy a seed or plant, pest prevention and control measures become a part of your gardening practices.

A Cultural Overview

You'll begin thinking about cultural controls in the dead of winter when you're paging through seed and plant catalogs, looking for disease-resistant cultivars of your favorite vegetables. Most gardeners are aware that selecting plants and cultivars that are resistant to certain disease and insect problems can help them avoid problems later in the season.

Cultural controls also play a role in how and when you plant, how you water and fertilize, and finally how you clean up your garden site in the fall when your plants have died off or have gone dormant for the winter.

Start with the Soil

A fundamental step in producing healthy plants is to promote healthy soil. Organically enriched, biologically active soil promotes plant health by encouraging root development, thus setting the stage for efficient use of water and nutrients. Healthy soil is also more likely to contain a natural, sustainable balance between beneficial and potentially destructive soil microorganisms.

Compost, a mix of decomposed organic materials, is one of the best things you can add to your soil. See "How to Compost" on page 425 for directions on making this valuable soil amendment. You can also encourage overall soil fertility by adding general-purpose organic fertilizers or tea made from compost (see "Fungus-Fighting Tea" on page 427). When using packaged organic fertilizers, follow the label directions for proper application. If your plants show nutrient deficiency symptoms, nourish them with the products listed in "Nutrients for Plant Deficiencies" on the opposite page.

If you suspect that your soil lacks certain nutrients, take a soil sample and have it analyzed by a reputable laboratory. Your local Cooperative Extension office or state land grant university can tell you how to take a soil sample; they may also be able to perform the tests for you or will recommend a private testing laboratory.

Choose Plants That Fight Back

Many diseases and even insects can be controlled by growing resistant or tolerant cultivars or species. By emphasizing disease resistance in your choice of cultivars, you can completely avoid some common pest problems. For example, some newer cultivars of shell pea, including 'Maestro', are resistant to mosaic virus as well as powdery mildew and Fusarium wilt. Similarly, 'Slicemaster' cucumber resists both downy mildew and powdery mildew, plus leaf spot, anthracnose, and mosaic virus.

Disease resistance also varies among species within the same genus. Among salvias, for example, the annual species, *Salvia splendens,* sold as a bedding plant, is seldom bothered by pests other than whiteflies. However, many of the perennial *Salvia* species that are native to arid climates have problems with mildews, rots, and insects. Whenever possible, get information (from extension agents, local nurseries, or fellow gardeners) about common pest and disease problems that occur in your area; use this knowledge to select plants that resist such problems. You'll find recommendations for resistant cultivars and species in individual plant entries in the "Problem-Solving Plant Guide" beginning on page 11.

Consider Timing and Placement

You can avoid certain pest problems by planting when the pest isn't around. Some pests, such as carrot rust flies, lay eggs only at a certain time in spring. By delaying planting, you reduce the chance that your crop will be infested by that pest.

Where you put plants in the garden and in relation to one another can also have an effect on pests. Changing the planting site of crops from one year to the next in your vegetable garden—known as crop rotation—can prevent pest populations from building up by denying them a food source from season to season. When practiced thoughtfully, crop rotation can keep nonmobile pest populations from reaching damaging thresholds. Problems that are particularly easy to control through rotation include onion and cabbage root maggots, black rot of broccoli, and most diseases caused by soilborne bacteria.

Companion planting—planting two or more types of plants in close proximity—is a popular gardening technique that can reduce pest problems in a variety of ways. Some good companion plants, such as dill, fennel, and Queen Anne's lace, will attract beneficial insects. Other types of companion plants repel pests or confuse and confound insects or disease organisms in search of their preferred host plants. To learn more about how diversity helps reduce pest problems, see "Encourage Diversity" on page 6.

Nutrients for Plant Deficiencies

If your plants can't get the nutrients they need from the soil, they may show a wide range of deficiency symptoms, from yellowing leaves to stunted growth. Once you identify what is lacking (by close observation of the plant or by a soil or plant tissue test), you can choose from a wide range of organic fertilizers to supply the needed nutrient. The table below lists some commonly deficient plant nutrients and fertilizers that can supply them.

Nutrient	Sources
Nitrogen	Alfalfa meal Blood meal Fish emulsion Fish meal Guano Soybean meal
Phosphorus	Bonemeal Colloidal phosphate Rock phosphate
Potassium	Granite meal Greensand Sul-Po-Mag Wood ashes
Magnesium	Epsom salts Dolomitic lime
Calcium	Gypsum Calcitic lime Oyster shells
Sulfur	Flowers of sulfur Gypsum
Boron	Borax
Other micronutrients	Kelp or seaweed extract Kelp meal

Plant Health Tonics

Although they are not pesticides, kelp, seaweed, and micronutrient sprays can help make plants more pest resistant by improving their overall health. Often described as growth enhancers, these products provide an extensive menu of nutrients such as iron and boron that plants need, in very small amounts, for proper growth and development.

Plants can absorb these micronutrients through their leaves. So if deficiency symptoms appear, applying a foliar spray such as a chelated iron product or seaweed will provide a quick remedy. (The long-term solution is to add soil amendments to correct the problem.) Those products made with seaweed or kelp can also be categorized as growth regulators, since they contain amino acids and enzymes that promote stronger growth and increase plant yields.

If you know from experience that certain plants will face challenges from insects and diseases, you can apply seaweed or kelp sprays to strengthen those plants and make them better able to defend themselves. These products, however, are not a substitute for good soil or nutrients taken up by plant roots.

If you have access to fresh seaweed, rinse it to remove the salt, then apply it to the garden as a mulch, or compost it. Gardeners who don't have access to fresh seaweed can buy liquid seaweed extract. Before spraying it on your plants, dilute this concentrated product according to the directions on the label. Spray plants every two to four weeks, depending on the label directions. Remember: A little is good for your plants, but too much can be toxic, so don't be tempted to spray more frequently or use stronger solutions. Dried seaweed, sold as kelp meal, is a good long-term soil conditioner; apply 1–2 pounds per 100 square feet of soil.

Keep Things Clean

Keeping a clean garden means preventing pest-infested material from getting in and doing as much as possible to get rid of plants or parts of plants that develop pest problems. Carefully inspect any plants you intend to buy for signs of insects or disease. See "Buying Healthy Plants" on page 4 for advice. And if pest problems do strike in your garden, use care in disposing of the diseased or infested material.

If you save your own seeds, you can try various heat or cold treatments to "clean" the seeds for next year's crops. Do not save seeds from plants that show symptoms of Fusarium or Verticillium wilts or any disfiguring virus. All seeds contain living cells capable of harboring these microscopic pathogens. If you save seeds from your healthiest, most robust plants, you are also practicing natural selection, and the resulting plants may be more tolerant of pathogens than preceding generations.

When the gardening season ends, you'll want to thoroughly clear any diseased plant material out of your garden beds. Otherwise, come the following spring, it can serve as a source of infection for new plants. In some cases, you can safely compost diseased plant material. In other cases, it's best to destroy it by burning or disposing of it in sealed containers with your household trash.

You may also want to till or cultivate the soil in garden beds that are cleared out in the fall to help reduce pest problems. Tilling exposes soil pests to the surface, where they may be killed by cold temperatures or spotted and eaten by predators.

Resistance

In 1859, Charles Darwin noted that in nature only the fittest will survive, describing the ongoing process of natural selection. All plants and animals are subject to a variety of

Pathogen-Free Potting Soil

Getting your seedlings and transplants off to a good start is a critical step in creating a healthy, productive garden. Seedlings and young plants are prone to a number of soilborne diseases, commonly lumped together under the term *damping-off.* Using a disease-free medium is a basic part of producing healthy young plants.

If you make your own potting medium, especially out of ingredients such as garden soil, you will probably want to pasteurize it. Exposure to temperatures from 160°-180°F for 30 minutes will kill most insects, weed seeds, and pathogenic bacteria and fungi. Higher temperatures, however, can also destroy beneficial organisms, deplete soil organic matter, and release toxic salts into the soil; monitor soil temperature carefully during treatment.

You can pasteurize your potting mix indoors in a conventional oven or microwave, or outdoors on a barbecue grill. Be aware, though, that heating soil will create a strong odor. If you're concerned about the smell lingering in your house, try the outdoor method. After treating the soil, let it cool before you use it, or store it in a closed container.

■ **Conventional oven.** Fill a clean, shallow pan with moist soil. Cover the pan with aluminum foil, insert a meat thermometer into the soil, and place the pan in an oven heated to 200°F. Start timing when the soil temperature reaches 140°F and continue heating for 30 minutes. Remember that the soil temperature should not exceed 180°F; if necessary, remove the pan from the oven and let it cool to below 180°F before returning it to the oven, or add a small amount of cool water to the soil to moderate the temperature.

■ **Microwave oven.** Be sure your soil is moist and free of stones and metallic particles. Put it in a polypropylene baking bag or a microwave-safe mixing bowl—in either case, the container should not be tightly sealed. At low power (30-40 percent), adequate heating may take 10-15 minutes of operation. At full power, heating will take about 2½ minutes for a 2-pound batch. If you see any sparks, stop the process immediately; this indicates that metallic particles are present in the soil. In this case, use another method.

■ **Outdoor barbecue grill.** Fill a clean, shallow pan with soil. Set the pan over the fire of an outdoor barbecue grill. Insert a meat thermometer into the soil. When the thermometer registers 140°F, begin timing, and keep the soil over the heat for 30 more minutes. If the soil temperature threatens to exceed 180°F, add a small amount of cool water to the soil to moderate the temperature.

hazards, including attack by insects and disease organisms. Those plants and animals that have natural mechanisms to cope with pests will live and reproduce more generations of healthy specimens. Those that can't cope don't survive to reproduce—a fact that is sometimes painfully evident in the garden.

The plants in your home and garden are the products of both natural and artificial selection. Plant breeders have accelerated natural selection. They select plants for resistance to damage caused by insects, diseases, and nematodes. The results are new plant cultivars with the ability to withstand some of the natural hazards in the garden.

Keep in mind that, in general, plants native to your area may have more natural pest resistance than exotic species. Plants will also

Resistance Ratings

Improved pest resistance is a strong trend in newer cultivars. It is among the most potent tools available for keeping plants free of insects and disease. When you are choosing plants for your garden, keep in mind that some plants are more resistant than others. The following are ratings for the level of defense plants exhibit when challenged by a specific problem.

- **Susceptible** indicates that the species or cultivar is at high risk to a specific pest or pathogen. When that insect or disease is present, it will attack a susceptible plant first, injuring or killing the plant and greatly reducing yield.
- **Tolerant** means that the species or cultivar can produce a good crop despite an attack by an insect or disease. The plant can, however, produce an even better crop if it is not subject to attack.
- **Resistant** species or cultivars can fight off attacks by specific enemies and are generally not infected or damaged. Choosing a resistant plant is a good first step in reducing the need for pest control. For example, if you know from past experience that early blight commonly attacks potatoes in your area, you can plant a resistant cultivar such as 'Kennebec' the following year. This greatly reduces the chance of early blight damaging your crop, and you probably won't have to do any spraying.
- **Immune** means the plant is incapable of being damaged by a certain pest or disease. If, for example, one of your garden beds contains the club root fungus, you can still grow many kinds of plants there. Only cabbage family plants can get club root; all other plants are immune.

have more natural resistance when they are grown in the appropriate site. For example, when azaleas or rhododendrons are grown in partial shade and acidic soil and are kept continuously mulched, they are more pest-resistant than when grown in full sun surrounded by moisture-grabbing grass. The latter situation subjects the plants to problems that do not occur in their natural setting—and for which they have developed no inherent resistance.

Host Resistance

Some plants have true resistance, which means they have an active physiological or mechanical means of fending off pests. Plants with physiological resistance produce antibiotics or other toxic compounds to inhibit pests. Plants with mechanical resistance have physical features that make pest attack difficult. Their defense may be a thick outer coating that resists penetration by a disease organism or a coat of long and stiff hairs that deters insect feeding. For example, shell beans and peas that have tough pods are more difficult for curculios to penetrate; corn that produces a husk that reaches past the tip of the ear is less likely to be bothered by corn earworms and birds.

Some plants have apparent resistance. This means that they can withstand attack by pests by growing rapidly, maturing early, or being very vigorous. Others are merely tolerant of invasion—they continue growing even when besieged by the enemy, as long as other conditions remain favorable. Some tolerate more damage than others.

Breeders search for these defensive abilities among wild plants that are related to our crops. They may also search for desirable characteristics among cultivars of foreign or heirloom plants (those plants that have been selected and saved for many generations). Sometimes, breeders will induce mutations in the laboratory, then breed the desirable characteristics into a new cultivar.

Autocidal Resistance

Some forms of pest resistance involve changing the pest rather than the host plant. The best example of autocidal resistance is the sterile male technique used to control insects. Scientists rear thousands of males of a particular insect species under controlled conditions, then sterilize them with x-rays. These sterilized males are released in the wild to mate with females of the same species. Since the males are sterile, mated females do not produce eggs, causing populations to drop drastically over time. (The technique is restricted to insect species with females that only mate once.) This method is most successful when the sterile males compete aggressively with the natural population of fertile males. A working example of autocidal resistance occurs at the border of Mexico and California, where sterile males of the Mexican fruit fly are released to help control populations and to prevent the pests from entering California.

Resistance in the Garden

There are several points to remember for making the best use of resistance for avoiding pest problems.

■ When skimming through seed catalogs, select cultivars with resistance to the pests common in your garden. Seed suppliers indicate a cultivar's resistant characteristics with a series of abbreviated codes. Look for a key to the codes in each catalog for help in deciphering listings.

■ When saving seed from your garden plants, collect seed only from the healthiest and most vigorous specimens. Select plants that seemed to naturally resist or tolerate pest attack, and those that were unattractive to your key pests.

■ Keep records of the cultivars that were most successful in your garden. In subsequent seasons, include the same cultivars in your rotation for as long as they remain trouble-free.

■ Try new cultivars. Just as plants are constantly changing, so are the pests. Plants that were once resistant may be vulnerable to pests that have adapted in order to feed on formerly resistant plants.

■ Sample heirloom cultivars for resistance. Heirloom plants that have been carefully selected and cultivated for many generations also may contain high levels of pest resistance. Some heirloom tomato cultivars, for example, have been screened and selected by gardeners for such a long time that they are naturally resistant to diseases such as Verticillium and Fusarium wilts.

■ Check with your local Cooperative Extension office to find out which cultivars are recommended for your area. Resistance for particular plants works better in some locations than in others.

Crop Rotation

Crop rotation is one of the most effective ways to control insect and disease pests in an organic garden. The theory of crop rotation is simple. Disease organisms such as club root fungi attack and grow in a crop during the growing season. During the winter, the organisms may overwinter as spores in the soil or in plant litter. The following year, they reinfest a new planting. So if you plant cabbage in the same spot year after year, the pathogens can continue to build their populations and club root will become an increasingly serious problem. However, if you plant cabbage in a bed one year, corn there the next, onions the third year, and beans the fourth year, many of the club root fungi will die off due to the lack of food. When you plant cabbage there again, you'll have little problem with club root.

This theory unfortunately becomes more complex as the number of crops you wish to grow increases. If you grow a large variety of crops, planning your rotation may leave your mind spinning as well as your crops. Don't despair! Just follow the steps and tips below.

Steps in Rotation Planning

In order to plan a rotation, you need some working knowledge of how botanical names of plants indicate relationships. All plants have a two-part botanical name, consisting of a genus name and a species name. For example, the botanical name of cucumber is *Cucumis sativus*. *Cucumis* is the genus name, *sativus* is the specific epithet. Together these make up the species name, *Cucumis sativus*.

A genus is a group of fairly closely related organisms, a species is an individual organism in the group. Many disease organisms can attack more than one member of a genus. Following through with our example, muskmelons, *Cucumis melo*, belong to the same genus as cucumbers and are subject to many of the same pest problems.

Some pests attack even a broader spectrum of plants. Plant families are larger groupings of plants that contain several genera. Cucumbers and melons belong to the family Cucurbitaceae. This family also includes summer squash and winter squash, which belong to the genus *Cucurbita*. It's important to know and understand these relationships, because one of the basic principles in planning a crop rotation is to plant crops from different families in sequence in a particular location to prevent the buildup of pest populations in the soil.

Keep family relationships in mind as you follow these steps.

1. Take a crop inventory. On index cards or small scraps of paper, list the crops you plan to grow. Use one card for each crop. Next, place the cards in major groups according to botanical family. Related crops (such as cabbage, cauliflower, and broccoli in the cabbage family; onions, leeks, and garlic in the lily family; or potatoes, tomatoes, and peppers in the nightshade family) will be in the same group. Grouping by family is only one of several ways to plan your rotation.

2. Design your garden. Work with a map of your garden in front of you. Divide your garden into several sections of approximately equal size and shape. The number of sections should be the same as the number of plant groups in step 1. Although the sections need not all be the same size, your plans will be simplified if they are. Sections may consist of several rows, a series of beds, or separate corners of the garden area.

3. Arrange your cards. Using the rules below, decide on an annual ordering sequence of plant groups that will place each crop at it's most advantageous position. For example, if your major crop groups are legumes, grains, peppers/tomatoes, salad greens, and alliums, you could create the following yearly order: year 1, legumes; year 2, grains; year 3, peppers/tomatoes; year 4, greens; year 5, alliums. In this case, nitrogen-fixing legumes precede grains, such as corn, which benefit most from the additional nitrogen. Alliums, such as onions, were placed last in the five-year sequence because they're good nutrient scavengers. The placement of peppers/tomatoes and greens is less obvious and less restricted. Perhaps you know from experience that peppers and tomatoes do well following corn. Afterward, a fall application of compost will boost the greens the following year. Within each section, five years will pass before the crop is grown again, to discourage pests specific to that plant family.

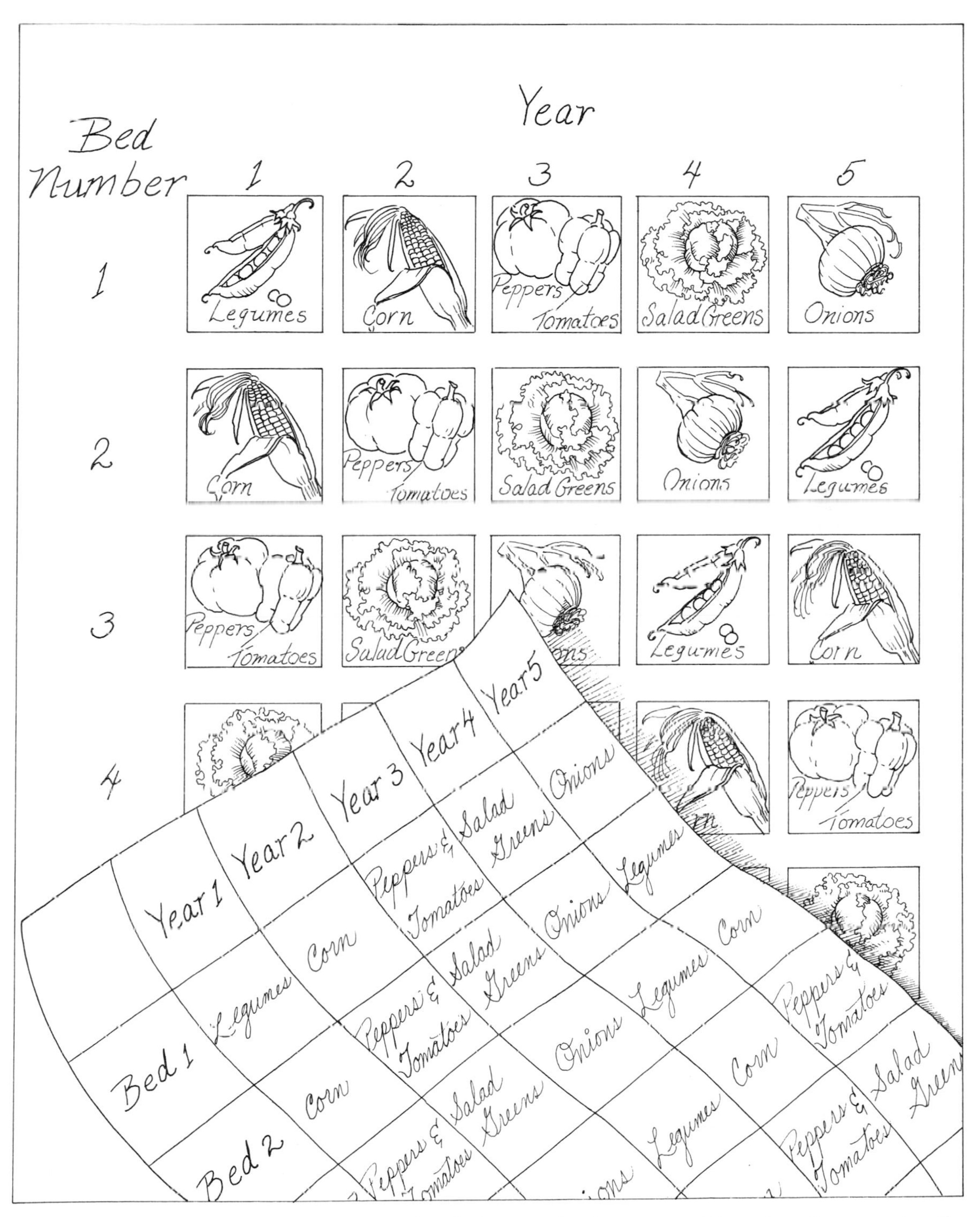

Gardeners use crop rotation to separate related plants by both space and time, thereby removing the host plants that pests and diseases need to survive. In this illustration, plants are rotated by family; other rotation plans might organize crops by growth habit, space requirements, or time of maturity.

4. Place crop groups in their garden sections. Assign a plant family to each of your garden sections. Then subdivide the sections to accommodate the plants in each family group. For example, designate space for snap beans, snow peas, and lima beans in the section reserved for legumes.

5. Make plans for future gardens. In each subsequent year, move on to the next plant group in each garden section. In year 2, for example, grains follow the legumes grown in section 1 in the first year, while section 3 is planted with greens.

Rules for Rotation

■ Alternate crops as many ways as possible. Plants may be grouped by families, as discussed earlier in the chapter. Alternately, you could choose to arrange your crops by the amount of space they require, by growth habit (root crops or vining crops), or by some other criterion.

■ Leave as much time as you can spare between related crops. Placing related crops three to six years apart in a rotation is important to effectively break disease and insect cycles.

■ Include soil-improving crops. Leave room for soil-builders such as buckwheat, rye, or oats. With a little more space, you can grow clover or alfalfa, which remains in place for more than one year, enriching the soil and attracting beneficial insects. Sod crops, such as lawns, cover crops, or even weeds that are intact for several years, however, may harbor soil pests such as grubs or wireworms, which can damage root crops. Avoid crops such as potatoes and onions the first year after sod is plowed under.

■ Grow legumes before grains. Nitrogen-fixing crops, including alfalfa, clover, beans, and peas, will boost the growth of most succeeding crops.

■ Apply compost or manure to those crops that need it most. Some crops prefer to feast on the leftovers of nutrients applied the previous year. This list includes members of the cabbage family, solanaceous plants such as tomatoes and potatoes, most root crops, and leafy crops like spinach and lettuce. Other crops, including squash family members, corn, and other grains, and legumes such as peas and beans prefer a fresh application of compost or manure each year.

■ Some crops, including onions, lettuce, and squash, seem to benefit any planting that follows them. In general, most crops grow poorly when planted after carrots, beets, and members of the cabbage family.

■ Plant heavy-feeding crops such as corn, tomatoes, and cabbage the season before light-feeders such as root vegetables, bulb crops, and herbs. Plant a soil-building crop in the third season.

■ Make efficient use of garden space by planting overwintering annuals such as spinach, parsnips, and garlic after crops that are harvested in late summer.

■ Although perennials such as asparagus and rhubarb remain in the same place each year, rotate the cover crops that surround them, or rotate different kinds of mulch.

■ If you find it impossible to rotate crops, at least rotate cultivars.

■ Certain soilborne diseases attack a broad spectrum of crops, and a more complex rotation is necessary to prevent infection. For example, the bacterium *Pseudomonas marginalis* can cause rots of lettuce, beans, cucumbers, and potatoes. If your garden is home to this disease organism, you may have to make it the sole focus of your rotation. Separate the susceptible crops by as much time as possible within your rotation.

■ Be prepared to change your plans if this year's problems threaten crops planned for next season's rotation.

■ Record your successes and failures, and review the list each time you plan a crop rotation.

Companion Planting

Organic gardeners often use companion planting to prevent pest damage and ensure a healthy harvest. The idea is that having a diverse grouping of plants growing side-by-side, where only one type grew before, will reduce the pest problems of one or more of the crops. Companion planting offers protection only from insect pests; there is little evidence that companion planting can control disease.

Mixed planting mimics a natural ecosystem. Your garden is like a mini-forest in which hundreds of organisms interact among the trees, usually out of sight. While plants remain stationary, insects and microbial organisms are moving in and out of the forest daily. By planting a diversity of crops in your garden forest, you can tip the scale in favor for vigorous growth and minimum damage by selectively attracting certain organisms and repelling others.

Gardens are good places in which to try companion planting, since gardeners may be less concerned with high yields and efficiency than are commercial growers. Flower beds often contain an array of species and cultivars, while stiffly regimented rows of identical plants fill most vegetable gardens. Combining the two creates a diverse planting that pleases the eye as well as the ecosystem. Mix companion plants in among your vegetables, or use them as a border.

The term *companion planting* describes several types of plant-to-organism relationships. These relationships, and how to use them, are described below. While examples are given, they are by no means a complete selection of possible companion plants. You'll probably want to do further reading and experimenting to discover the full range of companion planting possibilities for your garden.

Attractant Crops

Attractant crops tend to attract organisms, usually insects, in a way that benefits your yard and garden. Many beneficial insects that prey on garden pests supplement their diets with pollen and nectar produced by small-flowered plants; most beneficials are tiny and can't reach the food in larger blossoms. Plants in the carrot (Umbelliferae), daisy (Compositae), and mint (Labiatae) families are especially attractive to beneficial insects. Once good insects are attracted to your garden, they'll often stay to control garden pests. Mix the following attractant plants among your garden crops, or plant them in a border to greet the beneficials as they arrive: caraway, catnip, dill, fennel, hyssop, lemon balm, lovage, parsley, rosemary, thyme, and yarrow.

Repellent Crops

Some plants make good companions because of the company they discourage. While these relationships remain more a part of gardening folklore than scientific fact, many gardeners are convinced of their usefulness. Both catnip and tansy, for example, reportedly repel green peach aphids and squash bugs from susceptible garden crops. Wormwood *(Artemisia absinthium)*, southernwood *(A. abrotanum)*, and tomatoes are said to repel flea beetles from cabbage family plants. Plant radishes amid hills of cucumber and squash to repel cucumber beetles, and you'll also limit the diseases spread by these pests.

Good Neighbors

Certain plant combinations succeed because of differences in nutritional needs, space requirements, or harvest times. These associations may not directly retard pest damage, but they do promote plant health.

Interplanting certain crops fills precious

garden space efficiently without creating competition for light or nutrients. Corn and lettuce, for example, have different nutritional needs, rooting patterns, and pests. Lettuce fills the empty space between corn plants and is protected from the full glare of the summer sun by the towering corn stalks.

Other plants harmonize because they mature at different times. Radish and carrot seeds are often sown together; the larger radish seeds mark the row as you plant, and the quick-emerging radish seedlings hold the soil against rain for the slower-growing carrots. You'll harvest the radishes long before the carrots need the space to fatten.

Legumes such as alfalfa, clover, and vetch work with beneficial bacteria to convert atmospheric nitrogen gas into a form usable by plants, some of which is available to plants grown around them or after they are harvested. While often viewed as a weed in lawns, white clover actually helps supply nitrogen to the grass around it.

Still other plants seem to enhance the growth of plants growing near them. Basil is popular among gardeners who practice companion planting—they say that everything planted near it grows vigorously.

Patterns to Avoid

On the other hand, some crop combinations should be avoided in the interest of preventing pests. Crops that are closely related attract the same pests, and you may wish to separate them in your beds.

Some plants are allelopathic, which means they inhibit the growth of plants growing nearby. Their roots secrete substances that are toxic to a wide range of other plants. For example, a shallow-plowed cover crop of rye will inhibit the germination of small-seeded plants and weeds. Yellow and giant foxtail, nutsedge, quackgrass, sunflowers, and walnut trees all have allelopathic properties. Keep these relationships in mind if you're puzzling over a mysterious plant disorder that seems to be caused by invisible pests. The pest may actually be the large walnut tree growing nearby.

Timed Planting

Timed planting and harvesting are two of the oldest known pest-control methods. Using this method, gardeners either plant (or harvest) a crop earlier or later than usual in order to avoid the most damaging pest stages.

In earlier times, this type of information was passed from old to new generations within fairly stable communities. Modern gardeners, being part of a more mobile society, may find themselves in new environments where they don't know what planting schedule to follow to foil the local pests. Experimenting with new crops in a new setting, you may discover a whole new array of pests. If you're gardening in a new area, or if you haven't tried timing your growing schedule before, try the approaches described below.

Learn New Ways to Tell Time

The calendars in our homes and offices are useful tools for scheduling our time. Crop plants and pests, however, follow a different schedule and develop according to soil and air temperatures, which can vary yearly. In the northeastern states, for example, a snowstorm or sunny, warm weather is equally likely on the first day of spring.

Degree Days

To a large extent, temperature regulates pest development. Scientists use units of measurements called degree days to track heat accumulation over time and predict the local appearance of various pests. The degree day formula assigns a value to each day's maximum and minimum temperature and compares that value—the degree day or heat unit—to the temperature at which a particular pest begins to develop. Units accumulate daily, starting from a known or readily determined point in the pest's life cycle. If the pest is a caterpillar, for example, degree day units might be calculated from the first day its adult stage (moth) appears.

Using this formula, entomologists predict when each pest will emerge from hibernation and begin laying eggs on your plants. If temperatures are lower than normal, degree days accumulate more slowly and pests develop later. If temperatures are higher than normal, pests develop more quickly and you can expect to see them earlier than usual. Your Cooperative Extension office can provide you with local pest emergence predictions based on this formula. Use such information to schedule planting, harvesting, and preventive treatments.

Phenological Signals

Another way to tell time in the garden is by keeping track of local plant signals. Some gardeners can predict when a certain pest will emerge by watching the growth of flowering trees and shrubs and relating their development to that of the pest. This study of the timing of biological events and their relationship to one another is known as phenology. For example, you may observe that flea beetles appear when local lilacs begin to bloom. If you study and record these relationships over time, you may be able to use your observations to time plantings of crops to avoid pest attack.

Pheromone traps that mimic pests' own signals are useful tools for monitoring insect appearances. They can help you determine when the first pests arrive and when the population dwindles. The directions supplied with specific commercial traps explain how to interpret the population fluctuations. See "Pheromone Traps" on page 437 for details on using these traps.

Experiment for Success

To time planting dates accurately, begin collecting your own garden data. Each season, review your notes to detect patterns, then adjust your garden schedule accordingly. Keep track of the first date on which a particular pest appears, the coinciding botanical signals, and the last date you see it.

Experiment with successive plantings. Try several sowings of carrots, for example, to see which planting date is most effective for avoiding carrot rust fly infestation. Many gardeners find that delayed planting works best to control this pest. Sweet corn growers often plant in succession to ensure a weekly harvest and find that corn planted earlier escapes attack by corn earworms or European corn borers. As a gardener, you may decide to limit corn planting to the earliest part of the season in order to beat the pests in the race to the ears.

When arranging your planting and harvesting schedule, include cultivars that mature earlier or later than your usual crop selections. Pests arrive to find their favorite plant hosts have already been harvested or are not mature enough to produce the fruit on which they feed. Either way, damage to your garden is reduced or eliminated and pest populations dwindle in the absence of the appropriate hosts.

In areas where nematodes are common in the soil, plant susceptible crops extra early or extra late to take advantage of lower soil temperatures. Most nematodes can't penetrate plant roots when soil temperatures are below 64°F.

Here are some other examples of crops and pests for which timed planting is effective.

■ **Peas.** Cool, wet weather conditions encourage the development of many fungal diseases. If you live in an area with cool, wet fall weather, you may want to plant peas only in the spring. And if damping-off and root rot trouble your springtime seedlings, delay next spring's planting by a few weeks.

■ **Tomatoes.** If your plants tend to succumb to late blight near the end of the season, replace them in your next rotation with cultivars that mature earlier.

■ **Cabbage family crops.** In the North, plant radishes, broccoli, and other cabbage family crops early to avoid cabbage maggot damage to roots. If cabbage maggots persist, grow these crops for fall harvest to avoid the pest.

■ **Onions.** Plant them early to reduce damage caused by onion maggots. The first generation of adult flies usually appears as dandelions begin blooming. Onions that are already nearing maturity when pests arrive have greater resistance to injury than tiny seedlings.

Sanitation

Good sanitation is a crucial step in keeping most pests under control. Any pathogen that persists in soil from year to year, as do root knot nematodes, Fusarium wilts, tobacco mosaic virus, and many others, can be spread via contaminated tools, infected or infested plant debris, or even your hands. Make thorough cleaning of shovels, digging forks, or the tines of your tiller part of your gardening routine, especially if soil-dwelling pests and pathogens are present in sections of your garden.

Before You Begin

The temptation to get a quick start in the spring may lead gardeners to skip sanitation efforts that help prevent later problems. Before you take your first steps onto the newly awakening soil, stop and remember these important rules.

Keep the gardener clean. Because gardeners love to touch, smell, and investigate plants, they often unwittingly spread pest problems. They meander through their friends' gardens or visit local nurseries and farms brushing against infected plants and picking up soilborne spores on their boots. If you're among the ranks of such plant admirers, make sure your shoes and hands are clean *before* you enter your own garden. If you've been near potential sources of pathogens or insects, rinse your boots with a 10 percent bleach solution (1 part bleach to 9 parts water).

Start with clean tools. And keep them that way. Especially when working among diseased plants, rinse your tools with isopropyl alcohol or a 10 percent bleach solution before moving on to another part of the garden. When pruning infected trees or shrubs, disinfect pruners between cuts. When your work is finished, clean your tools again and let them air-dry, then coat them lightly with oil before storing.

Sterilize saved seeds. Disease organisms may linger invisibly on seeds you plan to save, ready to infect next year's emerging seedlings. Make a sterilizing dip to protect seeds from overwintering diseases. Place seeds on a large square of cheesecloth, gather the edges of the cloth together, and bind them with string or a rubber band. Dip the seeds in a solution of either vinegar and water (1 tablespoon apple-cider vinegar to 1 quart water) or a 10 percent bleach solution. Dry dipped seeds on newspaper or paper towels; make sure seeds are completely dry before storing.

You can also sterilize seeds with heat by immersing them in hot water. Take care if you use this method. Excessive heat can kill seeds and will usually reduce germination percentage. It's safer to use a large quantity—several gallons—of water to minimize the possibility of a sudden surge in temperature.

Heat water in a large pot to the correct temperature (see below). Place seeds in the center of a large square of cheesecloth, gather the edges together, and tie securely. Suspend the seeds in the water; don't let them rest on the bottom of the pot. Stir the water frequently during immersion. Use a good-quality thermometer to measure the water temperature, and time the immersion carefully. Don't try to store seeds after hot-water treatment; the treatment will have started the germination process. You can treat the following seeds at the specified times and temperatures.

- Brussels sprouts, cabbage, peppers, tomatoes—122°F, 25 minutes.
- Beets, broccoli, carrots, cauliflower, kale, kohlrabi, turnips—122°F, 20 minutes.
- Celery—118°F, 30 minutes.
- Eggplant—122°F, 30 minutes.

Choose healthy plants. Inspect new plants before introducing them to your yard or garden. Look for signs of disease and insects, and reject any that look suspicious. If available, buy certified disease-free plants and seeds. Inquire about seed sterilization practices at your seed source. Buy from suppliers that use sanitation procedures and heat-sterilizing techniques instead of synthetic chemicals to control disease on nursery stock and seeds.

Don't be afraid to buy sight-unseen from reputable mail-order nurseries. Most plants that are sold mail-order across state lines must be inspected and certified disease-free before they are packaged and shipped. When buying plants such as strawberries or raspberries, which often carry viral diseases, buy from a specialized producer who employs highly effective modern screening methods. Such growers are most likely to offer a wide selection of disease-free plants and are more likely to back up their sales with guarantees should your new plants show disease symptoms.

Solarize the soil. If you anticipate pest problems in a garden section that must be planted, you can use the sun's heat to kill some pathogens and insects in the soil. Stretch a sheet of clear plastic tightly over a smooth, moist soil surface and leave it in place for as long as several weeks. The heat that builds up in the covered soil will kill most soilborne pathogens and weed seeds.

Start a compost pile. A well-managed hot compost pile can be the perfect resting site for plant trimmings and other garden debris that may carry unseen pathogens and insect stages in waiting. See "Composting for Insect and Disease Control" on page 425 for instructions on composting.

Remove winter mulches. As temperatures begin to rise and spring rains start falling, soggy winter mulch provides a harbor for disease pathogens and hungry slugs or snails. Pull the mulch from beds a little at a time; pay special attention to clearing it away from tree trunks, stems, and crowns.

As You Garden

Maintain your sanitation standards consistently throughout the growing season. Cultural controls are especially useful in midsummer when beneficial insects are active and garden crops are nearing maturity. Keep problems at bay by adding a few more sanitation techniques to your garden plan.

Stay away when it's wet. Some plant disease organisms are just waiting for a free ride on your clothing and hands. Don't cooperate! Most diseases need only a thin film of water to spread from one plant to the next. During warm, rainy spells, disease transmission is at its greatest.

Rogue and prune. Pull plants with symptoms of disease or heavy insect infestation and prune diseased sections from perennial plants. A hot compost pile will kill pathogenic fungi and bacteria. If you suspect viral infections, bury the prunings in an out-of-the-

Soil Solarization

If soilborne pests are a big problem in your garden, why waste your energy trying to combat them? Let the sun do it for you! This process, known as soil solarization, involves covering the soil with clear plastic, so the heat can build up and destroy the soil organisms and weed seeds. Keep in mind that the high temperatures can also harm beneficial insects and organisms that live in the soil; only use this process when you have a serious pest or disease problem.

Soil solarization works best in midsummer, just before fall crops are planted. It is also most effective in areas that have long stretches of clear, hot weather. If you have a rainy or cool spell during the solarization period, soil temperatures may not stay high enough to kill off pathogens.

Prepare the site as you would for planting, removing old crop residues, working in any soil amendments, and raking it into planting rows or beds. Water the site thoroughly, and dig a trench a few inches deep around the edges of the site. Cover the area with a sheet of 1–4 mil clear plastic, and bury the edges in the trench to anchor them. Leave in place for three to four weeks.

Remove the plastic and plant your crops, disturbing the soil as little as possible. Solarization will only treat the top 3″–5″ of soil, so deeper cultivation may bring up pathogens and seeds that can reinfest the newly treated plot. Since solarizing may also harm the microorganisms that help make soil organic matter available to plants, you may want to topdress your crops with a microbe-rich material such as finished compost.

way area or put them in a sealed bag for disposal with the household trash. See "Pruning" on page 431 for specific suggestions on pest problems that you can combat by pruning.

Keep the site clean. Before planting a second crop, take the time to clean up and compost any plant remains. They'll be returned to the garden disease-free.

When You're Done

Some of us lose steam over the season and are happy to let things lie until next year, but this is no time to take it easy. End-of-season garden chores will make next year's garden that much healthier.

Prune perennial plants. It's time again for another round of pruning when growth has slowed and perennial plants have become dormant. Consult a pruning manual to help you determine the best time to trim and shape your plants.

Clean your equipment. Give tools a little tender loving care before storing them away. Clean them with a 10 percent bleach solution, wipe them dry, and coat them lightly with household oil to prevent rust.

Collect leaves and other plant residues. Continue adding to your compost pile for as long as you can. If temperatures dip low enough, decomposition will slow or stop, but will begin again when warmer temperatures return. Common fall compost ingredients include fruit you forgot to harvest, fallen leaves, and plant stems and flower stalks.

Remove row covers. If you're saving row covers for next season, dip them in a 10 percent bleach solution, then let them air-dry before storage. This is particularly important for covers used on related plants and for most early spring crops that are susceptible to damping-off pathogens.

Composting for Insect and Disease Control

Adding compost to your soil offers a simple, organic control for a variety of soil-dwelling pests. By improving the structure, moisture-holding capacity, and nutrient content of the soil, compost encourages healthy, balanced populations of soil organisms. Many of these organisms control plant pathogens by feeding on them, parasitizing them, or outcompeting them for food and water.

Foliar sprays of tea made from mature compost act as natural fungicides against the early stages of leaf blights, powdery mildew, and similar diseases by introducing competitive, beneficial microorganisms to infected plants. When compost is applied around plants as a thin layer of mulch, its beneficial microorganisms may benefit the lower sections of plants (when splashed up by rain) as well as the soil.

How to Compost

Recipes for preparing a compost pile abound, but only a few factors are essential for success. In addition to being a source for healthful, organic soil amendments, your compost pile can also serve as a place where diseased prunings and insect-infested plants can be converted back into usable materials for your garden. Follow these steps to create your pile.

1. Collect materials. Good raw materials for compost include garden wastes, grass clippings, kitchen scraps, manure, newspapers, and sawdust. Avoid meat scraps and oils, both of which will attract foraging animals and will slow decomposition. Save the materials in a separate pile outdoors, or layer them indoors in a 5-gallon bucket with a tight-fitting lid; sprinkle sawdust or other absorbent material between layers of kitchen waste to control odors.

2. Shred or chop large pieces. Shredding or chopping woody pruning debris, tree bark, and newspapers exposes more surface area to the decomposing organisms active in compost, which helps speed the process.

3. Layer the materials. When you've collected enough wastes to fill a space measuring approximately 4′ on each side, begin layering the materials. Your goal is to create a ratio of approximately 30 parts carbonaceous materials (dry, yellow or brown, plant-based) to 1 part nitrogenous materials (wet, green, or animal-based). This is achieved with roughly equal volumes of dry materials such as leaves, straw, sawdust, or paper, and wet materials, such as fresh grass clippings or manure.

4. Add a compost starter. While layering, include several shovelfuls of soil or finished compost—this serves to inoculate the pile with decomposer organisms naturally present in the soil. If no healthy soil or compost is available, a commercial compost starter can supply the necessary microorganisms. You can also add lime or mineral fertilizers, according to the needs of your soil. Keep the mixture damp but not soggy, sprinkling layers with water, as necessary. When finished, cover the pile to maintain the proper moisture level.

5. Turn the pile. Every few days, turn the pile with a garden fork, fluffing the materials as you go. If the pile seems dry, sprinkle with a little more water. Turning the pile works oxygen into the mix and hastens decomposition. As microbial activity increases, the temperature will rise. Use a soil or compost thermometer to monitor your compost's progress. Frequent turning helps maintain the proper temperature. Hot compost can reach 170°F, but try to keep your pile at 160°F. Higher temperatures tend to kill the organisms important for continuing decomposition, while lower temperatures allow insect pests and disease organisms to survive the composting process.

You'll know your compost is finished when the temperature returns to normal and original ingredients are no longer recognizable. This can take as little as two weeks, if you turn the pile regularly and if adequate nitrogen is supplied. Hot composting is necessary to kill pests, and only does so when the pile is actively heating. If your piles are built slowly and remain on the cool to warm side, refrain from adding pest-ridden garden wastes, since temperatures won't be high enough to control the pests and the resulting compost might reintroduce pests and diseases to your garden.

How Compost Controls Pests

The biological activity and heat in a compost pile work together to kill off disease organisms and insects.

Temperature. As microbial activity begins and increases, temperatures inside a compost pile increase until decomposition is complete. The pile is hottest at its center, where the intense heat can kill disease organisms and insects.

At 120°F, nematodes and pathogens causing Sclerotium rots are killed. At 140°F, most pathogenic bacteria, fungi, slugs and snails, and centipedes are dead; unfortunately, this temperature also kills earthworms. Many plant viruses and bacteria are killed once the temperature reaches 160°F. At temperatures greater than 160°F, however, important decomposer organisms begin to expire. Therefore, if you're composting for pest control, maintain the pile at 160°F.

Competition. Compost piles are like "end-of-the-week soup"—they contain a little bit of everything. This also applies to the organisms that inhabit the compost community. Rarely is any one species or group predominant, because the type and number of organisms present change along with the physical and chemical changes that occur as materials decompose.

The microbial organisms, insects, spiders, earthworms, and related decomposers tend to work together to produce finished compost. Along the way, however, friends quickly become foes and food for one another. In addition, researchers have discovered that many beneficial microorganisms are able to outcompete pathogenic fungi. Still other microorganisms may produce toxic substances that inhibit or kill the pathogens surrounding them. In a successful compost pile, the resulting compost contains a balance of organisms, large and small, beneficial and pathogenic, so that no one organism occurs in sufficient quantities to cause damage.

Using Compost for Disease Control

Garden clean-up should include composting plant residues and wastes that remain at the end of the season. Place debris that may carry pathogenic organisms in the center of the pile where temperatures are greatest. This also pertains to stalks, fruit, and leaves that may carry overwintering stages of insect pests and pathogens.

Mixing finished compost into your garden soil is an excellent way to distribute beneficial microorganisms that will continue working against plant pathogens over the season. Scientists also are researching ways to use compost extracts to control soilborne diseases. When liquid extracts are applied to soil, or when seedlings are started in specially prepared compost mixes, beneficial microorganisms are sufficiently aggressive to suppress the plant pathogens.

You can try this method at home to control damping-off and other common soilborne diseases. Mix 1 part finished compost with 6 parts water. Let the mix stand for one week, then strain it through a cloth, such as burlap or cheesecloth, and collect the liquid. Spray this extract undiluted on seedlings in the greenhouse or outdoors to help control many fungal pathogens and prevent infection. Research is now focused on creating "custom" compost mixtures that target specific pathogens. Some scientists recommend that compost ingredients include tree bark, animal manure, or a combination of the two, if disease control is the objective.

Soil Tillage

It makes sense that pests raise their families in garden soil. When larvae hatch from eggs or adults emerge from pupal cases, they're already seated at the dinner table. Gardeners may find this state of affairs very inconvenient, but at least they know where to start looking for problem pests.

How Tillage Helps

Tillage prepares the soil for planting. However, at the same time, it disrupts the life cycle of many soil organisms, both beneficial and pest. When you till the soil to prepare for planting, you disturb insects at their most vulnerable stages. Cultivation also upsets the nutrient cycles in the soil by mixing in debris and minerals and by adding oxygen. As a result, the rates of decomposition and nutrient release increase. This change in habitat results in many changes within the soil community.

Tillage physically destroys soil animals and their shelter, as egg-infested plant debris or hollow stalks with pupae inside are torn to smaller pieces. Food sources are removed and redistributed throughout the soil, starving some insects. Unfortunately, beneficial organisms are destroyed along with the rest. When you work the soil, you redistribute the pests themselves, often leaving them at the surface where they're exposed to predators. Atop the soil, eggs dry out, freeze, or overheat. Tillage may also bury pests so deeply that they cannot survive.

The scale of disruption and, therefore, the level of pest control, depends on how

Fungus-Fighting Tea

If your plants are weak and diseased, try perking them up with a nice cup of tea—compost tea, that is! Just as plant-based sprays can be used to repel insects, compost tea, made from manure-based compost, may help control fungal diseases. Apparently, the microorganisms present in compost tea attack or outcompete the problem fungi that cause such diseases as powdery mildew and Botrytis blight, thereby inhibiting their growth.

To make compost tea, place 1 gallon of well-aged compost in a 5-gallon bucket, and fill with water. Stir well and set in a warm place for three days. Filter the mixture through a screen or cloth, such as burlap or cheesecloth, and return the trapped solids to the compost pile or the garden. Place the strained liquid in a small sprayer or watering can. Pinch off heavily diseased leaves before applying the tea to the rest of the plant. For best results, use this treatment in the evening, when leaves are likely to remain damp for several hours.

Sometimes a single treatment will not stop the disease. Check affected plants every three to four days and repeat the application if the plants still show symptoms. Use any leftover tea to water plants growing in containers.

deeply you work the soil and the equipment you use. Rotary tilling is the most destructive method but is confined to shallow depths. Using hand tools creates much less disturbance.

Tillage for Insect Control

The best time to till is when the soil is fairly dry and crumbly. Maintaining soil structure is more important than tilling for pest control, so don't cultivate when the soil is too wet or too dry. Working the soil at these times can have disastrous effects on soil structure. When you till will depend on your local climate. If your area has lots of wet, overcast weather in fall, put off tilling until spring.

Review the following tilling tips and heed those that will help control pests that have been a problem in your garden.

■ Before planting, till to control weeds that may harbor armyworms, cutworms, and tarnished plant bugs, and to disturb corn root aphids.

■ Leave a tilled, dry strip of soil around your garden to discourage entry by slugs and snails.

■ After harvest, till to control pea weevils, root maggots, and tomato hornworms and pinworms. Fall cultivation also destroys plant debris that can shelter overwintering European corn borers, flea beetles, squash bugs, and other pests.

■ Summer tilling destroys grubs and exposes grasshopper eggs to unfriendly elements. However, your garden is not likely to have empty spaces that can be easily tilled. Leaving a garden fallow for the summer just to control insects in the soil is rarely feasible; in addition to the lost growing space, the extra tilling damages soil structure. Cultivating between successive crops is sometimes feasible. For example, till the soil between spring- and fall-planted crops to destroy lingering pests.

Tillage for Disease Control

Less attention has been given to controlling plant diseases with tillage. In the home garden, sanitation followed by hot composting or disposal of suspicious debris is probably more effective than cultivation in limiting disease organisms. However, on a larger scale of production, tillage offers some disease-control benefits. California lettuce growers, for example, till between crops, using special plows that penetrate the soil to a depth of at least 10″. This deep tillage buries a common fungus that causes a serious disease of lettuce and prevents the pathogen from infecting successive crops.

If You Don't Till

As tillage decreases, most soil organisms, including insect pests, are favored. Organic matter and nutrients become concentrated at a shallow depth, providing more food and cover. Animal life cycles remain undisturbed. Pests of seedlings tend to increase in numbers, along with earthworms, predaceous ground beetles, spiders, and slugs and snails.

Cultivated soil tends to warm up faster in the spring, which means that soil-dwelling pests emerge earlier. Researchers have found that pests may emerge from tilled soil one to two weeks earlier than from untilled soil. If your garden is just getting started, earlier pest emergence may increase injury to young plants. If pests emerge before your crops, however, they may look elsewhere for their meals.

PHYSICAL CONTROLS

When pests appear on plants, the most direct way to get rid of them is to pick them off. You can also handpick diseased leaves or other plant parts, limiting the ability of the infecting organism to further colonize them. In some cases this simple action effectively "cures" the host plants.

Handpicking and other physical controls are among the most straightforward and safe techniques for protecting plants. Most such controls rely on the gardener's ability to remove or exclude pests from plants or to lure pests away from plants and/or into traps.

The Physical Approach

In general, physical control methods are more often used to control insects than to control diseases. There is a wide variety of traps and barriers that thwart many common insect pests. Physical control of disease organisms is limited to removing diseased plant parts by hand or with pruning tools.

Keep Your Hand In

Very few insects have mandibles strong enough to bite into human skin, although a few, such as the striped blister beetle, can cause skin irritation if you squash them in your hands. Likewise, few plant diseases cause corresponding human illnesses. However, you can wear rubber gloves when handpicking pests if it makes you more comfortable with the job. You can also remove some types of insects with a small hand-held vacuum, while other types can be shaken off a plant onto a dropcloth.

Shut Pests Out

You will have fewer problems to handpick or otherwise control if you use physical barriers to keep pests, fungi, or bacteria from reaching their favorite hosts. Barriers range from a band of diatomaceous earth or wood ashes on the soil surface to deter slugs, to floating row cover draped over young cucumber plants to prevent cucumber beetles from feeding—and transmitting bacterial wilt.

Create a Diversion

Mulches sometimes help keep pests from finding the plants they seek. They may keep soil-dwelling insects from coming to the soil surface to feed, or prevent other pests from entering the soil to damage plant roots. Other mulches reflect light in ways that disorient pests and reduce their feeding. Scientists have experimented with different colors of plastic mulch and have found that white or aluminum foil mulches are particularly confusing to thrips.

Set a Trap

Besides the obvious benefits of trapping pests, you can use traps to warn of impending infestations. If you get early warning of the appearance of apple maggots, it will prompt you to quickly hang apple maggot traps in your trees. Traps like yellow sticky traps attract insects with color; other traps use chemicals produced by insects (pheromones) or plants to lure the insects to the trap.

An older method of capturing destructive insects is to plant crops that are known to be attractive to pests, and then harvest them before the pests can move on to other plants.

This approach works best when pest populations are monitored very closely.

While traps and trap crops hold little allure for disease organisms, you can use susceptible plants to tell you if a disease is present. For example, eggplant is very sensitive to Verticillium wilt and is often used to test for the presence of the disease. When planted in heavily infested soil, eggplant dies within three weeks. Okra is likewise used to help detect the presence of root knot nematodes, and susceptible tomato cultivars are grown to check for Fusarium wilt. When growing susceptible plants on purpose, remove them as soon as a disease diagnosis is made to keep the number of pathogens present in the site from increasing.

Manual Controls

Manual controls are simple and inexpensive pest-control strategies, since the only tools required are your time and patience. Besides controlling pests, manual controls require that you closely examine plants, increasing the likelihood that you'll detect the early symptoms of insect or disease problems.

Handpicking

When pest populations are low and time is on your side, handpicking pests and their eggs from plants is simple and effective. Removing adult pests before they've had a chance to lay eggs prevents pest buildup and the resulting damage.

Protection Offered: The best candidates for handpicking are large, slow-moving pests, such as caterpillars, Colorado potato beetles, and slugs and snails, that aren't quick enough to escape. It's also easy to find and smash mealybugs, scales, and squash bugs. Scrape away easily identifiable egg masses, such as those of gypsy and tussock moths, from tree trunks or other surfaces, before the hungry larvae emerge and begin feeding. Pests that fly away when disturbed, like leafhoppers, are better controlled with other methods.

How to Use: Monitor plants several times each week. Be on the lookout for pests that fly or crawl into your garden and for pests that emerge from the soil. It's important to learn to distinguish pests from beneficial organisms so you don't accidentally destroy the "good guys"; see "Insect Impostors" on page 260. Remember to look for pest eggs and immature stages. Smash pest egg masses. Squeamish gardeners can wear plastic gloves (thin surgical gloves are best) or carry long tweezers or tongs. If you're after cutworms, slugs and snails, weevils, and other nighttime diners, wait until just after sunset and search plants with a flashlight.

Some bugs are difficult to handpick because they hide or fly away. Before handpicking, it may help to first spray the infested plants with water to get the pests moving. For example, if you wet down squash plants before collecting squash bugs, they will either run to the topmost leaves or hide beneath those closest to the ground. In either hiding place, they are easy to find and pick up. For flying insects, use a butterfly net to catch them before and after spraying the plants with water.

After you handpick, you have to follow through and make sure that the captured insects don't escape to feed anew. Here are some suggestions for disposal.

- Carry a container of water mixed with soap or isopropyl alcohol (a 5 percent solution works well). Drop pests into the water; the soap helps to break the surface tension, causing pests to sink to the bottom.
- Kill pests by leaving them overnight in a bag or jar in the freezer.
- If you keep geese and chickens, they will quickly consume any insects tossed their way.

■ Compost captured pests in the center of a hot pile.

■ Flush them down the toilet.

■ Put them in your household trash—just make sure they can't escape.

■ Rub or scrape infestations of scale or mealybugs from plants, or swab them with cotton soaked in alcohol or liquid dish soap.

If you're uncertain as to the identity of a captured insect, keep it for identification. The staff at your local Cooperative Extension office or state land grant university can help you identify preserved captives.

You can preserve soft-bodied caterpillars by immersing them in boiling water for several seconds to kill them and to stop the enzymes that would cause them to break down (this is like blanching vegetables before freezing them). Large pests like tomato hornworms will take longer to boil than small cutworms. Don't let them boil too long or they will harden and lose their color. Place the boiled insects in vials or jars filled with rubbing alcohol. Use containers that seal tightly, since alcohol evaporates and loses its effectiveness as a preservative over time. For noncaterpillar pests, skip the boiling water bath.

Lawn Aeration Sandals

It's time-consuming, but some researchers effectively manage root-damaging grubs (most are Japanese beetle larvae) in lawns by using special sandals.

Protection Offered: Lawn aeration sandals are equipped with 3″ spikes that penetrate turf, piercing grubs that feed near the soil surface in spring and late summer.

How to Use: Put the sandals on and walk over the lawn several times, trying for about 2 penetrations per square inch of lawn—this is a lot of walking! You can purchase lawn aeration sandals from garden supply catalogs.

Mowing

Just as good pruning practices help maintain the health of trees, shrubs, and other garden plants, good mowing techniques encourage a healthy lawn.

Each time you mow, a freshly cut surface is available for pathogens to colonize. Repeated short mowing, under stressful conditions, makes plants smaller and fewer and allows weeds to easily move in. When plants are stressed by heat and drought in midsummer, it's best to raise the height of the blade and mow less frequently.

Protection Offered: Mowing to the proper height and mowing less frequently when plants are stressed will help prevent a broad range of pest problems.

How to Use: Proper mowing height is the top priority for controlling lawn pest problems without pesticides. To control lawn diseases, adjust the height of your mower as high as possible and mow infrequently. Prostrate, low-growing grasses do well at a height of ¾″-1″; fescues and ryegrasses should be maintained at 1½″-3″.

Pruning

In some situations, handpicking pests isn't enough. You may have to prune away leaves or branches when pests, their eggs, and debris are heavily concentrated on 1 or several plant parts. Selective pruning also removes leaves infected with foliar diseases, such as powdery mildew, and keeps pathogens from spreading to other parts of the plant. Added disease control occurs when pruning exposes more of a plant's foliage to air and light, reducing the conditions that encourage many fungal diseases.

Protection Offered: Control aphids, garden webworms, leafminers, and tent caterpillars by pruning infested plant parts. Pruning also limits the spread of foliar diseases and conditions that promote fungal growth.

How to Use: There are many times when pruning for insect and disease control is effective. Try the following:

■ Prune egg masses attached to foliage as you monitor your plants.

■ Pick spinach leaves infected with leafminers and destroy them to avoid future generations.

■ To control garden webworms, prune away webbed leaves.

■ Pruning may also work well to remove aphid infestations from plants when they're too numerous to handpick, and if they're concentrated in a small area that you can remove without harming the plant.

■ When powdery or downy mildew first appears on foliage, prune to prevent further infection and to improve air circulation.

■ Remove tomato leaves infected with blight or leaf spots.

■ In fruit trees, prune infested wood to control flatheaded borers.

■ Prune tent caterpillars and their webs from plants and burn, crush, or hot compost them.

■ Fruit trees infected by fire blight benefit from pruning that removes blighted twigs and disease cankers and stimulates air flow through the branches.

Pruning for Pest Control

For some insect and disease problems, a little careful pruning serves as a simple but effective control measure. The kind of pruning you'll do depends on the plant and the problem. For example, many kinds of perennials and woody plants form dense clumps of foliage that benefit from thinning. This involves cutting out some of the competing shoots, promoting better air circulation, and allowing more sunlight to reach the center of the plant. Removing water sprouts and root suckers from woody plants also reduces the amount of succulent growth attractive to insects and pathogens.

On all plants, handpicking or cutting off diseased or infested parts is quite effective for pest control. You can easily prune cedar-apple rust galls off cedar trees, for instance, or bagworm cocoons off evergreen trees. If your tree has a disease that starts at the branch tips and spreads inward (such as fire blight or twig blight), you can minimize the spread of disease by pruning off the affected parts during dry weather. Cut off the infected area of the branch, along with at least 6″ of healthy tissue. As in any pruning situation, prune back to a healthy bud or main limb; if you leave a stub of bare wood, it will die and become an entry point for other pests.

As you work with diseased plants, disinfect your tools between cuts by dipping them in a 10 percent bleach solution (1 part bleach to 9 parts water). Thoroughly clean up all trimmings as soon as you finish pruning, and place them in a hot (160°F) compost pile or in a sealed container for disposal. Be sure to wash your hands and disinfect tools before moving from diseased to healthy plants. And always use sharp tools, which will make clean, smooth cuts that are less prone to pest attack. After any pruning, disinfect your tools, rinse them with water, and lightly oil all metal parts to prevent corrosion.

Shaking

Shaking pests from plants is a method professionals use to monitor pest populations. In small gardens, shaking the pests from infested foliage takes less time than handpicking. Shaking pests from hard-to-reach tree limbs may remove enough pests to reduce damage.

Protection Offered: Shaking will dislodge any pest that crawls freely about on plants without clinging, including Colorado potato beetles, cucumber beetles, earwigs, Japanese beetles, and weevils.

How to Use: Shaking pests from foliage works best early in the morning when insects are cool and sluggish. Capture night-feeding insects just before sunset when temperatures begin to drop. Spread a white sheet or dropcloth under trees or plants, then shake or agitate the foliage with your hands or tap woody stems and limbs with a padded stick. Scoop up fallen pests and destroy them.

Professional scouts use beating trays to monitor pest populations. You can construct your own beating tray to catch and control pests. Sew 4 small triangular pieces of material at the 4 corners of a piece of canvas or plastic roughly 3′ square. Cut two 1 × 2's to the diagonal length of the canvas and fasten them together in the middle with a bolt and wing nut (to form an X). Insert the ends of these crosspieces into the corner pockets formed by the triangular pieces of fabric to make a rigid tray. Place the beating tray directly underneath foliage in the row, then shake the plants to knock pests onto the tray. The pests will be easy to spot against the white background.

Vacuuming

Commercial growers in California use tractor-mounted vacuums to remove bugs from strawberry plants, vacuuming twice each week when the insects are at their peak. In the home garden, careful use of a portable, rechargeable vacuum cleaner offers similar control advantages. Skim just the tops of infested plants and you'll avoid sucking up fragile beneficial wasps, which tend to remain on lower foliage. Beneficial predatory mites cling tightly to leaves and also remain unharmed.

Protection Offered: Vacuuming removes large leaf-eating pests like Japanese beetles and Colorado potato beetles as well as faster cucumber beetles and whiteflies. You can also vacuum up earwigs, sowbugs, and other pests.

How to Use: Lightly move your portable vacuum over the tops of vegetable and ornamental plants, taking care to avoid damaging tender leaves and shoots. Remove pests from the bag and dispose of them (see page 430 for disposal options). Don't give them a chance to escape!

Water Sprays

The physical action of a spray of water knocks plant-eating pests off plants and damages them sufficiently so that they are unlikely to resume feeding. Commercial grape growers control spider mites with overhead irrigation applied in late June and July when mites thrive in the heat.

Protection Offered: Controls small, soft-bodied insects like aphids, leafhoppers, spider mites, and thrips.

How to Use: To spray pests away with water, adjust the nozzle on your hose to provide a forceful spray that covers a large area. Hard and steady streams directed at 1 spot can damage foliage and flowers; move the spray back and forth among plants, being sure to reach under leaves.

To avoid encouraging diseases while controlling insects, refrain from spraying water during humid, wet weather. Excessive spraying of young, tender seedlings promotes damping-off disease; use other controls on seed beds. In hot, sunny weather, spray early in the morning or at dusk to avoid leaving droplets on the leaves, which could cause foliage to burn.

Baits, Traps, and Lures

Use the pest-control strategies in this section to lure pests into situations from which they can't escape. Pest baits and lures can be visual (using color or light to attract the pest), sexual (using pest-specific pheromones), or dietary (using preferred insect foods or kairomes, which are feeding stimulants).

When using sticky traps, cover them with a commercial sticky coating, such as Tangle-Trap, Stickem, Stiky Stuff, Tanglefoot, or Bug Gum, or make your own. Mix equal parts of petroleum jelly or mineral oil and liquid dish soap to make an inexpensive sticky coating that is easy to remove.

To capture small, lightweight pests like aphids, thrips, and whiteflies, dilute the heavier ointments to make application easier (use 1 part sticky coating to 2 parts paint thinner). Or, buy easy-to-apply, brush-on formulas. For heavier pests, such as caterpillars, that may resist being stuck, use sticky coatings at full strength.

Apple Maggot Traps

Apple maggot traps are red, apple-size spheres covered with sticky coating. The traps attract the attention of adult flies when they are ready to lay eggs. Zooming in for a landing on what they think is an apple, they become entrapped in the adhesive. You can trap enough flies on red sphere traps to control the damage they cause to fruit.

Protection Offered: Use sticky red spheres in your apple orchard to monitor the arrival and departure of apple maggot flies; this knowledge is useful if you choose to spray your trees to control apple maggots. You can also control these pests by concentrating several traps in small areas.

How to Use: You can buy inexpensive, reusable red spheres, designed to attract apple maggot flies, from garden suppliers. They come with attached hangers for easy placement in apple trees. Or, make your own traps by painting any apple-size red ball with sticky coating. Use old red croquet, rubber, or plastic balls. Install an eye screw and insert a support wire or string for hanging. You can also buy apple-scented lures to attach to your traps, but their effectiveness is controversial. The lures attract more apple maggot flies, but also appeal to flies of other closely related species. The concentration of flies may gum up the trap without improving apple maggot control. Some orchardists recommend the addition of a yellow sticky trap behind the red sphere. The red ball attracts adult flies ready to lay eggs, while the yellow trap lures newly emerged flies, which feed on apple foliage.

Hang traps in mid-June and leave them in place until after harvest. In small orchards of 10–15 trees, hang 1 trap for every dwarf tree, 2 or 3 traps for every semi-dwarf, and up to 6 for each full-size tree. In large orchards place 1 trap every 100′ among the perimeter trees in an orchard block. Also use 1 or 2 traps per acre within the block, and put a trap in every wild or abandoned apple and crab apple tree within 400 yards of the orchard. Renew the sticky surface of the balls every 2 weeks by scraping off the accumulated insects and applying new coating. In dusty locations, renew traps more frequently.

Commercial Products: BioLure trap with attractant dispenser card attached (also attracts blueberry maggots and walnut husk flies), Ladd apple maggot traps with synthetic apple odor bait

Bait Traps

In small gardens, you can lure pests away from susceptible crops by placing attractive vegetable baits near your crops. By regularly checking the bait, you can capture and destroy pests before they move on to your crops.

Protection Offered: Carrot and potato baits trap wireworms that live in garden soil and damage plant root systems. Wireworms are a common pest in areas that were covered by sod (such as a lawn, alfalfa, cover crops, or even weeds) the previous year. Use damaged squash to attract cucumber beetles, which damage foliage and transmit disease. Plant cull onions to distract onion maggot flies.

How to Use: Use carrots or potatoes to trap wireworms in garden soil. Insert several mature carrots every 2½′-3′ into the soil throughout the garden where sod was grown the previous year. Or, skewer raw potato chunks with bamboo stakes and bury them several inches deep every few feet, leaving the stakes protruding as a marker and handle. Several times each week throughout the season, pull these traps from the soil and remove the wireworms. Replace the same trap until a fresh vegetable is needed.

To attract adult cucumber beetles, cut chunks of fresh squash and place in recycled pint-size cartons sunk into the soil. Place 1 trap per hill of squash or related crops. Check the traps once a day during peak populations to collect and kill the pests; replace the bait when it loses effectiveness.

To control onion maggots in onions, plant cull onions (onions that have sprouted in storage) about 2″ deep between the rows where you've sown seeds for your main crop. The bulbs will grow much faster than the seedlings and attract the egg-laying flies. Two weeks after the bulbs sprout, pull them out and destroy them to prevent the next generation of onion maggots from developing. This trap works best when used along with repellents such as powdered ginger or hot pepper sprinkled on the seed rows to repel flies.

Cherry Fruit Fly Traps

Traps for cherry fruit flies differ from other yellow sticky traps in their shape and the addition of an ammonia lure.

Protection Offered: Use homemade sticky traps to catch and control adult cherry fruit flies in cherry orchards. You can also use them as monitoring tools to help you decide if and when to apply botanical insecticides.

How to Use: Paint a 10″ × 6″ piece of plywood bright yellow (use Federal Safety Yellow No. 659 from Rustoleum Company or Saturn Yellow from Day-Glo Colors), then cover it with sticky coating. Below it, hang a small, screen-covered jar filled with a mixture of equal parts ammonia and water, or a commercial apple maggot lure. Instead of a flat piece of plywood, you can paint the bell-shaped top half of a plastic, 2-liter soda bottle. In one study, this bottle trap was found superior to commercial designs for controlling Western cherry fruit flies.

For a small orchard, use at least 4 traps. Hang them 6′-8′ high among the leaves, preferably on the south side of the trees. Renew the bait weekly and check that the glue is still sticky. When it loses its stickiness, scrape off the accumulated insects and apply more glue.

Chinch Bug Traps

Chinch bugs are tiny lawn pests that destroy stems and leaves of lawn grasses. To monitor or control them in your lawn, drive them out of the sod with soapy water and into a dropcloth for counting and disposal.

Protection Offered: This type of protection works best for small areas. If you have a large lawn that is infested with chinch bugs, treatment with neem may provide more effective control.

How to Use: Mix 1 ounce of liquid dish soap in 2 gallons of water. Pour the mixture over 1 square yard of lawn. For large areas, you can use a garden hose with a siphon mixer attachment. Lay a large piece of white cloth, preferably flannel, over the treated area for 15-20 minutes, then pick it up. The bugs are driven out of the sod by the soapy soaking and then catch their feet on the flannel in their attempt to escape. Count them if you're monitoring the pest population, then kill them by rinsing them off the flannel into a bucket of soapy water. If there are fewer than 10-15 bugs per square foot, it is unlikely that they'll cause serious damage to your lawn. Water well to rinse the soap from the grass after you remove the flannel.

Earwig Traps

The same traps used to capture slugs and snails work equally well for trapping earwigs, but a different bait is used.

Protection Offered: If your garden has heavy infestations of earwigs, this simple trap may help control the population. Before you try to trap earwigs, remember that they are predators of aphids and other small garden pests; they do sometimes damage young plants, but more often are blamed for injuries caused by other night-feeding pests. Although they are scary-looking, earwigs pose no threat to people.

How to Use: A low container, such as a tuna can or pie plate, set into the soil so the rim is even with the soil surface and baited with fish oil or oil-soaked bread crumbs will attract large numbers of earwigs.

Japanese Beetle Traps

Japanese beetle traps rely on a combination of floral and fruit scents to attract females and a sex pheromone to trap the males. Most traps consist of some type of baffle with a bait container attached, hanging above a funnel that leads to a collection container. The beetles fly into the baffle and slide down the funnel into the container. The container can be removed so you can dispose of the beetles.

Protection Offered: You can protect small areas from adult Japanese beetles by surrounding them with traps baited with floral lures and sexual pheromones. They attract lots of beetles—the problem is that they may attract more than would normally arrive. Japanese beetle controls work best when used by entire communities, since the adults easily migrate from one yard to the next. Unfortunately, traps that capture adult beetles don't appear to result in a corresponding drop in the number of grubs found in lawns. Adults that fail to land in a trap often wind up on the lawn, where they mate and lay eggs; the lawn area surrounding a beetle trap may have a much higher population of grubs. To control grubs, use a combination of biological and physical methods. See "Lawn Aeration Sandals" on page 431 and "Milky Disease Spores" on page 460.

How to Use: You can buy commercial traps with either disposable plastic bags or permanent collection reservoirs. Make your own traps by cutting a wide opening in the top of a 1-gallon plastic jug (leave the handle in place for carrying). Fill it one-third full with wine or sugar and water, pieces of mashed fruit, and some yeast to enhance fermentation. An appealing bait for Japanese beetles is made by mixing 1 pint water, 1 banana or any non-citrus fruit, ½ cup sugar or honey, ½ cup wine, and ¼ teaspoon baking yeast. Let the mixture sit in a warm place until it starts to ferment before placing it in the traps.

In the spring, set traps 1′-3′ above the ground in open, sunny areas. Place them 20′-30′ downwind from the beetle's favorite plants (these may include most of the plants in your garden). To trap beetles over large areas, place a trap every 200′ surrounding the garden or yard. Or, use 3 traps for each ⅛ acre of garden space.

The traps should be emptied of beetles daily since dead beetles repel the live ones flying in.

Commercial Products: Bag-a-Bug, Bio-Lure traps, Safer traps

Light Traps

With the exception of the apple maggot fly, insects cannot see the color red. Their vision is limited to the ultraviolet (UV) to orange-red range, which is why they are attracted to blacklights (ultraviolet). Entomologists often use light traps to monitor populations of moths, which helps the scientists predict crop damage and advise growers when to apply control measures. Such traps direct pests to a holding reservoir from which captives are collected and identified.

Protection Offered: Don't use UV light "bug zappers" in the home garden, since they attract and kill just as many beneficial insects as pests. They do provide good fly control inside barns and poultry houses and are a useful defense in orchards against codling moths when used for limited periods of time. Blacklight traps also help control the adult moths of corn earworms, European corn borers, and fall armyworms.

How to Use: Place 1 UV light trap near your fruit trees or corn planting. Use an automatic timer to activate the light from 11:00 P.M. to 3:00 A.M. Moth control is most effective during this time, while injury to beneficial insects is minimized.

Pheromone Traps

Insects produce chemicals called sex pheromones to help them locate suitable mates. Some male insects are able to detect female pheromones at extremely low concentrations from as far away as 5 miles. Professional pest scouts take advantage of this ability by using synthetic pheromones, placed in special traps, to attract and capture the males of a species. By monitoring daily capture rates to detect population fluctuations, they can advise growers if and when control measures are necessary.

Some growers use pheromone traps to capture male codling moths and other pests to prevent them from fathering future generations and to control their damage. This type of pest control using pheromones is called mass trapping.

Protection Offered: Pheromones for professional pest monitoring are available for more than 40 insect species. Gardeners have a more limited selection available from garden suppliers, and only certain pests are actually controlled with the traps. Mass trapping is most successful against moth species that mate only once, such as codling moths. You can purchase pheromones to monitor such specific pests as cabbage loopers, cherry fruit flies, corn earworms, leafrollers, oriental fruit moths, and peachtree borers.

How to Use: Purchase traps from garden suppliers or make your own from 1-quart ice cream containers. Cut 3 large holes in the sides of the container for insects to enter. Fill the bottom half with soapy water. Suspend the pheromone capsule from the lid using string or wire, then snap the trap closed. Attach the trap to a garden stake for low-growing crops like tomatoes, or hang them from branches in fruit trees.

Commercial traps are inexpensive and easy to use and come in several shapes and sizes. Their bottoms are coated with sticky adhesive that holds the insects tight. Permanent, plastic traps will last for several growing seasons. Disposable cardboard traps will last the season if cleaned periodically.

In early spring before pests emerge, or by the time buds begin to open, hang traps in fruit trees about 6′ from the ground and away from the trunk. To control codling moth and other fruit pests, place 2 pheromone traps in each large tree; use 1 trap for each dwarf tree. The objective is to trap all of the males before they mate, capturing them in the sticky ointment that coats the trap bottoms. Situate traps for different species at least 10′ apart in the garden. In trees, hang traps for different species as far from each other as possible in the tree.

During the period when the pests are most active, check the traps twice each week. Remove insects and other debris. When traps are full, replace disposable ones, and follow the supplier's recommendations for renewing pheromone lures in reusable types. Replace pheromone capsules according to package instructions.

Mass trapping works best if your orchard is isolated from other sources of codling moth and other fruit pests (at least 1 mile away from other apple, pear, or walnut trees), since new pests migrate from neighboring orchards.

If you're just using traps to monitor a pest, 2 or 3 traps are sufficient for a 3-5 acre garden or orchard. Follow the guidelines above and the suppliers' instructions.

Pheromone lures are available from garden suppliers and pest management firms. You can purchase them separately or in a package with the coated trap and hanger. Keep lures in a cool, dry place until they're needed. BioLures can be stored for a few months, unopened, in the freezer.

Commercial Products: BioLure traps, Pherocon lures, Phero Tech traps, Scentry traps

Slug and Snail Traps

Slugs and snails are difficult to control since they feed at night and hide during the day. You can identify slug and snail damage by looking for shiny trails of slime that surround the chewed holes in leaves. Beware of commercially available baits that are not organic; some are highly toxic. Cover traps to keep pets and wildlife from lapping up their contents.

Protection Offered: Use these traps to capture slugs and snails and, occasionally, other crawling creatures, such as sowbugs.

How to Use: Garden supply stores and catalogs offer traps that feature a reservoir for bait or a liquid in which the pests drown. Follow trap instructions and bury them partially in the soil to create a cool, damp environment that will attract these pests. You can bait the traps with beer; a mixture of yeast, molasses, and water; or a commercial attractant. Use 1 trap for every 10 square feet of garden space. Under heavy infestation, check the traps daily to remove pests and replace bait.

You can make traps from aluminum pie plates or other recycled containers; just sink them to the brim in the soil and fill with beer. Or, cut a 1″ hole in the side of an empty coffee can about halfway up. Bury the can so the hole is even with the soil surface and pests can enter, then fill the trap half-full with beer. The lid will keep rain away and allow you access to the pests trapped inside.

Slugs and snails seek out moist, shady spots in which to spend the daylight hours. Place overturned pots, boards, shingles, or even the rind from your morning grapefruit on soil near plants, then collect and destroy the pests early in the morning before the sun drives them elsewhere.

Commercial Products: Garden Sentry, Slug Saloon

Sticky Boards

Traps covered with sticky coating strategically placed among your plants will capture sufficient numbers of certain pests to control their damage. Originally used to monitor pest population fluctuations, sticky boards work well as controls when concentrated in small areas.

Protection Offered: Use bright yellow traps to control aphids, cabbage root flies, carrot rust flies, cucumber beetles, fungus gnats, imported cabbageworms, onion flies, thrips, and whiteflies in the garden and greenhouse. Some growers report that blue traps are best for monitoring thrips, but not for controlling them. Use white traps to monitor European apple sawflies, flea beetles, and tarnished plant bugs.

How to Use: Sticky traps are simple affairs to mass produce at home. To make traps that will last several years, cut boards from ¼″ plywood, masonite, or a similar material and fasten them with staples or nails to garden stakes. You can also design them to hang from wire supports. Either way, make sure the traps are at average plant height. Traps made from cardboard or paper plates will work well in a greenhouse, where they are protected from the elements.

Trap size is limited only by the materials on hand, since scientists haven't yet decided exactly what is "too big" or "too small." Naturally, larger traps will catch more pests. Commercial traps are available in a range of sizes from 3″ × 5″ and up. Long, narrow, rectangular, or oval shapes may attract the most pests.

Apply sticky coating directly to your painted boards, or for easier clean-up, staple waxed paper or plastic wrap over them. In either case, leave 1 corner uncoated for easier handling.

Try these trapping tips for common pests.

- To catch whiteflies and other pests, paint boards yellow (Federal Safety Yellow No. 659 from Rustoleum Company and Saturn Yellow from Day-Glo Colors are 2 colors most pests find attractive) before applying homemade or commercial sticky coating.

- Staple a cotton ball soaked in oil of allspice, bay, or clove to your sticky traps to capture cucumber beetles in cucumber and squash plantings.

- Paint your traps a bright blue, such as cobalt or royal blue, to monitor thrips.

- Use bright white traps to monitor flea beetles or tarnished plant bugs.

Use 1 trap for several plants. Wait several days, then add more traps if pests continue to cause damage. For whitefly control, don't spread the traps too thin in the garden, since these pests don't travel far from the host.

Place your traps near plants, but not so close that foliage gets stuck to the boards. You can form a guard of hardware cloth or chicken wire around each trap to keep it from touching you and your plants. Set traps with the sticky sides facing your plants, but out of direct sunlight. Try to avoid placing traps where the wind will blow dust and debris onto them. Occasionally disturb infested foliage with your hands to drive feeding pests into flight.

You can clean pest-covered traps with a paint scraper, or wipe insects away with a cloth soaked in baby oil or vegetable oil, before applying fresh sticky coating.

To monitor tarnished plant bugs in apple and peach trees, hang 4 traps per full-size tree, or 1 per dwarf tree. To capture flea beetles, place traps just above eggplant, tomato, or pepper foliage.

Commercial Products: BioLure traps, Chroma-line Card traps, Olson Stiky Strips, Trapstix

Trap Crops

Grow pests' favorite plants to lure them away from your garden crops. Plant these trap crops around your garden or between the rows. When a trap crop becomes infested, pull it and destroy the pests. For pests that disperse quickly when disturbed, cover the trap plants with a sheet or bag before you pull them. This control method works best for pests that produce only a few generations each summer or for crops that need protection for a short, critical period in the season. A drawback is that you have to sacrifice harvest from the trap crop and garden space to grow it.

Protection Offered: You can use trap crops to catch a variety of insect pests. Unfortunately, trap crops may end up attracting more pests than usual to your garden. And if they're not pulled at the right time, the trap crops can provide more food for future generations of pests. Also, destroying the trap crop may mean sacrificing the beneficial insects that were attracted by the pests.

How to Use: To lure flying pests from your favorite crops, plant an attractive alternative nearby. When pests are concentrated on the trap crop, capture and destroy them, or pull and destroy the plant. Try the following trap crops.

- Plant various species of flowering mustard to trap cabbageworms and harlequin bugs.
- Nasturtiums will attract flea beetles away from cabbage seedlings until the plants are large enough to withstand attack, but you will have to plant nasturtiums earlier than usual. Try starting the trap crop indoors in peat pots, along with the vegetable seedlings. Nasturtiums are also favorites of aphids, but, if not removed, will serve as a breeding ground from which these pests can infest the rest of your garden.
- You may want to sacrifice an early crop of radishes in order to control flea beetle and root maggot damage on broccoli, cabbage, and related vegetables.
- Tomato hornworms are attracted to dill and lovage; the trap crops lure the pest away from tomato plants.
- If you want to protect soybeans from Mexican bean beetles, plant green beans between the rows of soybeans as a trap crop.
- Chervil is irresistible to slugs; plant it among vegetables and ornamentals in slug territory.
- In the greenhouse, use pots of sprouted wheat to lure fungus gnats from your plants, then hot compost the soil and wheat when the females have laid their eggs in it. Start a new crop of wheat every 2 weeks.

Water Traps

Bright yellow pans or basins full of water with a few drops of liquid soap added are attractive to winged aphids and cabbage root flies. Adding soap breaks the surface tension of the water, so insects sink to the bottom. The yellow color attracts the pests, who try to alight and are drowned.

Protection Offered: Water traps are useful for monitoring winged aphids and cabbage root flies in the home garden.

How to Use: Paint the inside of a heavy plastic basin or metal pan bright yellow (Federal Safety Yellow No. 659 from Rustoleum Company and Saturn Yellow from Day-Glo Colors are 2 colors most pests find attractive). Add ½″ or more of water and a few drops of liquid dish soap to the pan.

Set out the traps anywhere in the garden. Be sure the trap is exposed enough that insects will see it and be able to fly into it. Put the trap on a box or stand if plants are growing densely. Monitor this type of trap carefully to make sure you are not catching beneficial parasitic wasps in the water. If you find parasitic wasps, stop using the traps. The wasps are probably providing sufficient aphid control.

Barriers

Since barriers are meant to keep pests away from your plants, they should be put in place before the pests arrive. Among the choices below are several that you can leave in place all season. You may have to remove the covers when the plants outgrow them, when pollination by insects is necessary, or when temperatures under the barriers become too high. When using barriers, fasten down the edges at the soil line to prevent pests from moving in.

Ant Barriers

Ants tend to gather where aphids are concentrated, since they feed on the aphid secretions. Tending them like livestock, the ants will actually carry the aphids to a food source, then tend them and protect them from predators.

Protection Offered: Ant barriers on the legs of greenhouse tables prevent ants from reaching your seedlings.

How to Use: Cut a hole that corresponds in size and shape to a cross-section of each table leg through the center of an aluminum pie plate. Invert the cut pans and slide 1 up each table leg. Fold each pan downward to form a cup and cover the inside of the cupped plate with a sticky coating. Seal the crack between the pie plate and the table leg with caulking or sticky coating. Renew the sticky coating when necessary. Leave the traps in place permanently to guard against ants.

Antitranspirants

Antitranspirants are commercial products usually sprayed on trees and shrubs in fall to help protect them from winter damage.

Protection Offered: Reports indicate that antitranspirants form a barrier on leaves to prevent disease organisms such as powdery mildew spores from germinating and infecting plants. However, these products are not approved and registered for use as fungicides.

How to Use: Follow label instructions for coating plants. Avoid spraying drought-stressed plants on hot, sunny days.

Commercial Products: Folicote, For-EverGreen, Vapor Gard, Wilt-Pruf

Bagging Fruit

In small orchards, you can take the time to protect individual fruit on apple and pear trees by enclosing them in paper bags. The bags prevent insect pests from finding and damaging the fruit.

Protection Offered: Bagging works well to control codling moth damage on apples and pears.

How to Use: After thinning, place paper bags over individual fruit. Avoid using plastic bags, which will overheat the fruit. Tie them securely to the limbs with string or twist ties. Remove the bags just before harvesting.

Copper Strips

Copper repels slugs and snails. Some scientific studies indicate that copper is effective because slugs and snails actually get an electric shock when they touch it. Their slimy coating may interact chemically with the copper, creating an electric current.

Protection Offered: Using copper strips as a permanent edging for borders or beds is an effective but expensive way to keep slugs and snails away from flowers and vegetables.

How to Use: There are several ways to use copper repellent products to prevent slug and snail damage.

■ Buy copper-backed paper from garden suppliers and staple it to 3″ wide boards placed around the garden in a continuous border.

■ Bury a 3″-4″ wide copper strip around the edge of the bed or border, leaving 2″-3″ of the copper exposed. Bend the top ½″ of the strip outward at a right angle to form a lip (like an upside-down L). Once the barrier is in place, you may have to trap and remove the slugs and snails that were already inside the barricaded bed.

■ In the greenhouse, tack strips of copper 2½″ wide around bench legs.

■ Place strips of copper sheeting around tree trunks to keep the slimy pests away. Cut a 2″-3″ wide strip of copper several inches longer than the tree circumference, so you can enlarge the band as the plant grows. Punch holes in the ends, then fasten the strip securely around the trunk by feeding a piece of wire through the holes and twisting it tight. Remove suckers, water sprouts, and nearby weeds that might provide alternate routes for the snails and slugs.

Commercial Products: Snail-Barr

Crawling Pest Barriers

Bands of unattractive or abrasive materials that circle garden plants or beds can be used for mild infestations to keep pests away. Dusts or powders such as diatomaceous earth, wood ashes, or crushed seashells scratch the insects' waxy coating, destroying their water balance and killing them. Slugs and snails may also be deterred by these materials, which can irritate the soft bodies of these creatures. Insects and other crawling pests generally avoid such barriers. When pest populations are high or when wet weather renders the barrier ineffective, you may have to supplement with other control measures.

For control methods for ants, see "Ant Barriers" on page 440; for slugs and snails, see "Copper Strips" on page 441.

Protection Offered: Barriers keep earwigs, slugs and snails, sowbugs, and soft-bodied crawling pests like caterpillars away from your plants. On tree trunks, dehydrating dusts repel ants and may help deter adult forms of borers from laying eggs on the bark.

How to Use: Place a 2″ wide strip of wood ashes, diatomaceous earth, sawdust, crushed seashells, cinders, or similar organic materials as a border around beds or individual plants. Or, spread a circle of the same materials around individual plants, covering the area out to the dripline or at least within a 6″ radius of the stem. Some materials will have to be replaced after rain. Borders like diatomaceous earth and ashes are most effective when kept dry.

Make a dehydrating dust paint by mixing ¼ pound diatomaceous earth with 1 teaspoon of pure liquid soap, like Ivory, and enough water to make a thick slurry. Apply this paint to the lower trunk of a tree to give double protection; it shields the bark from the sun and discourages pests.

Place a "fence" of aluminum foil around beds and gardens if you are trying to control caterpillars such as armyworms and cutworms. Bury one edge of the foil in the soil, leaving several inches extending beyond the soil surface. Fold the top edge away from the garden to form a lip (like an upside-down L) that caterpillars will find difficult to cross.

Cutworm Collars

Cutworm collars protect seedlings from caterpillar damage. Cutworms are night-feeders that spend the day just under the soil surface, resting near your plants.

Protection Offered: Barriers placed around susceptible seedlings such as cabbage, eggplant, and beans provide protection from cutworms and other pests that crawl up the stem.

How to Use: Recycle cardboard tubes from toilet paper and paper towels. Cut the tubes into 2″-3″ sections, then place them over small seedlings while transplanting, pushing the collars into the soil. Thin, soft cardboard holds up in the garden long enough to protect your plants until they are beyond the susceptible stage. Shallow cans with the bottoms cut out also make satisfactory cutworm collars, although they won't deteriorate like cardboard collars and will have to be removed.

Floating Row Covers

When placed over young plants, these row covers don't float, they rest lightly on the foliage, looking like flattened sheets of cotton candy. They make great pest barriers, and you can irrigate as usual since they're water- and light-permeable. You can use floating row cover to extend the growing season, since temperatures undercover tend to be a few degrees higher than outside in the spring and fall. In the Northeast, spring harvests are often 5-10 days earlier. During hot summers, however, heat-sensitive plants like lettuce and cole crops can suffer, since temperatures under the row cover may increase by as much as 30°F during the day. Floating row cover doesn't provide adequate frost protection during cold spells, and you'll have to remove it to control weeds.

Several types of floating row cover are offered in garden supply stores and catalogs.

■ Polyethylene is actually plastic; it's the heaviest and warmest material and offers the most cold protection.

■ Polyester, polypropylene, and polyvinyl alcohol are "spunbonded" (fine threads tangled and bonded together), lightweight, porous materials that make the best pest barriers.

■ Polystyrene is heavier than the spunbonded covers and is popular with nursery owners for protecting perennial plants from the extremes of winter.

Commercial brands vary in longevity and the degree of warmth provided, but these differences are usually negligible. You can make your own alternatives to commercial floating row cover with mosquito netting, cheesecloth, or sheer drapery fabric. Remember that larger pores mean a greater chance that pests will penetrate your barrier, and fabric with finer pores may not offer adequate air circulation.

Protection Offered: Floating row cover controls insects and other pests that fly or crawl toward plants. They also control the plant diseases that insects transmit and stop rabbits, birds, groundhogs, and other animal pests. Row covers won't control the pests that emerge from soil once the covers are in place; crop rotation is still necessary.

Plan to use floating row cover barriers if you have problems with aphids, asparagus beetles, cabbage root maggots, caterpillars like cutworms and armyworms, Colorado potato beetles, flea beetles, leafhoppers, and Mexican bean beetles. Leave row cover over carrots all summer to control carrot rust flies. On ornamental plants place floating row cover over seedlings to protect them from foliage damage early in the season; remove it once the plants begin to flower and are large enough to withstand damage.

How to Use: Before covering, thoroughly weed the site. Place covers over the row immediately after planting seeds or transplants, since some insects are able to locate seeds before the plants emerge. If pests have already arrived, you'll have to eliminate them before covering. These lightweight covers are most easily installed on quiet, still days, since the slightest breeze sends them flapping away.

Lay them over the row or bed, leaving enough slack to allow plants to grow underneath. Anchor the edges with soil, rocks, boards, bricks, or other heavy objects. You can make or buy hoop supports for delicate crops like baby lettuce that may be damaged by abrasion. Later, remove the covers temporarily to thin plants, weed, or harvest. In the spring, periodically check for aphid infestations.

Remove covers permanently when plants are flowering, when plants are large enough to withstand some damage, or when the pests are no longer a threat. After using, wash them with soapy water (add bleach if you have disease problems) and rinse well before drying and storing. Patch holes and tears with duct tape placed on both sides. For extra cold protection, use 2 covers together; light penetration will be reduced to about 65 percent. Floating row cover is available in 5′-50′ widths. Used for spring and fall season extension in the northern states, it should last several seasons. When used all season long in hot, sunny climates, it deteriorates more quickly.

Commercial Products: Agryl row covers, Kimberly Farms Floating Row Covers, Reemay

Mulches

Mulches provide a barrier between plant foliage and soil that may contain disease spores or pests. They also serve as a barrier for pests that are headed for the floor of your garden, either to lay eggs or to rest for the winter.

Besides being useful for pest control, mulches of various materials serve many other cultural purposes such as retaining soil moisture, adding organic matter to the soil, warming the soil for heat-loving plants, and controlling weeds. The list of possible mulch materials seems endless. However, certain materials have distinct advantages when used for controlling insect and disease pests. An organic mulch, replaced several times each year, will help prevent disease spores from building up and prevent splashing rain drops that carry spores.

Protection Offered: Control aphids, leafhoppers, and thrips on cabbages and peppers in the garden and greenhouse with a mulch of aluminum foil. You will also control the plant diseases that these pests carry. Use a black plastic mulch to discourage sowbugs and other crawling pests that can't stand the heat; black plastic also keeps leafminers from emerging from infested soil and prevents their return to the soil to pupate. By blocking sunlight from the soil, black plastic mulch provides weed control as well. A thick layer of organic mulch will prevent shallow-rooted plants from being heaved out of the soil by frost action during the winter, because it moderates changes in soil temperature.

How to Use: You can buy aluminum-coated paper to cover beds or rows before planting. Anchor the edges with soil or weights, then cut 3″-4″ diameter holes for transplants or seeds. Later in the summer, remove the mulch to prevent reflected heat damage to large plants. You can also use wide, heavy-duty aluminum foil. On wide beds, run 2 strips of foil the length of the bed, leaving a gap several inches wide between the strips; plant in the gap. Scientists are experimenting with black plastic mulch sprayed with aluminum-colored paint as a substitute for foil.

To control rose diseases, rake debris and old mulch away from the stems in the fall. Apply fresh compost or other organic mulch. In mid-winter prune as usual and apply more mulch to protect the plants from spring heaving. In the spring gradually remove the mulch as temperatures increase, then apply a fresh summer mulch.

Covering garden and greenhouse soil with materials like newspaper or brown paper prevents thrips from reaching the soil in order to pupate. Weight paper mulches down with soil, boards, or other items so they can't blow away. These materials break down quickly; replace them periodically.

Painting

Protect fruit trees from both flatheaded borers and sunburn by painting the trunk with diluted white latex paint or with whitewash.

Protection Offered: Besides preventing borer damage, the paint will reflect sunshine from the trunk, reducing the chance of sunburns or cracks from uneven trunk warming.

How to Use: Paint the tree trunk using white latex paint diluted with equal parts of water, or use whitewash. Extend your paint barrier from 1″ below the soil line to 25″-30″ up the trunk. To paint at the base of the trunk, use a hand trowel to remove the soil to a depth of several inches; replace the soil after the paint dries.

Rigid Plant and Row Covers

Before floating row covers became popular, innovative backyard gardeners built their own barriers from window-screening attached to wooden frames. The list of variations for homemade barriers is limited only by your imagination and skills. Well-constructed and sturdy homemade covers should last for many years.

Protection Offered: Rigid barriers protect crops from damage by animals and a variety of insects. They prevent insects from transmitting disease problems as well. You can use rigid barriers as defense against the same pests for which floating row cover is effective.

How to Use: Make screen cones from aluminum screening. Cut sections of screen 1½′ × 2′. Wrap the shorter end around to form a cone, overlapping the edges. For stability, staple a wooden strip to the edge, so that it extends several inches below the base of the cone. Place cones over individual plants, inserting the wooden stakes into the soil to hold the cones in place. Sink the cone bottom into the soil so pests can't crawl under. Remove the cones when plants outgrow them or when pests are no longer a threat.

You can also make individual plant tents from window screening by folding long sheets of screening in half, then attaching triangular sections with a stapler to seal the ends.

Make permanent covers by constructing wooden frames that fit over sections of your garden. Make them wide enough to fit over your beds or rows and tall enough to accommodate plant growth for as long as you intend to keep them over the plants. Frames with the top narrower than the bottom stack easily for winter storage. Staple screening or porous fabric to the outside of the frames. Each fall, repair frames as needed, rinse them with a 10 percent bleach solution (1 part bleach to 9 parts water) to prevent disease problems, and dry thoroughly before storing.

Seedling Protectors

Adult root maggot flies lay their eggs in soil at the base of young vegetable plants. When the maggots hatch, they don't have far to travel to find their host. You can place paper barriers around seedlings as they emerge or as you transplant, to keep egg-laying flies away from the soil around your plants.

Protection Offered: Seedling disks prevent root maggot flies from laying their eggs in the soil near susceptible seedlings like broccoli, cabbage, cauliflower, and onions. Tar paper repels flies, offering added protection.

How to Use: Using tar paper or other heavy, flexible paper, cut circles 6″–8″ in diameter. Make 1 cut from the edge to the center. You may want to cut a small hole in the center of the circle so the paper fits snugly around the stem but lies flat on the ground. When transplanting, place 1 disk on the soil around each plant so that the stem is in the middle of the disk. Circles aren't magic shapes—squares work just as well. Since there will be several generations of root flies during the season, leave the covers in place until harvest.

Shade Cloths

Placed over rows of cool-season crops, shade cloth protects them from the scorching rays of the summer sun. It effectively extends the season for plants that normally prefer spring or fall temperatures.

Protection Offered: Because of its open weave, shade cloth offers little in the way of insect protection. It does, however, prevent injury from birds and small animals. It's most often used to protect lettuce and other cool-season crops from extreme temperatures and damage caused by direct sunlight.

How to Use: When summer temperatures reach 80°F or higher, suspend shade cloth over the crop to be protected, using stakes or other supports. Shade cloth is available in varying shade percentages that indicate the amount of light reduction; choose one that shades your crop adequately without reducing light levels to the point that plants are stretching for sunlight.

Commercial Products: Handyshade, Sudden Shade, Weathashade

Trunk Bands

Trunk bands act as barriers and traps to control pests that crawl along tree trunks on their way to the foliage or to the soil.

Protection Offered: Use bands to protect trees, vines, and shrubs from ants, codling moth and gypsy moth caterpillars, cutworms, leaf beetles, snails and slugs, and other pests that make daily trips along the trunk.

How to Use: You can apply most sticky coating products as a band directly on the plant you wish to protect, but it's generally more desirable to apply them to a removable wrap. A sticky coating applied to a tree's bark may injure the tree or encourage fungal growth in the treated area. It is also much more difficult to remove the coating once it is clogged with pests; such coatings eventually wear off but are rather messy and unsightly until they do.

To control codling moth damage on apple trees, prevent mature larvae from returning to the soil to pupate. Apply bands when leaves unfold in the spring. Use a band of sticky ointment, or wrap corrugated cardboard around trunks at least 1½′ above the soil. Cut 2″ wide corrugated cardboard strips long enough to wrap around the trunk; staple or tape them in place. Large larvae will crawl into the corrugations to pupate. Remove the bands at least once each week and destroy the pupae inside.

Trunk bands can also prevent codling moths and other pests from crawling up the trunk in the first place. Apply bands of sticky coating to tree trunks and rose stems with a stick or paint stirrer; remove debris and pests weekly and apply fresh material as needed. You can also buy foam-backed tree-banding strips to wrap around trunks. The foam forms a tight seal against rough bark, preventing insects from crawling under. Apply sticky coating to the band; when soiled, just remove and replace with fresh strips.

Folded burlap strips can also trap pests as they travel up the trunk. Cut a 20″ wide strip of burlap fabric and wrap it around the trunk. Tie it to the tree with a cord in the middle of the strip, and let the top half fold down over the bottom, forming a pocket that will trap pests. You can apply a solution of parasitic nematodes to the burlap for added control. Seal bands at the edges with a layer of sticky ointment to prevent other pests from crawling under them.

Commercially prepared trunk bands are also available. Special adhesive tape with a slippery silicone finish prevents caterpillars from crossing when placed around a tree's trunk. For extra protection, apply sticky ointment at the edges and in a thin strip around the center of the tape. Snail-repellent tapes are also available to wrap around trunks. These have a sticky backing and are coated with cayenne pepper and salt to repel and kill slugs and snails.

Commercial Products: Slick'n'Stick

BIOLOGICAL CONTROLS

Biological pest control—controlling pests with their natural enemies—is a phenomenon as old as the pests themselves. It is also one of the most successful forms of pest control.

Garden pests and their natural enemies coexist in balanced populations in well-managed organic gardens. Unfortunately, natural predators and parasites don't always provide relief from pests as quickly or as thoroughly as we would like. Their success depends on several factors, including climate, soil type, alternate sources of food and water, and the level of pest infestation. Some of these factors, such as the weather, are beyond your control. However, there are several things you can do to enhance the effectiveness of beneficial organisms already present in your yard and garden. And in some cases, you can release commercially reared predators and parasites around your yard to combat pest problems.

This section includes specific information on animals, insects, and disease-causing microorganisms that you can use to help fight garden pests. If you're not familiar with the terms used to describe these organisms, refer to "Insects" beginning on page 254 for basic information on the life cycles of insects and to "Diseases" beginning on page 338 for a rundown on the types of organisms that cause disease.

The Biological Balance

Biological control ranges from encouraging birds to visit your garden and eat pests to spraying microorganisms that can cause pests to sicken and die. An important aspect of biological control is making your garden attractive to the beneficial animals and insects that can do much of the pest-control duty for you. There are also commercial insectaries that raise beneficial insects in mass quantities to sell to farmers and gardeners. One biological control product has become a best-seller—BT, the bacterium *Bacillus thuringiensis,* which infects and kills a broad range of chewing caterpillar pests.

Encouraging Beneficials

The natural enemies of garden pests are often overlooked when gardeners consider their pest-control options. In fact, more beneficial organisms visit your plants than do pests.

Beneficial organisms go unnoticed because they're inconspicuous in size and habit. They're so efficient, you may never have realized that pests were there. Even so, pest outbreaks are clouds with silver linings since they attract more beneficials to your garden. It's a good idea to encourage natural enemies whenever you can. There are several ways to attract them to your garden, and keep them there.

Food

Predators and parasites often rely on nonpest food sources in addition to pests. Many beneficials that eat pests supplement their diets with pollen, nectar, or other insects, enabling them to survive when their target pests are scarce. And while birds will feast on insect pests, they may first be attracted to your yard by a well-stocked bird feeder.

Alternate food sources will help keep beneficial insects in your garden and are often necessary in order for the beneficials to reproduce. Some parasitic wasps, for example, lay more eggs after feasting on plant nectar. Green lacewings eat aphid honeydew to boost

(continued on page 450)

Important Native Beneficial Insects

Name	Pests Attacked	Comments	How to Attract
Aphid midge *Aphidoletes aphidimyza*	Over 60 species of aphids.	Very common, globally distributed, and hardy nearly to the Arctic circle. Especially attracted to aphids in roses, shrubs, and orchard trees.	Plant nectar-producing plants; shelter garden from strong winds; provide water during dry spells.
Assassin bugs (Family Reduviidae)	Many insects, including caterpillars, flies.	Robust, voracious insects, with strong beaks to attack prey; will squeak when handled; can inflict a painful bite.	Provide shelter in permanent plantings.
Bigeyed bugs *Geocoris* spp.	Aphids, small caterpillars, leafhoppers, spider mites, tarnished plant bugs.	They may resemble tarnished plant bugs or chinch bugs; their big black eyes are a distinctive trait.	Collect them from pigweed or goldenrod stands and transfer them to your garden; plant alfalfa, clover, or soybeans as cover crops or borders.
Braconid wasps (Family Braconidae)	Aphids, armyworms, beetle larvae, codling moths, European corn borers, flies, gypsy moths, imported cabbageworms, other caterpillars and insects.	Rigidly mummified aphids or dying caterpillars with white cocoons stuck to their backs are signs that braconids have been at work.	Grow nectar-producing plants with small flowers.
Damsel bugs (Family Nabidae)	Aphids, small caterpillars, leafhoppers, plant bugs, thrips, treehoppers.	These ⅜″–½″, gray or brown bugs are common and important predators in orchards and alfalfa fields (where you can collect them for your garden).	Collect them from alfalfa fields and transfer them to your garden; plant an alfalfa border around the garden.
Ground beetles (Family Carabidae)	Most prey on soil-dwelling pests, such as cabbage root maggots, cutworms, slugs and snails; some pursue aboveground pests, such as Colorado potato beetle larvae, gypsy moths, tent caterpillars.	Exceptionally long-lived (adults live up to 2 years), most active at night.	Provide permanent beds, plant sod pathways, and allow some weeds, especially pigweed; minimize tillage during the growing season; plant white clover as a groundcover in orchards.

Name	Pests Attacked	Comments	How to Attract
Hover flies/flower flies (Family Syrphidae)	Many species of aphids.	These insects hover over flowers and dart away like miniature hummingbirds. They often lay eggs in young aphid colonies to ensure that larvae will have enough prey.	Plant pollen- and nectar-producing flowers and encourage weeds such as wild carrots and yarrow; avoid pesticide use.
Ichneumon wasps (Family Ichneumonidae)	Caterpillars, sawfly and beetle larvae, other insects.	Although most ichneumon wasps are very small, some are frighteningly large with long, threadlike ovipositors trailing behind; they cannot sting people.	Plant pollen- and nectar-producing flowers; grow flowering cover crops in orchards.
Lacewings *Chrysoperla (=Chrysopa)* spp.	Soft-bodied insects—including aphids, mealybugs, thrips—small caterpillars, mites, moth eggs, some scales.	The delicate adults flutter erratically in a zigzag flight through the garden at dusk; their voracious larvae are known as aphid lions.	Plant pollen- and nectar-producing plants and allow flowering weeds such as dandelions and goldenrod; provide a source of water during dry spells.
Lady beetles (Family Coccinellidae)	Aphids, mealybugs, soft scales, spider mites.	Lady beetles abound in many sizes and colors, including solid black, ash gray, and yellow or orange with black spots or irregular blotches.	Plant pollen- and nectar-producing flowers; allow weeds such as dandelions, wild carrots, and yarrow; protect eggs, larvae, and pupae on plants.
Minute pirate bug *Orius tristicolor*	Small caterpillars, leafhopper nymphs, spider mites, thrips, eggs of many insects.	These plentiful, black and white harlequin bugs are easy to spot. Look for them in corn silks and stinging nettles; shake them into a jar and release them in the garden.	Plant alfalfa or other pollen-producing plants; encourage goldenrod and yarrow.
Rove beetles (Family Staphylinidae)	Many prey on aphids, fly eggs, maggots, nematodes, springtails; some are parasites on cabbage root maggots or other fly larvae.	Often mistaken for earwigs, rove beetles are usually smaller and have no pincers; more than 3,000 species in North America.	Maintain permanent plantings to protect the local population; interplant with cover crops or mulch planting beds; make permanent pathways in the garden.

(continued)

Important Native Beneficial Insects — Continued

Name	Pests Attacked	Comments	How to Attract
Soldier beetles (Family Cantharidae)	Aphids, beetle larvae, including cucumber beetles, caterpillars, grasshopper eggs.	Unlike most beetles, soldier beetles have leathery rather than hard wing covers.	Keep pollen-rich plants and weeds around the garden for adult feeding; keep permanent plantings in the garden to provide refuge and protect pupating beetles.
Spined soldier bug *Podisus maculiventris*	Fall armyworms, hairless caterpillars, including tent caterpillars, sawfly larvae; beetle larvae such as those of Colorado potato beetle and Mexican bean beetle.	These resemble stink bugs, but spined soldier bugs have sharp points on the "shoulders" of the thorax.	Maintain permanent beds of perennials to provide shelter.
Tachinid flies (Family Tachinidae)	Many species of caterpillars, including armyworms, cabbage loopers, cutworms, gypsy moths, tent caterpillars; also Japanese beetles, May beetles, sawflies, squash bugs.	One of the largest and most beneficial groups of flies, they are often mistaken for houseflies.	Grow plants rich in pollen and nectar; leave some garden weeds, especially goldenrod, wild carrots, and pigweed; don't kill caterpillars with white eggs stuck to their backs (the eggs will become the next generation of flies).
Tiger beetles (Family Cicindelidae)	Both adults and larvae prey on ants, aphids, beetles, caterpillars, flies, grasshoppers, spiders, other insects.	These insects are slow developers; larvae spend 2–3 years in their burrows before becoming spectacularly beautiful adults with bright, iridescent colors.	Maintain permanent beds in the garden as refuge; don't leave outdoor lights on all night or use light traps, because tiger beetles are highly attracted to light.

their egg production, and syrphid flies require a meal of pollen before they can lay eggs.

Assist natural enemies by including small-flowered, food-bearing companion plants in your landscape. Choose from members of the carrot family (Umbelliferae), which includes caraway, dill, fennel, lovage, and parsley. Many mint family (Labiatae) members such as catnip, hyssop, and lemon balm are important nectar plants. Rosemary, thyme, and other herbs are attractive to both gardeners and beneficials. Members of the daisy family (Compositae), including coneflowers, daisies, and yarrow, are excellent sources of both pollen and nectar.

Plants otherwise considered weeds might be useful wild food sources: corn spurry, goldenrod, lamb's-quarters, wild mustard, and Queen Anne's lace attract beneficial insects.

Cover crops are good sources of food and shelter. Buckwheat is an excellent choice, since it's easily worked into the garden season and quickly provides masses of blooms and cover to attract natural enemies. Try using alfalfa, buckwheat, clover, or ivy (*Hedera* spp.) as borders around the garden.

You can purchase bug food from suppliers of beneficial insects or from mail-order garden supply companies. This artificial diet usually contains yeast, sugar, and added vitamins and is placed in sheltered areas among garden plants when natural food supplies are low.

Water

Like animals and plants, beneficial insects require water. If rain is plentiful locally, puddles of water or morning dew are probably sufficient. During a drought, provide water in shallow containers filled with rocks that serve as insect perches. Change the water often to discourage mosquito pests. Also consider siting a deeper water source somewhere in your yard to help encourage a toad to call your garden home.

Shelter

Beneficials need protection from wind and weather extremes. When other parts of the garden are being tilled, mowed, sprayed, or harvested, beneficial insects seek out alternative cover. Provide them with hedgerows, flowering shrubs, cover crops, perennial borders, and mulches like newspaper or compost as resting and hiding places. To encourage soil-dwelling beneficials like ground beetles and rove beetles, plant in permanent beds and put down stone mulches to provide lots of hiding places. Avoid excess tillage, because it will destroy or disturb beneficials. The dust cloud created by tillage eventually settles on plants and may harm the more delicate beneficials, such as parasitic wasps. Excessive tillage also alters the soil environment and may harm populations of beneficial soil microorganisms (see "Soil Tillage" on page 427).

Environment

Just as gardeners don't consume every kernel of corn but save some for next year's crop, biocontrols rarely kill 100 percent of the pests they prey upon. Don't be tempted to spray those few remaining pests—they will provide the next generation of beneficials with food. Spraying with broad spectrum insecticides upsets the balance of beneficials and their prey.

In the wild, beneficials and harmful organisms often follow a natural oscillating rhythm of population highs and lows. Pest numbers may rise suddenly in response to some environmental cue, such as an increase in their food supply. Populations of beneficial organisms will increase accordingly. As a result, pest numbers decline, followed by a corresponding drop in numbers of beneficials. The cycle is repeated continuously each season in your garden. If you notice a certain pest population is on the rise, it may be worth the wait to let the beneficials move in and perform rather than reaching for pesticides immediately. Become familiar with common natural enemies so you'll know when to rely on them.

Many botanical insecticides, including rotenone and pyrethrin, are broad-spectrum pesticides that kill a wide range of insects. Beneficials are often included among the dead. Also, spraying recklessly for one pest may result in the development of secondary pests. When you spray a broad-spectrum pesticide that kills beneficials, some pest that survives the spraying and is now unchecked by a natural enemy may become a new problem. Pests like aphids are often able to recover more quickly than their natural controls, so outbreaks of aphids are common after the application of a broad-spectrum insecticide. Frequent spraying can also encourage populations of pesticide-resistant insects to increase, and pest species usually develop resistance more quickly than beneficial species do.

Keeping the soil environment healthy can also be important in preventing plant disease. Organically enriched soils have high populations of beneficial microbes that feed upon or outcompete pathogens. Preserve a healthy balance in your soils by adding compost or other organic soil amendments on a yearly basis.

Buying Beneficial Insects

If your native natural controls need assistance, you may choose to buy beneficial insects. Several garden supply catalogs offer beneficial insects among their products, and private insectaries offer a wide assortment. Keep in mind that most types of beneficials are reared at only one or a few insectaries. The size of the crop of beneficials may vary depending on the success of the insectaries' breeding efforts. You'll be competing with commercial growers and garden supply companies for what may be a limited supply of beneficials. Availability of specific species can change from year to year. (For the most recent list of companies that specialize in raising and selling biocontrols, contact the California Department of Food and Agriculture, Biological Control Services Program, 3288 Meadowview Road, Sacramento, CA 95832.)

Purchased beneficials need special care or they can easily die before they get a chance to do their job. Follow these steps when purchasing and releasing beneficials.

■ Become familiar with the life cycle of the target pest; know where and when to find eggs and larvae in your garden. Use garden books with good illustrations or photographs to help you.

■ Identify the natural enemies available for control of your pest and locate a source. Consult with your supplier to be certain you're buying the best biological control for the pest.

■ When the beneficials arrive, study the instructions for storage and release; each beneficial organism is unique and must be treated differently. Release them as soon as you can; if you can't release them immediately, follow the instructions for proper storage.

■ Release them at the proper rate and location. Rate of release will depend on the size of your garden or greenhouse and the severity of your pest infestation. Take a good look at the beneficials before releasing, so you don't confuse them for a pest later.

■ Once you've released beneficials, monitor their progress. If you're releasing egg parasites, watch for discolored pest eggs with odd-shaped exit holes. Parasitized larvae often become discolored and inactive. If necessary, make several releases to keep a major pest under control. Once released, some natural enemies are vigorous enough to survive from year to year. Others may require annual releases if they aren't able to survive your area's climate year-round.

Beneficial Animals and Insects

We can rely on an assorted crew of large and small creatures to help with pest control. The list of beneficial animals ranges from birds and toads to spiders and predatory nematodes. Beneficial insects include the familiar lady beetle along with less well known predatory bugs and parasitic wasps.

Large Animals

Mammals are often overlooked for their pest-control abilities. For example, in Canadian forests, shrews control larch sawflies and other pests. Lizards and toads are important natural enemies of pest insects, and snakes can help control problems with mice and voles. In many communities in Europe and North America, farmers encourage birds to control pests by placing nesting boxes around their fields. Even domesticated animals can assist in pest control. If you have a small orchard and keep chickens, let the chickens feed in the orchard. They will peck pests such as plum curculios out of dropped fruit and other ground litter.

How to Encourage: Encourage or discourage these large and small animals as necessary in your garden. Naturally, you'll probably want to keep out those animals that cause more damage than benefit. But it may be time to reconsider some of your prejudices when assessing the harmfulness of certain animals. If you're the squeamish type, have patience with the snakes that cross your garden paths—they're great rodent controllers! Also remember that bats are generally harmless to people, but they can consume huge numbers of mosquitoes, flies, and other pests.

If water is scarce, place small containers filled to the brim in shady locations for thirsty animals. Build birdhouses to attract the insect-eating birds, and place the houses along the garden's border.

Insects

One of the best steps you can take to reduce pest insects is to encourage nonpest insects such as lady beetles and lacewings. You can also buy beneficial insects and release them in your garden. This approach works better with some species than others. For example, purchased convergent lady beetles released in your garden may not be there long because their instinct tells them to migrate elsewhere.

For information on many of the beneficial insects that may be present naturally in your garden, see "Important Native Beneficial Insects" on page 448. For details on how and when to release beneficials that are commercially available from garden suppliers and insectaries, see "Buying and Releasing Beneficial Insects" on page 454. If you'd like more information about particular types of insects mentioned in those listings, look them up in the "Insect Identification Guide" beginning on page 268.

Predatory Mites

Predatory mites are less than 1/25″, 8-legged creatures that are related to spiders. In your garden, predatory mites live in the soil, in compost piles, and on your plants. Soil-dwelling species usually reside in the upper 2″ of the soil, preying on other mites and insects. Species that live on garden plants consume other mites and their eggs; some types also feed on thrips. Still other species provide valuable pest control in stored grains.

Predatory mites are useful biological controls because they reproduce quickly, keeping pace with oscillating pest populations: Some are able to complete their life cycle in 1 week. If you observe mites through a magnifying glass, you can distinguish predatory mites from

(continued on page 457)

Buying and Releasing Beneficial Insects

Insectaries and garden supply catalogs offer a wide variety of beneficial insects and predatory mites, but the first step to getting good results is to be absolutely certain that the insect you purchase will prey on the pests you're trying to control. You also need to be sure that you buy the correct number of insects for your garden size. Here you'll find all the information you need to shop wisely for beneficial insects.

Beneficial Species	Pests Attacked	How and Where to Use	Comments
Aphid midge *Aphidoletes aphidimyza*	Over 60 species of aphids.	Apply 250 cocoons in a small garden or greenhouse or use 3-5 cocoons per plant; repeat in 2 weeks; use 5-20 cocoons per fruit tree.	Reliable except in dry, windy areas; not suitable for melon aphid control.
Aphid parasites *Aphidius matricariae, Lysiphlebus testaceipes,* other species	Apple aphids, green peach aphids, melon aphids, others.	Release 50-100 pupae of these parasitic wasps in a garden or small greenhouse early in spring.	Experimental; best effect if used with aphid predators.
Colorado potato beetle parasite *(Edovum puttleri)*	Parasitizes eggs of Colorado potato beetle.	Best effect seen against second generation of potato beetles. Become established faster when there is a source of honeydew from aphids nearby.	Experimental; not hardy in northern areas.
Lacewings *Chrysoperla (=Chrysopa)* spp.	Soft-bodied insects—including aphids, mealybugs, thrips—small caterpillars, mites, moth eggs, some scales.	Scatter 500-1,000 eggs throughout an average garden.	Unreliable due to poor survival during shipping, but survivors are effective.
Lady beetles *Hippodamia convergens*	Aphids, mealybugs, soft scales, spider mites.	Not recommended for outdoor release; use in greenhouses at a rate of 3-5 beetles per plant or 500 per home greenhouse.	Unreliable outdoors due to migration, unless released over community-wide area; other species good candidates for outdoor releases.

Beneficial Species	Pests Attacked	How and Where to Use	Comments
Mealybug destroyer, also called Australian lady beetle *Cryptolaemus montrouzieri*	Mealybugs indoors and in citrus orchards.	Release 2-5 beetles per infested plant in greenhouses, 1 or 2 times per year; for infested houseplants, confine 10-20 beetles to each plant for 4-6 weeks; in citrus and grapes, release 250-500 per ¼ acre.	Used for decades in citrus groves; reliable in summer; slow indoors in winter.
Minute pirate bug *Orius tristicolor, O. insidiosus*	Small caterpillars, leafhopper nymphs, spider mites, thrips, eggs of many insects.	Release 1 bug per 1-2 plants in greenhouses; try 50-100 in a home garden.	Experimental; studied in commercial greenhouses only.
Predatory mite *Phytoseiulus persimilis*	Spider mites, especially Pacific and two-spotted spider mite.	Release 2-10 mites per plant on strawberries or garden or greenhouse plants at first sign of spider mites; repeat if necessary in 2-3 weeks; not winter hardy in most of U.S.; doesn't perform well on tomatoes.	Reliable, except in hot, dry conditions.
Predatory mite *Amblyseius cucumeris*	Western flower thrips, cyclamen mites, onion thrips, spider mites.	Apply 100-200 per infested plant to control thrips; effect is slow because mites cannot kill adult thrips; cold-hardy and will become established in strawberries to control cyclamen mites.	Reliable thrips control on greenhouse peppers; unreliable on cucumbers.
Western predatory mite *Metaseiulus (=Typhlodromus) occidentalis*	Spider mites, especially citrus red mites and European red mites.	Establish in orchards with rates of 50-100 mites per tree in summer or early fall; apply 1,000 per tree to control infestation of spider mites in same season; hardy in most apple-growing regions.	Reliable; pesticide-resistant strains available.

(continued)

Buying and Releasing Beneficial Insects — Continued

Beneficial Species	Pests Attacked	How and Where to Use	Comments
Praying mantid *Mantis religiosa*	Any insect they can catch, including beneficial species.	Not advisable to release in garden because of destructive effect on native species.	Not effective as control for pest insects.
Scale predatory beetles *Chilocorus nigritus, Lindorus lophanthae*	Soft scales, especially soft brown scale.	Minimum order of several hundred sufficient for a greenhouse; confine 10–20 beetles for a month to an infested houseplant.	Reliable, especially indoors.
Soil mite *Geolaelaps* (=*Hypoaspis*) spp.	Fungus gnats, springtails, thrips pupae.	Sprinkle granular carrier containing mites on soil around plants; apply 1 quart per 1,000 square feet once early in season to establish population.	Reliable on fungus gnats when established early.
Spined soldier bug *Podisus maculiventris*	Hairless caterpillars, including fall army worms, sawfly larvae, tent caterpillars; beetle larvae such as those of Colorado potato beetle and Mexican bean beetle.	Release 2–5 per square yard of bean or potato patch to control beetle larvae.	Experimental; results best on tent caterpillars and bean beetles; not very promising on Colorado potato beetles.
Trichogramma wasps *Trichogramma minutum,* other species	Eggs of over 200 species of moths such as corn borers, corn earworms, codling moths, cotton bollworms, spruce budworms.	For codling moths, release 5,000–7,000 per tree, at 5–7 day intervals throughout month when moths are flying; try 5,000–7,000 per square foot to control corn borers; use pheromone traps to determine moth populations and time releases effectively.	Most effective in large numbers over large crop areas.
Whitefly parasite *Encarsia formosa*	Greenhouse whiteflies, sweetpotato whiteflies.	Release 1,000 wasps in a small greenhouse or 2–5 parasites per plant when the first whiteflies are seen; repeat in 1–2 weeks; double releases for sweetpotato whiteflies.	Reliable on greenhouse whiteflies if released while pest numbers are extremely low; unlikely to succeed in winter months; works best in warm, bright conditions.

spider mites by their faster movement and oval shape. Their beneficial effects are often hardly noticed until they are killed off by pesticide sprays; the subsequent population explosion in spider mites highlights their importance.

Several species are available from insect suppliers for controlling pests in the garden, orchard, vineyard, and greenhouse. *Phytoseiulus persimilis* is the most widely sold species for spider mite control. In hot greenhouses and gardens (temperatures greater than 90°F), try *Amblyseius californicus* or *Phytoseiulus longipes. Metaseiulus occidentalis* is an excellent hardy predator of European red mite in apple orchards and berry patches, surviving as far north as the Canadian apple-growing regions. Once established, it should only need to be re-released after a severe winter or other disruption to the population.

How to Release: Predatory mites are usually sold as a mix of several species to ensure complete control under a variety of conditions, because each species has overlapping humidity, temperature, and prey requirements. In the greenhouse or garden, order about 1,000 predatory mites for every 200-500 square feet. Shake the carrier and mites out among your plants. The mites are so small you may not be able to spot them in the mixture, but rest assured they will travel among your plants as necessary to find pests.

For fruit trees, if European red mite numbers are low and you just want to establish a predator population for future years, release 50-100 per tree; if you want to control an outbreak during the same season, release 1,000 per tree.

Nematodes

Mention nematodes and most gardeners picture the parasitic type that infests plant root systems. The beneficial nematodes such as *Steinernema (=Neoplectana) carpocapsae* and *Heterorhabditis heliothidis* almost make up for the damage their relatives cause. They're both known for quick control of insect pests that spend part of their lives in your garden soil.

Immature larvae of both species have a protective cuticle. They aggressively search for hosts, usually in or near the soil. Once the host is located, the nematode enters its body through natural openings, such as the mouth or spiracles, shedding the cuticle it no longer needs. Beneficial nematodes carry an intestinal bacterium (*Xenorhabdus* spp.) that is released inside the new host, paralyzing and killing it within 24 to 48 hours. Once the host is dead, the nematode completes several generations within the carcass until host tissues are depleted. Larvae then redevelop the protective cuticle and leave the carcass in search of a new host. Larvae in this protected state can survive in the soil without a host for as long as 1 year if moisture levels and temperatures remain favorable.

Nematodes require a moist, dark environment. As temperatures drop in the winter, they burrow deeper in the soil and begin hibernation; as temperatures rise in spring, they move closer to the soil surface. Their return lags behind the arrival of early spring pests, so for early pest control, gardeners must reintroduce the nematodes each year.

In the laboratory, over 250 species of pests fall prey to beneficial nematodes. Outdoors, however, uncontrolled soil temperatures and moisture reduce their efficiency. *Steinernema carpocapsae* lives closest to the soil surface and successfully controls carpenterworms, currant borers, earwigs, navel orangeworms, onion maggots, pillbugs, seedcorn maggots, sod webworms, sowbugs, and strawberry root weevils. *Heterorhabditis heliothidis* ranges deeper in soil, effectively controlling billbugs, black vine weevils, corn rootworm larvae, Japanese beetle grubs, masked chafers, mole crickets, and wireworms. Both types of nematodes show promise against armyworms, cabbage root maggots, codling moth larvae, Colorado potato beetle larvae, cutworms, and rose chafers.

How to Release: Rate of application depends on the susceptibility of the target pest, its location in reference to the damaged plant, and the particular species or strain of nematode supplied. Commercial suppliers vary in the strain of nematodes they raise; you may have a choice of several strains or species. Consult with specialists at your supply source for the most appropriate species and the most efficient method of application. Always follow the suppliers' label directions.

The nematodes arrive in the infective larval stage, either dehydrated or suspended in gels, sponges, or moist peat and vermiculite. It's best to release them when pest larvae are known to be present. Mix the formulations thoroughly with water, then apply to the soil with a watering can or pressure sprayer. At planting time, treat a 3″ wide band centered over the row. Control pests of perennial plants by spraying at the base of the plants. To control pests in sod, spray nematodes evenly over the grass with a watering can or sprayer, or pour them directly onto the grass: Use 100,000-500,000 nematodes per square yard. Water the nematodes in well; apply ½″-1″ of water after spraying to soak them into the sod. Control is most successful when nematodes are applied to moist soil in the late afternoon or evening. Treatment may be less successful in lawns with thatch, which impedes the progress of the nematodes into the root zone.

Since the nematodes are susceptible to drying, control is less effective when applied directly to plants. It may take more than 6 weeks before you know whether treatments were effective. Parasitized hosts will appear chalky white, reddish, or gray.

To control borers in trees, mix nematodes according to label directions, and use an oilcan to squirt the nematodes into the hole (about 17,000 nematodes per squirt). Or use a syringe to inject nematodes into squash vines to control squash borers or in corn silks to control corn earworms. Syringes are usually marked in cubic centimeters (cc); you'll be adding about 5,000 nematodes in 2 cc of the prepared mix.

Decollate Snails

Decollate snails *(Rumina decollata)* prey on the brown garden snail *(Helix aspersa)*. This predatory snail, with an elongated, cone-shaped shell, efficiently controls the pest species but will consume your young vegetable plants if other food is unavailable. They are available for release in certain parts of California, but they should not be released in home gardens. Commercial citrus growers release them in citrus groves, where they control pest snails without damaging the trees.

Spiders

The easiest way to tell a spider from an insect is by counting the number of legs: Spiders have 4 pairs while insects move about on only 3 pairs. Spiders come in a great variety of sizes, shapes, and colors. They're the predators you're likely to encounter most often in your garden, yard, and home.

Insects are the most popular items on the spider's menu. Spiders capture their insect prey with unusual tactics. They weave traps with the silk threads produced by spinnerets. The multipurpose silk is woven into coverings for eggs; parachutes for travel; and snares, drag lines, and webs for capturing prey. Some spiders, including common wolf spiders, are adequately large and fast to chase their prey down. As soon as they have hold of their prey, they inject digestive fluids that liquefy the tissues, then they suck the shell dry. Almost all spiders are poisonous, but only a few have venom toxic enough to threaten humans.

Daddy longlegs are closely related to spiders. Often called harvestmen, they'll eat just about anything, including scraps from the kitchen table. They're most active at night, patrolling garden plants for infestations of small insect pests. Like spiders, they have 8 legs. Their legs are easily shed but don't grow back. By the end of the summer, many daddy longlegs are getting about adequately on only a few legs.

How to Encourage: Depending on their size and stage of life, spiders and daddy longlegs are susceptible to the same pesticides used to kill insects, so refrain from spraying. Solicit their help in the garden by leaving them alone when you encounter them in webs strung between the plants or under stones or debris on the soil.

Beneficial Microorganisms

Unlike introduced beneficial insects, which require exact conditions if they are to work well, microbial and biological insecticides can tolerate a range of environmental conditions. However, they do require some understanding of both the pest and the remedy if they are to work well. For example, spraying BT (*Bacillus thuringiensis*) on your broccoli plants will effectively control imported cabbageworms. However, it will do nothing to stop aphids feeding on those same plants, because the bacteria cannot infect aphids. And while milky disease spores kill Japanese beetle grubs, they have no affect on adult Japanese beetles. Always read instructions and follow them exactly when using biological pesticides.

The specific nature of biological pesticides, like BT, means that you can use them without destroying the natural balance that exists between pests and predators. There is no danger of pollution or residual toxicity when such pesticides are properly used. Still, there may be occasions when it's best to wait and see what kind of pest-predator relationships develop before you reach for a biological pesticide. For example, if you place a bird feeder close to cabbage or broccoli, the birds may consume leaf-eating worms for you, eliminating the need for treatment with BT. Because biological insecticides are host-specific, they are not toxic to people. However, it still pays to follow proper spray procedures for best results. See "Spray Safely" on page 463 and "The Way to Spray" on page 465 for information on how and when to spray pesticides and on the best types of application equipment to use.

Remember to use cultural methods, including plant selection, soil care, timely cultivation, and crop rotation, before considering biological control. Biological pesticides are much gentler to the environment than chemical agents, but they are pesticides just the same. Although their impact may be small and finely tuned, they can still disturb the natural balance that exists on any site between soil, plants, pests, and predators. Using microbes to control garden pests is an active area of research; be on the lookout for new microbial insecticides.

Bacillus thuringiensis

The most widely used microbial biological control method is the application of *Bacillus thuringiensis,* or BT. Scientists have identified more than 35 varieties of this bacterium; several are available to home gardeners for controlling various larval insect pests. *B.t.* var. *kurstaki* (BTK) was one of the first bacterial pest controls offered on the market. BTK produces crystal toxins that poison, paralyze, and kill various common pest caterpillars. After BTK is applied to plants, caterpillar pests ingest it as they feed. Although the pest may live several more days, it stops feeding, darkens in color, and eventually drops to the soil. BTK effectively controls a wide range of caterpillars including cabbage loopers, codling moth larvae, diamondback moths, gypsy moth larvae, imported cabbageworms, spruce budworms, tomato hornworms, and others. Consult the label for a complete list of susceptible pests.

B.t. var. *israelensis* (BTI) attacks larvae of blackflies, fungus gnats, and mosquitoes when applied to standing water where these pests reproduce. *B.t.* var. *san diego* (BTSD) controls certain leaf-feeding beetles, including black vine weevils, boll weevils, Colorado potato beetles, and elm leaf beetles.

BT products are selective—they will not harm the great majority of beneficial insects in your garden. However, butterfly larvae are caterpillars and will be infected by BT. So don't spray BT indiscriminately throughout the garden: Limit applications to plants you know are infested by pests.

BT will not infect wormlike pests excluded from the label, like slugs or leafminers. BT is considered nontoxic to people.

Although it was once thought that insects would not develop resistance to microbial insecticides, scientists have confirmed cases of diamondback moths in commercial farms that showed resistance. This is another argument for using BT with care, and only after trying cultural or physical control methods for which resistance is not a factor.

How to Use: You can purchase BT in many forms, including liquids, powders, dusts, and granules. Follow label directions for either dusting on plants or preparing and applying a batch of spray solution. If sprays roll off leaves rather than adhering to them, add a few drops of liquid dish soap to the spray solution to enhance sticking.

Adding a commercial appetite stimulant to the BT spray solution may encourage pests to eat more of the treated leaves. While this increases the likelihood that the pests will ingest the BT, it could also lead to more severe plant damage; this is especially true if the appetite stimulant has a longer residual effect than the BT. Appetite stimulants have not been scientifically tested, but you may wish to experiment with them in your own garden.

To control mosquitoes and flies, add BTI to standing water in ponds, puddles, storm drains, and other wet areas. BTI has no harmful effects on fish or amphibians. To control fungus gnats in the greenhouse, apply BTI as a drench to soil on benches, floors, and in large pots.

You can prepare a BT/bran bait for controlling cutworms by moistening bran with a dilute solution of BTK. Sprinkle the moist bran on the surface of planting beds 2 weeks before you plan to plant.

Store BT products in a cool, dark place to retain viability for several years. Use a fresh batch for each application and use it within several days. BT products break down quickly in sunlight, so several applications may be necessary to continue to control new pests as they arrive. You can prolong the period of effectiveness by applying sprays in the evening.

Commercial Products: BTK: Bactur, Bactospeine, Caterpillar Attack, Caterpillar Killer, Dipel, Javelin, Larvo-Bt, SOK-Bt, Thuricide, Worm Ender; BTI: Bactimos, Mosquito Attack, Teknar, Vectobac; BTSD: Colorado Potato Beetle Attack, M-One; Appetite stimulants: Entice, Wheast

Milky Disease Spores

Bacillus popilliae and *B. lentimorbus* are 2 types of bacteria that are combined as a microbial insecticide product called milky disease or milky spore disease. The bacteria kill grubs of Japanese beetles and several related beetles but are harmless to other organisms. Beetle grubs eat spores that have been sprayed on lawns while feeding on grass roots. Infected grubs are filled with an opaque, milky white liquid full of bacterial spores, which will remain in the soil after the grubs die. Infected grubs inoculate the soil as they travel. Usually, only 1 application is necessary since the disease is carried over from year to year by new generations of beetles. In northern climates, however, annual applications may be necessary.

How to Use: Apply spores any time the ground is not frozen to control grub damage to your lawn. Treatment will not have much affect on populations of adult beetles unless it is applied on a community-wide basis, since adults are mobile and new ones will arrive daily from untreated areas.

Use 7-10 pounds of spore dust per acre or about 10 ounces per 2,500 square feet, applying the dust in spots roughly 4′ apart. This is about 1 level teaspoon of powder per spot. You can make an old-fashioned applicator from a 1-pound coffee can and a discarded broom handle. First, punch holes in the bottom of the can with a 10-penny nail. Then bolt the can to the broom handle so that the bottom of the can is 4″ from the bottom of the handle.

Fill the can ¾ full with milky disease spores. To apply the dust, sharply tap the end of the stick against the lawn. One rap will release the right amount of powder for each treatment spot. Try to apply the dust just before a rain, or water the area lightly after application.

You can apply granular formulations with a broadcast seeder or fertilizer spreader. Use about 4 pounds per 2,500 square feet, and water immediately after applying.

Commercial Products: Doom, Japademic

Fungi

Beauveria bassiana has long been recognized as a biological control of many insect pests that spend part of their lives in the soil, particularly Colorado potato beetles. This fungal pathogen of insects exists naturally in garden soil, making a few insects sick each year. When environmental conditions are right and susceptible insect populations are high, outbreaks of this and other insect pathogens occur that can cause pest populations to crash almost overnight. Unfortunately, the concentrated pest population damages many plants before being controlled.

A commercial product containing *B. bassiana* is available as an insecticide in the Soviet Union and China. Scientists in the United States are currently studying the fungus. A commercial version may be available in the United States in the near future. Meanwhile, to encourage native populations, refrain from using fungicides in the garden.

Many other specific fungal pathogens of insect pests are being studied and developed for their potential use as biological control agents. *Coelomomyces, Lagenidium, Entomophthora,* and *Culicinomyces* are fungal pathogens that infect whiteflies and mosquitoes and show promise for future development.

Viruses

Like the human cold virus, insect viruses tend to thrive when their hosts are under stress. They occur naturally but don't severely impact pest populations except under stressful conditions, such as high numbers of pests with a limited food supply. They're specific to insects and mites and are safe for other organisms and the environment.

Nuclear polyhedrosis viruses (NPV) and granulosis viruses (GV) have received the most attention. Both of these belong to a general group of viral diseases known as baculoviruses. Both of these types of viruses form a capsule that protects them in the environment until they are eaten by an insect. Afterward, the capsule dissolves in the insect's gut, releasing the particles, which spread to other organs. When the insect dies, its body is full of a liquid containing millions of virus capsules. Infected caterpillars are discolored and have a soggy, limp appearance. When disturbed, the body breaks open releasing more capsules to infect other insects.

NPV strains control armyworms, cabbage loopers, cotton bollworms, gypsy moths, soybean loopers, tobacco budworms, and velvetbean caterpillars. GV controls cabbage butterflies, codling moths, oriental fruit moths, and potato tuberworms.

While there are currently no viral products available for home use, several are in the experimental stages. Some are in use for forest and field crop applications.

Protozoans

Protozoans are tiny, single-celled organisms that may have complex life cycles. *Nosema locustae* is a protozoan pathogen that infects the fat tissue of most grasshopper and cricket species. Once ingested, spores germinate in the insect's gut. The pest becomes sluggish and slowly dies or remains sufficiently ill that it no longer reproduces or causes damage. Grasshoppers that survive lay infected eggs that serve as inoculum during the following season. *Nosema* is recommended for long-term control, since mortality may take as long as 4–6 weeks, although some species die within a few hours. Once applied, each successive generation is reinfected by the preceding one. *Nosema* is an effective grasshopper control in large fields. It is not effective for backyard use, because grasshoppers are highly mobile and new adults will continue to migrate in. To be effective, community-wide applications are necessary.

How to Use: Apply *Nosema* products to control grasshoppers when the late nymphs are emerging. In areas with severe community-wide grasshopper problems, Cooperative Extension personnel can help you time the application. In your yard, broadcast the bait on lawns and garden beds in early summer when you see immature grasshoppers (less than ¾″ long).

Commercial Products: Grasshopper Attack, Grasshopper Control, Nolo Bait, Semaspore

Antagonistic Organisms

Microbial antagonists are pathogens that don't directly attack pest organisms. Instead, they secrete antibiotics that are toxic to other microbial organisms. They also compete aggressively with disease-causing pathogens for food, often winning the race without harming the host plant. This special group of microbials occurs naturally in the soil. Scientists are only recently learning how to manipulate them to control plant diseases.

Agrobacterium radiobacter is a fungal antagonist of crown gall diseases that affect fruit and nut trees. *Gliocladium virens* is a fungal organism approved for greenhouse control of the pathogens that cause damping-off (*Rhizoctonia solani* and *Pythium ultimum*). *Trichoderma* species fungi show promise in controlling damping-off diseases of seedlings.

How to Use: Binab-T, a mixture of several species of the beneficial fungus *Trichoderma*, is currently registered for use on fresh tree wounds. Used before sealing, it establishes residence at the site and prevents other pathogenic organisms from moving in. Galltrol-A and Norbac 84-C are commercial sources of *Agrobacterium radiobacter*. Mix these products according to label directions, and use as a disinfectant dressing for pruning cuts or as a dip for bare-root plants. Look for antagonistic microbials in garden supply catalogs that feature least-toxic pest controls.

Commercial Products: Binab-T, Galltrol-A, Norbac 84-C

CHEMICAL CONTROLS

In isolated instances, even a combination of cultural, physical, and biological controls won't be enough to remedy a pest problem. In these cases, you'll face the choice of losing your harvest or valuable plant or resorting to organically acceptable chemical controls.

Not all natural chemical controls are true insecticides or fungicides like rotenone or sulfur. Pest-plagued gardeners have long resorted to homemade sprays with ingredients such as hot peppers, garlic, and strongly flavored herbs. The general idea is to make a pungent potion that will fool pests into turning away from their favorite plants. Some, such as homemade hot pepper and garlic sprays, deter insect feeding and possibly confuse the pests' smell receptors (often located on their feet).

Organically acceptable chemical controls such as horticultural oils make pest habitats inhospitable; they also coat and suffocate scales. Soap sprays kill soft-bodied pests, including many beneficials, but are safer for the gardener than botanical pesticides such as rotenone or sabadilla. Gentle, nonpesticidal, chemical controls such as homemade herbal or baking soda sprays are much less toxic than botanical pesticides; select them first when you seek chemical remedies to pest problems.

Handle with Care

Use organically acceptable chemical control methods with restraint. Some, especially botanical insecticides, present the same risks to the dynamic living community in your garden as do synthetic poisons. Use them too freely, and you may face the problem known as pest resurgence. In other words, you kill off some pests and most of their predators, and the pests come back stronger than ever with no predators to slow them down. And, although organically acceptable botanicals like pyrethrin and rotenone break down quickly after application, they are highly toxic initially. During the first few hours after they are applied, these and other naturally derived chemicals can kill large numbers of honeybees, spiders, and other beneficial organisms.

Spray Safely

Proper application of pesticides and fungicides is time consuming, but necessary. A lackadaisical approach at the treatment stage throws away the time you've spent caring for your garden, identifying pests, and selecting appropriate controls. It also endangers you, the environment, and the plants you wish to protect.

A few of the chemical products covered in this section are available as dusts or premixed sprays, but many others are liquid concentrates or wettable powders. The following guidelines will help you apply controls safely and correctly; in addition, always follow label instructions.

1. Wait for calm weather. Applying any pesticide under windy conditions causes the poison to blow or drift where it is not wanted and greatly increases the risk that you'll inhale it or get it in your eyes.

2. Pick ripe fruits or vegetables, prune plants if appropriate, and remove weeds that provide pests cover. Although fruits and vegetables are safe to eat within a few days after application of most botanical pesticides, it makes sense to gather up everything that's ready to eat before you spray. This also reduces your need to enter recently treated areas while chemicals are actively working.

3. Dress properly. Wear long pants, shoes and socks, and a long-sleeved shirt. If you are applying a liquid pesticide, do not wear leather shoes, which readily absorb spillage. Wear a face mask, such as a disposable dust mask. However, if the product label advises you not to inhale the spray vapor or mist (you'll see this warning on some botanical pesticide products), wear a respirator. It's a good idea to wear goggles as well, because the tissues of the eye can absorb pesticides fast and easily. Wear rubber gloves when handling any pesticide.

4. Precheck your equipment. Make sure spray nozzles are open and pump mechanisms are working properly before you load the reservoir with a pesticide. Otherwise, you can end up with a serious disposal problem.

5. Mix up only as much as you will need. Work slowly and carefully so you don't spill chemicals during the mixing process.

6. Apply the pesticide when beneficials are relatively inactive. You may need to observe insect activity patterns for a few days to discover the best time. Generally, mid-morning should be avoided, as this is prime time for honeybees and many winged beneficial insects. If you're only going to treat a few plants, put a row cover over them for a day or two after the pesticide is applied to keep beneficials out of the treated area.

7. Spray and dust early in the morning or in the evening. Many pesticides, such as horticultural oils and insecticidal soaps, can injure plants when they are applied during the hottest and brightest part of the day. Others, like pyrethrin, lose their toxicity when exposed to high temperatures. Most botanicals begin to break down immediately when exposed to light; to get the best control, don't apply these in hot sun.

8. Make a thorough application. Treat upper and lower leaf surfaces as well as places in mulch where pests might run and hide.

9. The best way to dispose of any pesticide is to apply it properly, but if you do have any leftover solution, dilute it with water and place in a bucket or other container. Rinse out equipment and allow the rinse water to drip into the bucket. Place the bucket of contaminated water in sunlight for a day or two to let the pesticide degrade. Make sure it's inaccessible to children, pets, and wildlife while it sits—cover it with a screen if it's in an outdoor spot where animals might find it and drink from it. After it has degraded, dump the solution away from all water sources, including ponds.

10. Change your clothes and wash your hands and face. If you spilled some solution on yourself, take a shower.

11. Return the container to storage, which should be a cool, dark place out of the reach of children. Always store pesticides in their original containers. Most botanical dusts are mixed with diatomaceous earth or a fine clay powder and are easily confused with plain diatomaceous earth. Liquid concentrates also should stay in their original containers—never store concentrates or solutions in beverage containers.

12. Apply botanicals and other organically acceptable pesticides only to the plants specified on the label and at the recommended rates.

Chemical Control Products

Organically acceptable chemical pesticides have three characteristics that distinguish them from their synthetic counterparts: They are derived from natural substances, they are generally less toxic to humans than synthetic pesticides, and they break down relatively quickly in the environment to harm-

The Way to Spray

Finding the right dust or spray for pest-ridden plants is only the first step in pest control. The next step is actually applying the substance exactly where it's needed. For best results, choose an application method suited to the substance you are applying and to the size of the job. Here are some practical guidelines.

■ For small jobs, use a hand-held trigger sprayer. Keep one sprayer for nonpoisonous liquid sprays, like those made from soap, garlic, peppers, or kelp, and label a separate sprayer for biological or botanical pesticides. Remember to wear rubber gloves while applying biological or botanical controls.

■ For larger spray jobs, select a pressure sprayer that you can pressurize manually through pumping. Choose a sprayer large enough to handle the area you want to cover without frequent refilling. The more times you must refill the spray reservoir, the more often you will be exposed to the spray. Carefully clean and dry the nozzle and any filters after each use. For complete information on cleaning your sprayer and disposing of unused pesticides, see "Spray Safely" on page 463.

■ When spreading nonpoisonous dust, such as diatomaceous earth, punch small holes in the sides and bottom of a paper bag, place a cup of the dust inside, and shake it over the areas you want to treat. Or place the dust in an old sock, tie the end closed, and beat the sock above the plants with a stick or broom handle. Use caution, and wear protective gear to avoid inhaling dusts.

■ With biological or chemical pesticide dusts, such as sabadilla or rotenone, use a duster designed for this purpose. Besides keeping the dust away from your face, dusting appliances push out puffs of air along with the pesticide, resulting in much better coverage. Pesticide dusts sprinkled onto plants are much less effective than those applied so that they temporarily form a cloud around affected plant leaves.

less substances. However, there is no easy way to group chemical controls. Some are truly insecticidal or fungicidal, while others work by deterring pests. Some substances kill both insects and disease organisms.

Many chemical controls are known by more than one name. For example, highly refined horticultural oils used to kill a variety of pests are also called supreme, summer, or superior oils. And some gardeners refer to spraying pyrethrin, while others call that botanical pesticide pyrethrum.

This listing of chemical controls is in alphabetical order. If you have trouble finding a control measure that you are interested in, look it up in the index for a quick reference to the appropriate listing.

Alcohol

Sprays of 70 percent isopropyl alcohol (also known as rubbing alcohol) control a variety of pests on garden plants. Alcohol also works well against pests on houseplants that have waxy foliage.

Protection Offered: Alcohol sprays can control aphids, flea beetles, scales, thrips, and whiteflies.

Precautions: Undiluted alcohol and solutions of water and alcohol may injure foliage. Test sprays on a few leaves before applying to an entire plant. It may take several days for damage symptoms to appear.

How to Use: Spot treat scale infestations by wiping them from your plants with a cotton ball soaked in 70 percent isopropyl alcohol. To make a whole-plant spray, dilute 1-2 cups of isopropyl alcohol with 1 quart water. Test this solution on a small area before spraying to be sure the plants will not be damaged.

All-Purpose Insect Pest Spray

Since its first issue, editors of *Organic Gardening* magazine have collected the pest-control remedies and recipes mailed in by readers from around the world. Several ideas were repeatedly offered as safe, effective insect controls. The editors combined several of these home remedies to make an all-purpose spray out of ingredients found in most kitchens. This spray combines the repellent effects of garlic, onion, and hot pepper with the insecticidal and surfactant properties of soap.

Protection Offered: Try this spray against any leaf-eating pests in the garden, and make a note of what pests are successfully controlled.

Precautions: When spraying, be careful to keep the mixture well away from your eyes and nose and wear rubber gloves, since it could cause painful burning.

How to Use: Chop, grind, or liquefy 1 garlic bulb and 1 small onion. Add 1 teaspoon of powdered cayenne pepper and mix with 1 quart of water. Steep 1 hour, strain through cheesecloth, then add 1 tablespoon of liquid dish soap to the strained liquid; mix well.

Spray your plants thoroughly, being sure to cover the undersides of leaves. Store the mixture for up to 1 week in a labeled, covered container in the refrigerator.

Ammonia

Household ammonia is a common cleaning product that also provides control for several common plant pests.

Protection Offered: Ammonia can control aphids, flea beetles, scales, thrips, and whiteflies.

Precautions: Ammonia solutions may injure the leaves of some plants. Do not apply in hot weather or on drought-stressed plants.

How to Use: Mix 1 part ammonia with 7 parts water and apply as a spray, after you've tested a few leaves to be sure the plants will not be damaged.

Avermectins

Streptomyces bacteria are familiar to us as important sources of antibiotics used in medicine. Avermectins, including abamectins and ivermectins, are naturally occurring antibiotics that have insecticidal properties that are useful in the garden as well as on pets and livestock.

Protection Offered: Avermectins are registered to control spider mites and leafminers on flowers and ornamental foliage crops. They are also available as a fire ant bait, but other garden uses await federal approval.

Precautions: Avermectins have been found to be extremely toxic to mammals, aquatic invertebrates, fish, bees, and other beneficial insects. They also produce developmental toxicity in mice. Soil microorganisms readily degrade avermectins; studies indicate they are unlikely to leach through most soils.

How to Use: You can buy avermectins in spray form to control leafminers and spider mites. Spray both upper and lower surfaces of leaves thoroughly to ensure good coverage. To control fire ants, broadcast the poison bait or apply it to individual ant mounds, according to label instructions. Placing the bait in the shade prolongs its effectiveness.

Commercial Products: Avid, Fire Ant Ender (formerly Affirm)

Baking Soda

Baking soda, or sodium bicarbonate, has fungicidal properties when used as a spray on garden plants. It is an effective protectant as well as an eradicant that will kill some organisms that have already infected your plants.

Protection Offered: Research has shown that a 0.5 percent solution of baking soda will help prevent roses from being damaged by black spot. It may also control other fungal diseases, such as powdery mildew.

How to use: Make a baking soda spray by dissolving 1 teaspoon of baking soda in 1 quart warm water. Add up to 1 teaspoon of liquid dish soap or insecticidal soap to make the solution cling to leaves for a longer period of time. Spray infected plants thoroughly, being sure to cover the undersides of leaves.

Bleach

Sodium and calcium hypochlorite, otherwise known as bleach, is a common disinfectant in the garden and greenhouse.

Protection Offered: Use bleach to disinfect greenhouse tools and benches and as a dip for cuttings taken from plants with disease symptoms. You can also use it to disinfect shovels, pruning shears, and seed-starting equipment to prevent the spread of plant pathogens.

Precautions: In diluted form, bleach is relatively safe; undiluted, it is a toxic and caustic eye and skin irritant. It breaks down rapidly in the soil, but may drive soil pH into the alkaline range. It is not selective, so it kills the beneficial organisms along with the bad.

How to Use: In the greenhouse, dip plant cuttings, before rooting, in a 10 percent bleach solution (1 part bleach to 9 parts water). Prevent damping-off disease in seedling flats by drenching the soil after planting with a 2 percent solution (2½ tablespoons bleach to 1 gallon water). This drench has no residual effect; avoid contaminating flats once they have been treated (wash your hands and tools in a 10 percent bleach solution before touching the soil or plants in the flat). In the garden, disinfect tools such as shovels, trowels, and clippers with a 10 percent bleach solution between uses to avoid spreading diseases from plant to plant. After each treatment and at the end of the season, rinse all tools thoroughly and coat the metal parts with household oil. You can also disinfect plant cages, insect traps, and row cover with a bleach solution before storing them.

Bordeaux Mix

Bordeaux mix combines copper sulfate and hydrated lime into a wettable powder that may be dusted onto plants or mixed with water and applied as a spray.

Protection Offered: Bordeaux mix acts as a fungicide with insecticidal and insect repellent properties. Use it to control common plant diseases like anthracnose, bacterial leaf spots and wilts, black spot, fire blight, peach leaf curl, powdery mildew, and rust.

Precautions: Bordeaux mix can burn plant foliage. One of the safer times to apply it is just before plants leaf out in the spring. Injury is most common at temperatures below 50°F and when humidity is high. Read the product label carefully before applying bordeaux mix.

How to use: You can buy bordeaux mix in powdered form, and apply it as a dust or mix it with water for spraying.

Boric Acid

Boric acid is a crystalline material that is derived from the mineral borax. It is effective as a stomach poison against several kinds of insects, including ants and cockroaches.

Protection Offered: In the garden, boric acid is commonly used as a bait to control ants. Such baits typically combine boric acid dust and a food that attracts ants, such as sugar, jelly, or pet food.

Precautions: Boric acid can be toxic when ingested in high doses. Keep baits out of the reach of children and pets. Wear a dust mask, eye protection, and gloves when handling boric acid dust.

How to Use: Buy commercial baited traps, or make your own bait by mixing a few teaspoons of boric acid or borax with foods attractive to ants. Sugary or high-protein foods are usually most effective. Enclose the bait in a container with small holes to admit ants, and place in problem areas.

Commercial Products: Bait: Drax; Powders: Roach Prufe, Roach Kill

Botanical Pesticides

Farmers and homeowners used botanical pesticides to control both indoor and outdoor pests long before synthetic pesticides were formulated in the mid-1900s. Some of the earliest insecticides were extracted from plants growing in tropical forests and other natural environments. These special plants are now grown commercially to produce botanical insecticides. Many of these products are useful for controlling pests in organic gardens and in commercial production. You can expect to see expanded crop registrations for some botanical insecticides as well as the development of new types of botanicals, as interest in environmentally sound pest control increases.

Some botanical insecticides are made by grinding raw plant materials, such as flowers, roots, stems, or seeds. Others are extracted from plant materials, then refined, purified, and packaged for sale.

Botanical pesticides have broad-spectrum activity, meaning that they kill a wide variety of insects, including beneficials. Still, you will find that certain botanicals work better than others to control certain pests. Although they break down quickly in the environment, most botanicals are toxic to wildlife when first applied. They are also toxic to humans and their pets. Use these insecticides with caution and always follow label directions.

For information on specific botanical pesticides, see the following entries: "Citrus Oils" on page 470, "False Hellebore" on page 472, "Neem" on page 477, "Nicotine" on page 478, "Pyrethrin" on page 480, "Quassia" on page 481, "Rotenone" on page 481, "Ryania" on page 482, "Sabadilla" on page 482, and "Tomato-Leaf Spray" on page 485.

Bug Juice

Long ago, backyard gardeners discovered that some insect pests wouldn't feed on plants if dead members of their species were on those plants. In the 1960s, this observation caught on as a way to control pests, and gardeners began experimenting with solutions of pulverized pests as insecticidal sprays. Soon after, the Environmental Protection Agency warned that the bug juice sprays might contain pathogens that could harm people, and they discouraged gardeners from using them.

The reasons why bug juice works to control pests are unknown, but there are several possible explanations. One is that some of the pests you collect and pulverize are infected with an insect disease. By making the bug juice, you're actually culturing the disease pathogen that will kill other pests when applied to infested plants. Like people, insects suffer from diseases that can spread quickly through the community.

Another explanation is that the odor of the crushed pests sprayed on the plants repels new arrivals. For example, the odor or alarm pheromones of dead beetles may discourage other beetles of the same species. Also, the

The Other Ingredients

As if the words *sabadilla* and *diatomaceous* weren't tongue-twisting enough, you'll find other even less familiar words in the fine print on pesticide labels. Here's an explanation of some common label lingo.

Synergists

If you find the words *piperonyl butoxide* on a label, you've just discovered that your pesticide contains a synergist. Synergists enhance the effects of the pesticide's active ingredients, such as pyrethrin or rotenone. The addition of a synergist reduces the amount of active ingredients needed to do the job, and since active ingredients tend to be expensive, synergists help keep the final product cost low. Some synergists also help to deactivate protective enzymes that insects carry in their bodies; without the synergist, these enzymes often degrade a pesticide before it has a chance to work.

Piperonyl butoxide, also called PBO, is a common, synthetically formulated synergist used with botanical insecticides such as pyrethrin, rotenone, ryania, and citrus oils. PBO is sufficiently toxic that it was recently recognized as a pesticide itself. When used frequently at high concentrations, PBO may affect the human nervous system. Many organic certification programs prohibit the use of pesticides that contain PBO.

Inert Ingredients

On most labels you'll also find the words *inert ingredients,* a catchall term for additives that make pesticides easier to use. Inert ingredients do not enhance a product's toxicity. Rather, they are nonreactive substances that help to keep a product stable in storage and act as "filler" material. If inert ingredients weren't added to pesticides, it would be almost impossible to measure the small amount of active ingredient actually required, and even more difficult to achieve the proper coverage.

Inert ingredients in liquid pesticides include water, alcohol, and other solvents. Pesticide powders and dusts often contain added diatomaceous earth, powdered seeds, or talc for added bulk, or fuller's earth to help prevent caking. Other inert ingredients, known as surfactants, help powders mix easily with water or help sprays adhere to foliage so that pests are more likely to contact it. In the absence of surfactants, many pesticide sprays would simply roll off waxy leaf surfaces. Soap is one of the most common surfactants added to pesticide sprays.

Some pesticides, like sulfur or boric acid, are packaged in their pure form without inert ingredients. Other pesticides combine an active ingredient, such as pyrethrin, with an active diluting agent, such as insecticidal soap, which also acts as a synergist. Most pesticides, however, are a single toxic ingredient mixed with one or several inert ingredients.

To figure out just how much active ingredient you are buying, read the product label for the total percentage of inert ingredients. For example, a typical insecticidal soap in a ready-to-use formula contains about 98 percent inert ingredients. Insecticidal soap in concentrated form contains a lower percentage of inert ingredients since you will be adding your own inert ingredient—water—prior to using the product.

odors may attract the natural enemies of the pests.

Protection Offered: Gardeners report that bug juice will control cabbage loopers and Colorado potato beetle and Mexican bean beetle larvae. Larger pests are easier to collect than small, fast-moving types.

Precautions: There hasn't been much research conducted on bug juice. Since you're dealing with the unknown, wear a filter mask, long-sleeved shirt, rubber gloves, and long pants when mixing and spraying the juice to avoid skin contact. Don't use a kitchen blender to prepare the spray—you could contaminate your food with pathogens that may sicken you as well as the pests. Wash sprayed produce thoroughly before eating it.

How to Use: Collect about ½ cup of the pests, especially ones that appear to be sluggish or sick. Put the insects in an old blender with about 2 cups of water and liquefy. Strain the mix, then dilute ¼ cup of this concentrate with 1-2 cups of water in a sprayer. You can store leftover mix in the freezer, but be sure to label it.

Use the spray to control the same pest species that is in the mixture. Spray the plants thoroughly on both sides of leaves. You can reapply once or twice a week, although some bug juice sprays are reported to be effective for as long as 2 months.

Chitin

Chitin is a protein complex found in shellfish shells. When shells of lobsters, crabs, shrimp, and other shellfish are added to the soil, the chitin stimulates population growth of beneficial soil microorganisms. These microorganisms produce the enzyme chitinase, which destroys nematode eggs and larvae. As chitin breaks down, nitrogen and potassium become available to plants.

Protection Offered: You can use chitin on all ornamentals, fruits, nuts, vegetables, trees, lawns, and row crops to control nematodes.

Precautions: Adding chitin alone will not be effective, because the microorganisms need extra nitrogen to support their increased populations. If you add a chitin source such as shellfish shells, also add dried blood or some other organic nitrogen source to the soil. The only commercially available chitin source, ClandoSan, contains urea formaldehyde (a synthetic nitrogen fertilizer) as its nitrogen source and is prohibited under some organic certification programs.

How to Use: To control nematode pests on lawns and under trees, apply crushed shells at a rate of 5 pounds per 200 square feet, then water them in. Be sure to include a source of nitrogen. In vegetable gardens, till chitin into the soil at the same rate. Or, apply it in bands along the row at one-half the rate. Chitin breaks down slowly and provides nematode control for 1 year with a single application.

Commercial Products: ClandoSan

Citrus Oils

Citrus peels contain oils with insecticidal properties. The botanical insecticide-containing citrus oil has the potential to control many garden and home pests and has an orangy fragrance. Linalool and d-limonene are the active ingredients extracted from citrus waste. Linalool is an insect nerve poison that kills pests on contact; the mode of action of d-limonene remains unknown. Look for these extracts among the list of active ingredients in new pesticide products that claim to have a citrus base.

Precautions: Although generally not harmful to humans, exposure to citrus oils may cause some animals to experience tremors and salivation.

Protection Offered: One manufacturer has combined citrus extracts with insecticidal soap for effective aphid and mite control. Citrus extracts are also toxic to leaf-eating caterpillars, Colorado potato beetles, fire ants, flies, and wasps, although they aren't yet registered for those uses. Expect to see more garden pesticides made from citrus oils as new products are developed and registered.

How to Use: To control aphids and mites in the garden, prepare the spray solution according to label directions. Cover the leaves thoroughly, making sure to coat the undersides where many pests hide. As plants grow and rain washes away the oils, you may need to spray at 1-2 week intervals to maintain the protection.

Commercial Products: Aphid-Mite Attack (combines insecticidal soap and citrus oils)

Compost Tea

A solution made by soaking finished compost in water not only provides important nutrients to plants, it also serves as a natural fungicide. See "Composting for Insect and Disease Control" on page 425 for details on how to make compost tea and use it as a fungicide.

Copper

Copper has been used as a pesticide since the 1700s. Copper sulfate has herbicidal and fungicidal properties: It inactivates critical enzyme systems in fungi, algae, and other plants.

Protection Offered: Copper is a broad-spectrum protectant fungicide used to protect vegetables, ornamentals, fruits, and nuts from plant pathogens. Use it in the garden to control anthracnose, bacterial leaf spot, black rot, blights, downy mildew, peach leaf curl, and Septoria leaf spot.

Precautions: Copper products are toxic to humans and other mammals; they also irritate the eyes and skin, so use adequate protection when you use these products. Copper is highly toxic to fish and aquatic invertebrates. Its toxicity is reduced when mixed with lime (this product is called bordeaux mix), but repeated applications of any copper product will stunt a plant. Although copper does not become concentrated in plant tissues, it persists indefinitely in the soil.

How to Use: You can buy fixed copper as a dust or sprayable solution from organic garden suppliers. Copper sulfate, also called bluestone, is available as a dust, wettable powder, or liquid concentrate. Since copper is a protectant, cover the entire plant surface to prevent invasion by disease organisms. Spray in the early morning in dry, bright weather so that plants have time to dry. If the solution remains on leaves too long, it may penetrate the cuticle and kill the tissue. Copper may be applied up to 1 day before harvest.

Commercial Products: Bonide Liquid Copper, Kocide, Top Cop, Top Cop with Sulfur

Diatomaceous Earth

Diatomaceous earth (DE) is a nonselective, abrasive dust commonly used to kill pests in the home and the garden. It is unique among the insecticidal dusts, since it works by physical rather than chemical action. DE is the fossilized silica shells of algae called diatoms; the microscopic shells are covered with sharp needlelike projections that penetrate an insect's cuticle, allowing vital liquids to leak out. DE also absorbs the waxy coatings on insects' bodies; both actions cause pests to die of dehydration.

Protection Offered: Dust plants and soil with DE to control crawling pests like slugs and snails. On plant foliage, DE will kill soft-bodied pests like aphids, caterpillars, leafhoppers, and thrips; hairy-bodied pests may be somewhat resistant. It is an excellent product for use in stored grain and seeds, and for indoor flea and louse control. Farmers often add DE to animal feed to control internal parasites of livestock.

Precautions: DE is considered nontoxic to mammals, but the same properties that make it lethal to pests also cause it to irritate mucous membranes. Wear a dust mask when applying DE to avoid inhaling the particles. Don't apply the dust where children are likely to encounter it. Don't confuse pool-grade DE with the garden variety; the type used in pools is chemically treated and poses a severe respiratory hazard. DE is nonselective and will kill beneficial insects. Rain will dilute or wash away DE and mix it into the soil, but the DE will retain its insecticidal properties.

How to Use: Purchase natural-grade DE as a dust and apply only in problem areas to minimize harm to beneficials. When applied around the base of susceptible seedlings like cabbage, onions, and other transplants, DE helps control root maggots and other soil-dwelling pests. Dust foliage to control chewing pests, but avoid dusting flowers. For stubborn thrips infestations, dust only the undersides of affected leaves and in a circular band on the soil beneath each plant. Apply DE when plants are wet from dew or after overhead watering, to help keep the dust on the plants. Mix diatomaceous earth with liquid dish soap and water to make a thick slurry; paint on tree trunks to protect them. DE is also available mixed with pyrethrin.

Commercial Products: Perma-Guard

False Hellebore

Several species of false hellebore (*Veratrum* spp.) contain highly poisonous alkaloids with insecticidal properties. Harvested and dried, the powdered roots of these lilylike plants yield an insecticide that poisons chewing insects.

Protection Offered: False hellebore controls garden insects with chewing mouthparts such as beetles, caterpillars, European corn borers, grasshoppers, and sawflies. It has also been used to control flies, imported currantworms, and mosquitoes.

Precautions: Once prepared, false hellebore loses its effectiveness quickly when exposed to air and sunlight. Store it in a cool, dry place to retain its potency. False hellebore was once used to make poison arrows—keep this in mind and use extreme caution when working with hellebore; it is highly toxic if ingested.

How to Use: You can grow your own false hellebore and collect the roots for drying and grinding into powder, but be aware of its toxicity. To make a spray for controlling insect pests on garden plants, mix 1 ounce with 2 gallons of water. Or, mix the powder with flour or hydrated lime to make an insecticidal dust. Since it loses effectiveness quickly, repeated applications may be necessary.

Fungicides

The 2 most commonly used organically acceptable fungicides are sulfur and copper. (Synthetic fungicides also contain sulfur and copper, but they are chemically combined with other elements such as carbon.) Even in their pure and simple form, they are deadly poisons that should be used with discretion. These and other fungicides act as protectants that inhibit the germination and growth of fungal spores; unfortunately, they won't stop the spread of plant disease once infection has occurred, so it is important to apply them prior to periods of wet and humid weather, when disease organisms can spread and grow easily. Follow label directions for frequency of application; spray or dust only when the threat of disease is great.

For specific fungicidal product information, see the following entries: "Baking Soda" on page 467, "Bordeaux Mix" on page 467,

"Compost Tea" on page 471, "Garlic Oil" below, and "Soap Sprays" on page 483.

Garlic Oil

Herbal folklore recommends garlic for its antibiotic, antifungal, and insecticidal properties. When combined with mineral oil and pure soap, it becomes an effective insecticide with fungicidal properties as well.

Protection Offered: Garlic oil kills insects, but not selectively. Use it to control aphids, imported cabbageworms, leafhoppers, larval mosquitoes, squash bugs, and whiteflies. It also works against some fungi and some nematodes. Some gardeners report that it has little effect against Colorado potato beetles, grapeleaf skeletonizers, grasshoppers, red ants, or sowbugs. Adult lady beetles seem unharmed by garlic oil sprays.

Precautions: Since garlic is consumed daily in some cultures with no ill affects, we assume it is a safe product to use in the garden and that excessive precautions are unnecessary. Some foliar injury may occur when garlic sprays include oil and/or soap. Since it is nonselective, garlic spray can kill beneficials as well as pests.

How to Use: Soak 3 ounces of finely minced garlic cloves in 2 teaspoons of mineral oil for at least 24 hours. Add 1 pint of water that has ¼ ounce of liquid dish soap mixed into it. Stir well and strain into a glass jar for storage. Combine 1-2 tablespoons of this concentrate with 1 pint of water to make a spray. Test your mixture on a few leaves to check for injury caused by the oil and soap; damage may not appear for 2-3 days. Spray plants thoroughly to ensure good coverage.

Growth Regulators

Insect growth and reproduction is controlled by hormones that must be present in the right place at the right time and at the right concentration. Most insects shed their external skeletons periodically as they grow and develop. Juvenile hormone (JH) is one of the chemicals that help to regulate this process. However, if insects are exposed to JH at the wrong time, they do not mature. Scientists have used this and similar concepts to develop insecticides that control several important insect pests. Most available products are synthetic versions of naturally occurring insect growth-regulating hormones.

Protection Offered: Methoprene is a growth regulator that prevents the immature blackflies, fire ants, fleas, fungus gnats, mosquitoes, and some midges and cattle flies from maturing. You can expect to see commercial products released containing growth regulators that control codling moth eggs and larvae, Colorado potato beetle larvae, armyworms, and whitefly larvae.

Precautions: Since humans do not have the same chemical processes as insects, insect growth regulators are considered among the safest of pest-control products. They don't irritate the skin or eyes, and since the growth-regulating hormones must be eaten by the pests to be effective, they are not as likely to affect beneficial and nontarget insects.

How to Use: To control greenhouse fungus gnats with methoprene products, follow label instructions and spray thoroughly to get good coverage on all leaf surfaces. The mosquito control Altosid is available as a slow-release liquid and in briquette form floating in wet areas where mosquitoes are likely to breed.

Commercial Products: Altosid (methoprene), Enstar (kinoprene, for whitefly control), Precor (for flea and gnat control)

Herbal Sprays

Many of the aromatic herbs are well-known for their hardiness and for the lack of attention that insect pests pay them. Several scientific studies confirm the repellent effect of herbal extract sprays, long relied upon by organic gardeners.

People and Pet Protectors

While tending your garden, you may find that the pests that most plague you are not on your plants but on you or your pet. In addition to protecting your garden from pests, some of the control methods in this book offer safe alternatives to dousing yourself or Fido with synthetic chemicals.

Once you've exhausted (or been exhausted by) physical controls such as swatting, handpicking, and using barriers—long pants, mosquito netting, and so on—you may find a solution among the biological and chemical controls listed below.

■ ***Bacillus thuringiensis* var. *israelensis* (BTI).** This variety of BT provides safe, effective control of mosquito larvae in water, reducing the population of irritating, disease-carrying adults.

■ **Avermectins.** These naturally occurring antibiotics produced by the bacterium *Streptomyces avermitilis* play an important role in modern agriculture. The group of avermectins known as ivermectins are used to control parasitic worms and stable and horn flies in livestock.

Pet owners use avermectins to control heartworms and other internal parasites in dogs. Avermectins are toxic to collies; consult your veterinarian before treating your dog with avermectins.

■ **Growth regulators.** A variety of products that interfere with pests' ability to molt and mature may become the first line of defense against annoying, hard-to-control pests such as mosquitoes and fleas. Methoprene, sold as Altosid for mosquito control, is available as a slow-release liquid and in briquette form for floating in wet areas where mosquitoes are likely to breed. Precor, another methoprene product, prevents larval fleas and gnats from maturing.

■ **Citrus oils.** A pleasant-smelling alternative to synthetic insect repellents, citrus oils repel and kill fleas on dogs and cats and in their bedding; look for products with the active ingredients linalool and d-limonene. Products containing citrus oils are also available to repel flies, mosquitoes, and other pests of humans. In general, citrus oils are safe for humans, pets, and other mammals. However, exposure to citrus oils may cause some animals to experience tremors and salivation. Test your pets for sensitivity by treating a small area of skin before bathing or dipping them. As with any pesticide, an overapplication can poison your pet.

There are many other home remedies for controlling the pests that plague you and your pets. Commercial products make use of botanicals such as citronella and cedar chips, to name a few; a broad spectrum of herbal concoctions are also effective and worthy of further investigation.

Protection Offered: The essential oils of sage and thyme and the alcohol extracts of such herbs as hyssop, rosemary, sage, thyme, and white clover can be used to reduce the number of pest eggs laid and the amount of feeding damage caused by a broad range of pests. Sprays made from tansy repel imported cabbageworms on cabbage, reducing the number of eggs laid on the plants. Teas made from wormwood or nasturtiums may repel aphids from fruit trees. Some gardeners have used extracts of catnip, chives, feverfew, marigolds, or rue against leaf-feeding pests. Experiment with your own herbal extracts to control leaf-eating pests in the garden.

Precautions: The medicinal properties of many herbs are well-known; make sure you don't give yourself an overdose when applying herbal sprays. Follow the same precautions as for other pesticides: Wear a mask and protective clothing to avoid inhaling the spray or getting it on your skin.

How to Use: You can buy essential herbal oils and dilute them with water to make sprays. Experiment with proportions, starting with a few drops of oil per cup of water.

Make your own herbal extracts by mashing or blending 1-2 cups of fresh leaves with 2-4 cups of water and leaving them to soak overnight. You can also make concentrated herbal teas by pouring the same amount of boiling water over 2-4 cups fresh or 1-2 cups dry leaves and leaving them to steep until cool. Strain the mixture through cheesecloth and dilute the resulting liquid with 2-4 cups of water to make a spray. Add a few drops of liquid soap to help the spray stick to leaves.

You can also make alcohol extracts of pest-repelling herbs. Follow the directions in the Pyrethrin entry on page 480 for this procedure.

Homemade Sprays

In the past, a lack of organically acceptable commercial pesticides forced organic gardeners to use innovation and creativity in solving their pest problems. The result is an array of homemade pesticides that you can prepare at home. Follow the precautions included with the recipes throughout this controls encyclopedia, since even the safest ingredients can be used in the wrong way. Some of the remedies are used for controlling both insects and diseases.

Unlike commercial pesticides, the homemade versions tend to be simple and unrefined, with a lot of room for experimenting. Yet many work just as well as commercial products. You may find that some concentrations work better than others; note your observations for future reference. You may also discover that some of the mixtures have the potential to injure your plants, so be sure to test your concoctions on a few leaves before making a garden-wide application. Any damage should be apparent within a few days.

You'll find recipes for homemade sprays in the entries: "Alcohol" on page 465, "All-Purpose Insect Pest Spray" on page 466, "Ammonia" on page 466, "Baking Soda" on page 467, "Bleach" on page 467, "Bug Juice" on page 468, "Compost Tea" on page 471, "False Hellebore" on page 472, "Garlic Oil" on page 473, "Herbal Sprays" on page 473, "Horsetail Spray" below, "Hot Dusts" on page 476, "Neem" on page 477, "Nicotine" on page 478, "Pyrethrin" on page 480, "Quassia" on page 481, "Soap Sprays" on page 483, "Starch Spray" on page 484, and "Tomato-Leaf Spray" on page 485.

Horsetail Spray

Some organic and biodynamic gardeners use the common weed horsetail *(Equisetum arvense)* as a botanical fungicide to prevent and control plant diseases.

Protection Offered: In general, horsetail is used as a protectant fungicide. Although its use is popular among backyard gardeners, it hasn't been scientifically tested for fungicidal properties.

How to Use: You can buy dried horsetail from organic and biodynamic garden suppliers. In a glass or stainless steel pot, mix ⅛ cup of dried leaves in 1 gallon of unchlorinated water. (If you have chlorinated tap water, either collect rainwater or let the tap water sit uncovered for 2 days so the chlorine will volatilize.) Bring to a boil, then let simmer for at least ½ hour. Cool and strain through cheesecloth. This mixture will keep for 1 month, stored in a glass container. Be sure to label it!

Dilute the horsetail concentrate by adding 5-10 parts unchlorinated water to every 1 part concentrate. Spray infected plants once every week to 2 weeks. If you anticipate a disease outbreak because of prevailing weather conditions, begin a preventive spray program before you spot any symptoms. Experiment to see which plants and diseases it works best on, and keep records for future reference.

Hot Dusts

Many of the culinary seasonings we use, like black pepper, chili pepper, dill, ginger, paprika, and red pepper, owe their hot stuff to capsaicin, a compound shown to repel insects. Researchers have found that as little as 1/25 ounce of capsaicin sprinkled around an onion plant reduced the number of onion maggot eggs laid around the plant by 75 percent, compared to untreated plants.

Protection Offered: Use capsaicin dusts to repel onion maggots and other root maggots in vegetable seedlings. Pepper dusts will help repel ants, thus controlling the aphids they protect.

Precautions: Avoid inhaling the dust or getting it in your eyes. Contact with hot peppers can irritate sensitive skin, so wear rubber gloves when handling.

How to Use: Sprinkle the dust along both sides of each seeded row of cabbages, carrots, or onions. Make the application band at least 6″ wider than the planted row. Or sprinkle around the base of each plant as far out as the largest diameter of the plant top. A light application should be effective, but the more you use, the better it works. Renew after irrigation or a heavy rain.

Lime

Many of our grandparents probably dusted plants with agricultural lime to control a variety of insect pests in the garden.

Protection Offered: This old-fashioned remedy is effective against a wide range of insects. Experiment to see what it controls best in your garden.

Precautions: Wear a protective mask to avoid inhaling the dust.

How to Use: You can apply lime as often as necessary, shaking it through an old strainer, window screen, or mechanical duster onto plants wet with dew. Reapply it after rain or high winds.

Lime-Sulfur

Adding lime to sulfur enhances its fungicidal properties. The lime causes a chemical change that allows the sulfur to penetrate the leaf tissue. This is an important change; since once sulfur penetrates the leaf tissue, it becomes an eradicant that can kill recently germinated spores. Unfortunately, the boost in fungicidal properties also means a greater risk of damaging the plant.

Protection Offered: You can use lime-sulfur to help control diseases like anthracnose, brown rot, leaf spot, mildew, and scab as well as scales on dormant perennials, roses, foundation evergreens, and many fruit crops. It also kills mites and scales.

Precautions: Lime-sulfur is extremely toxic to mammals and can cause severe eye damage and skin irritation. When using it, wear rubber gloves and goggles or other face protection. Lime-sulfur is more caustic than pure sulfur, so it is more likely to damage the host plant if used improperly. If you have sprayed plants with horticultural oil, wait 2-3 weeks before using a sulfur product. Avoid spraying during periods when temperatures will exceed 85°F, to prevent plant damage. Lime-sulfur discolors wood and painted surfaces, so use with caution around structures.

How to Use: Buy lime-sulfur as a liquid concentrate. In early spring, spray it on dormant shrubs, like lilacs and roses, and evergreens, such as junipers. You can control powdery mildew on roses by applying lime-sulfur when the buds break in spring and by repeating the application 1 week later. Spray raspberries infected with anthracnose or blight when the buds first show silver. Spray currants and gooseberries infected with anthracnose at bud break, and repeat 10-15 days later.

Mineral Oil

Mineral oil has insecticidal properties and is sometimes mixed with other materials such as soap or garlic to control garden pests. See "Garlic Oil" on page 473 for a recipe for a homemade insecticidal spray containing garlic and mineral oil.

Protection Offered: Mineral oil works physically, trapping pests on plant surfaces and gumming up their wings and body coatings, causing suffocation. Placed on wilted corn silks, mineral oil prevents corn earworms from crawling down into the husks to feed.

How to Use: To use mineral oil to control corn earworms, apply a drop or 2 to the tip of each ear of corn after the silks have wilted.

Neem

Neem (also known as azadirachtin) is an insecticide extracted from the seeds of the neem tree *(Azadirachta indica)* common in most of Africa and India. It is closely related to the chinaberry tree *(Melia azadarach),* common in the southern and southeastern United States. Extracts of both trees have insecticidal properties. Neem is unique among pesticides since it has so many uses: It acts as a broad-spectrum repellent, growth regulator, and insect poison. It discourages feeding by making plants unpalatable to insects; if they still attack, it inhibits their ability to molt and lay eggs.

Protection Offered: Use neem to kill a wide range of pests, including aphids, gypsy moths, leafminers, loopers, mealybugs, thrips, and whiteflies. It kills the difficult pests, like Colorado potato beetles, corn earworms, cucumber beetles, flea beetles, Mexican bean beetles, and pest mites. However, use of neem is currently restricted to nonfood plants like ornamentals, trees, and shrubs in and around greenhouses, nurseries, and homes. The manufacturer is seeking approval for its use on food crops.

Precautions: Neem is almost nontoxic to mammals and beneficial insects and is biodegradable. It is used in India as an ingredient in toothpaste, soap, cosmetics, pharmaceuticals, and cattle feed. The seeds and extracts of both neem and chinaberry trees, however, are poisonous if consumed. Because neem's chemical structure is so complex, scientists hypothesize that it will take a long time for pests to develop resistance to it.

How to Use: You can purchase a commercial neem product for spraying on ornamental plants. Mix according to label instructions, then apply it to plants twice, at weekly or longer intervals. To enhance effectiveness, spray in early morning or late afternoon.

If you live in the South, you may be able to find neem or chinaberry tree seeds, and extract the insecticide from them yourself. Wrap 1 pound of depulped, cleaned, dried seeds in a cloth bag, crush them, and suspend the bag overnight in a bucket containing 1 pint of water. After 12 hours, remove the bag and squeeze out the liquid. Discard the crushed seeds, wrapping them securely so children and animals cannot eat them. Add ⅛ teaspoon liquid dish soap or commercial insecticidal soap to the liquid extract to make it stick to plants. Dilute 1 part of this concentrated insecticidal solution with 17 parts water to make a spray.

Alternately, you could shred 1 pound of whole neem or chinaberry fruit in an old blender or food processor (one that is no longer used for food preparation). After shredding, add enough water to cover the pulp, and let stand overnight. The next day, strain and discard the pulp and add water to the strained liquid to make 4-5 quarts of spray mixture. Add 1/16 teaspoon liquid dish soap or insecticidal soap to improve sticking. The mix should remain effective for 3-4 days if stored in a dark place. Keep it in a tightly sealed, labeled container.

Commercial Products: Bioneem/Neemisis

Nicotine

Nicotine is a highly poisonous alkaloid, extracted from special tobacco selected for its high nicotine content. When used as an insecticide, concentrations of 0.05 to 0.1 percent will kill most soft-bodied insects (like aphids) on contact. When poisoned, insects convulse and quickly die. Its potency increases when mixed with alkaline or soapy water.

Protection Offered: Nicotine products are poisonous to most plant pests, including aphids, mealybugs, scales, and spider mites. Use it to control pests that spend part or all of their life cycle in the soil, such as root aphids or fungus gnats.

Precautions: Nicotine is highly toxic to mammals when taken internally or absorbed through the skin, so wear protective clothing including gloves, goggles, and a respirator. To be safe, use nicotine products only on young plants and only up to 1 month before harvest, since it remains toxic on leaf surfaces for several weeks. Since nicotine products may contain the pathogen that causes tobacco mosaic virus, avoid using them on eggplant, peppers, tomatoes, and potatoes.

How to Use: You can purchase tobacco dust (0.5 percent nicotine) made from ground stems and leaves to apply directly to plants. Or, purchase liquid nicotine sulfate and mix with water according to label directions. To control soil pests, pour the mixture onto the soil in the area of the stem base and root zone. For leaf pests, spray or dust leaves thoroughly.

You can make your own nicotine tea by soaking tobacco leaves or cigarettes in water to make a spray. Soak 1 cup of dried, crushed tobacco leaves, or an equivalent amount of cigarettes, in 1 gallon of warm water with 1/4 teaspoon liquid dish soap added. Strain the mixture through cheesecloth after 1/2 hour. You can store the tea for several weeks in a tightly closed container; be sure to label it.

Commercial Products: Black Leaf 40

Oil Sprays

Prior to the 1970s, orchardists sprayed their fruit trees each spring with heavy petroleum oils known as dormant oils, which killed insect pests and their overwintering stages before the next season began. It was important to spray before leafing out, since the heavy oil damaged the leaves of plants.

Most of today's horticultural oil sprays are lighter and contain fewer of the impurities that made the heavy dormant oils phytotoxic, so they can be used year-round on a variety of plants. These products are called superior, summer, or supreme oils. They are especially effective at controlling pests because they spread thoroughly over the leaf surface. They work physically to smother and kill pests and their eggs, and may also have some repellent properties.

Superior oils are designated by their purity. If in doubt, read the label: They must contain at least 92 percent unsulfonated residues and no more than 8 percent sulfonated aromatics or impurities. Most of the horticultural oils available today are this new, lighter version, although the product names may have remained the same. Use the new oils as both a dormant and summer spray to control pests on garden and orchard plants.

Protection Offered: Superior oils are unique because they control a broad variety of insect pests while going easy on the beneficial insects. Use superior oils to control aphids, mealybugs, mites, and scales on a variety of fruit, nut, ornamental, and shade trees.

In addition, the new oils may have fungicidal properties (although they aren't yet registered for this use). Some of the oils have shown promise as herbicides. These products have been formulated with a greater concentation of the phytotoxic impurities, to kill weeds by dissolving their cell walls. On contact, the plant sap leaks from the cells, killing the plant.

Precautions: The light superior oils developed for controlling insect pests break down quickly and are less toxic to beneficials and the environment than other insecticides. They are slightly toxic to mammals and humans. They do not harm most garden plants when mixed according to label directions. (An exception is blue spruce, *Picea pungens:* The oil removes the bluish frost and it may take 2-3 years for the normal color to return.) If your plant is not listed on the label, it is a good idea to test the oil on a few leaves before treating the whole plant.

Don't spray water-stressed plants unless you irrigate them thoroughly before spraying. Avoid using the oils on plants weakened by disease, drying winds, or high-nitrogen applications. Don't apply if daytime temperatures are likely to exceed 85°F or night temperatures are expected to fall below freezing. Never apply oils within 1 month before or after applying sprays containing sulfur.

How to Use: Horticultural oils are concentrated and must be mixed with water. The oil and water mix forms an emulsion: tiny droplets of oil suspended evenly in the water. Use a 3 percent solution for a dormant application in early spring before buds appear. Spray a 2 percent solution against insects and mites on plants in full leaf if environmental conditions are right. To make a 2 percent solution, pour ⅓ cup oil into a 1-gallon container, then fill with water to make 1 gallon of solution. For a 3 percent solution, start with ½ cup of oil. Apply successive sprays at least 6 weeks apart.

You can apply a 1 percent oil solution (2½ tablespoons of oil mixed with enough water to make 1 gallon of spray) on plants that cannot tolerate higher concentrations, on rutabagas, and on cucumber and tomato plants in the greenhouse. Spray plants until the leaves are well-coated and some solution starts to drip from the leaves. Apply at weekly intervals as needed. A 3-day period between the last spray and harvest is required for greenhouse crops; leave 21 days between spraying and harvesting rutabagas. Spray early in the morning or in the evening to avoid direct sunlight. Wait 24 hours before using other sprays.

Before spraying oil on citrus crops, check with local Cooperative Extension Service agents for the best timing of sprays for your area and cultivars. On lemons, oils are usually applied in April and May; for other citrus crops, oils are used in late summer or fall. (Navel oranges are very susceptible to damage from oil, so do not spray them after September 1.) Avoid treating citrus trees while mature fruit is present, because fruit may drop or the color and quality may suffer.

Commercial Products: SunSpray Ultra-Fine Oil, Volck Oil Spray

Pheromones

When insects are ready to reproduce, they depend on chemical signals, called sex pheromones, to help find mates. Mature females emit the pheromones, and males of the same species are able to detect them in extremely low concentrations from far away. The males follow the chemical signal in order to find receptive females. You can make or buy pheromone lures to intercept and trap pests before they reach your garden. Some products use pheromones as mating disruption lures. These products work by flooding the air with female sex pheromones, making it difficult for male insects to find the females for mating. Pheromones have been used extensively in commercial horticulture to reduce the amount and frequency of use of conventional pesticides. Refer to "Pheromone Traps" on page 437 for more information about the use of pheromones in pest control.

Protection Offered: Mating disruption lures can reduce the damage caused by pests such as codling moths, grape berry moths, and oriental fruit moths. Lures for more pests may be available in the future.

How to Use: Lures are available as twist ties or patches; apply according to label directions. Make the first application in early spring before the moths emerge; reapply 3 months later to continue protection throughout the summer. Mating disruption lures seem to work best when applied over a large area, such as an orchard. If you want to use them to protect one tree or a few trees in your backyard, apply the lures to other plants or structures (such as fences) throughout your property.

Commercial Products: Isomate ties

Pyrethrin

The pyrethrum daisies *Chrysanthemum (=Tanacetum) cinerariifolium* and *C. coccineum* contain several compounds that kill insect pests on contact. You can purchase this insecticide in several forms: The dried flower heads are called pyre*thrum,* while the extracted active ingredients are called pyre*thrin.* (Pyre*throids* are synthetic versions of the insecticide that are even more toxic to insects.)

Protection Offered: Pyrethrin products are broad-spectrum insect nerve poisons approved for controlling pests on flowers, fruits, and vegetables in the garden and greenhouse. Use them to control many chewing and sucking insects, including most aphids, cabbage loopers, celery leaftiers, codling moths, Colorado potato beetles, leafhoppers, Mexican bean beetles, spider mites, stink bugs, thrips, tomato pinworms, and whiteflies. Pyrethrin products are less effective on diamondback moths, flea beetles, imported cabbageworms, pear psyllas, and tarnished plant bugs.

Precautions: Pyrethrin insecticides are moderately toxic to mammals. They will kill lady beetles but do not appear to be harmful to bees. Be aware that many commercial products contain the synergist piperonyl butoxide to enhance the toxicity of the pyrethrin. (See "The Other Ingredients" on page 469 for more information.)

If you suffer from hay fever, avoid contact with pyrethrum flowers; some people have allergic reactions to them. Pyrethrin and related products tend to work best at the lower end of summertime temperatures and are less effective when temperatures exceed 80°F.

How to Use: You can purchase pyrethrin products alone or combined with fungicides, like copper or sulfur, or with rotenone or other botanical insecticides. Follow label directions for mixing, spraying, or dusting. Two applications may be necessary for complete control.

Or, you can grow your own pyrethrum daisies. Pick the blossoms in full bloom (when the concentration of pyrethrin is at its peak) and hang them in a sheltered, dark and dry place. When the flowers are thoroughly dry, store them in a tightly sealed, labeled container in a dark, cool spot. To use, grind the flowers with a mortar and pestle, then mix the powder with a little liquid dish soap and enough water to make a sprayable solution. You'll have to experiment with proportions, because the concentration of pyrethrin in home-grown flowers is variable. If your spray doesn't seem to kill the pests, try a more concentrated solution the next time.

You can also extract the pyrethrin in alcohol. Soak 1 cup of packed fresh flower heads overnight with ⅛ cup of 70 percent isopropyl alcohol; cover the container. Strain the mixture through cheesecloth, then store the extract in a tightly sealed and labeled container. When you need to use the pyrethrin, add 3 quarts water to the extract to make a spray.

Commercial Products: Entire (soap and pyrethrins, without piperonyl butoxide), Red Arrow (rotenone and pyrethrins), Safer Yard and Garden Insect Killer

Quassia

Bitterwood *(Quassia amara)* is a Latin American tree. The wood and bark contain quassia, an insecticide used for controlling several garden pests.

Protection Offered: Quassia controls aphids, caterpillars, Colorado potato beetle larvae, sawflies, and several species of flies.

Precautions: Quassia was once used as a substitute for hops in making beer, and it has long been used as a medicinal herb, so it appears to be fairly safe to use. It is one of the few botanical insecticides that seems to go easy on the beneficials like lady beetles and honeybees.

How to Use: Purchase quassia bark chips and shavings in natural food stores. To make a spray, crush, grind, or chop ¼ cup of bark chips and add them to 1-2 quarts of boiling water. Allow the mixture to cool, then strain and use the liquid to spray pest insects. Or, steep the bark chips overnight in 1-2 quarts cool water, then strain and spray the liquid. Spray plants thoroughly, including the undersides of leaves. Spray as often as twice a week to control pests once they begin damaging your plants.

Some sources indicate that the extract has less insecticidal power than the wood chips used alone. Try spreading the wood chips on soil at the base of plants for long-lasting control.

Rotenone

Rotenone occurs naturally in more than 65 species of plants; however, most commercial supplies come from Peruvian cubé, Malaysian derris, or Brazilian tembo plants. When fewer pest-control options were available, rotenone was a common item on the organic gardener's shelf since it was considered safer to use than the synthetic insecticides. Its importance has lessened since new research highlighted the deadliness of this natural poison, and since safer organic practices have been discovered. You can also buy products that contain a mixture of rotenone, ryania, and pyrethrin.

Protection Offered: All forms of rotenone are highly toxic to most insects with chewing mouthparts, like beetles. You'll notice less control on aphids, which have sucking mouthparts, and on immature stages of moths like cutworms and cabbageworms. You can use rotenone to control pests on vegetables, berries, fruit trees, nut trees, ornamentals, turf, and shade trees.

Precautions: Rotenone's insecticidal property is broad spectrum, so beneficial insects will be killed on contact along with the pests. It is moderately toxic to people and most animals and very toxic to swine, birds, and fish. Residues will remain on plant parts for at least 1 week after application, so be sure to schedule your harvest well afterward. Some people are highly allergic to rotenone on food crops and suffer from violent reactions even after the week-long post-spray period, sometimes even after the food has been cooked. New evidence suggests that rotenone may cause growth abnormalities in laboratory animals.

When using rotenone in the home or garden, wear protective clothing and a respirator to avoid contact, and carefully follow label instructions. Never use rotenone near waterways or ponds. Purchase only enough rotenone to last the season since old dust is ineffective.

How to Use: You can purchase rotenone as a dust, wettable powder (WP), or emulsifiable concentrate (EC). The wettable powder form, which is mixed into a spray with water, is preferable to the dust formulation because less active ingredient is needed to get good coverage and there is less likelihood of inhaling the dust. Rotenone dust residues may last nearly twice as long as residues from the wettable powder formulation. Mix wettable powders carefully to avoid contact. Don't try to mix dust formulations with water, because they are nearly impossible to dissolve. Spray plants thoroughly for best control. Rotenone is available alone or mixed with organic fungicides, other botanical insecticides, or the synergist piperonyl butoxide (see "The Other Ingredients" on page 469 for more information).

Commercial Products: Deritox

Ryania

Ryanodine is the active, insecticidal ingredient of a tropical shrub *(Ryania speciosa).* You can purchase ryania through many garden supply catalogs. You can also buy products that contain a mixture of ryania, pyrethrin, and rotenone. Ryania dust can be stored at least 3 years, and some gardeners report good results with powder stored 5 years. Keep the powder in a cool, dry place.

Protection Offered: Ryania is a broad-spectrum insecticide, but is most effective against lepidopterous larvae (the immature, wormlike stages of moths and butterflies). Use it to control citrus thrips; corn earworms and European corn borers on sweet corn; and codling moth on apples, walnuts, and pears. Ryania also controls many chewing and sucking insects, such as aphids, Colorado potato beetles, Japanese beetles, Mexican bean beetles, and squash bugs, although it is not labeled for use against these pests.

Precautions: Ryania is more toxic to pests than it is to beneficial insects. It is even more toxic to mammals and water life; wear protective clothing when applying ryania and do not use it near waterways and ponds. Ryania is more stable than rotenone once applied to plants; schedule your applications several weeks before harvest.

How to Use: To control codling moths, apply sprays 10-14 days apart starting when petals begin to fall. For corn borers and corn earworms, use dust and apply it to the ears at 5-day intervals. For most other pests, 2 applications 10 days apart will provide control. Follow label instructions for mixing and applying; cover foliage and stems thoroughly to reach hiding pests.

To control thrips on citrus, make a baited poison spray to attract and kill them. Mix ¼ pound Ryan 50 with ¼-½ pound sugar and 2½ gallons water. Ryania is most effective against citrus thrips in hot, sunny weather. (To control citrus thrips in cool, cloudy, wet weather, sabadilla is a better choice.)

Commercial Products: Ryan 50

Sabadilla

Like nicotine and ryania, sabadilla owes its insecticidal powers to several poisonous alkaloids. It is made from the seeds of *Schoenocaulon officinale,* a lilylike plant found in Venezuela.

Protection Offered: Use sabadilla to control aphids, cabbageworms, diamondback moths, flea beetles, grasshoppers, green stink bugs, harlequin bugs, leafhoppers, loopers, squash bugs, striped cucumber beetles, tarnished plant bugs, and thrips. When other botanical insecticides fail, sabadilla often does the job.

Precautions: Sabadilla is moderately toxic to mammals and causes violent allergic reactions in susceptible individuals. It is toxic to honeybees, but does not seem to harm predators and parasites such as lady beetles, predatory mites, and armored scale parasites. Sabadilla becomes stronger as it ages when kept in dry, dark storage. It breaks down quickly with sunlight.

How to Use: You can apply the dust to plants directly or mix it with water to make a solution for spraying on plant foliage. Sabadilla is most effective when applied at weekly intervals to plants wet with dew or moist from showers. Prepare the spray according to label directions. Screen the solution through a fine mesh strainer or nylon stocking before adding it to your sprayer to prevent blockage from seed hulls. Shake the liquid frequently as you spray to keep the powder from settling out of the solution. To control citrus thrips in areas with cool, cloudy, wet weather, make a baited spray by mixing ¼ pound Veratran D with ¼-½ pound of sugar and 2½ gallons of water. (To control thrips in hot, sunny weather, ryania is a better choice.)

Commercial Products: Necessary Organics Sabadilla Pest Control, Veratran D

Salt

Sodium chloride rock salt (NaCl) has been shown to help improve asparagus production while helping plants to resist disease.

Protection Offered: Sodium chloride rock salt helps asparagus resist crown and root rot diseases caused by *Fusarium* fungi.

Precautions: Iodized table salt (NaCl) and rock salt made from calcium chloride (CaCl) are not recommended. Older plants are more salt-tolerant than young ones, so wait a year after planting a new bed before applying salt.

How to Use: Add 2 pounds of sodium chloride rock salt (also sold as pickling salt) per 100 square feet of asparagus bed. Apply the salt in early spring, or later in summer (early July).

Soap Sprays

For many years, organic gardeners used solutions of water and natural soaps for controlling garden pests. This old home remedy involved simmering a pot of water with chunks of Fels Naptha soap, straining it, and coating garden plants with the suds to control aphids and thrips. Since then, scientists have discovered that the salts of fatty acids found in many soaps act as selective insecticides. Insecticidal soaps control insect pests by penetrating their cuticles, which causes their cell membranes to collapse and leak, resulting in dehydration. While some insects can overcome the effects of a soap spray, others are immediately affected and die.

Protection Offered: Use insecticidal soaps to control soft-bodied insects like aphids, mealybugs, and whiteflies. It also works on chiggers, earwigs, fleas, mites, scales, thrips, and ticks. The soap is less effective on chewing insects like caterpillars and beetles.

You can also purchase a fungicidal version to control powdery mildew, black spot, brown canker, leaf spot, and rust on ornamental and food plants. Still another soap product contains high levels of the chemicals that make soap poisonous to plants; it is available as an organic herbicide.

Precautions: All soaps have phytotoxic properties, so test the kind you plan to spray on a few leaves before treating whole plants. Plants with thin cuticles, like beans, Chinese cabbage, cucumbers, ferns, gardenias, Japanese maples, nasturtiums, and young peas, are easily damaged by soap sprays. Tomatoes and potatoes are less susceptible to damage, and thick-leaved cabbages seem virtually impervious, although heavy soap use may reduce yields. A good rule is to use no more than 3 successive sprays on any plant.

For all of their pesticidal properties, insecticidal soaps are nontoxic to humans and to other test animals, and they biodegrade rapidly in the soil. They will, however, kill beneficial insects along with the pests, so limit their use to problem areas. Herbicidal soaps will affect any plant that they are sprayed on, so use with care around valuable ornamentals.

How to Use: You can use household soaps such as Ivory Snow, Ivory Liquid, or Shaklee's Basic H to make your own insecticidal soap solution. Since manufacturers often make small changes in their soaps' contents, the effectiveness of homemade solutions may vary more than that of commercial insecticidal soaps.

Be aware that many products commonly known as soap also contain impurities, such as perfumes and whiteners, that can damage plants. For pest control, it is important to use pure soap; avoid detergents or soaps with additives.

To prepare a homemade soap solution, mix from 1 teaspoon to several tablespoons of soap per gallon of water. Start at the lower concentration and adjust the strength to maximize pest control while avoiding plant damage.

To control aphids, spray when the first aphid colonies develop in early spring and again when winged females arrive. Use yellow sticky traps to monitor pest populations; see "Sticky Boards" on page 438 for instructions on making yellow sticky traps. If aphids are a continuous problem, make several applications about 2 weeks apart. Spray during cool, humid, or foggy weather to improve control. To control plant bugs, spray as soon as nymphs begin feeding. To control mites, use a high-pressure spray, like that of a hose-end sprayer, that washes away many pests and kills those that remain. Repeat 7-10 days later to kill newly hatched mites.

You can mix soaps with other insecticides like BTK, horticultural oil, pyrethrin, and rotenone to boost their toxicity. You can also mix homemade soap or liquid dish soap with cooking oil to boost the effectiveness of both insecticidal ingredients.

Mix 1 tablespoon of liquid dish soap and 1 cup of oil (peanut, safflower, corn, soybean, or sunflower). When you're ready to spray, mix 1 to 2½ teaspoons of the prepared base to 1 cup of water. The mixture has been used successfully to control a variety of pests on carrots, celery, cucumber, eggplant, lettuce, and peppers. However, some plants may be injured by the oil, so test the spray on a few leaves and wait several days before spraying all of your plants.

Herbicidal soaps are most effective on young weeds with tender leaf tissue. They are not very effective against mature perennial weeds, especially those with tap roots. Spray on weeds according to label directions.

Commercial Products: Aphid-Mite Attack, Safer Garden Fungicide (soap and sulfur), Safer Insecticidal Soap, Savona, SharpShooter (herbicidal)

Starch Spray

Starches like ordinary baking flour and potato starch dextrin work to control insect pests by gumming up the leaf surfaces, trapping and holding the critters until they die. Flour is an old-fashioned pest control, but dextrin is the object of recent attention by researchers looking for new and safe insecticides. Dextrin is actually a sticky sugar extracted from potato starch.

Protection Offered: You can use potato starch dextrin to control aphids, spider mites, thrips, and whiteflies. It also controls powdery mildew on cucumbers. Other uses will likely be recommended once a commercial product reaches the market. Flour is a good control for larger pests like imported cabbageworms and loopers on cabbage family crops.

How to Use: Commercial starch insecticides aren't yet available in the United States, but you can make your own potato starch spray by mixing 2-4 tablespoons of potato flour in 1 quart water, and adding 2 or 3 drops of liquid dish soap. Shake the mixture, and spray to cover the leaves thoroughly. You can also apply the flour as a dust. If a residue remains on ornamentals, simply wash it away with water a few days after the application.

Sulfur

Sulfur is one of the oldest pesticides known. It has been used for centuries to control both plant pathogens and pests like insects and mites on contact. Sulfur is mined from natural deposits as a yellow solid that is almost insoluble in water. Wettable sulfur is finely ground sulfur mixed with a wetting agent to help it go into solution. Sulfur is also formulated as a finely ground dust with 1-5 percent clay or talc added to enhance dusting qualities.

Protection Offered: Sulfur can be purchased alone or mixed with other insecticides and inorganic fungicides to control a wide variety of garden pests. Use sulfur as an insecticide to control insects and mites on fruit trees and citrus. You can also use sulfur as a protectant fungicide. It is effective against apple scab, brown rot of stone fruits, powdery mildews, rose black spot, rusts, and other plant diseases on many crops, including grapes, potatoes, strawberries, and tomatoes.

Precautions: Sulfur is moderately toxic to humans and other mammals. It can irritate or damage the lungs, skin, or eyes if not used carefully. Wear protective clothing when applying sulfur. Although it is more toxic to mites than to insects, sulfur is nonspecific and can kill beneficial insects, soil microorganisms, and fish. Do not apply it within 1 month of using an oil spray; use a copper fungicide instead. Sulfur spray may cause plant injury if applied when temperatures exceed 80°F. It is corrosive to metal, so use a sprayer with plastic parts; rinse equipment thoroughly after use.

How to Use: You can buy sulfur as a dry powder for dusting on plants or as a wettable formulation (also known as flowable sulfur) that mixes readily with water. Some sulfur products are colloidal, which means they have very fine particles that disperse over the leaf surfaces to provide excellent protection. Noncolloidal sulfur is made of larger particles that will leave unprotected areas on the leaf surfaces. Mix sulfur solutions according to the instructions on the label, agitating it frequently since it tends to settle out of solution.

Commercial Products: Bonide Liquid Sulfur, Safer Garden Fungicide, That Flowable Sulfur

Tomato-Leaf Spray

Tomatoes and potatoes have significant amounts of poisonous compounds called alkaloids in their leaves. Instead of acting as an insecticide, however, sprays made from tomato leaves appear to reduce pest damage by attracting natural pest enemies searching for their prey.

Protection Offered: Use tomato-leaf sprays to protect plants from aphids and to reduce corn earworm damage.

Precautions: Since alkaloids tend to be toxic to mammals, use care in handling this spray and avoid getting it on your skin. Some individuals are extremely allergic to plants in the nightshade family. Don't use tomato-leaf spray on other nightshade family crops because of the risk of spreading mosaic virus.

How to Use: Finely chop 1-2 cups of tomato leaves, then soak them overnight in 2 cups of water. In the morning, strain the slurry through cheesecloth, add about 2 more cups of water to the strained liquid, and spray, covering leaves thoroughly.

USDA PLANT HARDINESS ZONE MAP

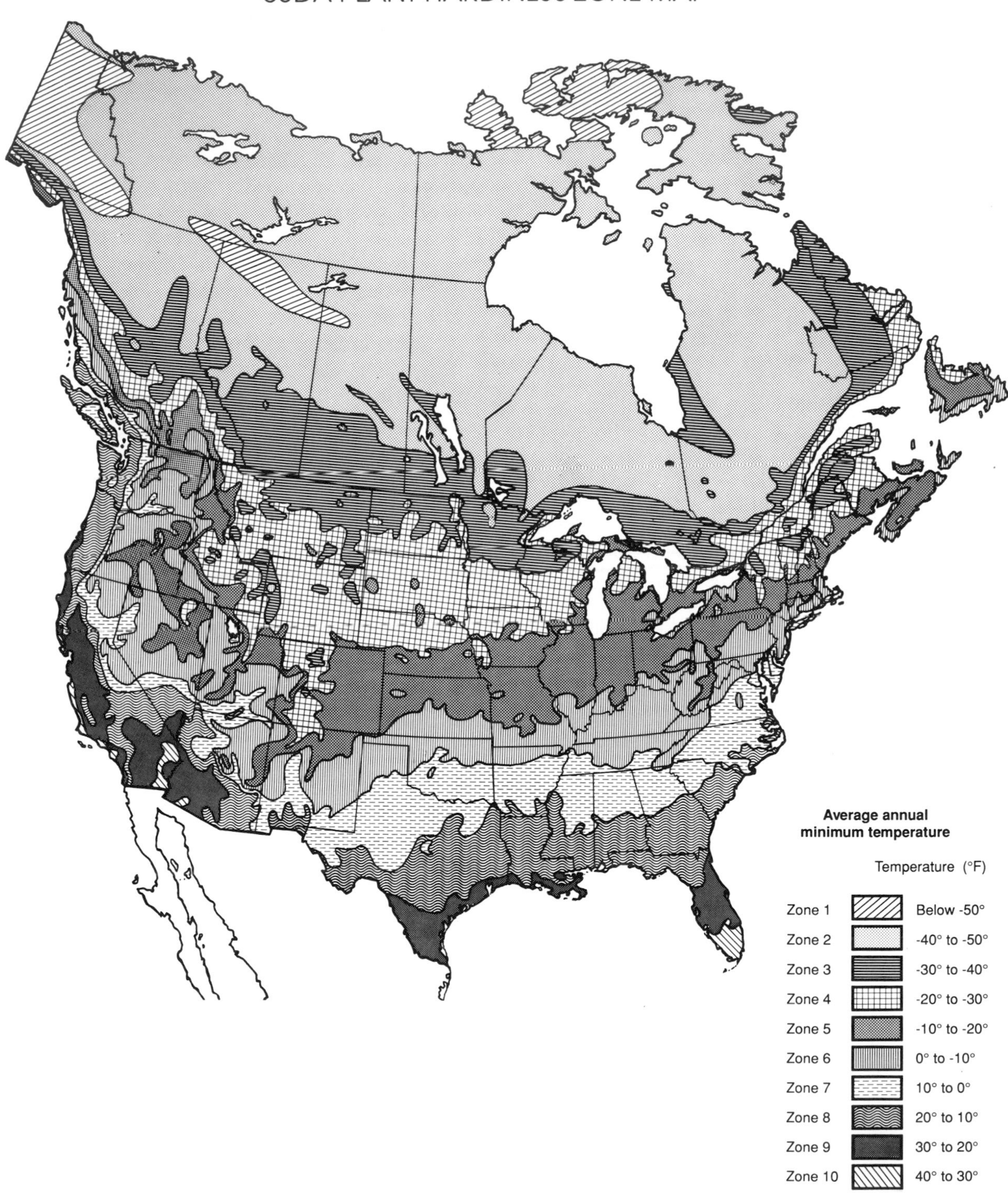

COMMON AND BOTANICAL NAMES OF PLANTS

Adam's-needle, *Yucca*
Ageratum, *Ageratum*
Ajuga, *Ajuga*
Albizia, *Albizia*
Allium, *Allium*
Alumroot, *Heuchera*
Amaranth, *Amaranthus*
Anemone, *Anemone*
Arborvitae, *Thuja*
Artemisia, *Artemisia*
Ash, *Fraxinus*
Aster, *Aster*
Astilbe, *Astilbe*
Autumn crocus, *Colchicum*
Azalea, *Rhododendron*
Baby's-breath, *Gypsophila*
Bachelor's-button, *Centaurea*
Balsam, *Impatiens*
Barberry, *Berberis*
Bee balm, *Monarda*
Beech, *Fagus*
Begonia, *Begonia*
Bellflower, *Campanula*
Bergenia, *Bergenia*
Birch, *Betula*
Bittersweet, *Celastrus*
Black-eyed Susan, *Rudbeckia*
Blazing-star, *Liatris*
Bleeding heart, *Dicentra*
Bluebells, *Mertensia*
Boston ivy, *Parthenocissus*
Boxwood, *Buxus*
Brooklime, *Veronica*
Browallia, *Browallia*
Bugbane, *Cimicifuga*
Bugleweed, *Ajuga*
Butterfly bush, *Buddleia*
Calendula, *Calendula*
Camellia, *Camellia*
Candytuft, *Iberis*
Canna, *Canna*
Carnation, *Dianthus*
Cedar, *Cedrus*
Celosia, *Celosia*
China aster, *Callistephus*
Chrysanthemum, *Chrysanthemum*
Clematis, *Clematis*
Cockscomb, *Celosia*
Coleus, *Coleus*
Columbine, *Aquilegia*
Coneflower, *Echinacea, Rudbeckia*
Coral bells, *Heuchera*
Coreopsis, *Coreopsis*
Cornflower, *Centaurea*
Cosmos, *Cosmos*
Cotoneaster, *Cotoneaster*
Crab apple, *Malus*
Cranesbill, *Geranium*
Crape myrtle, *Lagerstroemia*
Crocus, *Crocus*
Daffodil, *Narcissus*
Dahlia, *Dahlia*
Daphne, *Daphne*
Daylily, *Hemerocallis*
Delphinium, *Delphinium*
Deutzia, *Deutzia*
Dogwood, *Cornus*
Dusty miller, *Senecio*
Elm, *Ulmus*
Epimedium, *Epimedium*
Euonymus, *Euonymus*
False spirea, *Astilbe*
Fir, *Abies*
Firethorn, *Pyracantha*
Flag, *Iris*
Flossflower, *Ageratum*
Flowering quince, *Chaenomeles*
Flowering tobacco, *Nicotiana*
Forsythia, *Forsythia*
Foxglove, *Digitalis*
Fritillary, *Fritillaria*
Garden mum, *Chrysanthemum*
Garden portulaca, *Portulaca*
Gayfeather, *Liatris*
Geranium, *Pelargonium*
Glad, *Gladiolus*
Gladiolus, *Gladiolus*
Gypsophila, *Gypsophila*
Harebell, *Campanula*
Hawthorn, *Crataegus*
Hellebore, *Helleborus*
Hemlock, *Tsuga*
Hibiscus, *Hibiscus*
Holly, *Ilex*
Holly grape, *Mahonia*
Hollyhock, *Alcea*
Honey locust, *Gleditsia*
Honeysuckle, *Lonicera*
Hornbeam, *Carpinus*
Horse chestnut, *Aesculus*
Hosta, *Hosta*
Hyacinth, *Hyacinthus*
Hydrangea, *Hydrangea*
Impatiens, *Impatiens*
Iris, *Iris*
Ivy, *Hedera*
Juniper, *Juniperus*
Lagerstroemia, *Lagerstroemia*
Lantana, *Lantana*
Larch, *Larix*
Larkspur, *Delphinium*
Lilac, *Syringa*
Lily, *Lilium*
Lily-of-the-valley, *Convallaria*
Linden, *Tilia*
Lobelia, *Lobelia*

Lupine, *Lupinus*
Magnolia, *Magnolia*
Mahonia, *Mahonia*
Maple, *Acer*
Marigold, *Tagetes*
Meadow rue, *Thalictrum*
Michaelmas daisy, *Aster*
Mimosa, *Albizia*
Mock orange, *Philadelphus*
Morning glory, *Ipomoea*
Moss rose, *Portulaca*
Mountain ash, *Sorbus*
Mountain laurel, *Kalmia*
Myrtle, *Vinca*
Narcissus, *Narcissus*
Nasturtium, *Tropaeolum*
Oak, *Quercus*
Pachysandra, *Pachysandra*
Pansy, *Viola*
Pelargonium, *Pelargonium*
Peony, *Paeonia*
Periwinkle, *Vinca*
Petunia, *Petunia*
Phlox, *Phlox*
Pieris, *Pieris*
Pine, *Pinus*
Pink, *Dianthus*
Planetree, *Platanus*
Plantain lily, *Hosta*
Poplar, *Populus*
Poppy, *Papaver*
Pot marigold, *Calendula*
Primrose, *Primula*
Privet, *Ligustrum*
Purple coneflower, *Echinacea*
Pyracantha, *Pyracantha*
Red cedar, *Juniperus*
Rhododendron, *Rhododendron*
Rock cress, *Arabis*
Rose, *Rosa*
Rose-of-Sharon, *Hibiscus*
Salvia, *Salvia*
Sapphire flower, *Browallia*
Sedum, *Sedum*
Shrub verbena, *Lantana*
Silk tree, *Albizia*
Snakeroot, *Cimicifuga*
Snapdragon, *Antirrhinum*
Speedwell, *Veronica*
Spindle tree, *Euonymus*
Spirea, *Spiraea*
Spruce, *Picea*
Spurge, *Euphorbia, Pachysandra*
Stonecrop, *Sedum*
Strawflower, *Helichrysum*
Sunflower, *Helianthus*
Sweet alyssum, *Lobularia*
Sweet gum, *Liquidambar*
Sweet pea, *Lathyrus*
Sycamore, *Platanus*
Thunbergia, *Thunbergia*
Tickseed, *Coreopsis*
Tulip tree, *Liriodendron*
Tulip, *Tulipa*
Verbena, *Verbena*
Vervain, *Verbena*
Viburnum, *Viburnum*
Vinca, *Vinca*
Virginia creeper, *Parthenocissus*
Weigela, *Weigela*
Willow, *Salix*
Windflower, *Anemone*
Wisteria, *Wisteria*
Witch hazel, *Hamamelis*
Woodbine, *Parthenocissus*
Wormwood, *Artemisia*
Yarrow, *Achillea*
Yellow poplar, *Liriodendron*
Yew, *Taxus*
Yucca, *Yucca*
Zinnia, *Zinnia*
Zonal geranium, *Pelargonium*

Seeds and Plants for Disease- and Insect-Resistant Cultivars

The following companies sell seeds and plants of many disease- and insect-resistant fruit and vegetable cultivars. Many of these companies also offer flower seeds as well as trees, shrubs, and perennials.

Adams County Nursery, Inc.
P.O. Box 108
Aspers, PA 17304

Ames' Orchard and Nursery
Rte. 5, Box 194
Fayetteville, AR 72701

Bear Creek Nursery
P.O. Box 411
Northport, WA 99157

W. Atlee Burpee & Co.
300 Park Ave.
Warminster, PA 18974

C & O Nursery
P.O. Box 116
Wenatchee, WA 98807

Country Heritage Nursery
P.O. Box 536
Hartford, MI 49057

Cumberland Valley Nurseries, Inc.
P.O. Box 471
McMinnville, TN 37110

Farmer Seed and Nursery Co.
818 N.W. 4th St.
Faribault, MN 55021

Henry Field's Seed & Nursery Co.
P.O. Box 700
Shenandoah, IA 51602

Gurney's Seed & Nursery Co.
110 Capital St.
Yankton, SD 57079

Harris Seeds
P.O. Box 22960
Rochester, NY 14692

H. G. Hastings Co.
P.O. Box 115535
Atlanta, GA 30310

HollyDale Nursery
P.O. Box 68
Pelham, TN 37366

Johnny's Selected Seeds
2580 Foss Hill Rd.
Albion, ME 04910

J. W. Jung Seed Co.
335 S. High St.
Randolph, WI 53957

Kelly Nurseries
1800 Hamilton Dr.
Bloomington, IL 61701

Orol Ledden & Sons
P.O. Box 7
Sewell, NJ 08080

Liberty Seed Co.
P.O. Box 806
New Philadelphia, OH 44663

Mellinger's, Inc.
2310 W. South Range Rd.
North Lima, OH 44452

J. E. Miller Nurseries, Inc.
5060 W. Lake Rd.
Canandaigua, NY 14424

New York State Fruit Testing Cooperative Association, Inc.
P.O. Box 462
Geneva, NY 14456

Nichols Garden Nursery
1190 N. Pacific Hwy.
Albany, OR 97321

North Star Gardens
2124 University Ave.
St. Paul, MN 55114

Northwoods Nursery
28696 S. Cramer Rd.
Molalla, OR 97038

Park Seed Co.
P.O. Box 31
Greenwood, SC 29647

Pinetree Garden Seeds
Rte. 100
New Gloucester, ME 04260

Raintree Nursery
391 Butts Rd.
Morton, WA 98356

Seeds Blüm
Idaho City Stage
Boise, ID 83706

Shepherd's Garden Seeds
6116 Hwy. 9
Felton, CA 95018

Southern Exposure Seed Exchange
P.O. Box 158
North Garden, VA 22959

Southmeadow Fruit Gardens
15310 Red Arrow Hwy.
Lakeside, MI 49116

Stark Bro's Nurseries & Orchards Co.
Hwy. 54
Louisiana, MO 63353

Stokes Seeds, Inc.
Box 548
Fredonia, NY 14063

Territorial Seed Co.
P.O. Box 157
Cottage Grove, OR 97424

Thompson & Morgan, Inc.
P.O. Box 1308
Jackson, NJ 08527

General Gardening Equipment and Supplies

Many organic gardening products are not readily available at local garden centers. Fortunately, there are many fine mail-order suppliers. The following companies offer a wide range of merchandise, including botanical pesticides, biological controls, composting equipment, copper and sulfur fungicides, floating row covers, insect traps, organic soil amendments and fertilizers, soil test kits, sprayers, tillers, tools and accessories, and watering equipment.

Agri-Diagnostics Association
2611 Branch Pike
Cinnaminson, NJ 08077

Bio-Dynamic Preparations
P.O. Box 133
Woolwine, VA 24185

The Clapper Co.
1121 Washington St.
West Newton, MA 02165

Gardener's Supply Co.
128 Intervale Rd.
Burlington, VT 05401

Gardens Alive
5100 Schenley Pl. NE
Lawrenceburg, IN 47035

Green Earth Organics
12310 Hwy. 99 S, #119
Everett, WA 98204

Harmony Farm Supply
P.O. Box 460
Graton, CA 95444

The Kinsman Co., Inc.
River Rd.
Point Pleasant, PA 18950

A. M. Leonard, Inc.
P.O. Box 816
Piqua, OH 45356

Mantis Manufacturing Corp.
1458 County Line Rd.
Huntingdon Valley, PA 19006

The Natural Gardening Co.
217 San Anselmo Ave.
San Anselmo, CA 94960

Necessary Trading Co.
703 Salem Ave.
New Castle, VA 24127

North Country Organics
R.R. 1, Box 2232
Bradford, VT 05033

Ohio Earth Food, Inc.
5488 Swamp St. NE
Hartville, OH 44632

Peaceful Valley Farm Supply Co.
P.O. Box 2209
Grass Valley, CA 95945

Smith & Hawken
25 Corte Madera
Mill Valley, CA 94941

The Urban Farmer Store
2833 Vicente St.
San Francisco, CA 94116

Insect Predators and Parasites and Other Biological Controls

These companies offer microbial disease products, pheromone traps, and insect predators and parasites. Insectaries specialize in rearing pest predators and parasites. Some rear several species at their facilities, while others raise only one or two and buy the other species they offer from other insectaries. When you place an order, ask whether the insectary raises the predator or parasite you're interested in on site. If not, try to find a direct source from which to order. The less shipping involved, the better for the health and viability of the organisms. If you live in the United States and plan to order insects or any other organisms from a Canadian firm, you must apply for a permit from the U.S. Department of Agriculture. Write to U. S. Department of Agriculture, Plant Protection and Quarantine/APHIS, Federal Building, 6505 Belcrest Road, Hyattsville, MD 20782; ask for application form #526. There is no fee, but it may take up to a month to process the necessary paperwork before the permit is issued. The permit allows you to import the organisms you specify in your application for up to one year from the date of issuance.

Applied Bionomics
(for U.S. customers)
11074 W. Saanich Rd.
Sidney, B.C.
Canada V8L 3X9

OR

Westgro Sales, Inc. and Agrico Sales, Ltd.
(for Canadian customers)
7333 Progress Way
Delta, B.C.
Canada, V4G 1E7

Beneficial Insectary
14751 Oak Run Rd.
Oak Run, CA 96069

Biofac
P.O. Box 87
Mathis, TX 78368

BoBiotrol
54 S. Bear Creek Dr.
Merced, CA 95340

Buena Biosystems
P.O. Box 4008
Ventura, CA 93007

W. Atlee Burpee & Co.
300 Park Ave.
Warminster, PA 18974

Henry Field's Seed & Nursery Co.
P.O. Box 700
Shenandoah, IA 51602

Foothill Ag. Research
510½ W. Chase Dr.
Corona, CA 91720

Gardener's Supply Co.
128 Intervale Rd.
Burlington, VT 05401

Gardens Alive
5100 Schenley Pl. NE
Lawrenceburg, IN 47035

Growing Naturally
P.O. Box 54
149 Pine Ln.
Pineville, PA 18946

Harmony Farm Supply
P.O. Box 460
Graton, CA 95444

Hydro-Gardens, Inc.
P.O. Box 9707
Colorado Springs, CO 80932

Mellinger's, Inc.
2310 W. South Range Rd.
North Lima, OH 44452

The Natural Gardening Co.
217 San Anselmo Ave.
San Anselmo, CA 94960

Nature's Control
P.O. Box 35
Medford, OR 97501

Necessary Trading Co.
703 Salem Ave.
New Castle, VA 24127

Orcon Organic Controls, Inc.
5132 Venice Blvd.
Los Angeles, CA 90019

Peaceful Valley Farm Supply Co.
P.O. Box 2209
Grass Valley, CA 95945

Richters
Goodwood, ONT
Canada, L0C 1A0

Rincon-Vitova Insectaries, Inc.
P.O. Box 95
Oak View, CA 93022

Ringer Corp.
9959 Valley View Rd.
Eden Prarie, MN 55344

RECOMMENDED READING

General Reference Books

Many Cooperative Extension Services offer excellent photo identification guides for diseases and pests of fruits and vegetables. Often these are published by the Cooperative Extension Service of a state land grant university and are available to the general public through a local Cooperative Extension office. To find out which Extension publications are available for food crops in your area, call or visit your local Cooperative Extension office or contact an Extension specialist at your state land grant university.

An asterisk (*) next to a title indicates that while the book offers valuable information, such as good photos and descriptions of various insects and plant diseases, it may also describe pest and disease remedies and gardening methods that are not organic and thus not endorsed by Rodale Press.

Bradley, Fern Marshall, ed. *Rodale's Chemical-Free Yard and Garden.* Emmaus, Pa.: Rodale Press, 1991.

Bradley, Fern Marshall, and Barbara W. Ellis, eds. *Rodale's All-New Encyclopedia of Organic Gardening.* Emmaus, Pa.: Rodale Press, 1992.

Campbell, Stu. *Let It Rot: The Gardener's Guide to Composting.* Charlotte, Vt.: Garden Way Publishing, 1975.

Carr, Anna. *Good Neighbors: Companion Planting for Gardeners.* Emmaus, Pa.: Rodale Press, 1985.

Coleman, Eliot. *The New Organic Grower: A Master's Manual of Tools and Techniques for the Home and Market Gardener.* Chelsea, Vt.: Chelsea Green Publishing Co., 1989.

Cox, Jeff, and the Editors of Rodale's *Organic Gardening* Magazine. *How to Grow Vegetables Organically.* Emmaus, Pa.: Rodale Press, 1988.

*Damrosch, Barbara. *The Garden Primer.* New York: Workman Publishing, 1988.

*Darr, Sheila, Helga Olkowski, and William Olkowski. *Common-Sense Pest Control.* Newtown, Ct.: Taunton Press, 1991.

Ellis, Barbara W., ed. *Rodale's Illustrated Encyclopedia of Gardening and Landscaping Techniques.* Emmaus, Pa.: Rodale Press, 1990.

*Flint, Mary Louise. *Pests of the Garden and Small Farm: A Grower's Guide to Using Less Pesticide.* Oakland, Calif.: ANR Publications of the University of California, 1990. (Available from Publications, Division of Agriculture and Natural Resources, University of California, 6701 San Pablo Ave., Oakland, CA 94608.)

Gershuny, Grace, and Joseph Smillie. *The Soul of Soil: A Guide to Ecological Soil Management.* 2nd ed. St. Johnsbury, Vt.: Gaia Services, 1986. (Available from Gaia Services, R.F.D. 3, Box 84, St. Johnsbury, VT 05819.)

Hall-Beyer, Bart, and Jean Richard. *Ecological Fruit Production in the North.* Trois-Rivières, Quebec: Jean Richard, 1983. (Available from Bart Hall-Beyer, 163 McNamee, Scotstown, Quebec, Canada J0B 3B0.)

Halpin, Anne Moyer, and the Editors of Rodale Press. *Foolproof Planting: How to Successfully Start and Propagate More Than 250 Vegetables, Flowers, Trees, and Shrubs.* Emmaus, Pa.: Rodale Press, 1990.

Hamilton, Geoff. *The Organic Garden Book.* New York: Crown Publishers, 1987.

Maltas, Michael. "Orchard Pest Management and Spray Schedule." (Available from Northwoods Nursery, 28696 S. Cramer Rd., Molalla, OR 97038.)

Martin, Deborah L., and Grace Gershuny, eds. *The Rodale Book of Composting.* rev. ed. Emmaus, Pa.: Rodale Press, 1992.

*Page, Stephen, and Joseph Smillie. *The Orchard Almanac.* 2nd ed. Rockport, Maine: Spraysaver Publications, 1988. (Available from Spraysaver Publications, P.O. Box 392, Rockport, ME 04856.)

Schultz, Warren. *The Chemical-Free Lawn: The Newest Varieties and Techniques to Grow Lush, Hardy Grass.* Emmaus, Pa.: Rodale Press, 1989.

Shigo, Alex L. *Tree Pruning: A Worldwide Photo Guide.* Durham, N.H.: Shigo & Trees Assoc., 1989. (Available from Shigo & Trees Assoc., 4 Denbow Rd., Durham, NH 03824.)

*Smith, Michael D., ed. *The Ortho Problem Solver.* 2nd ed. San Francisco: Chevron Chemical Co., 1984.

Stebbins, Robert L., and Michael MacCaskey. *Pruning: How-To Guide for Gardeners.* Los Angeles: HPBooks, 1983.

Identification Guides

Borror, Donald J., and Richard E. White. *A Field Guide to the Insects of America North of Mexico.* The Peterson Field Guide Series. Boston: Houghton Mifflin Co., 1970.

Carr, Anna. *Rodale's Color Handbook of Garden Insects.* Emmaus, Pa.: Rodale Press, 1979.

*Davidson, Ralph H., and William F. Lyon. *Insect Pests of Farm, Garden, and Orchard.* 8th ed. New York: John Wiley & Sons, 1987.

*Johnson, Warren T., and Howard H. Lyon. *Insects That Feed on Trees and Shrubs.* 2nd ed. Ithaca, N.Y.: Cornell University Press, 1988.

MacNab, A. A., A. F. Sherf, and J. K. Springer. *Identifying Diseases of Vegetables.* University Park, Pa.: The Pennsylvania State University College of Agriculture, 1983. (Available from The Publications Distribution Center, 112 Agricultural Administration Building, University Park, PA 16802.)

Milne, Lorus, and Margery Milne. *The Audubon Society Field Guide to North American Insects and Spiders.* New York: Alfred A. Knopf, 1980.

Muenscher, Walter Conrad. *Weeds.* 2nd ed. New York: Macmillan Publishing Co., 1955. (Reprint, with forward and appendixes by Peter A. Hyypio. Ithaca, N.Y.: Comstock Publishing Assoc., 1980.)

*Sinclair, Wayne A., Howard. H. Lyon, and Warren T. Johnson. *Diseases of Trees and Shrubs.* Ithaca, N.Y.: Cornell University Press, 1987.

Smith, Miranda, and Anna Carr. *Rodale's Garden Insect, Disease, and Weed Identification Guide.* Emmaus, Pa.: Rodale Press, 1988.

*Westcott, Cynthia. *The Gardener's Bug Book.* Garden City, N.Y.: Doubleday & Co., 1973.

*———. *Westcott's Plant Disease Handbook.* 5th ed., rev. by R. Kenneth Horst. New York: Van Nostrand Reinhold Co., 1990.

Periodicals

**Common Sense Pest Control Quarterly,* Bio-Integral Resource Center (BIRC), P.O. Box 7414, Berkeley, CA 94707.

**HortIdeas,* Greg and Patricia Y. Williams, Rt. 1, Box 302, Black Lick Rd., Gravel Switch, KY 40328.

**National Gardening,* National Gardening Association, 180 Flynn Ave., Burlington, VT 05401.

Organic Gardening, Rodale Press, Inc., 33 E. Minor St., Emmaus, PA 18098.

**The IPM Practitioner,* Bio-Integral Resource Center (BIRC), P.O. Box 7414, Berkeley, CA 94707.

PHOTOGRAPHY CREDITS

Agricultural Research Service, U.S. Department of Agriculture: p. 270, bottom right; p. 283, top right; p. 309, top left; p. 315, top right; p. 335, bottom.

Max E. Badgley: p. 260, top right, bottom right; p. 261, top left, center left, and center right, bottom; p. 269, top, bottom left; p. 270, bottom left and center; p. 271; p. 272, top; p. 273; p. 275, top, bottom left; p. 276, top; p. 277, top center; p. 280, bottom center; p. 282, top left and center; p. 283, top center, bottom; p. 284, top left, bottom; p. 285, top left, bottom; p. 286, top right, bottom; p. 287, bottom left; p. 289, top left and center; p. 292, top left and center; p. 293, top right; p. 296, top left, bottom left and center; p. 297, top left; p. 300, bottom center and right; p. 301, top left, bottom left; p. 302, top left; p. 303, top right, bottom; p. 304; p. 305, top left and right, bottom left; p. 306, top; p. 307, bottom; p. 308, top left, bottom; p. 316; p. 317, top right; p. 320, top, bottom right; p. 321, top, bottom left; p. 322, top left and center, bottom left and right; p. 323, bottom; p. 324, top left and right; p. 327, bottom left and right; p. 330, top center and right, bottom left; p. 331, top left and center; p. 332, bottom right; p. 333; p. 334, bottom; p. 335, top left; p. 337.

Charlton Photos, Inc.: p. 291, top center, bottom; p. 312, left; p. 315, top left; p. 317, bottom; p. 325, bottom; p. 328, top; p. 390, right col., bottom; p. 393, left col., bottom.

Clemson University Extension Service and the U.S. Department of Agriculture: p. 277, bottom left; p. 312, right; p. 313, right; p. 329, bottom; p. 386, right col., left.

Crandall & Crandall Photography: p. 330, bottom right.

Davey Tree Expert Co. (by D. L. Caldwell): p. 274, top; p. 298, top; p. 301, bottom center and right; p. 302, top center and right, bottom left; p. 319, top center, bottom right; p. 321, bottom center; p. 397, right col., top right.

Dr. John A. Davidson, Department of Entomology, University of Maryland at College Park: p. 291, top right; p. 318, top.

Department of Plant Pathology, University of Illinois: p. 372, right col.; p. 373, left col., top; p. 376, left col.; p. 378, left col., top; p. 380, left col., right; p. 381, right col., left; p. 384, right col., right; p. 388, right col.; p. 389, right col.; p. 394, left col.; p. 395, left col., p. 396, left col.; p. 400, right col.

Entomological Society of America/Ries Memorial Slide Collection: p. 325, top left and center; p. 334, top left.

The Fertilizer Institute: p. 379, right col., top.

Tom Gettings, Rodale Press Photography Department: p. 381, right col., right; p. 390, right col., top.

Dr. Gerald M. Ghidiu, Rutgers Research and Development Center: p. 280, top.

L. Gilkeson: p. 260, top left; p. 268, center.

Courtesy of the Ken Gray Collection, Oregon State University (available through Eugene Memmler): p. 277, bottom center and right; p. 279; p. 291, top left; p. 312, center; p. 314, top left, bottom.

John Hamel, Rodale Press Photography Department: p. 274, bottom right.

Dr. George W. Hudler, College of Agriculture and Life Sciences, Cornell University: p. 371, left col., left; p. 375, right col.; p. 377, left col.; p. 383, left col., bottom; p. 385, right col.; p. 394, right col., bottom; p. 397, right col., top left, bottom; p. 398, right col.

Dr. Stephen A. Johnston, Rutgers Research and Development Center: p. 399, right col.

Reprinted by permission from Dr. Alan L. Jones: p. 391, left col.

Reprinted by permission from Dr. Alan MacNab: p. 380, left col., left; p. 384, left col., bottom, right col., left.

Bob Mulrooney: p. 374, right col., top; p. 376, right col., top.

Nematology Laboratory, U.S. Department of Agriculture: p. 309, top right.

New York State Agricultural Experiment Station: p. 270, top; p. 282, top right, bottom; p. 287, bottom right; p. 289, top right; p. 311; p. 313, left;

p. 314, top center and right; p. 329, top left; p. 371, right col.; p. 373, left col., bottom; p. 374, left col., right col., bottom; p. 386, left col., top, right col., right; p. 387, right col., bottom; p. 389, left col.

Joe Ogrodnick: p. 260, bottom left; p. 294, top; p. 298, bottom right.

Pam Peirce: p. 322, bottom center; p. 386, left col., bottom.

Thomas M. Perring, University of California at Riverside: p. 268, left and right.

Potash & Phosphate Institute: p. 379, left col.

Lee Reich, Ph.D.: p. 388, left col.

Robert Reise, Nematology Laboratory, U.S. Department of Agriculture: p. 309, bottom.

Ann F. Rhoads: p. 272, bottom left; p. 372, left col.; p. 373, right col.; p. 376, right col., bottom right; p. 378, left col., bottom; p. 380, right col.; p. 381, left col.; p. 384, left col., top; p. 387, left col.; p. 390, left col., top right, bottom; p. 391, right col.; p. 392, left col.; p. 397, left col.; p. 399, left col.; p. 401, left col.; p. 403, right col.

Rodale Press Photography Department: p. 278; p. 286, top left; p. 293, top left; p. 295, center; p. 298, bottom center; p. 315, bottom right; p. 331, top right.

E. S. Ross: p. 272, bottom right.

Barbara Rothenberger: p. 290, right; p. 382, left col.; p. 385, left col.; p. 393, left col., top.

David Shetlar, Landscape Entomologist, The Ohio State University: p. 275, bottom center; p. 281, right, p. 292, bottom; p. 294, bottom center; p. 298, bottom left; p. 310; p. 315, bottom left; p. 319, top left, top right, bottom left; p. 325, top right; p. 327, top; p. 329, top right; p. 334, top right.

Patrick Temple, University of California at Riverside: p. 378, right col.; p. 379, right col., bottom.

U.S. Department of Agriculture: p. 284, top right; p. 295, left and right; p. 322, top right; p. 390, left col., top left; p. 398, left col.; p. 402, right col.

Ron West: p. 260, top center left and center right, bottom center left and center right; p. 261, top right; p. 269, bottom center and right; p. 274, bottom left and center; p. 275, bottom right; p. 276, bottom; p. 277, top left and right; p. 280, bottom left and right; p. 281, left; p. 283, top left; p. 284, top center; p. 285, top right; p. 286, top center; p. 287, top; p. 288; p. 289, bottom; p. 290, left; p. 292, top right; p. 293, bottom left and center; p. 294, bottom left and right; p. 296, top right, bottom right; p. 297, top right, bottom; p. 299; p. 300, top, bottom left; p. 301, top center and right; p. 302, bottom right; p. 303, top left and center; p. 305, top center, bottom right; p. 307, top; p. 308, top center and right; p. 309, top center; p. 317, top left; p. 318, bottom; p. 320, bottom left; p. 321, bottom right; p. 323, top; p. 324, top center, bottom; p. 326; p. 327, bottom center; p. 328, bottom; p. 329, top center; p. 331, bottom; p. 332, top, bottom left; p. 335, top center and right; p. 336; p. 376, right col., bottom left; p. 377, right col.; p. 382, right col.; p. 383, left col., top, right col.; p. 400, left col.; p. 402, left col.; p. 403, left col.

Katharine D. Widin, Ph.D.: p. 293, bottom right; p. 306, bottom; p. 330, top left; p. 371, left col., right; p. 375, left col.; p. 387, right col., top; p. 392, right col.; p. 393, right col.; p. 394, right col., top; p. 395, right col.; p. 396, right col. p. 401, right col.

INDEX

Note: Page references in *italic* indicate tables.
Boldface references indicate photographs or illustrations.

A

B

D

E

F

K

L

N

O

P

S

T

U

V

X

Y

Z

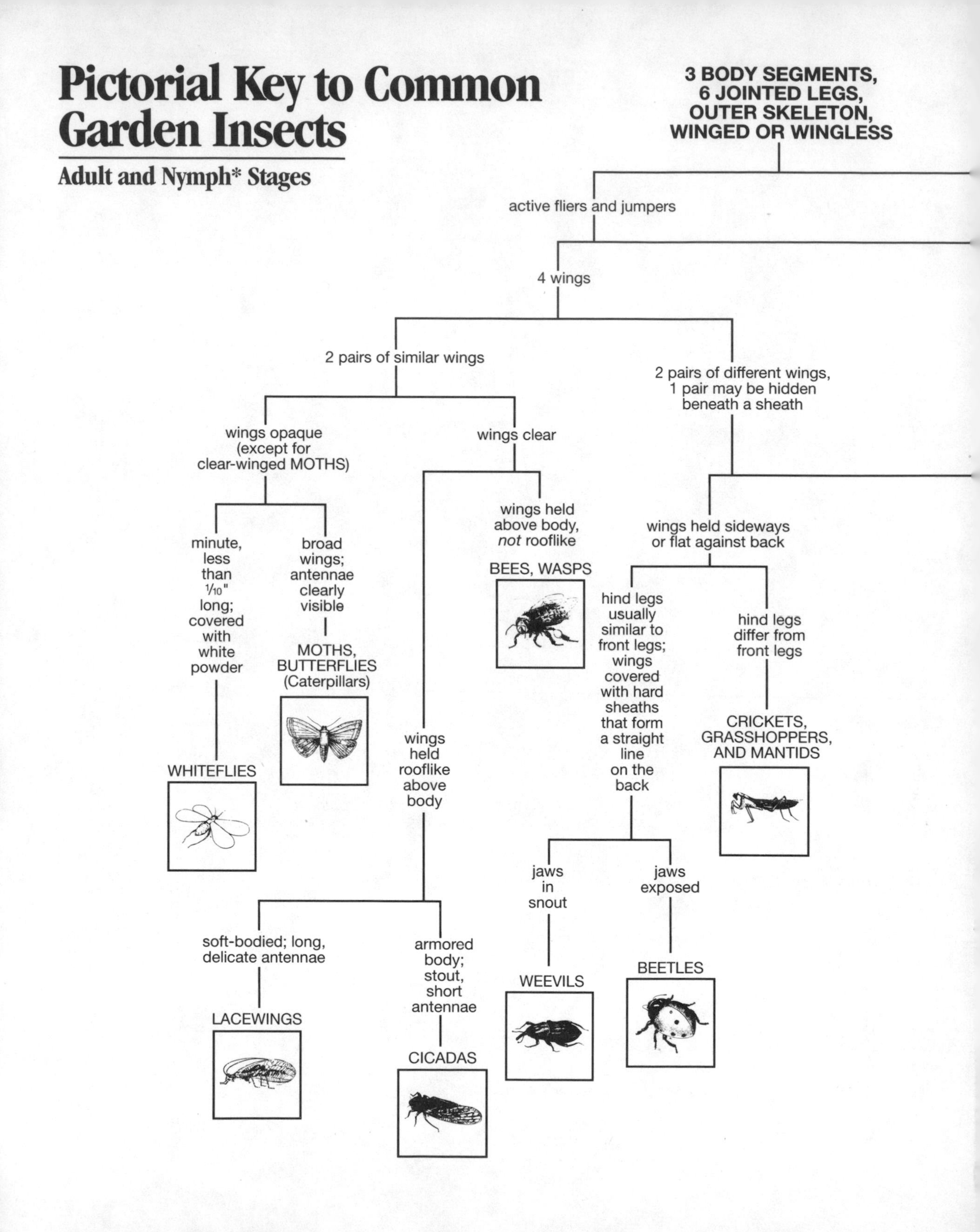

Pictorial Key to Common Garden Insects
Adult and Nymph* Stages
3 BODY SEGMENTS, 6 JOINTED LEGS, OUTER SKELETON, WINGED OR WINGLESS
active fliers and jumpers
4 wings
2 pairs of similar wings
2 pairs of different wings, 1 pair may be hidden beneath a sheath
wings opaque (except for clear-winged MOTHS)
wings clear
minute, less than 1/10" long; covered with white powder
broad wings; antennae clearly visible
WHITEFLIES
MOTHS, BUTTERFLIES (Caterpillars)
wings held above body, *not* rooflike
BEES, WASPS
wings held rooflike above body
soft-bodied; long, delicate antennae
LACEWINGS
armored body; stout, short antennae
CICADAS
wings held sideways or flat against back
hind legs usually similar to front legs; wings covered with hard sheaths that form a straight line on the back
hind legs differ from front legs
CRICKETS, GRASSHOPPERS, AND MANTIDS
jaws in snout
WEEVILS
jaws exposed
BEETLES